Android 4.X 手机/平板电脑程序设计
入门、应用到精通
（第二版）

孙宏明　著

中国水利水电出版社
www.waterpub.com.cn

内 容 提 要

本书是著名 Android 技术专家孙宏明老师经典之作。

本书从 Eclipse 的操作技巧、强大的程序代码编辑辅助功能、程序的调试排错技术等基本功开始，到 Android 程序架构详解、各种接口组件用法介绍以及 Android 程序的高级功能和应用，带领读者从菜鸟一路晋升成为 Android 技术牛人。

本书不仅内容丰富完整，更重要的是笔者根据教学经验整理出一条由浅入深的学习路径，搭配主题单元的学习方式和清晰明了的步骤讲解，再加上精心设计的实战案例，让读者在学过每一个单元之后都能立即上手，达到最高的学习效率。

除了完整的 Android 基础知识，本书还包含了 Android 的最新高级技术，所以本书不仅适合于 Android 技术的初学者，还可用于中高级读者参考学习。

本书为经台湾碁峰资讯股份有限公司独家授权发行的中文简体版。本书中文简体字版在中国大陆之专有出版权属中国水利水电出版社所有。在没有得到本书原版出版者和本书出版者书面许可时，任何单位和个人不得擅自摘抄、复制本书的一部分或全部以任何方式（包括资料和出版物）进行传播。本书原版版权属碁峰资讯股份有限公司。版权所有，侵权必究。

北京市版权局著作权合同登记号：图字 01-2011-5269 号

图书在版编目（CIP）数据

Android 4.X手机/平板电脑程序设计入门、应用到精通 / 孙宏明著. -- 2版. -- 北京：中国水利水电出版社，2012.9
 ISBN 978-7-5170-0122-5

Ⅰ. ①A… Ⅱ. ①孙… Ⅲ. ①移动终端－应用程序－程序设计 Ⅳ. ①TN929.53

中国版本图书馆CIP数据核字(2012)第207071号

策划编辑：周春元　　责任编辑：陈 洁　　封面设计：李 佳

书　　名	Android 4.X 手机/平板电脑程序设计入门、应用到精通（第二版）	
作　　者	孙宏明　著	
出版发行	中国水利水电出版社	
	（北京市海淀区玉渊潭南路1号D座　100038）	
	网址：www.waterpub.com.cn	
	E-mail：mchannel@263.net（万水）	
	sales@waterpub.com.cn	
	电话：（010）68367658（发行部）、82562819（万水）	
经　　售	北京科水图书销售中心（零售）	
	电话：（010）88383994、63202643、68545874	
	全国各地新华书店和相关出版物销售网点	
排　　版	北京万水电子信息有限公司	
印　　刷	北京市蓝空印刷厂	
规　　格	184mm×240mm　16开本　41.75印张　937千字	
版　　次	2012年3月第1版　2012年3月第1次印刷	
	2012年8月第2版　2012年8月第1次印刷	
印　　数	0001—3000 册	
定　　价	68.00元（赠1CD）	

凡购买我社图书，如有缺页、倒页、脱页的，本社发行部负责调换

版权所有·侵权必究

推荐序

近年来由于3G网络普及、硬件运算效能提升，再加上各种应用软件带来的灵活便利性，智能手机渐渐取代传统移动电话融入日常生活，包括电子邮件、网页浏览、影音播放、地图查询、拍照摄影，各种功能随手呈现出实时互动带来的无比方便性。这也反映在许多手机品牌的亮丽销售数据上。我们可以预期在通讯成本持续降低、云端运算逐步成熟、网络早已无所不至的时代，通过不断发布优秀产品，增进各项周边功能，以及提出更多元种类的应用服务，智能手机终将成为数字生活不可或缺的重要帮手。

Android平台基于Linux开放核心架构，对硬件设计制造充满弹性，容易使销售对象自行规划产品，自从2007年底正式公布以来深受各家手机制造商支持，结合最新处理器与周边组件，不断推出各种智能手机，并且依照功能与价格提供消费者多样化的选择。在应用上，Android平台充分发挥新型处理器运算效能，呈现高质量图形与音效，更直接提供各种Google服务，包括网页浏览搜寻、地图定位、语音处理等。Android Market在线软件商店提供了无数程序：软件工具、商务应用、游戏、电子书、新闻气象、交通旅游……，供使用者直接以手机分类浏览、检索、在线购买、下载安装或更新，不断延伸手机功能。按照现行发展趋势，Android手机应用将会更多元化、更丰富，并且具备更完整的商务应用支持。如果说智能手机的出现重新定义了移动通讯，那么在此时此刻与即将来临的年代，各种应用软件功能所提供服务质量的深度与广度也将重新书写手机平台的价值。

有别于早期移动平台开发经验，Android应用程序开发资源取得的方式十分简单，相关工具皆可在网站上自行下载，并且轻易适用于Windows、MAC、Linux操作系统。这种配置的便利性简化了软件系统开发成本，对于学习者而言更是友善亲切。此外，Android架构在先天规划上区分了操作系统核心与应用接口层，开发者使用Java语言专注于应用程序本身的设计编写，再通过标准架构类函数调用与系统互动，这种方式不但降低设计与调试复杂度，也免除许多不必要的负担。从开发工具的获得到最后应用程序的发布方式，其中开放与标准化架构的精神，更使我们的设计研发无须受限于单一品牌厂商与特定商品，就数字文明历史来看，这应该也是应用领域长期发展持续繁荣的重要基础。

无论是对于初学者还是专业软件工程师，备妥合适的参考工具都将会事半功倍，虽然我们可通过网络找到丰富而开放的在线源代码范例或各讨论区的技术解决方案，一本精心规划妥善编排的参考书所带来的实时帮助仍是无可取代的。本书作者孙宏明老师具有丰富的教学与实践经验，熟悉各种计算机语言、程序设计、图像处理、计算机图形学、计算机视觉、计算机游戏设计等领域，指导学生参加程序设计竞赛取得丰硕成果。本书编排方式由浅入深、循序渐进，将相关知识与技巧详细解说呈现，搭配各种实例练习，又可通过章节主题直接寻找特定解决方案，巧妙涵盖学习与实战不同层次需求，相信能为读者带来有效的帮助，轻松提升程序开发技能，解决各种疑难状况。

黄文良教授

推荐序

我和孙老师认识已有十年的时间，当时孙老师在软件公司负责新技术的研发，他和许多大学的教授及研究生一起合作相关的研究计划。虽然当时是我们与他首次合作，但是孙老师发挥他负责及亲切的优点，非常详细及精确地为我们提供整个计划的目标及要求，孙老师不仅在专业领域有很深厚的基础，其沟通与协调的能力也是很令人赞赏的。现在孙老师愿意针对"手机程序设计"这个新领域，将自己累积多年的程序开发技术和经验，结合目前当红的 Android 手机平台程序设计，编写成适合入门学习的书籍，实在是学习 Android 手机程序设计者的一大福音。

孙老师后来离开业界到大学教书，转眼之间教授程序设计相关课程已超过六年的时间，他擅长各种程序语言包括 C/C++、Java 和 VB，以及许多应用领域，例如图像处理与辨识、计算机图形学、游戏程序设计以及 UML 软件工程，所授课程受到学生的喜爱。由于他在应用程序设计方面具有深厚的专业背景，所以编写的这本书内容非常精彩丰富，不仅内容的广度涵盖 Android 程序技术的各种学习主题，同时还包括目前热门的应用，比如游戏程序所需的 2D 和 3D 绘图、Google 地图服务、拍照录像、网页处理和 NFC 程序。

孙老师将他亲切及讲解仔细的特色融入于文字叙述中，使得本书不仅逻辑及描述非常清楚，而且在程序范例的设计和编排上也顾及初学者的需要，读来非常轻松流畅，相信阅读本书必能提高读者的学习兴趣和成效。如果读者想要加入 Android 程序开发的行列，我相信这本书必能给予很大的帮助！

清华大学　赖尚宏教授

推荐序

Android 在两年前还是一个大众不太熟悉的名字，虽然当时在业界已经是个成熟的环境。很感谢 Google 的大力支持，这两年来让以 Linux 为基础的操作系统能够在移动设备中运行，同时也照顾了广大的 Java 开发与爱好者。

欣闻孙老师邀请我帮忙写 Android 书籍的推荐序，我与孙老师这两年间参与了 Android 相关课程的教授，除了 Android 一般的课程外，孙老师在图像处理与游戏软件设计与开发方面有长足的经验，在书中他也分享了重要的心得与整理了重要的数据。孙老师平日与同学有良好的互动，多年来的认真教学让我相信他编写的书一定具有良好的水平与质量。

这两年间，坊间也出了不少 Android 的教学书籍。这些书籍适合不同层次的学习者与开发者，有的例子较少，着重案例与概念的讲解，适合初学者，而有的例子很多，多到像是一本电话簿，对于进阶的应用与开发多有着墨。在此，我要推荐给大家这本由孙老师精心编写的 Android 教学书籍，其中包含许多重要主题单元，内容非常丰富，讨论了非常多的例子。个人认为这是一本非常好的学习手册与工具书，非常适合在已经具备 Java 语言基础后，直接进行 Android 程序开发时放在案头的一本学习与参考范例。

书中的内容除了教导初学者如何使用 Android 窗口组件开发手机程序之外，也分享了重要的程序调试经验与技巧，对于初学者是非常重要的。对于多运行序如何配合窗口组件一起使用、如何在不同的 Activity 间互动、传递与交换数据，都有深入的探讨与解释，这些单元对于学习如何在 Android 平台开发一个快速且有效的程序将会非常受用。

书中对于 App Widget 程序开发也有所着墨，这些内容对于进阶手机程序开发者而言是一个重要的参考。Part 10 之后的单元讨论了许多进阶的 Android 程序设计主题，如 SQLite、多语系和屏幕支持、2D 与 3D 绘图、Google 地图服务、多媒体功能、网页处理以及 NFC，读者可以充分运用本书的内容开发出专业且互动接口良好的程序。相信通过这本书清楚且详尽的解说，您可以完整地学习与欣赏 Android 的美，欢迎您加入 Android 的行列。

台湾大学信息工程系　李盛安助理教授

前　言

昔日的绿芽已经长成大树

本书的第一版是针对 Android 2.X 手机程序设计所编写的，虽然其中包含 Android 3.X 的内容，可是就功能上来说，3.X 版是给平板电脑专用而并非通用的平台，因此等到 2011 年 10 月 Google 发表 Android 4 通用版本之后，笔者才结合手机和平板电脑的功能进行内容的更新。回顾 2011 年，Android 系统在市场上可谓铺天盖地。虽然新闻上看似 Apple 的 iPhone 和 iPad 抢尽风头，可是如果仔细观察市场销售数据，Apple 产品的市占率只是与 Android 持平，反观 Android 系统却是以超越线性的方式快速成长。本书前一版形容 Android 是"信息科技的绿芽"，如今看来这个昔日的绿芽已经长成大树。

Android 对于程序开发人员的致命吸引力

安装 Android 程序开发工具需要到不同的网站下载相关软件，虽然过程有些繁复，但是读者只要依照书上的说明操作就可以顺利完成，而且这些网站一年 365 天，每天 24 小时开放，只要连上 Internet 就可以下载，更棒的是"完全免费"。另外相较于其他软件来说，Android 程序开发工具可说是非常"环保"（对于用户的计算机而言），只要将安装好的文件夹复制到另一台计算机，再设定好文件夹路径就可以使用。日后如果不想用，只要删除它的文件夹即可。

Android 系统的功能也超越传统的 PC 操作系统，像是 Google 地图服务、定位功能、拍照、录音录像、近场通讯（NFC）、语音和人脸辨识等，只要加上您的创意，就可以发展出比一般软件更生活化的应用。而且 Google 还针对全世界 Android 应用程序开发人员架设了一个 Android Market 网站，让他们可以在上面贩卖或是免费下载自己的作品。由于以上这些吸引人的条件，让世界各地加入开发 Android 应用程序的人员快速地增加。如果过去我们已经错过研发 PC 软件的先机，现在岂能再错失成为手机和平板电脑应用程序领头羊的大好机会！

学习 Android 程序设计的方法

开发 Android 应用程序需要使用 Java 语言，而且 Android 系统的功能非常丰富，程序项目的架构也和传统的 PC 程序不同，因此学习时如果没有适当的规划，只靠东拼西凑的方式恐怕效率不佳。为了让读者能够有效率地学习 Android 程序设计，笔者对于本书内容的编排花费很大的心思，希望从五花八门的 Android 程序技术中理出一条由浅入深、适合初学者的快捷方式。最后笔者决定摒弃传统程序设计书籍惯用的章节编排模式，改成以教学单元的方式，搭配切合主题的实作范例，再辅以详细的操作步骤说明，让读者能够确实了解每一个单元主题的技术和用法。在学习 Android 程序设计的过程中，除了知道 SDK 相关的知识之外，开发工具的操作技巧也很重要，善用辅助功能不但可以减少打字的时间，同时也能够避免打错字的情况，因而缩短程序除错的过程。另外当遇到开发工具异常时，如何迅速排除问题也是实务上很重要的技能，否则徒耗时间在非程序技术的问题上往往令人为之气结。这一次改版笔者特别将重要观念、操作技巧和相关知识等用"补充说明"的小文框提示，一来可以达到更好的提醒功能，二来也方便日后的查阅。

开发 Android 应用程序大致上需要三个条件。第一是具备程序语言的基础（Java、C/C++、Visual Basic 皆可），也就是说必须知道变量、数组、判断式、循环等基本语法。第二是了解面向对象的观念和用法，因为 Android 程序是使用 Java 语言撰写，它采用面向对象的架构。面向对象的基本观念并不难，当然进阶的用法需要比较多的经验和技术，但是本书的程序范例是针对入门学习者所设计，因此只要依照书上循序渐进的内容安排来阅读就可以了解。如果读者可以配合书上的操作步骤动手实作，学习效果更能倍增。再者 Android 程序中使用事件处理程序和系统 callback 函数的机制，这种机制并不是 Android 程序的专属特性，任何图形操作接口的操作系统如 Windows 也都是采用这种方式，基本上这也是一个观念问题，只要读者了解它的运作原理就会知道如何使用，本书会在适当的时候加以解说，因此只要读者了解程序语言的基本语法，就可以由本书的说明和范例开始学习 Android 程序设计。

如何使用这本书

Android 程序设计是一个新兴的领域，早期介绍相关技术的书籍并不多，近来陆续有新的著作问市，目前市面上已经有许多入门学习或是进阶应用的书籍。有些入门书籍涵盖的技术范围有限，无法满足实战上的需要，反观进阶应用的著作虽然包含比较完整的技术内容，但是解说的方式可能不适合初学者。笔者编写这本书的目的就是希望在内容的广度和解说的细节上取得更好的平衡。为了达到这个目的，笔者将 Android 系统的功能加以分类，然后根据由浅入深的原则进行编排，再搭配许多精心设计的范例程序贯穿相关的学习单元，让读者在学习单一主题的时候也能够了解相关的功能。本书使用的单元编排方式是希望将每一次的学习时间做适当的切割，再以切合主题的程序项目为范例，让读者能够充分了解学习的内容并知道如何使用，如此一来自然能够达到更好的学习效果。采用教学单元编排的另一个好处是方便日后查询，由于每一个单元的内容长度适中、主题明确，读者可以根据目录快速找到需要的数据。另外 Android 4 的特色之一就是同时适用不同类型的装置包括手机和平板电脑，因此如何让程序的操作接口可以动态配合不同的屏幕尺寸是很重要的技术主题。由于每个程序的操作接口复杂度不同，因此使用的技术也有所差异，本书特别针对这方面的考虑提供完整的解决方法。另外像 fragment、action bar、action item/view、property animation、NFC beam 等新功能在书中都有完整的介绍和操作范例。

由于 Android 3 和 4 增加了许多新技术，为了让学习更加流畅和完整，笔者将这些新功能和原来的内容重新整理编排，而不是单纯附加在原来的内容之后，以达到更好的整体性。另外为了方便读者区分不同 Android 版本的功能差异，在每一个学习单元开头都特别以表格的方式注明适用的版本。这些精心的安排无疑是希望用最有效的方式传达知识，以提升读者的学习效率。

　　本书的第一部分是介绍 Android 的发展史、系统架构、市场趋势等背景知识，以及程序开发工具的安装和使用。第二部分是介绍程序项目的架构、接口组件的用法、除错技术、仿真器的操作和设定等程序开发的基本技巧，这个部分是后续学习的重要基础，建议读者配合书上的解说动手实作。学习程序设计的秘诀无他，就是多动手，然后想一想、改一改、试一试自然就能理解。完成第二部分的学习之后就可以根据自己的兴趣或需要学习特定的单元，由于本书的内容编排是基于由浅入深的原则，而且部分程序范例具有前后连贯的关系，因此笔者在解说的过程中会视需要提示参考相关的单元。另外如果读者在学习上遇到障碍，可以先回到前面相关的单元研读之后，再依关联性往后续单元继续学习，依照这种方式就能够让本书对于读者学习 Android 程序设计发挥最大的帮助。坐而言不如起而行，现在就让我们一起踏上 Android 程序设计的学习之旅吧！

　　最后感谢我最亲爱的家人 Maysue、小 D 和小 M 在本书编写期间的宽容和体谅，虽然因为我的忙碌而疏忽了你们，但是有你们的陪伴，让一切的付出和努力都更有意义！

孙宏明
于 玫园

目 录

推荐序
前言

PART 1 拥抱 Android

UNIT 1 Android 造时势或是时势造 Android 2
    1-1 Android 从何而来 3
    1-2 Android 的功能、应用和商机 4
    1-3 先睹为快——Android 手机和平板电脑模拟器 6

UNIT 2 安装 Android 程序开发工具 9
    2-1 不同操作系统的开发工具版本 9
    2-2 安装 Android 程序开发工具的步骤 10
    2-3 Android 程序开发工具的维护和更新 20

UNIT 3 建立 Android 应用程序项目 22
    3-1 修改程序的接口 26

UNIT 4 Eclipse 程序项目管理技巧 29
    4-1 根据已经写好的程序文件来建立项目 29
    4-2 根据 Android SDK 中的程序范例来建立项目 30
    4-3 把建立好的 Android 程序项目加载 Eclipse 30
    4-4 程序项目的管理和维护 31

PART 2 开发 Android 应用程序的流程

UNIT 5 Eclipse 程序项目管理技巧 35

UNIT 6	使用 TextView、EditText 和 Button 接口组件	40
6-1	TextView 接口组件	41
6-2	EditText 接口组件	42
6-3	Button 接口组件	43
6-4	链接接口组件和程序代码	44
6-5	设置 Button 的 click 事件 listener	44
6-6	取得 edtSex 和 edtAge 接口组件中的字符串	47
6-7	将结果显示在 txtResult 接口组件	47
6-8	在模拟器中输入中文	49

UNIT 7	程序的错误类型和除错方法	51
7-1	程序的语法错误和调试的方法	51
7-2	程序的逻辑错误和调试的方法	52
7-3	运行时期错误和调试的方法	54

UNIT 8	使用 Android 模拟器的技巧	57
8-1	启动模拟器的时机	57
8-2	Eclipse 选择不同版本 AVD 的规则	60
8-3	同时运行多个 AVD	60
8-4	使用 AVD 的调试功能	61
8-5	AVD 的语言设置、时间设置和上网功能	63
8-6	把实体手机或平板电脑当成模拟器	64

UNIT 9	良好的程序架构是程序开发和维护的重要基础	66

UNIT 10	升级 Android 手机程序成为平板电脑程序	71
10-1	针对 Android 平板电脑作优化	72
10-2	将程序升级成为 Android 平板电脑专属程序	73

PART 3　学习使用基本接口组件和布局模式

UNIT 11	学习更多接口组件的属性	76
11-1	match_parent 和 wrap_content 的差别	78
11-2	android:inputType 属性的效果	79
11-3	控制文字大小、颜色和底色	80
11-4	控制组件四周的间隔距离以及组件内部的文字和边的距离	81

UNIT 12	Spinner 下拉式菜单组件	82
12-1	建立 Spinner 下拉式菜单的第一种方法	82

	12-2	建立 Spinner 下拉式菜单的第二种方法	87
UNIT 13		使用 RadioGroup 和 RadioButton 组件建立单选清单	91
	13-1	将"婚姻建议"程序改成使用 RadioGroup 菜单	92
UNIT 14		CheckBox 多选清单和 ScrollView 滚动条	98
UNIT 15		LinearLayout 界面编排模式	105
UNIT 16		TableLayout 接口编排模式	109
UNIT 17		RelativeLayout 布局	115
UNIT 18		FrameLayout 布局和 Tab 卷标页	126
	18-1	建立 Tab 标签页的步骤	127
	18-2	范例程序	130

PART 4　学图像接口组件与动画效果

UNIT 19		ImageButton 和 ImageView 接口组件	134
UNIT 20		Gallery、GridView 和 ImageSwitcher 接口组件	140
	20-1	Gallery 组件和 GridView 组件的使用方法	141
	20-2	ImageSwitcher 组件的使用方法	146
	20-3	完成"图像画廊"程序	148
UNIT 21		使用 Tween 动画效果	151
	21-1	建立动画资源文件	152
	21-2	建立各种类型的动画	154
	21-3	使用随机动画的"图像画廊"程序	157
	21-4	在程序代码中建立动画效果	159
	21-5	应该使用动画资源文件还是在程序代码中建立动画对象	160
UNIT 22		Frame animation 和 Multi-Thread 游戏程序	161
	22-1	建立 Frame animation 的两种方法	161
	22-2	Multi-Thread "掷骰子游戏"程序	163
	22-3	使用 Handler 对象传送信息	164
	22-4	实现"掷骰子游戏"程序	164
UNIT 23		Property animation 初体验	170
	23-1	Property animation 的基本用法	171
	23-2	范例程序	173
UNIT 24		Property animation 加上 Listener 成为动画超人	178
	24-1	使用 AnimatorSet	178

24-2 加上动画事件 listener 180
24-3 ValueAnimator 182
24-4 范例程序 183

PART 5 Fragment 与进阶接口组件

UNIT 25 使用 Fragment 让程序界面一分为多 189
25-1 使用 Fragment 的步骤 190
25-2 为 Fragment 加上外框并重设大小和位置 191
25-3 范例程序 192

UNIT 26 动态 Fragment 让程序成为变形金刚 200
26-1 Fragment 的总管——FragmentManager 200
26-2 范例程序 202

UNIT 27 Fragment 的进阶用法 212
27-1 控制 FrameLayout 的显示和隐藏 213
27-2 使用 Fragment 的 Back Stack 功能和动画效果 215

UNIT 28 Fragment 和 Activity 之间的 callback 机制 220
28-1 检查"计算机猜拳游戏"程序架构 220
28-2 实现 Fragment 和 Activity 之间的 callback 机制 222
28-3 范例程序 224

UNIT 29 ListView 和 ExpandableListView 232
29-1 使用 ListActivity 建立 ListView 菜单 232
29-2 帮 ListView 加上小图标 236
29-3 ExpandableListView 二层式选项列表 239

UNIT 30 AutoCompleteTextView 自动完成文字输入 243

UNIT 31 SeekBar 和 RatingBar 接口组件 248

PART 6 其他接口组件与对话框

UNIT 32 时间日期接口组件和对话框 255
32-1 DatePicker 日期接口组件 255
32-2 TimePicker 时间接口组件 256
32-3 范例程序 257
32-4 DatePickerDialog 和 TimePickerDialog 对话框 259

UNIT 33　ProgressBar、ProgressDialog 和 Multi-Thread 程序 261
　　33-1　Multi-Thread 程序 263
　　33-2　使用 Handler 对象完成 Thread 之间的信息沟通 263
　　33-3　第一版的 Multi-Thread ProgressBar 范例程序 264
　　33-4　第二版的 Multi-Thread ProgressBar 范例程序 269
　　33-5　ProgressDialog 对话框 270
UNIT 34　AlertDialog 对话框 271
　　34-1　使用 AlertDialog.Builder 类别建立 AlertDialog 对话框 271
　　34-2　使用 AlertDialog 类别建立 AlertDialog 对话框 273
　　34-3　范例程序 274
UNIT 35　Toast 消息框 279
UNIT 36　自定义 Dialog 对话框 282

PART 7　Intent、Intent Filter 和传送数据

UNIT 37　工程中的 AndroidManifest.xml 程序功能描述文件 289
UNIT 38　Intent 粉墨登场 296
　　38-1　Eclipse 的 DDMS 功能以及模拟器的 Linux 命令行模式 299
UNIT 39　Tab 标签页接口——使用 Intent 对象 303
UNIT 40　Intent Filter 让程序也能帮助别人 308
　　40-1　设置 AndroidManifest.xml 文件中的 Intent Filter 309
　　40-2　Android 系统检查 Intent 和 Intent Filter 的规则 311
　　40-3　程序接收到 Intent 对象的工作 311
　　40-4　范例程序 312
UNIT 41　让 Intent 对象附带数据 318
　　41-1　传送数据的 Activity 需要完成的工作 318
　　41-2　从 Intent 对象中取出数据 320
　　41-3　范例程序 320
UNIT 42　要求被调用的 Activity 返回数据 327

PART 8　Broadcast Receiver、Service 和 App Widget

UNIT 43　Broadcast Intent 和 Broadcast Receiver 334
　　43-1　程序广播 Intent 对象的方法 334

  43-2 建立 Broadcast Receiver 监听广播消息 335
  43-3 范例程序 336

UNIT 44 **Service 是幕后英雄** 341
  44-1 Service 的运行方式和生命周期 341
  44-2 在程序项目中建立 Service 342
  44-3 启动 Service 的第一种方法 345
  44-4 启动 Service 的第二种方法 346
  44-5 范例程序 347

UNIT 45 **App Widget 小工具程序** 351

UNIT 46 **使用 Alarm Manager 增强 App Widget 程序** 358
  46-1 建立增强版的 App Widget 程序 358
  46-2 取得并更新 App Widget 程序的界面 362

UNIT 47 **App Widget 程序的其他两种执行模式** 365
  47-1 预定运行时间的 App Widget 程序 365
  47-2 用按钮启动 App Widget 程序 367

PART 9 Activity 的生命周期与进阶功能

UNIT 48 **Activity 的生命周期** 371

UNIT 49 **帮 Activity 加上菜单** 375
  49-1 onCreateOptionsMenu()的工作 376
  49-2 onOptionsItemSelected()的工作 376
  49-3 建立 xml 格式的菜单定义文件 377
  49-4 范例程序 379

UNIT 50 **使用 Context Menu** 384
  50-1 Context Menu 的用法和限制 384
  50-2 范例程序 385

UNIT 51 **在 Action Bar 加上功能选项** 389
  51-1 控制 Action Bar 390
  51-2 在 Action Bar 加上 Action Item 391
  51-3 在 Action Bar 加上 Action View 392
  51-4 范例程序 392

UNIT 52 **在 Action Bar 上建立 Tab 标签页** 398

UNIT 53 **在状态栏显示信息** 404

PART 10　储存程序数据

UNIT 54　使用 SharedPreferences 储存数据 411
- 54-1　储存数据的步骤 411
- 54-2　读取数据的步骤 412
- 54-3　删除数据的步骤 412
- 54-4　清空数据的步骤 413
- 54-5　范例程序 413

UNIT 55　使用 SQLite 数据库储存数据 417
- 55-1　进入模拟器的 Linux 命令行模式操作 SQLite 数据库 417
- 55-2　SQLiteOpenHelper 类 419
- 55-3　SQLiteDatabase 类 420
- 55-4　范例程序 421

UNIT 56　使用 Content Provider 跨程序存取数据 428
- 56-1　Activity 和 Content Provider 之间的运行机制 429
- 56-2　范例程序 431

UNIT 57　使用文件储存数据 438
- 57-1　将数据写入文件的方法 438
- 57-2　从文件读取数据的方法 439
- 57-3　范例程序 440

PART 11　程序项目的整备工作和发布

UNIT 58　支持多语系和屏幕模式 446
- 58-1　让程序支持多语系的方法 447
- 58-2　让程序支持多种屏幕模式 448
- 58-3　范例程序 449

UNIT 59　开发不同 Android 版本程序的考虑 453
- 59-1　利用 Fragment 控制分页或单页显示 455

UNIT 60　取得屏幕的宽度、高度和 分辨率 465
- 60-1　取得屏幕的宽高和分辨率 465
- 60-2　取得程序界面的宽和高 466
- 60-3　利用 AndroidManifest.xml 文件设置程序运行的屏幕条件 471

UNIT 61　将程序安装到设备或在网络上发布 474

61-1 利用 Export Wizard 帮程序加上数字签名和完成 zipalign ... 475
61-2 将程序上传到 Google 的 Android Market 网站 ... 477

PART 12　2D 和 3D 绘图

UNIT 62　使用 Drawable 对象 ... 479
　62-1　从 res/drawable 文件夹的图像文件建立 Drawable 对象 ... 479
　62-2　在 res/drawable 文件夹中建立 xml 文件格式的 Drawable 对象定义文件 ... 480
　62-3　在程序中建立 Drawable 类型的对象 ... 481
　62-4　范例程序 ... 482

UNIT 63　使用 Canvas 绘图 ... 485

UNIT 64　使用 View 在 Canvas 上绘制动画 ... 489
　64-1　程序绘制动画的原理 ... 489
　64-2　范例程序 ... 490

UNIT 65　使用 SurfaceView 进行高速绘图 ... 494
　65-1　使用 SurfaceView 的步骤 ... 494
　65-2　范例程序 ... 495

UNIT 66　3D 绘图 ... 500
　66-1　3D 绘图的基本概念 ... 500
　66-2　3D 绘图程序 ... 502

PART 13　Google 地图程序

UNIT 67　使用 Google 地图 ... 509
　67-1　开发 Google 地图应用程序的准备工作 ... 509
　67-2　建立 Google 地图应用程序的步骤 ... 513
　67-3　范例程序 ... 515

UNIT 68　Google 地图的进阶用法 ... 519
　68-1　地图的缩放和拖曳功能 ... 519
　68-2　加上键盘控制功能 ... 521
　68-3　切换地图显示模式 ... 522

UNIT 69　帮地图加上标记 ... 526
　69-1　地图程序使用 Overlay 的步骤 ... 526

	69-2 范例程序	531
UNIT 70	加上定位让地图活起来	533
	70-1 移动设备的定位技术	534
	70-2 第一种定位方法——使用 MyLocationOverlay	535
	70-3 第二种定位方法——使用 LocationManager	538

PART 14 拍照、录音、录像与多媒体播放

UNIT 71	使用 MediaPlayer 建立音乐播放器	544
	71-1 音乐播放程序的架构	544
	71-2 MediaPlayer 类的用法	545
	71-3 范例程序	548
UNIT 72	播放背景音乐和 Audio Focus	555
	72-1 用 Service 的方式运行 MediaPlayer	555
	72-2 使用状态栏信息控制 Foreground Service	557
	72-3 使用 Audio Focus 和 Wake Lock	559
	72-4 播放不同来源的文件	560
	72-5 范例程序	561
UNIT 73	录音程序	571
	73-1 MediaRecorder 类的用法	571
	73-2 范例程序	574
UNIT 74	播放影片	579
	74-1 Android 支持的图像和影片文件格式	579
	74-2 使用 VideoView 和 MediaController	580
UNIT 75	拍照程序	585
	75-1 Camera 和 SurfaceView	585
	75-2 范例程序	587
UNIT 76	录像程序	594
	76-1 Camera 和 MediaRecorder 通力合作	594
	76-2 在接口布局文件中建立 SurfaceView	595
	76-3 范例程序	596

PART 15　WebView 与网页处理

UNIT 77　WebView 的网页浏览功能 606
77-1　WebView 的用法 606
77-2　范例程序 608

UNIT 78　自己打造网页浏览器 611
78-1　WebView 的网页操作方法 612
78-2　设置 WebViewClient 和 WebChromeClient 613
78-3　范例程序 615

UNIT 79　JavaScript 和 Android 程序代码之间的调用 622
79-1　从 JavaScript 调用 Android 程序代码 622
79-2　从 Android 程序代码调用 JavaScript 的 function 624
79-3　使用 WebView 的 loadData() 625
79-4　范例程序 625

PART 16　开发 NFC 程序

UNIT 80　NFC 程序设计 632
80-1　Android 处理 NFC tag 数据的方式 633
80-2　开发 NFC 程序 635

UNIT 81　把资料写入 NFC tag 638
81-1　Android 4.X 的 Android Application Record（AAR） 639
81-2　Android Beam 640

UNIT 82　NFC 的进阶用法 642
82-1　让运行中的程序优先处理 NFC Intent 644

附录　本书光盘内容与使用说明 647

PART 1 拥抱 Android

UNIT 01　Android 造时势或是时势造 Android
UNIT 02　安装 Android 程序开发工具
UNIT 03　建立 Android 应用程序项目
UNIT 04　Eclipse 程序项目管理技巧

UNIT 1
Android 造时势或是时势造 Android

信息科技之所以能有今日的成就，计算机产业的不断创新和进步是幕后最大的功臣。最早期的计算机不仅体积庞大而且计算速度缓慢，但是随着半导体技术的发展，以及计算机软硬件技术的进步，计算机的体积逐渐缩小，同时运算效能却不断地提升。如果今天有一个可以随身携带的电子设备，它的功能就像计算机一样具有方便的图形操作接口，而且可以安装新的程序，也提供程序开发工具让程序开发人员编写应用程序，同时还能够上网以及具备强大的影音功能，那么它未来就有很大的发展潜力，这就是我们的主角 Android 平台，而它的势力正快速地从智能手机蔓延到平板电脑。

有人将智能手机和平板电脑的兴起称为后 PC 时代，意思是说计算机已经开始走出以往我们熟知的桌上型和笔记本电脑的框架，进一步深入我们的日常生活当中。这些后 PC 时代的小型计算机设备在显示接口和操作方式上的特色，让新型态的操作系统有崭露头角的机会，像是 Symbian、iOS（Apple）、BlackBerry OS、Palm OS、MS Windows Phone（以前称为 Windows Mobile）以及 Google 的 Android（Android 单字的意思是"机器人"，因此 Android 系统的代表图腾就是一个可爱的机器人，如图 1-1 所示）。在这些众多的竞争者中，虽然 Android 是后起之秀，但是却已经后来居上成为目前占有率最高的智能手机平台。Android 平台之所以能够成功，除了因为它是国际信息科技大厂 Google 的产品之外，另外一个重要的原因是采取开放原始码的策略，有关 Android 手机平台的技术资料与开发工具完全公布在以下的网址：

图 1-1　Android 系统的代表图案

- http://android.git.kernel.org　　Android 手机平台原始码
- http://developer.android.com　　Android 手机程序开发工具

不论是厂商或是个人都可以自由取得相关的数据或是软件。

1-1　Android 从何而来

虽然现在 Android 已经和 Google 划上等号，但其实 Android 的诞生地并不是在 Google，它是由一位名字叫做 Andy Rubin 的美国人所开发出来的软件系统，2005 年 Google 公司收购了 Android 系统并网罗 Rubin 先生进入 Google 继续发展 Android 直到现在。2007 年 11 月 Google 结合 33 家手机相关软硬件厂商组成开放手持设备联盟（Open Handset Alliance，简称 OHA），并共同对外公开 Android 智能手机平台。Google 在 OHA 联盟所扮演的角色是研发 Android 系统核心程序，并提供手机软硬件厂商免费使用。除此之外，Google 也提供开发 Android 程序的软件开发工具（Software Development Kit，简称 SDK），并将它公布在网络上供大家免费下载。为了提高大家开发 Android 程序的动机以及速度，Google 先后举办过二次 Android 程序开发挑战赛（Android Developer Challenge，ADC），以下我们将 Android 平台版本和功能的演进整理成表 1-1。

表 1-1　不同 Android 版本的功能演进

Android 版本（名称）	发布日期	功能说明
Android 1.0	2008 年 9 月	Android 智能手机平台诞生
Android 1.5 (Cupcake)	2009 年 4 月	<ul><li>Android 智能手机平台诞生</li><li>在视频模式下可以观看和拍摄影片</li><li>可以直接将影片上传至 YouTube</li><li>可以直接将照片上传至 Picasa</li><li>虚拟键盘</li><li>Bluetooth 功能</li><li>动画转场效果</li></ul>
Android 1.6 (Donut)	2009 年 9 月	<ul><li>改良拍照、摄影和浏览接口</li><li>更新文字转语音核心</li><li>支持 WVGA 屏幕模式</li><li>改善搜寻功能</li><li>改良运行速度</li></ul>
Android 2.1 (Éclair)	2010 年 1 月	<ul><li>运行速度优化</li><li>支持更多屏幕模式</li><li>改良虚拟键盘功能</li><li>支援 Bluetooth 2.1</li><li>支援 HTML 5</li></ul>
Android 2.2 (Froyo)	2010 年 5 月	<ul><li>系统优化</li><li>SD 卡支援能力</li><li>支援 Flash 10.1</li><li>支援 Microsoft Exchange</li><li>快捷键盘语系转换</li></ul>

续表

Android 版本（名称）	发布日期	功能说明
Android 2.3.3	2010 年 12 月 2011 年 2 月	● 改良用户操作接口 ● 支持更大屏幕和分辨率 ● 支援 WebM/VP8 影片 ● 支持 AAC 音频编码 ● everb、equalization、headphone virtualization、bass boost 音讯效果 ● 游戏的声音和绘图功能强化 ● 支持多摄影机
Android 3.0 (Honeycomb)	2011 年 2 月	专门给平板电脑使用的版本，由于手机和平板电脑的屏幕尺寸和使用方式有相当程度的差异，因此 Android 3.0 较以往的版本做了很大的改变，包括： ● 用户接口 ● 键盘操作方式 ● 2D 和 3D 绘图效能 ● 网页浏览器 ● 影片播放和管理 ● 复制粘贴功能
Android 2.3.4	2011 年 5 月	● 新增 Open Accessory Library 提供对 USB 设备的联机能力
Android 3.1	2011 年 5 月	● 提供对 USB 设备的联机能力 ● 支援 PTP（Picture Transfer Protocol） ● 新增不同输入设备的支持，例如 mice、trackballs、joysticks、…… ● 支持 RTP（Real-time Transport Protocol） ● 强化 Animation framework、UI framework、Network 的能力
Android 3.2	2011 年 7 月	● 支持更多型号的平板电脑 ● 新增应用程序界面的放大模式
Android 4.0	2011 年 10 月	● 适用手机、平板电脑和其他设备 ● 改良的操作接口 ● 针对多核心 CPU 进行优化 ● 强化多媒体处理能力 ● 新增人脸侦测功能 ● 新增 Android Beam 功能（NFC） ● 加强 text-to-speech engine

1-2 Android 的功能、应用和商机

从技术面来看，Android 是一个采用 Linux 为核心的手机操作系统。Linux 在 Android 平台中所扮演的角色是系统资源管理，像是内存、网络、电源、驱动程序等，读者可以参考图 1-2 的 Android 平台架构图。在 Linux 核心的上一层则是各种功能的链接库，包括 C 链接库、SGL 和 OpenGL ES 绘图和多媒体链接库、SQLite 数据库链接库等。另外还有一个和链接库具有同样重要功能的组件就是 Android Runtime，也就是这个组件让 Android 平台可以支持 Java 程序语言，它专门负责转换 Java 运行码成为底层的硬设备机器码。在链接库和 Android Runtime 的上一层是所谓的 Application

Framework。这一层定义了 Android 平台的应用程序架构,最后在 Application Framework 之上才是 Android 平台的应用程序。

图 1-2　Android 平台架构图

虽然 PC 和 Android 系统的架构相似,但是二者的使用在时、地、物等方面都有所不同。在先进国家中,手机几乎已经成为每个成年人的随身必备物品,手机小巧玲珑,随手就可使用,但缺点是运算能力不及笔记本强大。笔记本的优点是计算能力强、扩充性较高,但缺点是携带不便。就是因为这些时、地、物的差别,让手机的应用程序可以发展出和笔记本不一样的特色。以 2008 年举办的 "Android 程序开发挑战赛" 来说,获得评审青睐的程序多半是和日常生活的应用有关,例如 CompareEverywhere 程序的功能是根据手机用户的位置(利用 GPS 定位功能),查询附近商店的商品价格,让用户可以用最便宜的价格买到所需的物品。Life360 则是利用 BBS 聊天室的功能,让用户可以和邻近的朋友分享信息,或是寻求生活上的协助,此外还有许多其他功能的程序,有兴趣的读者可以到下列网址取得更多的信息:

http://code.google.com/android/adc

为了吸引程序开发人员加入编写 Android 程序的行列,Google 建立了一个 Android Market 网站,让程序开发人员辛苦编写的 Android 程序可以上传到该网站上供人查询、浏览、免费下载或是付费下载。Google 要求程序的作者必须在 Android Market 网站上建立一个账号(和 Gmail 账号共享),并支付 25 美元的注册费(只在第一次建立账号时收取)。日后如果程序作者在 Android Market 上贩卖程序,必须将所得的 30%支付给 Google。Android Market 会记录每一个程序的下载次数和使用者意见。藉由 Android Market,Google 希望可以和程序开发人员一起达到双赢的结果。

另外一个读者一定也很关心的问题是 Android 平台的前景究竟如何?我们可以通过观察市场的一些数据来回答这个问题。首先是知名手机系统的市占率,目前能见度比较高的是以下 4 种:RIM 黑莓机、Apple iPhone、Microsoft Windows Mobile 手机和 Android 手机。从 2009 年第 4 季到 2011 年第 2 季它们的市占率消长状况如图 1-3 所示。从变化趋势中可以发现 Android 手机不但呈现高度的成长，而且造成其他 3 者的成长停滞甚至下滑。另外一项值得注意的数据是 Android 系统平台的应用程序数量，图 1-4 是从 2009 年到 2011 上半年的 Android Market 网站的应用程序数量，从它的增加趋势来看，未来的前景也毋须怀疑。

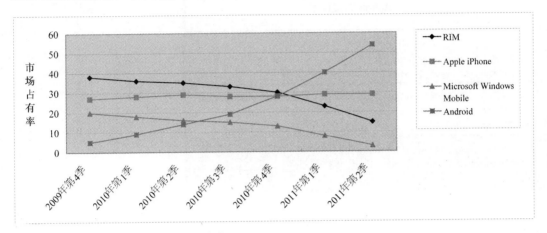

图 1-3　不同品牌的智能手机市占率的变化

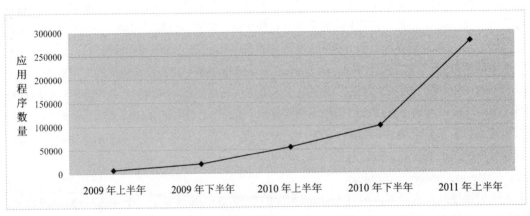

图 1-4　Android Market 网站上的应用程序增长趋势

1-3　先睹为快——Android 手机和平板电脑模拟器

以 Android 版本的应用来说，1.X 和 2.X 版是提供手机使用，3.X 版则是平板电脑专用，而最新发布的 Android 4 则是将平板电脑和手机两个不同的版本合而为一，甚至加上未来可应用到其他不同设备的能力，像是网络电视或是其他智能型家电，由于 Android 4 同时支持手机和平板电脑，因此它的模拟器有两种模式，一种是手机模拟器，另一种是平板电脑模拟器。模拟器的运行模式是由选择的屏幕大小决定，图 1-5 是选择手机型态的屏幕尺寸后的模拟器运行界面，用鼠标按住屏幕下方的解锁区域往右拉至锁头图标位置就会出现图 1-6 的 Home screen。界面下方有 5 个按钮，由左至右分别为"拨打电话"、"个人资料设置"、"浏览应用程序"、"网络聊天"和"网页浏览器"，按下中间的"浏览应用程序"按钮后会出现图 1-7 所示的界面。

图 1-5　Android 4 手机模拟器的初始界面

图 1-6　Android 4 手机模拟器的操作首页

图 1-8 是 Android 4 平板电脑模拟器的启动屏幕，我们把界面上的解锁按钮按住往右拖曳到解锁位置就会看到图 1-9 的 Home screen。Home screen 的右上方有一个 Apps 按钮可以用来检视已经安装的应用程序，如图 1-10 所示，Home screen 左下方的三个按钮由左至右分别为"回上一页"、"回到 Home screen"和"浏览最近使用的程序"。看完这些 Android 模拟器的界面读者是不是觉得跃跃欲试，迫不急待想要亲自试看看？下一个单元就让我们一起动手安装 Android 程序开发工具！

图 1-7　Android 4 手机模拟器中的应用程序

图 1-8　Android 4 平板电脑模拟器的启动屏幕

图 1-9　Android 4 平板电脑模拟器的 Home screen　　图 1-10　按下 Apps 按钮后进入应用程序浏览界面

UNIT 2
安装 Android 程序开发工具

在开始学习设计 Android 应用程序之前，必须先打造一个 Android 应用程序的开发环境。建立 Android 应用程序开发环境需要安装一些工具软件，而学习 Android 应用程序设计的一大好处是这些工具软件完全免费，只要上网就可以下载。而且这些免费下载的软件，在质量和功能上相较于市面上贩卖的商业程序也毫不逊色，并且对于各种操作系统都百分之百支持，包括 Windows XP、Windows Vista、Windows 7、Linux、Mac OS 等。

2-1 不同操作系统的开发工具版本

因为 Android 是整合目前世界上许多开放源代码技术而成的平台，并不是单一厂商的独家产品，因此在安装过程中需要到不同的网站下载软件，并且软件的版本也要谨慎挑选（因为考虑到软件间的兼容性），而不像一般商业软件只要运行一个安装程序就大功告成，因此安装过程中需要多一点的耐心。不过也毋须太过担心，因为这些开放源代码软件都已经发展得非常成熟，只要依照后续的说明一步一步操作即可顺利完成。以下我们就以 Windows 操作系统为例，详细介绍 Android 程序开发工具的安装过程。

在开始安装之前让我们先检查一下运行 Android 程序开发工具所需的软硬件条件。就硬件而言必须考虑计算机的运行速度和硬盘容量，因为 Android 程序开发工具都是使用 Java 语言编写，Java 程序的运行速度比较慢，而且需要占用较多的内存，因此如果 CPU 不够快，或是内存不够大，在操作的过程中会经常处于等待的状态，为了能够平顺地运行 Android 程序开发工具，所使用的计算机必须符合以下所列的最低需求。

计算机硬件的最低需求

- CPU 运作速度（频率）：2.5GHz（建议双核以上）

- 内存：2GB
- 硬盘剩余空间：4GB

如果读者的计算机是在 3 年内购买的新型计算机，应该能够符合以上的条件，如果是旧型的计算机，也可以先尝试安装看看，然后再视情况决定是否需要升级。

软件需求

软件需求包括操作系统的种类和开发工具的相关软件。

1. 操作系统

    Android 程序开发工具支持的 Windows 操作系统版本包括 Windows XP、Windows Vista 和 Windows 7。

2. Java Development Kit（JDK）

    请读者注意一定要安装 JDK 而不能只安装 JRE（JDK 包含 JRE），JDK 的版本必须是 JDK5 或 JDK6。

3. Eclipse

    Eclipse 是由 IBM 捐赠的开放源代码软件，它是一个功能超强的程序开发平台，经过全世界开放源代码程序设计人员的通力合作，目前已经发展出支持多种程序语言开发的版本，包括 Java、C/C++、PHP、Software modeling 等。我们需要的版本是 Eclipse IDE for Java Developers，目前最新版本是 3.7.1，旧版的 3.6 也可以和 Android 程序开发工具兼容。

4. Android Development Tools（ADT）plugin for Eclipse

    这是用来编写 Android 程序的工具软件，它是一个 Eclipse 的 plugin（插件），也就是必需安装在 Eclipse 中和 Eclipse 一起运作。

5. Android Software Development Kit（Android SDK）

    Android SDK 包括开发 Android 程序的过程中需要用到的资源，像是链接库、程序调试工具、平板电脑和手机模拟器等。

由于 Android 程序开发工具会持续地更新，在安装之前读者可以先到下列网址查看最新的信息：

http://developer.android.com/sdk/installing.html

在了解计算机的软硬件需求之后，接下来就让我们到相关网站下载所需的软件开始进行安装吧！

2-2 安装 Android 程序开发工具的步骤

第一步 安装 JDK

JDK 是编写 Java 程序必备的工具，如果读者曾经学过 Java 程序设计，那么计算机中应该已经安装好 JDK。假如不确定或者不知道计算机中安装的 JDK 版本，可以运行 Windows 的"控制台>新增或移除程序"，找找看其中是否有一项叫做"Java SE Development Kit"（旧版的 JDK 名称叫做"Java 2 SDK, SE"）。如果找到了（注意版本必须是 5 或 6）代表计算机中已经安装好 JDK，那么就可以跳过这一个步骤。如果找不到或是版本比 5 或 6 还低，请先将旧版的 JDK 移除，然后依照下列说明安装新的版本。

请开启网页浏览器在网址列输入下列网址，或是利用 Google 搜索 JDK 再从中选择下列网址：
http://www.oracle.com/technetwork/java/javase/downloads/index.html

开启以上网址后会看到如图 2-1 所示的网页，如果读者看到的网页编排和书上的图有些不同也不用担心，因为网页上的数据随时都有可能更新，但是内容基本上是相同的。如果读者仔细看一下目前最新的版本号码，会发现已经是 JDK 7，与 Android 网站建议的版本不同，如果将网页往下拉就可以找到 JDK 6 的版本，然后单击 JDK Download 按钮，之后就会出现图 2-2 的界面。在网页下方有各种操作系统的 JDK 安装文件，请先单击同意遵守版权的项目，再选择适合的操作系统版本进行下载，等下载完成之后直接运行下载的文件就可以完成安装。

图 2-1　Java 的官方网页，请单击 JDK Download 按钮下载 JDK

第二步 安装 Eclipse

Eclipse 是一个通用的程序开发平台，不仅功能强大，还有一个优点就是不需要安装，只要把它解压缩到一个新的文件夹就可以运行。这就是所谓的绿色软件，它不会随意变更计算机的设置，也不会在移除之后留下任何额外的文件。首先我们到 Eclipse 的官方网站，请读者开启网页浏览器

在网址列输入下列网址,或是利用 Google 搜索 Eclipse 再从中选择下列网址:

http://www.eclipse.org/

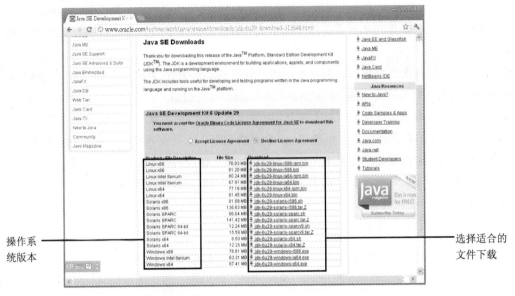

图 2-2　选择符合操作系统的 JDK 版本

　　开启该网址后就会看到如图 2-3 的界面,请单击右上方的 Download Eclipse 按钮就会进入图 2-4 的网页,请单击其中的 Eclipse IDE for Java Developers 项目进入图 2-5 的界面,最后从右边的版本列表中选择适合的操作系统版本进行下载。

图 2-3　Eclipse 开发工具的官方网站

图 2-4　下载 Eclipse 的网页界面

图 2-5　选择我们需要的 Eclipse IDE for Java Developers 版本

　　完成下载后请建立一个新的文件夹，例如可以在 C:\Program Files 中新增一个名为 Eclipse 的文件夹，然后把下载的 zip 文件全部解压缩到该文件夹即可完成 Eclipse 的安装。当要启动 Eclipse 时，只要运行其中的 eclipse.exe 即可。为了日后使用上的方便，读者可以在桌面上建立一个 eclipse.exe 文件的快捷方式。

> **补充说明**　**安装 Android 程序开发工具的注意事项**
>
> 在建立 Eclipse 文件夹的路径中（包括从驱动器号开始）不要有任何的中文名称，否则在运行开发工具的过程中可能会出现不明原因的错误。

第三步 安装 Android Development Tools (ADT) plugin for Eclipse

ADT plugin for Eclipse 可以在 Eclipse 主程序中完成下载和安装。首先运行上一个步骤安装好的 eclipse.exe，然后选择 Help>Install New Software 就会出现图 2-6 所示的对话框，按下对话框右上方的 Add 按钮并输入下列信息。

- Name: Android plugin
- Location: https://dl-ssl.google.com/android/eclipse/

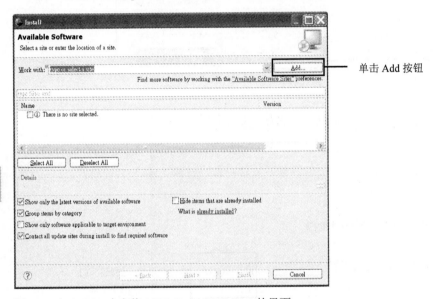

图 2-6 在 Eclipse 中安装 ADT plugin for Eclipse 的界面

单击 OK 按钮，在对话框中央的列表会出现 Developer Tools 项目。如果没有出现，请读者再重新单击 Add 按钮，把 Location 改为 http://dl-ssl.google.com/android/eclipse/（也就是把原来的 https 改成 http）应该就会出现 Developer Tools。把该项目打勾然后依照说明操作就可以完成安装。安装后会显示一个对话框要求重新启动 Eclipse，单击 Restart Now 重新启动即可，当 Eclipse 重新启动之后会显示一个对话框询问安装 Android SDK 的事项，请读者参考下一个步骤继续操作。

第四步 安装 Android Software Development Kit（Android SDK）

完成上一个步骤之后会显示图 2-7 的对话框让我们继续安装 Android SDK，请读者选择 Install new SDK，然后勾选 Install the latest available version of Android APIs，在下方的 Target Location 字段中设置好 Android SDK 的储存路径，然后单击 Next 按钮就会出现图 2-8 所示的对话框，左边会列出要安装的项目，确认无误后单击 Install 按钮开始安装。

拥抱 Android **PART 1**

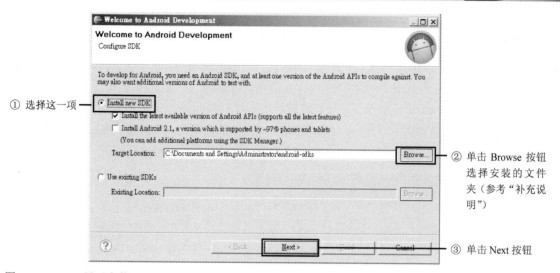

图 2-7　Eclipse 显示安装 Android SDK 的对话框

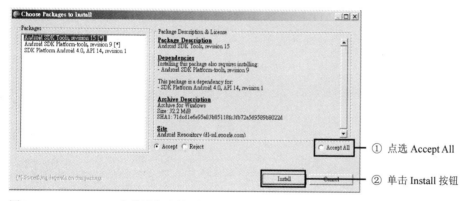

图 2-8　Android SDK 安装过程中的项目说明

补充说明　**Android SDK 的储存路径**

1. 在 Android SDK 的储存路径中不要出现任何中文文件夹名称，否则在开发 Android 程序时会出现错误信息。
2. 我们可以将 Android SDK 存放在 Eclipse 文件夹中，这样安装好开发环境之后就可以将整个 Eclipse 文件夹压缩复制到 U 盘，然后在任何一台含有 JDK 的计算机上将 Eclipse 文件夹解压缩到任何一个文件夹，再依照后续的说明完成两项设置就可以开始开发 Android 程序。

如果读者顺利完成上述的操作就可以跳到下一个步骤，如果安装过程中出现错误，则可以改用下列方式安装。请开启网页浏览器在网址列输入下列网址，或是利用 Google 搜索 Android SDK 再从中选择下列网址。

15

http://developer.android.com/sdk/index.html

开启该网址后会出现图 2-9 的网页，网页上会列出不同操作系统的 Android SDK 版本。以 Windows 操作系统来说，官方网站推荐使用 installer_r15-windows.exe，因此我们单击该文件进行下载，然后运行该文件。首先会显示一个确认 JDK 版本的对话框，单击 Next 按钮会显示图 2-10 的界面让我们设置 Android SDK 的储存路径，设置好之后单击 Next 按钮决定快捷方式的位置，如果不想建立快捷方式可以勾选下方的 Do not create shortcuts，然后单击 Install 按钮进行安装。安装完成后会询问是否运行 SDK Manager，勾选该项目继续运行 SDK Manager。

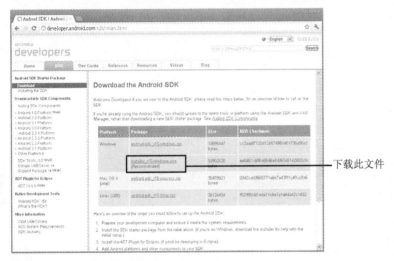

图 2-9　Android SDK 的官方网站

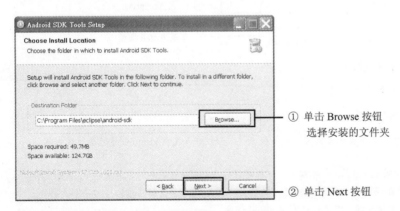

图 2-10　设置 Android SDK 的储存路径

启动 SDK Manager 之后会看到图 2-11 的对话框，其中有许多可以展开的项目，包括一些相关工具程序和组件，以及不同版本的 Android SDK，每一个项目的后面都会显示安装状态。读者可以勾选需要安装的项目，再按下右下方 Install packages 按钮进行安装，由于这个步骤需要下载大量的数据，如果网络速度不够快，可能需要等待一段时间，完成后关闭对话框。

补充说明 使用 Android SDK Manager

图 2-11 是 Android SDK Manager 的操作界面，往后如果我们需要安装其他版本的 Android SDK 或是删除已经安装的版本都必须通过它来完成。需要运行 Android SDK Manager 时先启动 Eclipse，然后选择主选单的 Window > Android SDK Manager。

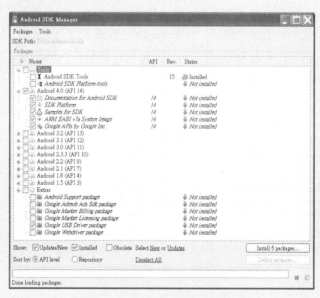

图 2-11 SDK Manager 操作界面

最后还必须在 Eclipse 中设置 Android SDK 的路径才可以使用，请读者运行 Eclipse，单击选单上的 Window>Preferences 便会出现图 2-12 的对话框，在对话框的左边单击 Android 项目，然后在右边的 SDK Location 中填上 Android SDK 的储存路径（也可以使用右边的 Browse 按钮来选择），最后单击下方的 Apply 按钮，再单击 OK 按钮。

第五步 建立 Android Virtual Device（AVD）

AVD 就是 Android 的设备模拟器，在编写 Android 应用程序的过程中，我们必须让程序在模拟器中运行才可以看到结果。请先运行 Eclipse，单击主选单的"Window>AVD Manager"就会出现图 2-13 的对话框，单击对话框右上方的 New 按钮便会出现图 2-14 的对话框，在 Name 字段中帮这个虚拟设备取一个名称，在 Target 字段中选择一个 Android 版本，在下方的字段还可以指定 SD 卡的容量、设备的屏幕大小和分辨率。如果在 Target 字段挑选的是 Android 4.0 以上的版本，那么所设置的屏幕大小和分辨率决定究竟这是手机模拟器还是平板电脑模拟器。在 Skin 字段中我们可以选择一个内建的屏幕型号，或是自行设置屏幕的宽和高，然后在 Hardware 字段中可以设置 Abstracted LCD density，这二者的搭配结果决定模拟器的屏幕尺寸。请读者参考表 2-1，它是 Android SDK 用来分类屏幕尺寸的方式，上方第一列就是 Abstracted LCD density 的值，表格中就是 Skin 字段设置的屏幕型号或是自行输入的宽和高。假设在 Skin 字段选择 WVGA800（480×800），然后将 Abstracted LCD density 设置成 160，那么模拟器的屏幕尺寸分类就是 Large，可是如果将 Abstracted LCD density 换成 240，那么模拟器的屏幕尺寸分类就变成 Normal。对于 Android 4.0 以上来说，屏幕尺寸分类为 Small、Normal 和 Large 的都是手机模拟器，屏幕尺寸分类为 Extra Large 的才是平板电脑模拟器。如果 Target 字段是设置 Android 3.X 则一律是平板电脑模拟器，如果 Target 字段是设置 Android 2.X 或 1.X 则一律是手机模拟器。设置好模拟器的属性之后，按下 Create AVD 按钮就完成新增模拟器的动作。视情况需要我们可以建立多个不同型态的模拟器，以便测试程序在各种设备的运行结果。

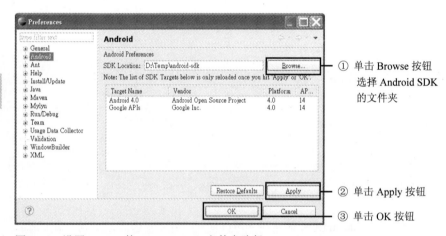

图 2-12 设置 Eclipse 的 Android SDK 文件夹路径

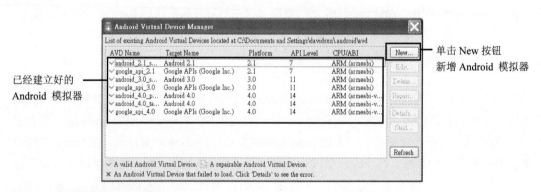

图 2-13 新增模拟器的对话框

拥抱 Android **PART 1**

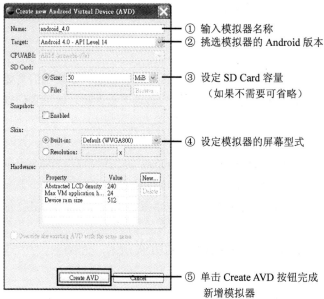

① 输入模拟器名称
② 挑选模拟器的 Android 版本
③ 设定 SD Card 容量
　（如果不需要可省略）
④ 设定模拟器的屏幕型式
⑤ 单击 Create AVD 按钮完成新增模拟器

图 2-14　设置模拟器的名称和属性

表 2-1　模拟器屏幕尺寸的分类

屏幕尺寸的分类	Low density (120), ldpi	Medium density (160), mdpi	High density (240), hdpi	Extra high density (320), xhdpi
Small	QVGA（240×320）		480×640	
Normal	WQVGA400（240×400） WQVGA432（240×432）	HVGA（320×480）	WVGA800（480×800） WVGA854（480×854） 600×1024	640×960
Large	WVGA800（480×800） WVGA854（480×854）	WVGA800（480×800） WVGA854（480×854） 600×1024		
Extra Large	1024×600	WXGA（1280×800） 1024×768 1280×768	1536×1152 1920×1152 1920×1200	2048×1536 2560×1536 2560×1600

完成上述的所有步骤之后，安装 Android 程序开发工具的任务就大功告成了。如果读者的计算机是使用 Linux 操作系统或是 Mac 系列的计算机，安装 Android 程序开发工具的过程和以上介绍的步骤类似，只是在软件版本的选择上要挑选适合的操作系统版本，请读者同样依照上述步骤操作即可顺利完成安装。

2-3　Android 程序开发工具的维护和更新

Android 程序开发工具包含许多不同来源的软件，这些软件会不断地推陈出新。如果我们需要更新其中某一个软件，或是觉得系统有问题想移除某一个软件再重新安装该怎么办呢？以下我们以 Windows 版本的开发工具为例来加以说明。

1. 移除 JDK 再重新安装

    移除前如果要先检查计算机中安装的 JDK 版本可以运行 Windows 的"控制面板>添加或删除程序"，寻找其中名为 Java SE Development Kit 的项目。该项目名称的后面会有一个数字代表版本号码（例如 5 或 6），如果想要移除旧的版本，可以先用鼠标单击它，然后单击"移除"按钮。

    要安装新版的 JDK，首先要连到 JDK 的官方网站下载想要的 JDK 版本，然后运行它的运行文件即可完成安装。

2. 移除 Eclipse 再重新安装

    Eclipse 是一个绿色软件，只要把它复制到计算机磁盘中就可以运行。如果想要移除它，也只要删除它所在的文件夹即可。但是请注意，如果 Android SDK 是放在 Eclipse 文件夹中，记得先把它复制出来，等重新安装好 Eclipse 之后再复制回去，并依照前面第四步骤最后的说明设置好 Android SDK 的路径。

3. 移除 ADT plugin for Eclipse 再重新安装

    运行 Eclipse 然后单击菜单 Help>About Eclipse，在出现的对话框中单击 Installation Details 按钮就会出现图 2-15 的对话框。在对话框中会列出安装在 Eclipse 中的工具，请找出"Android DDMS"、"Android Development Tools"、"Android Hierarchy Viewer"和"Android Traceview"4 个项目，选择它们后（可以同时按住键盘上的 Ctrl 键进行多重选择）单击下方的 Uninstall 按钮就可移除。如果是想要更新，就单击 Update 按钮。移除后要重新安装请参阅前面的安装步骤重新做一次即可。

4. 移除 Android SDK 再重新安装

    先删除储存 Android SDK 的文件夹，然后依照前面介绍的安装步骤重新安装 Android SDK。如果是要更新，可以参考前面的"补充说明"运行 Android SDK Manager，在对话框的 Status 字段会列出可以更新的项目，勾选需要更新的项目后单击右下方的 Install packages 按钮进行安装。

5. 移动 Eclipse 程序的文件夹

    安装好 Android 程序开发工具之后，如果有一天突然想把 Eclipse 程序的文件夹搬到其他地方也可以，只不过如果 Android SDK 是放在 Eclipse 文件夹中，在移动 Eclipse 程序文件夹之后必须重新设置 Android SDK 的文件夹路径，请读者依照前面第四步骤最后的说明操作即可。

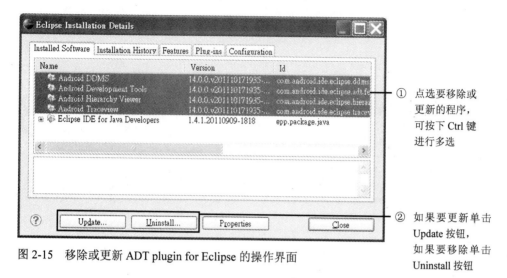

① 点选要移除或更新的程序，可按下 Ctrl 键进行多选

② 如果要更新单击 Update 按钮，如果要移除单击 Uninstall 按钮

图 2-15　移除或更新 ADT plugin for Eclipse 的操作界面

Android 版本	1.X	2.X	3.X	4.X
适用性	★	★	★	★

UNIT 3
建立 Android 应用程序项目

在完成程序开发工具的安装之后,读者是不是迫不及待想试用看看呢？第一次使用新东西总是让人感到特别地新鲜和期待,现在就让我们来体验一下 Android 应用程序的开发流程吧！

建立 Android 应用程序项目的过程和建立一般 PC 程序项目的过程大致相同,如果读者已经具备编写计算机程序的经验应该很容易就能够上手,如果读者还不熟悉计算机程序的开发也无妨,只要依照下列说明一步一步操作,就能够顺利完成第一个 Android 应用程序。

Step 1 运行 Eclipse 程序。

Step 2 从选单中选择 File > New > Project 就会出现图 3-1 所示的对话框。

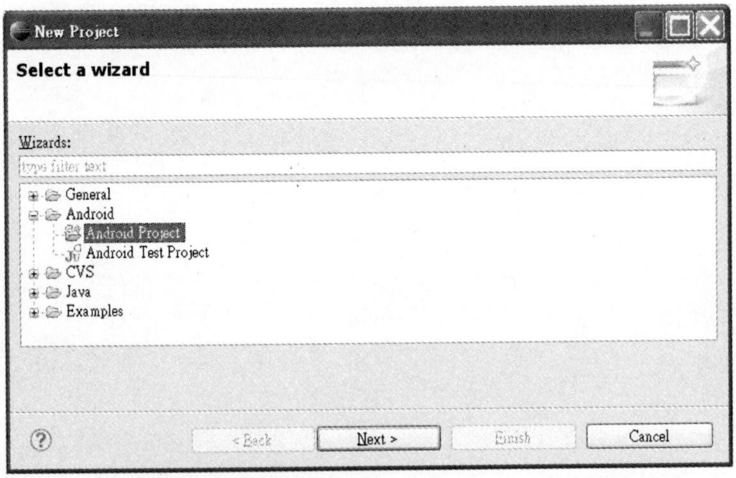

图 3-1 建立新程序项目的对话框

Step 3 在对话框中间的清单中单击 Android > Android Project，再单击 Next 按钮就会出现图 3-2 所示的对话框。

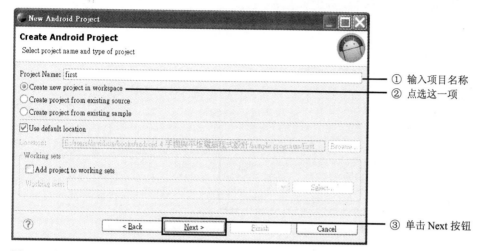

图 3-2 设置 Android 应用程序项目的属性－步骤一

Step 4 输入以下信息：

1. Project Name：自己帮此项目取一个名字，例如 first。
2. 单击 Create new project in workspace 项目。
3. 勾选 Use default location。

单击 Next 按钮就会出现图 3-3 的对话框。

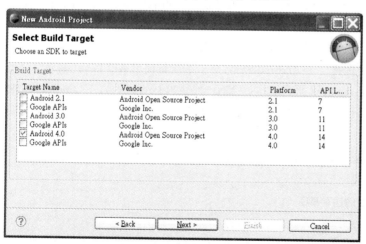

图 3-3 设置 Android 应用程序项目的属性－步骤二

Step 5 选择此程序项目使用的 Android 平台版本（Build Target），例如勾选 Android 4.0，然后单

击 Next 按钮就会出现图 3-4 的对话框。

图 3-4　设置 Android 应用程序项目的属性 - 步骤三

Step 6 继续输入以下信息：

1. Application Name：当此程序运行时，显示在屏幕上方的标题，例如可以填上"我的第一个 Android 应用程序"。
2. Package Name：就字义上来说是组件名称，但其实这个字段是决定程序文件在项目文件夹中的储存路径，它是用网址的格式代表，但是是从大区域到小区域而不是网址惯用的小区域到大区域，例如可以填上 tw.android，注意至少要有两层，也就是 xxx.xxx。
3. Create Activity：这是程序的主类名称，主类就是程序开始运行的地方，这个主类是继承 Activity 类，例如读者可以输入 Main。
4. Minimum SDK：限制这个程序项目必须在哪一个版本以上的 Android 平台上运行，读者可以从下拉式菜单中选择一个 Android 版本，这个字段是使用 API Level 编号也就是 3, 4, 5, …代表，每一个 Android 平台版本都会有一个 API Level 编号，在 Android SDK 技术文件中经常以 API Level 标号来描述 Android 平台的版本，因此建议读者留意它们之间的对应关系。

最后单击 Finish 按钮就完成建立 Android 程序项目的步骤。

补充说明　建立 Android 程序项目的注意事项

1. 程序项目的 Project Name 不可以使用中文。
2. 程序项目的 Application Name 可以使用中文。
3. Android 3.0（API Level 11）是 Android 平台发展过程中的重要分界，在 Android 3.0 以前都是手机程序，Android 3.X 则是平板电脑程序，Android 4.0 以后才是手机和平板电脑的通用版本，因此如果我们将 Minimum SDK 字段设置为 11 以下，就代表此程序项目是属于旧 Android 版本的手机程序，如果设置成

11～13（13是Android 3.2的API Level编号）就是平板电脑程序，设置成14以上才是手机和平板电脑的通用程序。

建立好程序项目之后单击上方工具栏的 Run 按钮就会出现图3-5所示的对话框，它是用来设置程序的运行模式，每一个程序项目第一次运行的时候都会出现这个对话框，请读者选择 Android Application 再单击 OK 按钮。如果读者依照上一个单元的说明分别建立了 Android 4 的手机和平板电脑模拟器，可以让这个程序项目分别在这两个模拟器上运行，就可以看到如图3-6和图3-7所示的运行界面。如果读者没有事先启动 Android 模拟器，第一次运行程序时会需要比较长的时间（可能数分钟），因为 Eclipse 必须先启动 Android 模拟器并等待启动完成，然后才能在上面运行程序。

图 3-5　选择 Android 程序项目的运行模式

图 3-6　在 Android 4 手机模拟器上的运行界面

图 3-7　在 Android 4 平板电脑模拟器上的运行界面

3-1 修改程序的接口

虽然我们已经完成了第一个 Android 应用程序，也看到了运行结果，可是还是觉得有些意犹未尽，因为我们并没有写任何程序代码，只是按几下鼠标输入几个文字罢了，整个程序从哪里开始运行到哪里结束还是没有头绪！没关系，接下来就让我们为程序加油添醋一下吧。

Step 1 请用鼠标单击 Eclipse 左边项目检查窗口中的项目名称将项目展开，如图 3-8 所示。

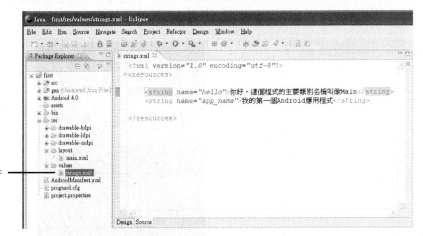

图 3-8　在 Eclipse 左边的项目检查窗口中展开项目内容

Step 2 项目名称下面有几个文件夹，请打开其中的 res 文件夹，然后打开里头的 values 文件夹，再用鼠标双击 strings.xml 文件，该文件便会出现在 Eclipse 中央的程序代码编辑窗口（请参考图 3-8）。

Step 3 strings.xml 是程序项目的字符串资源文件，它有两种编辑模式，读者可以在程序代码编辑窗口左下方找到两个小小的标签并用它们进行切换，第一种是 Design 模式，这是利用类似表格字段的方法，配合阶层状的展开模式进行编辑。第二种是 Source 模式，它是用原始的文本文件进行编辑。请读者使用 Source 模式将 hello 字符串的内容改成"你好，这个程序的主要类名称叫做 Main"，如下：

```
<?xml version="1.0" encoding="utf-8"?>
<resources>

    <string name="hello">你好，这个程序的主要类名称叫做 Main</string>
    <string name="app_name">我的第一个 Android 应用程序</string>

</resources>
```

拥抱 Android　　PART 1

Step 4　仿照第二步的方法打开 res/layout/main.xml 文件。

Step 5　在最后一行的上方，也就是</LinearLayout>的上一行插入下列的程序代码，输入时请注意遵守原来文件的锯齿状编排格式以方便阅读程序代码。如果程序项目是使用 Android 2.2 以后的版本，可以把 fill_parent 改成 match_parent。

```
<Button
    android:layout_width="fill_parent"
    android:layout_height="wrap_content"
    android:text="这是按钮"
    />
<ImageButton
    android:layout_width="fill_parent"
    android:layout_height="wrap_content"
    android:src="@drawable/ic_launcher"
    />
```

> **补充说明**　**接口布局文件的编辑技巧**
>
> res/layout/main.xml 是程序运行界面的定义文件，我们可以称它为"接口布局文件"，其中可以加入各式各样的接口组件。在设置接口组件的属性时，每一次输入"android:"之后稍停半秒钟就会自动弹出一个属性列表（如果没有请先关闭文件，再用鼠标右击该接口布局文件，从快捷菜单中选择 Open With > Android Layout Editor），我们可以继续输入属性名称的前几个字母，列表中会自动过滤出符合的属性，如果单击列表中的属性，还会出现另一个窗口说明该属性的效果。善用这项功能不但可以减少打错字的机会，还可以学到许多不同属性的用法。

完成后请读者按下上方工具栏中的 Run 按钮，再分别选取手机模拟器和平板电脑模拟器来运行程序，就可以看到如图 3-9 和图 3-10 的结果。读者可以发现程序显示的信息改变了，而且多了 2 个按钮。读者可以按下其中任何一个按钮，只是不会有任何效果，因为我们并没有加上处理按钮的程序代码，这一部分我们留待后续单元再作介绍。根据以上的操作，读者是不是对 Android 程序开始有一些感觉了呢？

图 3-9　修改后的程序在 Android 4 手机模拟器上的运行界面

27

图 3-10　修改后的程序在 Android 4 平板电脑模拟器上的运行界面

> **补充说明**　**Android 模拟的使用技巧**
>
> 如果想要结束在模拟器上运行的程序，可以按下模拟器的"回上一页"按钮，"回上一页"按钮是一个类似"回转"的半圆型箭头。如果我们结束模拟器，等一下再运行程序时必须重新将它启动，这将会很耗时间。

Android 版本	1.X	2.X	3.X	4.X
适用性	★	★	★	★

UNIT 4
Eclipse 程序项目管理技巧

俗话说"工欲善其事，必先利其器"，在学习 Android 程序设计的过程中，我们最常使用的软件就是 Eclipse，在实现不同的程序设计主题时，我们经常需要建立新的程序项目，或是修改之前已经完成的项目。除了自行建立程序项目之外，网络上也有别人已经写好的 Android 程序项目源代码可供下载，但是这些网络上的程序文件并不是程序项目的形式，我们必须在下载后将它们建立成程序项目。另外在 Android SDK 中也有许多程序范例，这些程序范例也必须通过适当的方式把它们变成程序项目。还有如果我们想要修改已经写好的程序项目，但又想保留原来的项目，该如何完成？为了满足以上种种程序项目的操作需求，这个单元就让我们来学习 Eclipse 的程序项目操作技巧。

4-1 根据已经写好的程序文件来建立项目

一个完整的 Android 应用程序包括程序文件、图形文件、接口布局文件、字符串资源文件等，我们将这些文件称为项目原始文件。这些项目源文件是以特定的文件夹结构加以储存，我们可以将网络上别人分享的 Android 程序项目源文件下载到计算机中，并按照它们原来的文件夹结构放到一个新的文件夹里头，然后再利用以下的步骤建立程序项目。

Step 1 运行 Eclipse，单击 File > New > Project。

Step 2 在对话框中间的清单中选择 Android > Android Project 再单击 Next 按钮。

Step 3 单击 Create project from existing source，然后单击 Browse 按钮选取程序源文件所在的文件夹，最后单击 OK 按钮回到原来的对话框，再单击 Next 按钮。

Step 4 选择此程序项目使用的 Android 平台版本（Build Target），然后单击 Next 按钮。

Step 5 Application Name 和其他字段都会自动填入数据，确定无误后单击 Finish 按钮。

Step 6 如果在 Eclipse 下方的 Console 窗口中出现错误信息（红色文字），请参考后面"程序项目的管理和维护"小节中的说明进行处理。

完成以上步骤之后，我们就可以使用前一个单元学过的方法来运行这个 Android 程序了。

4-2 根据 Android SDK 中的程序范例来建立项目

Android SDK 中提供许多程序范例，这些程序范例是学习 Android 程序设计很好的参考数据，我们可以把它们建立成程序项目来测试运行结果，请读者依照下列步骤操作。

Step 1 运行 Eclipse，单击 File > New > Project。

Step 2 在对话框中间的清单中选择 Android > Android Project 再单击 Next 按钮。

Step 3 单击 Create project from existing sample 再单击 Next 按钮。

Step 4 选择想要使用的 Android 版本然后单击 Next 按钮。

Step 5 在对话框中间会列出 Android 程序范例列表，请从中选取一个再单击 Finish 按钮。

Step 6 如果在 Eclipse 下方的 Console 窗口中出现错误信息（红色文字），请参考后面"程序项目的管理和维护"小节中的说明进行处理。

完成以上步骤之后就可以运行该程序范例了。

> **补充说明**
>
> 其实这些 Android SDK 中的程序范例也可以使用前一个小节的方法来建立程序项目，只是我们要知道这些程序范例文件夹的位置，它们是放在 Android SDK 文件夹中的 samples 子文件夹下面，如果读者有兴趣不妨自行试看看。

4-3 把建立好的 Android 程序项目加载 Eclipse

如果使用前面 2 种方法建立程序项目的过程中出现图 4-1 上方的警告信息，那就代表所指定的文件夹已经是一个完整的程序项目，这时候我们要改用 import 方式来加载。另外如果我们想要修改一个已经完成的程序项目，但又想保留该项目目前的内容，这时候可以先用 Windows 文件管理器复制该项目的文件夹，然后修改复制文件夹的名称，接着再利用以下介绍的 import 方法加载复制后的程序项目。

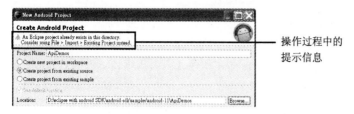
操作过程中的
提示信息

图 4-1　建立程序项目时出现警告信息

Step 1　运行 Eclipse，单击 File > Import…。

Step 2　在对话框中间的清单中选择 General > Existing Projects into Workspace 再按下 Next 按钮。

Step 3　在对话框中单击 Select root directory（请参考图 4-2），再按下右边的 Browse 按钮，然后选取项目所在的文件夹。

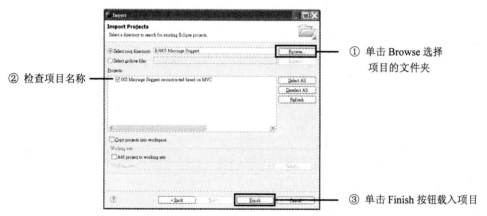
① 单击 Browse 选择
项目的文件夹
② 检查项目名称
③ 单击 Finish 按钮载入项目

图 4-2　使用 Import 方式加载 Android 程序项目

Step 4　在对话框中间的 Projects 清单中会列出找到的项目名称，确定无误后按下 Finish 按钮。完成之后在 Eclipse 左边的项目检查窗口中就会出现新加载的程序项目名称。

4-4　程序项目的管理和维护

在 Eclipse 中操作程序项目时偶尔会出现错误信息，遇到这种状况时要如何排除？另外我们也许想要删除某一个程序项目，有关这些程序项目的管理和维护整理说明如下。

1. 解决程序项目出现的错误信息

如果我们之前已经完成程序项目并经过测试无误，可是不知何故突然在程序项目名称的前面出现红色打叉符号，此时有可能是加载程序项目时出错。遇到这种情况请先在项目检查窗口中单击该项目名称，然后从 Eclipse 上方的主菜单中选择 File > Refresh，如果错误还是存在，可以尝试关闭 Eclipse 再重新启动。如果错误还是没有解决，请参考下列第 3 点，先移除该程序项目再重新 import。

2. Fix Project Property 错误信息

有时候项目的属性也会出现错误,这时候 Eclipse 会在程序代码编辑窗口下方的 Console 窗口中,显示信息要求运行 Fix project property。如果读者看到这个信息,请在项目检查窗口中用鼠标右键单击该项目,然后在快捷菜单中选择 Android Toos > Fix Project Properties 就可以更正错误,请参考图 4-3。

图 4-3　使用 Fix Project Properties 选项修正程序项目的错误

3. 删除程序项目

如果要删除某一个程序项目,请在项目检查窗口中用鼠标右击该项目,然后在快捷菜单中选择 Delete。在出现的对话框中有一个"连同文件一并删除"的选项,如果勾选它,该项目的数据会从磁盘中删除,否则 Eclipse 只会从项目清单中移除该项目,往后读者可以再利用 import 方式重新将它加载。

4. 指定程序项目使用特定的模拟器

如果我们在 Eclipse 中建立了多个模拟器(例如我们可以建立不同 Android 版本的模拟器,以便测试程序在不同平台的运行结果),那么可以为每一个程序项目指定所使用的模拟器。请在 Eclipse 左边的项目检查窗口中用鼠标右击项目名称,在弹出的快捷菜单中选择 Run As

> Run Configurations…就会出现图 4-4 所示的对话框。接着单击 Android Application，再利用上方工具栏的新增按钮加入一个启动设置，在右边的 Android 标签页中指定项目名称，再切换到 Target 标签页指定使用的模拟器名称，然后单击 Apply 按钮再单击 Run 按钮。

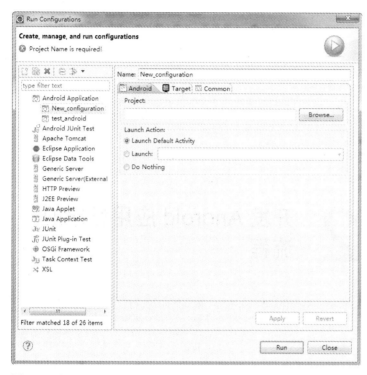

图 4-4　设置 Android 程序项目使用的模拟器

　　学会本单元介绍的程序项目管理技巧之后，相信读者对于 Eclipse 开发环境的使用已经更加熟悉，这些操作步骤请读者务必熟练，因为在后续的学习过程中经常需要使用。另外在编写程序的过程中也可以善用 Eclipse 的辅助功能，像是"语法错误修正提示"、"对象方法候选列表"、"类方法清单对话框"……，这些好用的功能将在后续的学习单元中再伺机介绍。如果读者能够熟练地使用 Eclipse 这个功能强大的开发平台，就能够有效提高程序开发的效率。

补充说明　Android 模拟器的操作技巧

在开发程序时我们可以先利用 Eclipse 主菜单的 Window > AVD Manager 启动需要使用的模拟器，甚至也可以同时启动多个模拟器。当开始运行程序时，如果 Eclipse 发现有超过一个以上的模拟器可以运行该程序，它会显示一个对话框让我们选择想要使用的模拟器。

PART 2　开发 Android 应用程序的流程

UNIT 05　了解 Android 程序项目的架构和查询 SDK 说明文件
UNIT 06　使用 TextView、EditText 和 Button 接口组件
UNIT 07　程序的错误类型和除错方法
UNIT 08　使用 Android 模拟器的技巧
UNIT 09　良好的程序架构是程序开发和维护的重要基础
UNIT 10　升级 Android 手机程序成为平板电脑程序

Android 版本	1.X	2.X	3.X	4.X
适用性	★	★	★	★

UNIT 5
Eclipse 程序项目管理技巧

在单元三中我们学会了建立 Android 程序项目的方法,并且知道如何修改程序界面的提示字符串,也加上了 2 个按钮。这些都只是针对程序资源(放在项目的 res 文件夹中的文件)进行修改,还没有真正进入程序代码的部分。在这个单元中就让我们来打开潘多拉的宝盒,一起检查项目的程序代码吧。

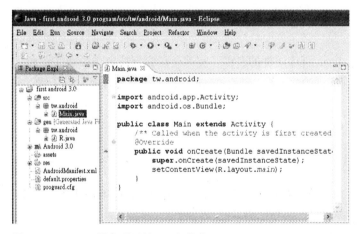

图 5-1　Android 程序项目的 src 文件夹

Android 程序项目的程序文件是放在项目的 src 文件夹(请参考图 5-1),该文件夹中会根据我们在建立项目时输入的 Package Name 建立一个 Package,再把程序文件置于其中,而程序文件的名称就是在建立项目时输入的 Activity 名称。根据上一个单元建立的程序项目,我们的程序文件是 src/tw.android/Main.java。请读者从项目检查窗口中找到该文件(依序展开项目的树状结构即可找到),再用鼠标双击把它打开,就可以看到如下的程序代码:

```
package tw.android;

import android.app.Activity;
import android.os.Bundle;

public class Main extends Activity {
    /** Called when the activity is first created. */
    @Override
    public void onCreate(Bundle savedInstanceState) {
        super.onCreate(savedInstanceState);
        setContentView(R.layout.main);
    }
}
```

我们在上一个单元看到的程序运行界面就是这一段程序代码的运行结果，接下来我们就用它来解说 Android 程序的运行流程。

> **补充说明** **Eclipse 操作技巧**
>
> 在程序代码编辑窗口中如果看到某一行的最前面有一个小圆圈里面有一个加号，代表可以用鼠标单击那个加号把程序代码进一步展开。如果是小圆圈里面有一个减号，则代表可以单击把程序代码收起来。

第一行程序代码：

```
package tw.android;
```

是指定这个程序文件属于哪一个组件。Java 程序组件的名称其实就是对应到特定路径的文件夹，如果读者使用 Windows 文件管理器找到项目的文件夹，再沿着 src 文件夹往下展开就可以了解，而一个组件中可以包含多个程序文件。

接下来的二行 import 程序代码：

```
import android.app.Activity;
import android.os.Bundle;
```

是要求编译程序加载指定的类或是组件以供此程序文件使用。最后一段程序代码则是建立一个类：

```
public class Main extends Activity {
    /** Called when the activity is first created. */
    @Override
    public void onCreate(Bundle savedInstanceState) {
        super.onCreate(savedInstanceState);
        setContentView(R.layout.main);
    }
}
```

这个类的名称叫做 Main，也就是我们在建立项目时输入的 Activity 名称，同时也是这个程序文件的名称。这个 Main 类是继承自 Activity 类，它就是我们这个项目的主类，也就是程序开始运行的地方。@Override 指令是说下一行开始的 onCreate()方法要取代原来基础类 Activity 中的 onCreate()方法，这个 onCreate()方法是当程序启动的时候会由 Android 系统调用运行。onCreate()方法中的第一行程序 super.onCreate(savedInstanceState)是调用基础类 Activity 的 onCreate()方法，并且把 Android 系统传进来的 savedInstanceState 对象也传给它处理。因为 Android 程序有许多内定要处理的工作，这些工作我们可以根据调用基础类的方法帮我们完成即可。下一行程序代码是调用 setContentView()方法并指定使用程序接口布局文件 R.layout.main，于是程序就会显示我们所指定的接口。关于 R.layout.main 请读者在项目检查窗口中把项目中的 gen 文件夹展开（请参考图 5-1），就可以看到其中有一个 package，里头包含一个 R.java 文件，请用鼠标双击打开就可以看到如下的程序代码：

```
/* AUTO-GENERATED FILE.   DO NOT MODIFY…

package tw.android;

public final class R {
    public static final class attr {
    }
    public static final class drawable {
        public static final int icon=0x7f020000;
    }
    public static final class layout {
        public static final int main=0x7f030000;
    }
    public static final class string {
        public static final int app_name=0x7f040001;
        public static final int hello=0x7f040000;
    }
}
```

第一行是此程序文件的批注，提醒我们这是由程序编译程序自动产生的文件，不要修改。这个文件的内容就是定义一个名称叫做 R 的类，这个类整合了程序项目中所有的资源，包括 drawable（图文件）、layout（程序接口）和 string（字符串）等，每一个型态的资源都是一个独立的子类，其中的子类 layout 中的 main 属性就是对应到项目下的 res/layout/main.xml 文件。

根据以上的说明，读者应该对于这个项目的架构和运行流程有了初步的概念。只是程序不能只有操作接口，还必须要有实际功能的程序代码。截至目前为止我们只加了两个按钮，可是按下按钮后却没有任何效果，下一个单元我们将建立一个具有实际功能的程序项目。不过在开始编写程序之前，必须先知道如何查询程序设计相关数据，以便解决随时可能出现的问题。

在安装 Android SDK 的文件夹中有许多程序设计相关的参考数据，前面已经介绍过 Android SDK 的程序范例，除此之外还有类相关的技术文件，这些类相关的技术文件是放在 docs 文件夹中，请读者用网页浏览器开启其中的 index.html 就可以看到 Android developers 的首页。这个首页的内容和读者安装的 Android SDK 版本有关。此外读者也可以利用因特网连到 Android developers 网站，该网站同样也可以查询 Android SDK 的说明文件。计算机中的数据和 Android developers 网站上的数据的差异是计算机上的数据不会自动更新，但是 Android developers 网站上的数据会由 Google 公司负责持续地更新。

利用网页浏览器打开 Android SDK 文件夹中的 docs 子文件夹里头的 index.html 文件，或是利用因特网连到 Android developers 网站之后，在网页上会显示许多卷标页，包括 Home、SDK、DevGuide、……，当单击其中的 Reference 标签页就会出现图 5-2 的类说明网页，这个网页的使用方式如下：

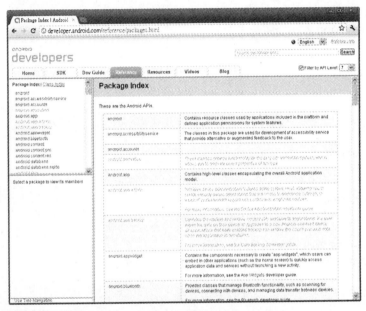

图 5-2　使用 Reference 标签页查询类说明文件

1. 如果是要查特定组件的数据，请单击网页左边上方的 Package Index，在下方就会出现所有组件的列表，用鼠标在组件名称上点一下，就会在下方的窗口中出现该组件的所有类，找到想要的类后用鼠标点一下就会在右边的窗口显示该类的详细说明。

2. 如果想要直接查询特定的类，可以单击网页左边上方的 Class Index，然后在右边窗口的上方有 A 到 Z 的字母。请读者依照类名称的第一个字母单击就会跳到该位置，然后再往下找到想要查询的类，最后点一下类名称就会出现该类的说明文件。

3. 查询特定类的数据还有一个更快速的方法，就是利用界面右上方的 Search 按钮。请读者在该按钮左边的字段输入类名称，输入的同时在下方会出现参考清单，我们可以继续输入直到出现想要查询的类后，用鼠标在该类名称上单击就会切换到该类的说明文件。

4. 在上述 Search 功能的下面有一个 Filter by API Level 选项，如果勾选它，然后在右边的下拉式清单中选择一个 API Level 版本，就会发现网页左边的组件清单中有些变成灰色。这项功能是说如果把程序项目的 Minimum SDK 属性设置成这个 API Level 版本，这些灰色的组件就不能使用。

> **补充说明**　**查询类别说明文件的技巧**
>
> 由于 Android SDK 的所有类都是以面向对象技术的继承方式产生，在每一个类说明页的最上方都会有一个继承的阶层图。如果在目前的类说明页中查不到想要的属性或是方法，就代表该属性或方法是定义在它所继承的基础类中，这时候可以依照继承的阶层图查阅它的基础类。

Android SDK 文件夹中除了类说明文件之外，还有讨论 Android 程序设计的技术文章，请读者单击 Android SDK 网页上方的 Dev Guide 就会出现图 5-3 所示的网页，网页的左边列出各种程序设计主题的分类，读者可以单击有兴趣的项目就会在网页右边出现说明文件。这些文件都是以英文编写，目的是让程序开发人员了解如何开发各种 Android 应用程序。只是这些文章对于程序设计经验尚浅的人来说帮助有限，即使是经验丰富的程序设计人员，也还需要配合许多测试才能够完全掌握其中的技术。但是这些文件毕竟是 Google 官方的第一手数据，因此还是具备相当重要的参考价值，读者如果有兴趣可以尝试阅读其中的内容。

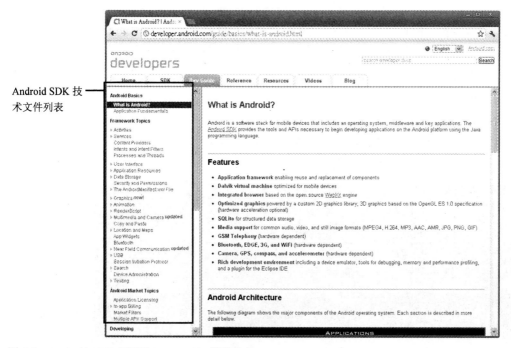

Android SDK 技术文件列表

图 5-3　Android SDK 网页的 Dev Guide 说明文件

Android 版本	1.X	2.X	3.X	4.X
适用性	★	★	★	★

UNIT 6

使用 TextView、EditText 和 Button 接口组件

TextView、EditText 和 Button 这三个接口组件是最常使用的程序操作接口。TextView 是用来显示信息,例如域名或是程序的处理结果。EditText 则是让用户输入数据,Button 可以让用户按下以启动处理程序。在这个单元中我们要利用这 3 种接口组件完成如图 6-1 所示的操作界面,它是在平板电脑模拟器上的运行结果,读者可以改变程序项目使用的 Android 版本让它在手机上运行。我们必需在 res/layout/main.xml 文件中建立 3 个 TextView 组件、2 个 EditText 组件和 1 个 Button 组件。建立接口组件时,我们必须设置好每一个接口组件的属性。首先请读者依照单元三中介绍的方法建立一个新的程序项目,然后依照后续说明进行操作。

图 6-1　使用 TextView、EditText 和 Button 接口组件的"婚姻建议"程序

6-1 TextView 接口组件

TextView 接口组件的功能是显示信息，用户无法编辑其中的文字。在 res/layout/main.xml 文件中我们可以依照如下的语法格式加入 TextView 组件：

```
<TextView android:id="@+id/自定义组件名称"
    android:属性="属性值"
    ...
/>
```

以上其实就是 xml 的语法，TextView 是卷标名称，后面则是它的属性。第一个属性 android:id 是设置这个 TextView 组件的名称，它的值"@+id/自定义组件名称"中的"@+id /"是一个指令，指示要将后面的组件名称加入程序资源 R 中的 id 类，这样程序才能够找到此组件。如果程序运行过程中不会再去更改这个组件的内容，那么我们可以省略 android:id 这个属性。接下来的每一行都是设置一个属性值，最常用的属性包括 android:layout_width、android:layout_height 和 android:text，它们分别决定组件的宽、高和组件内显示的文字，例如以下范例：

```
<TextView android:id="@+id/txtResult"
    android:layout_width="fill_parent"
    android:layout_height="wrap_content"
    android:text="程序运行结果"
/>
```

以上程序代码的功能是增加一个名为 txtResult 的 TextView 组件，它的宽度设置为 fill_parent，也就是填满它所在的外框，高度设置为 wrap_content，也就是由文字的高度来决定，组件中会显示"程序运行结果"这个字符串。

> **补充说明 程序设计技巧**
>
> 在决定接口组件的 id 名称时，为了在程序代码中能够清楚地知道该接口组件的种类，我们可以在组件名称的前面加上小写的组件型态的缩写，例如我们把上述 TextView 组件取名为 txtResult，前面的 3 个小写英文字母 txt 代表这是一个 TextView 接口组件，本书的所有范例都将采用这种方式来命名以方便阅读程序代码。

除了以上介绍的属性之外，还有许多其他的属性，我们会在后续单元补充说明。如果读者想先自行了解，可以用 Google 搜寻，例如在 Google 搜索页中输入 Android TextView 就可以找到官方对于 TextView 的说明网址，例如图 6-2 就是 TextView 组件的官方说明数据，其中有非常详细的说明包括所有属性的列表和解释。读者在阅读时要注意，Android 程序的所有接口组件都是以面向对象技术的继承方式产生，在每一个接口组件说明页的最上方会有一个继承的阶层图，如果在目前的组件说明页中查不到想要的属性，就代表该属性是定义在它所继承的基础类中，我们可以依照继承的阶层图查阅它的基础类。

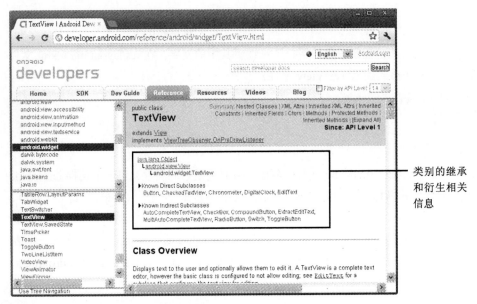

类别的继承和衍生相关信息

图 6-2　TextView 接口组件的官方说明数据

6-2　EditText 接口组件

EditText 组件可以让用户在上面输入一串文字，再让程序读取该字符串。我们可以在 res/layout/main.xml 文件中依照以下的语法加入 EditText 组件：

```
<EditText android:id="@+id/自定义组件名称"
    android:属性="属性值"
    …
/>
```

读者可以和前面的 TextView 组件的语法比较，两者除了标签名称不同以外其余的格式都一样，例如我们可以利用以下程序代码建立一个 EditText 组件：

```
<EditText android:id="@+id/edtSex"
    android:layout_width="fill_parent"
    android:layout_height="wrap_content"
    android:inputType="text"
    android:text=""
/>
```

上面的范例中出现了一个新的 android:inputType 属性，它是用来限制这个组件可以接受的字符类型，text 代表任何字符都可以被接受，如果设置成 number 则只能输入 0～9 的数字字符。

6-3 Button 接口组件

Button 组件是让用户单击以启动程序的某一项功能，它的语法格式如下：

```
<Button android:id="@+id/自定义组件名称"
    android:属性="属性值"
    …
/>
```

除了标签名称换成 Button 之外，其他都和前面介绍的组件语法相同，例如以下程序代码会在程序的界面建立一个按钮，按钮上面显示"运行"：

```
<Button android:id="@+id/btnDoSug"
    android:layout_width="fill_parent"
    android:layout_height="wrap_content"
    android:text="运行"
/>
```

现在请读者尝试阅读以下的接口布局文件，它就是本单元范例程序的操作界面：

```
<?xml version="1.0" encoding="utf-8"?>
<LinearLayout xmlns:android="http://schemas.android.com/apk/res/android"
    android:orientation="vertical"
    android:layout_width="fill_parent"
    android:layout_height="fill_parent"
    >
<TextView
    android:layout_width="fill_parent"
    android:layout_height="wrap_content"
    android:text="性别："
    />
<EditText android:id="@+id/edtSex"
    android:layout_width="fill_parent"
    android:layout_height="wrap_content"
    android:inputType="text"
    android:text=""
    />
<TextView
    android:layout_width="fill_parent"
    android:layout_height="wrap_content"
    android:text="年龄："
    />
<EditText android:id="@+id/edtAge"
    android:layout_width="fill_parent"
    android:layout_height="wrap_content"
```

```
        android:inputType="number"
        android:text=""
    />
    <Button android:id="@+id/btnDoSug"
        android:layout_width="fill_parent"
        android:layout_height="wrap_content"
        android:text="婚姻建议"
    />
    <TextView android:id="@+id/txtResult"
        android:layout_width="fill_parent"
        android:layout_height="wrap_content"
        android:text="结果"
    />
</LinearLayout>
```

其中<LinearLayout>卷标是指定接口组件采线性顺序排列，它是一种接口组件的编排模式，我们留待单元十五再详细介绍如何控制接口组件的编排方式。接下来的工作是如何链接接口组件和程序代码，也就是让程序可以读取用户输入的数据，并在按下按钮后开始运行。

6-4　链接接口组件和程序代码

建置好 res/layout/main.xml 文件中的接口组件之后，下一步就是编写控制接口组件的程序代码。图 6-1 范例程序的操作方式是在使用者输入性别和年龄之后按下按钮，程序就会读取用户的性别和年龄，然后显示判断结果，因此接口组件和程序代码之间必须能够互动以完成下列 3 件事：

1. btnDoSug 按钮被按下时，程序会开始运行判断的程序代码；
2. 程序代码必须能够读取 edtSex 和 edtAge 这两个 EditText 组件中的字符串；
3. 程序代码必须将最后的判断结果显示在 txtResult 组件中。

接着就让我们依序说明如何完成以上的工作。

> **补充说明**
> 我们将建立在接口布局文件中的卷标像是<TextView>、<EditText>、<Button>和<LinearLayout>称为"接口组件"，为了在程序中使用这些"接口组件"，我们必须在程序代码中建立对应到它们的"对象"，这些用来对应到"接口组件"的"对象"必须和它们所对应的"接口组件"具备相同的型态，后续会有实际范例。

6-5　设置 Button 的 click 事件 listener

事件 listener 就是用来处理某一个组件的事件的处理程序。举例来说当我们单击某一个 EditText 组件时，这个单击的动作就会引发 EditText 组件的 click 事件，该事件会传到 EditText 组件的 click

事件 listener 中进行处理。当在 EditText 组件中输入文字时，又会引发另一个事件，该事件又会传到另一个对应的事件 listener 中处理。像 Android 这种图形操作接口的程序就是利用各式各样的事件处理程序来运行用户的操作，而事件处理程序在 Android 程序中就称为事件 listener。要设置 Button 组件的 click 事件 listener 需要完成下列步骤：

Step 1 建立一个 OnClickListener 对象。

OnClickListener 本身是一个接口，当我们要建立一个 OnClickListener 的对象时，必须实现其中的 onClick() 方法，例如以下程序代码就是建立一个 OnClickListener 对象，并取名为 btnDoSugOnClick，因为它将被设置成 btnDoSug 的 click 事件 listener。

```
Button.OnClickListener btnDoSugOnClick = new
Button.OnClickListener() {
    public void onClick(View v) {
        // 按下按钮后要运行的程序代码
        …
    }
};
```

补充说明　Eclipse 操作技巧

当要建立 OnClickListener 对象时，我们可以先输入以上范例中的粗体字部分，把对象内部的程序代码留白，这时候程序代码编辑窗口会在这一段程序标示一个红色波浪底线代表语法错误，将鼠标光标移到该红色波浪底在线，就会弹出信息窗口其中有建议修正项目，单击该修正项目就会自动修正程序代码的错误，善用这个功能可以提高程序编写的效率。

Step 2 在程序代码中取得定义在 res/layout/main.xml 接口布局文件中的 Button 组件，所有定义在接口布局文件中的组件经过编译之后都会放在 gen/(组件名称)/R.java 资源文件中，我们可以利用 findViewById() 从 R.java 文件中取得接口组件，例如：

```
Button btnDoSug = (Button)findViewById(R.id.btnDoSug);
```

我们把取得的 btnDoSug 组件存入一个同名的 Button 对象。其实这两个对象的名称也可以不同，但是因为它们都是负责相同的功能，只是一个定义在接口布局文件，一个宣告在程序代码，所以我们使用相同的名称。

Step 3 把在第一个步骤中建立的 OnClickListener 对象，设置给第二个步骤中的 Button 对象成为的 click 事件 listener。

```
btnDoSug.setOnClickListener(btnDoSugOnClick);
```

> **补充说明** 程序代码编辑技巧
>
> 在编写程序代码的时候，每一次输入"对象名称."之后稍停半秒钟就会自动弹出一个方法清单，我们可以继续输入方法名称的前几个字母，列表中会自动过滤出符合的方法，如果单击清单中的某一个方法，还会出现另一个窗口说明该方法的功能。
>
> 当输入变量名称、对象名称或方法名称时，输入到一半同时按下键盘上的 Alt 和 / 键，程序编辑器会立即提供适当的协助。善用这项辅助功能可以提高程序编写的效率，也可以减少打错字的机会，建议读者多加利用。

我们将以上 3 个步骤的程序代码整合到程序项目的程序代码得到如下的结果：

```java
package tw.android;

import android.app.Activity;
import android.os.Bundle;
import android.view.View;
import android.widget.*;

public class Main extends Activity {

    private Button btnDoSug;
    private EditText edtSex, edtAge;
    private TextView txtResult;

    /** Called when the activity is first created. */
    @Override
    public void onCreate(Bundle savedInstanceState) {
        super.onCreate(savedInstanceState);
        setContentView(R.layout.main);

        // 从资源类 R 中取得接口组件
        btnDoSug = (Button)findViewById(R.id.btnDoSug);
        edtSex = (EditText)findViewById(R.id.edtSex);
        edtAge = (EditText)findViewById(R.id.edtAge);
        txtResult = (TextView)findViewById(R.id.txtResult);

        // 设置 button 组件的事件 listener
        btnDoSug.setOnClickListener(btnDoSugOnClick);
    }

    private Button.OnClickListener btnDoSugOnClick = new Button.OnClickListener() {
        public void onClick(View v) {
            // 按下按钮后要运行的程序代码
            ...
        }
    };
}
```

针对上述程序代码我们再补充几点说明：

1. 要在程序中使用接口组件必须在程序的前面 import 下列组件，这个工作可以利用前面"补充说明"的 Eclipse 操作技巧让程序代码编辑器帮我们完成。

    ```
    import android.widget.*;
    ```

2. 在 OnClickListener 对象中会用到一个 View 对象，所以还要另外 import 该对象如下：

    ```
    import android.view.View;
    ```

3. 除了 Button 对象 btnDoSug 以外，我们也一起宣告了其他会用到的接口对象包括 edtSex、edtAge 和 txtResult。这些接口对象都宣告在 onCreate()方法之前，也就是让这些对象可以被所有方法共享，这样如果未来程序中又增加其他方法时才可以使用这些对象。另外 Button 的 OnClickListener 对象也是宣告在 onCreate()方法外面。

4. 在 onCreate()方法中我们从资源类 R 取得所有会用到的接口组件（利用 findViewById()方法），并设置给步骤 3 中宣告的接口对象，最后完成设置 Button 对象的 click 事件 listener。

6-6　取得 edtSex 和 edtAge 接口组件中的字符串

在前面的程序代码中我们已经从资源类 R 取得 edtSex 和 edtAge 接口组件，也把它们设置给同名的对象，接下来就可以利用 EditText 对象本身提供的 getText()方法来取得字符串，然后把字符串数据转换成 String 型态存入一个 String 对象中如下：

```
String strSex = edtSex.getText().toString();
```

我们也可以利用同样的方法取得 edtAge 中的字符串，不过因为年龄应该是数值型态，所以我们再将得到的 String 对象转成 int 型态的数据，这些动作可以全部写成一行程序代码如下：

```
int iAge = Integer.parseInt(edtAge.getText().toString());
```

6-7　将结果显示在 txtResult 接口组件

txtResult 接口组件也在前面的程序代码中被取得并设置给一个同名的对象，我们只要将程序要显示的信息存入一个 String 对象，再把该 String 对象设置给 txtResult 对象即可（利用 setText()方法）。

```
String strSug = "程序的运行结果";
txtResult.setText(strSug);
```

我们将上述的程序代码范例整合到程序文件的 onClick()方法中，再加上我们对于性别和年龄的判断条件，最后得到如下的结果：

```
private Button.OnClickListener btnDoSugOnClick = new Button.OnClickListener() {
    public void onClick(View v) {
        // 按下按钮后要运行的程序代码
        String strSex = edtSex.getText().toString();
        int iAge = Integer.parseInt(edtAge.getText().toString());

        String strSug = "结果：";
        if (strSex.equals("男"))
            if (iAge < 28)
                strSug += "不急";
            else if (iAge > 33)
                strSug += "赶快结婚";
            else
                strSug += "开始找对象";
        else
            if (iAge < 25)
                strSug += "不急";
            else if (iAge > 30)
                strSug += "赶快结婚";
            else
                strSug += "开始找对象";

        txtResult.setText(strSug);
    }
};
```

到此我们已经完成整个项目的所有程序代码。这个单元可能会让读者觉得比较辛苦一点，但是付出总是会有回报，根据这个单元的范例，我们可以了解整个 Android 程序的架构和运行流程，这是后续学习的重要基础。现在读者可以尝试运行这个程序，如果顺利的话就可以看到运行界面，如图 6-3 所示。如果启动程序的过程中出现错误信息，就代表程序代码还有错误，我们将在下一个单元学习如何帮程序调试。

图 6-3 "婚姻建议"程序的运行界面

6-8　在模拟器中输入中文

手机和平板电脑输入中文的方式与一般计算机不同，初学者常常不知道从何处下手，以下我们做个说明。如果是在 Android 2.X 的手机模拟器上操作，当用鼠标单击任何一个 EditText 组件准备输入数据时，手机屏幕下方就会自动出现一个虚拟键盘，如图 6-4 所示。我们可以单击该键盘左下方的"中文（英文）"按钮切换中英文输入模式，或是利用该按钮右边的"符号数字"按钮切换到符号数字键盘。要提醒读者的是 Android 平台的中文输入是使用"谷歌拼音"，也就是所谓的"汉语拼音"，它是利用 26 个英文字母代替注音符号，详细的对照表可以参考下列网址：

http://www.cccla-us.org/pinyin.htm

如果读者想要自己选择输入法，可以先用鼠标单击任何一个 EditText 组件，然后再按一次不要放开就会出现一个选单，请选择其中的 Input method 就会出现所有输入法的清单，如图 6-5 所示。

图 6-4　Android 2.X 手机模拟器的虚拟键盘

图 6-5　Android 2.X 手机模拟器的输入法列表

如果是 Android 4.X 的手机模拟器，当用鼠标单击任何一个 EditText 组件准备输入数据时，屏幕左下方会出现一个键盘图标，我们可以用鼠标单击它按住不放再往下拉就会出现如图 6-6 的界面，单击 Select input method 便可进入输入法选单，如图 6-7 所示，我们可以单击"谷歌拼音输入法"来输入中文。如果是 Android 3.X/4.X 的平板电脑模拟器，当用鼠标单击任何一个 EditText 组件准备输入数据时，屏幕右下方会出现一个键盘图标，当我们用鼠标单击它时就会弹出输入法选择窗口如图 6-8，我们可以单击"谷歌拼音输入法"来输入中文，并且把最上面的 Use physical keyboard 项目右边的 ON 按下（变成只看到 OFF）。设置好输入法之后，单击屏幕的空白处就会进入键盘输入界面，如图6-9所示。单击左下方的"中英文切换"按钮切换成中文模式就可以利用"谷歌拼音"来输入中文。

图6-6　Android 4.X手机模拟器的虚拟键盘设定画面

图6-7　Android 4.X手机模拟器的输入法列表

图6-8　Android 3.X/4.X的平板电脑模拟器的输入法清单

图6-9　"谷歌拼音"中文输入界面

Android 版本	1.X	2.X	3.X	4.X
适用性	★	★	★	★

UNIT 7
程序的错误类型和除错方法

　　一般初学者学习程序设计通常必须克服 3 道关卡，第一就是必须想出解决问题的详细步骤，也就是所谓的"算法"。第二道关卡是编写程序代码，由于计算机程序是非常严谨的逻辑和语法，因此输入程序代码的时候要谨慎小心，否则很容易出错。最后一道关卡是调试，不管程序代码是长是短，发生错误是常有的事。当程序出现错误的时候，如何根据目前程序运行的情况，推敲出可能出现错误的地方然后加以测试并更正，这个过程必须依赖经验、工具和技巧。其实程序的调试过程就像是现在流行的在线游戏一样，必须依赖技术和智慧去突破每一个障碍。为了能够提高程序调试的效率，花些时间学习调试工具和使用方法是非常值得的，因此这个单元就让我们来介绍程序的错误类型和学习相关的调试技巧。

　　程序的错误一般可以分成三种：语法错误、逻辑错误、运行时期错误（又称为例外），以下我们依序说明。

7-1　程序的语法错误和调试的方法

　　所谓语法错误是指程序代码不符合程序语言的语法。语法错误是最明显也是最容易解决的错误类型。目前许多程序代码编辑器都有实时检查语法的功能，只要有错误就会立刻标示出来，包括我们使用的 Eclipse 也是如此。读者可以参考图 7-1 的界面，当某一行程序的前面出现红色打叉符号就代表该行有语法错误，在该行的某处会有红色波浪底线，把鼠标光标移到波浪底线上面时，会弹出一个说明窗口并且提供修正的建议。但是读者要了解的是在该处标示波浪底线的真正原因是该处的程序代码不符合语法，而实际出错的地方有可能就是该处，但也有可能是前面的程序代码错误所造成的。因此如果读者在标示错误的该行程序反复检查却仍然找不出错误，就要检查前面的程序代码才能找出真正错误的地方。如果看了许久还是找不出错误，可以把可能有问题的那一段程序代码先删除看看语法错误还在不在。如果已经没有语法错误，代表问题就在那一段程序代码，可以按下 Ctrl+Z 回复到原来的状态，再依此法继续缩小寻找范围，直到找出错误的地方。

```
private Button.OnClickListener btnDoSugOnClick = new Button.O
    public void onClick(View v) {
        String strSex = edtSex.getText().toString();
        int iAge = Integer.parseInt(edtAge.getText().toStrin

        String strSug = "结果:"; // getString(R.string.promptR
        if (strSex.equals("男"))
            if (iAge < 28)
                strSug += "不急"; //getString(R.string.sugNotH
            else if (iAge > 33)
                strSug += "赶快结婚"; //getString(R.string.sug
            else
                strSug += "开始找对象"; //getString(R.string.sug
        else
            if (iAge < 25)
                strSug += "不急"; //getString(R.string.sugNotH
            else if (iAge > 30)
                strSug += "赶快结婚"; //getString(R.string.sug
            else
                strSug += "开始找对象"; //getString(R.string.sug

        txtResult.setText(strSug);
    }
};
```

图 7-1　Eclipse 程序代码编辑窗口标示语法错误的界面

7-2　程序的逻辑错误和调试的方法

所谓逻辑错误是说程序已经可以运行，但是运行的结果不对，例如要计算输入的成绩总分，却发现加总的结果不对。这有可能是数据读取错误，或是成绩累加的时候出错。遇到这种错误我们可以采取下列步骤：

Step 1　首先思考可能在哪一段程序代码出错。

Step 2　在可能出错的程序代码的第一行设置一个断点。所谓断点就是当程序在调试模式下，运行到该处就会暂停。设置断点的方式是单击 Eclipse 上方选单的 Run > Toggle Breakpoint 就会在目前编辑光标所在的那一行程序代码标示一个断点，也就是那一行的最前面会出现一个小圆点。要取消断点时，直接在小圆点上双击即可。

> **补充说明**　**Eclipse 操作技巧**
> 我们同样可以在小圆点出现的地方用鼠标双击来设置断点。

Step 3　单击工具栏上的小虫按钮（称为 Debug 按钮），让程序以调试模式运行。

Step 4　操作程序让程序进入设置断点的程序代码，当程序运行到断点时会弹出一个如图 7-2 所示的信息窗口，通知我们即将切换到调试界面，这时候请单击 Yes 按钮。

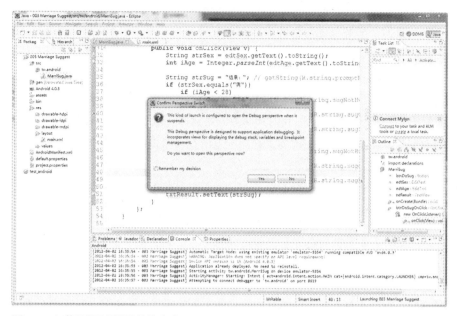

图 7-2　切换到调试界面的信息窗口

Step 5　Eclipse 切换到调试（Debug）界面，如图 7-3 所示。

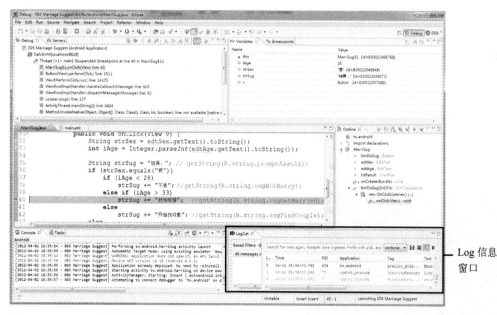

图 7-3　Android 应用程序的调试界面

左边中央的窗口就是程序代码窗口，在断点那一行前面会有一个箭头代表目前程序正要运行该行。界面的右上方窗口则会列出程序中的变量以及每一个变量的值，也可以点一下某一个变量的值然后将它修改，程序就会用修改后的值继续运行。

在图 7-3 界面的主要工具栏的下方（也就是 Debug 窗口的上方）有一个如图 7-4 的调试工具栏可以控制程序的运行，每一个调试工具栏上的按钮代表不同的运行方式，例如继续往下运行、单行运行、进入调用函数、离开这个函数。例如我们可以用"单行运行"按钮以一次一行的方式运行程序，并持续观察变量的值，以了解程序是否依照我们预期的方式运作。读者可以视情况使用适合的运行方式，也可以加入新的断点。根据交互运用以上介绍的方法找出错误的程序代码。

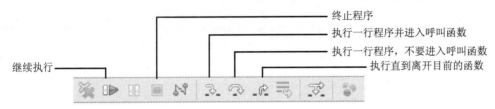

图 7-4　调试界面中的调试工具栏

Step 6　要终止调试模式时，先用鼠标单击左上方 Debug 窗口中的程序项目名称，然后按下调试工具栏上的停止按钮，最后再单击主工具栏最右边的 Java 按钮就会回到原来的程序编辑界面。

7-3　运行时期错误和调试的方法

如果程序在运行过程中弹出非预期的终止信息（如图 7-5 所示）就代表出现运行时期错误。要找出发生运行时期错误的原因也可以利用前面介绍的设置断点的方式，或是利用以下介绍的 log 方法来调试。log 是程序在运行过程中输出的信息，例如许多软件的安装程序在安装过程中都会产生一个 log 文件，以了解安装的整个过程。我们可以根据在程序中加入产生 log 的程序代码，来追踪究竟程序运行到哪一段程序代码才出现运行时期错误。要让程序产生 log 必须先在程序前面 import 以下组件。

图 7-5　Android 应用程序出现运行时期错误的界面

```
import android.util.*;
```

然后再使用以下指令：

```
Log.d("这个 log 信息的分类卷标", "信息");
```

每一个 log 都可以设置一个标签以方便分类和搜寻，但是目前 log 只能使用英文，因为中文会显示乱码。由于 Android 操作系统在运行的过程中会自动产生许多 log 信息（在调试模式右下方的 LogCat 窗口中，请读者参考图 7-3），为了能够找出我们的程序所产生的 log，我们可以设置一个专用的 log 标签名称，例如在上一个单元的婚姻建议程序中的每一个判断结果都增加一个 log 指令如下：

```
private Button.OnClickListener btnDoSugOnClick = new Button.OnClickListener() {
    public void onClick(View v) {
        String strSex = edtSex.getText().toString();
        int iAge = Integer.parseInt(edtAge.getText().toString());

        String strSug = "结果：";
        if (strSex.equals("男"))
            if (iAge < 28) {
                strSug += "不急";
                Log.d("MarriSug", "man, don't hurry");
            }
            else if (iAge > 33) {
                strSug += "赶快结婚";
                Log.d("MarriSug", "man, hurry to get married!");
            }
            else {
                strSug += "开始找对象";
                Log.d("MarriSug", "man, start to find girlfriend!");
            }
        else
            if (iAge < 25) {
                strSug += "不急";
                Log.d("MarriSug", "woman, don't hurry!");
            }
            else if (iAge > 30) {
                strSug += "赶快结婚";
                Log.d("MarriSug", "woman, hurry to get married!");
            }
            else {
                strSug += "开始找对象";
                Log.d("MarriSug", "woman, start to find boyfriend!");
            }
```

```
        txtResult.setText(strSug);
    }
};
```

修改程序代码之后请启动程序（用一般模式或是调试模式皆可），在运行过程中可以单击主工具栏右边的 Debug 按钮切换到调试界面，在调试界面右下方的 LogCat 窗口中就可以看到全部的 log 信息。为了过滤出我们程序所产生的 log，读者可以单击 LogCat 窗口左边 Saved Filters 工具栏中的 "+" 按钮，然后在出现的 Log Filter 对话框中（请参考图 7-6）输入 filter 名称，在 by Log Tab 字段输入要查看的标签名称（依我们的程序范例请输入 MarriSug），最后在 LogCat 窗口中就会列出我们的程序的所有 log 信息。

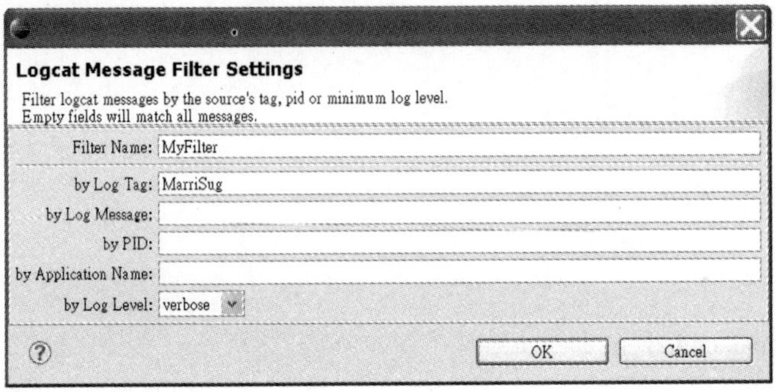

图 7-6　Log Filter 对话框

当程序出现运行时期错误的时候，先在可能出错的程序代码前面和后面加上产生 log 的指令，你也可以把相关变量的值一起显示在 log 信息中，以便了解程序运行的状况。当程序发生运行时期错误并终止之后，再切换到 Debug 界面检查 log 信息，就可以知道究竟是哪一段程序代码造成的。

以上介绍的 log 方法可以和设置断点的方法互相配合使用。举例来说，如果是在程序运行循环的过程中发生错误，可以先在循环中加上 log。因为如果循环运行的次数很多，一步一步追踪程序代码非常耗时，我们先利用 log 找出究竟是哪一次循环造成错误，再利用设置断点的方法来追踪那一次循环的运行过程就可以比较快地找出错误的原因。

在这个单元我们学到了程序的调试工具和使用技巧。程序调试是学习程序设计一个很重要的主题，如果不能帮程序调试，就算学会再多的语法也无法完成一个完整的程序项目。只要能够多做，程序设计的功力自然就会提升，调试的技巧和效率也会随之提高。

Android 版本	1.X	2.X	3.X	4.X
适用性	★	★	★	★

UNIT 8
使用 Android 模拟器的技巧

不管是开发 Android 平板电脑程序或是手机程序都需要先在模拟器（AVD）上进行测试，等到程序能够正确且稳定地运行再进行真机测试，因此在编写 Android 应用程序的过程中，使用 AVD 的时间占了很大的比例。如果读者能够熟悉 AVD 的运行，并且学会排除偶尔可能发生的问题，就能够更顺利地开发和测试程序，避免被一个突然发生的问题耽误了宝贵的时间，因此这个单元就让我们来学习如何操作 AVD 并解决运行上可能出现的问题。

8-1 启动模拟器的时机

当我们从 Eclipse 个工具栏上单击 Run 按钮时，Eclipse 会自动检查是否现在有运行中的 AVD 可以使用，如果没有就会自动挑选一个我们以前建立好的 AVD 并将它启动。启动 AVD 通常需要数十秒甚至是数分钟的时间，为了避免等待模拟器的启动时间，我们可以在运行 Eclipse 之后先单击工具栏上的 AVD Manager 按钮（旧版的开发工具是 Android SDK 和 AVD Manager 共享同一个按钮），就会看到图 8-1 的对话框。在对话框的 AVD 清单中选择要启动的 AVD，再单击 Start 按钮就会弹出一个 AVD 属性对话框，单击其中的 Launch 按钮就会开始启动 AVD。

> **补充说明** **Eclipse 操作技巧**
>
> Eclipse 工具栏上有许多按钮，如果把鼠标光标停在按钮上就会弹出按钮的名称，读者可以根据这个方式找到需要的按钮。

在启动 AVD 的过程中，我们可以同时进行程序代码的编辑工作，完成后就可以直接运行程序，

Eclipse 会自动把程序安装到已经启动好的 AVD 并开始运行。在运行程序的过程中，读者可以注意 Eclipse 下方的窗口中有一个叫做 Console 的标签页（参考图 8-2），其中会显示程序启动的过程。如果程序启动的过程中出现错误（红色字符串）并终止，就必须仔细检查错误的说明并加以排除才能继续测试程序，如果只是显示 warning 信息而且启动过程还在持续进行则可以不用在意。就笔者的经验而言，偶尔会发生下列两种错误：

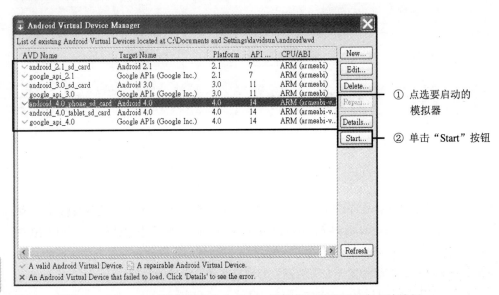

图 8-1　利用 AVD Manager 对话框预先启动 AVD，旧版的对话框要先在左边选择 Virtual devices 才会在右边显示 AVD 列表

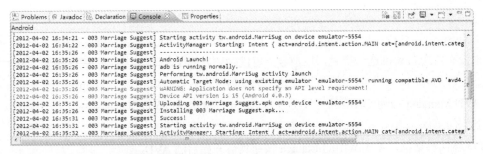

图 8-2　启动程序的过程中在 Eclipse 的 Console 卷标页显示的信息

1. 已经启动 AVD 但是 Eclipse 却尝试再启动一次 AVD，最后显示 AVD 无法启动而终止。这是因为 Eclipse 没有侦测到已经启动的 AVD，排除这个错误的第一步是找到 Eclipse 工具栏最右边有一个名称叫做 Open Perspective 的按钮（把鼠标光标停在 Eclipse 工具栏上的按钮时，会弹出一个按钮名称的信息，读者可以利用这个方法找到指定的按钮）。单击它然后选择其中的 DDMS 选项，Eclipse 的操作界面会变成如图 8-3 所示的模式。在左边的窗口会显

示目前运行中的模拟器的名称,从模拟器名称上方的工具栏中找到最右边有一个向下的三角形按钮,单击该按钮后选择最下面的 Reset adb(参考图 8-3),就可以让 Eclipse 重新和 AVD 取得连接,然后再单击 Eclipse 工具栏最右边的 Java 按钮切换成原来的程序编辑界面,重新运行程序。

2. Eclipse 有找到 AVD 但是在安装程序的过程中失败,并显示要 Uninstall 程序。这是因为目前要运行的程序和之前测试过的程序互相冲突,要解决这个问题请切换到模

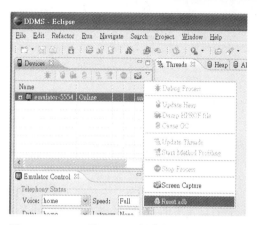

图 8-3　Eclipse 的 DDMS 操作界面

拟器并回到 Home screen 界面,然后单击 Apps 按钮(就是由多个小方格组成的那一个按钮)进入应用程序浏览界面,然后单击其中的 Settings 选项,再从显示的项目中选择 Apps 就会出现我们之前测试过的程序,图 8-4 是平板电脑模拟器的界面,读者可以在手机模拟器上用同样的方式操作。在图 8-4 的界面中单击我们测试过程序并选择 Uninstall,完成后利用"回上一页"按钮回到 Home screen,然后切换到 Eclipse 重新运行程序。

图 8-4　检查系统中安装的程序

> **补充说明** 模拟器操作技巧
>
> 在我们编写程序的过程中需要重复在 AVD 上测试程序，笔者建议在每一次运行程序后单击"回上一页"按钮退出程序，以便进行下一次测试。

8-2 Eclipse 选择不同版本 AVD 的规则

我们可以在 Eclipse 中建立多个不同 Android 版本的模拟器，而每一个 Android 程序项目也可以设置自己使用的 Android 版本和 minSdkVersion 属性（在 AndroidManifest.xml 文件中设置，代表程序需要的最低 Android 版本运行环境）。那么当运行一个程序项目时，Eclipse 如何决定要启动哪一个 Android 模拟器呢？Eclipse 在选择 AVD 时会同时考虑程序使用的 Android 版本和 minSdkVersion 属性的设置值，它会以两者中比较低的版本决定要启动的 AVD。举例来说，如果我们有三个程序项目，它们使用的 Android 版本和 minSdkVersion 属性设置如下：

- 方案一：Android 2.3.1　　minSdkVersion = 7（也就是 Android 2.1）
- 方案二：Android 2.3.1　　minSdkVersion = 9（也就是 Android 2.3.1）
- 方案三：Android 2.3.1　　minSdkVersion = 11（也就是 Android 3.0）

另外在 Eclipse 中我们已经建立好以下的 AVD：

- AVD1：Android 2.2
- AVD2：Android 2.3.3
- AVD3：Android 3.0

则当运行项目一的程序时 Eclipse 会启动 AVD1，运行项目二的程序时 Eclipse 会启动 AVD2，运行项目三的程序时 Eclipse 也会启动 AVD2。以上的情况是说当运行程序时，计算机里没有正在运行中的 AVD。如果计算机中已经有正在运行的 AVD，那么挑选的方式就不一样！

8-3 同时运行多个 AVD

如果程序需要在不同版本的 Android 平台上测试，我们可以同时启动多个 AVD。当程序运行时，Eclipse 会自动从运行中的 AVD 挑选条件适合的来使用，所谓条件适合是说只要 AVD 的版本大于或等于程序使用的 Android 版本或是 minSdkVersion 属性的设置值即可，如果有多个运行中的 AVD 符合条件，Eclipse 会弹出一个如图 8-5 所示的对话框，要求使用者挑选其中一个 AVD 来使用，举例来说，如果我们有三个程序项目，它们使用的 Android 版本和 minSdkVersion 属性设置如下：

- 方案一：Android 2.3.1　　minSdkVersion = 7（也就是 Android 2.1）
- 方案二：Android 2.3.1　　minSdkVersion = 9（也就是 Android 2.3.1）

■ 方案三：Android 2.3.1　minSdkVersion = 11（也就是 Android 3.0）

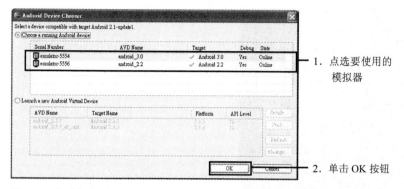

1. 点选要使用的模拟器
2. 单击 OK 按钮

图 8-5　启动程序后 Eclipse 要求用户挑选一个正在运行中的 AVD 来使用

另外在计算机中我们已经运行以下的 AVD：

■ AVD1：Android 2.2
■ AVD2：Android 3.0

则当运行项目一的程序时 Eclipse 会弹出图 8-5 的对话框要求我们从以上 2 个 AVD 中选择一个使用，运行项目二的程序时 Eclipse 就会直接使用 AVD2，运行项目三的程序时 Eclipse 也会直接使用 AVD2。

8-4　使用 AVD 的调试功能

如果读者单击模拟器屏幕上的 Apps 按钮，然后选择其中的 Dev Tools > Development Settings 就会出现图 8-6 的项目列表，这些选项可以改变 Android 模拟器的运行或是显示额外的信息，像是屏幕的重绘区域、单击的位置等，帮助我们了解程序运行的过程，我们针对其中比较重要的项目说明如下：

1. Debug App

   用来选择要调试的程序，当我们从 Eclipse 中启动程序之后，该程序会先被加载到模拟器中并完成安装，安装后的程序就会在这个项目的列表中出现供我们选择。基本上要进行程序调试并不需要设置这个项目，只要直接在 Eclipse 中设置程序代码的断点即可，只是如果同时加上这个项目的设置会有以下两个效果：

   i. 如果程序代码的断点停留太久，Android 系统也不会发出例外信息（exception error）；

   ii. 配合下一个 Wait for debugger 选项让程序先等待模拟器的 debugger 完成启动之后再开始运行程序。

2. Wait for debugger

   必须设置上一个项目才能使用，它是要求程序先等待模拟器的 debugger 完成启动之后再开始运行，设置这个项目可以帮助我们对程序的 onCreate()方法调试。

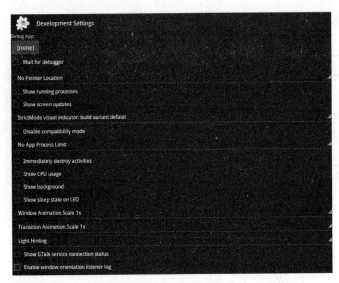

图 8-6　Android 模拟器的程序开发功能设置

    3. No Pointer Location

可以根据设置这个项目（从下拉式菜单中单击 Pointer Location）让模拟器用水平和垂直线的交叉方式显示目前用户单击的屏幕位置。

    4. Show running processes

设置这个项目后会在屏幕右上方显示目前系统正在运行的程序，如图 8-7 所示。

图 8-7　设置 Show running processes 功能后在屏幕右上方显示的程序运行信息

    5. Show screen updates

像 Android 这种使用图形接口的操作系统，需要不断地重绘界面，以反映程序最新的运行的状态。设置这个项目会让模拟器的界面在重绘之前，先显示一个矩形代表即将重绘的区域，让程序开发者观察程序界面重绘的情况。这项功能对于绘图相关的应用程序来说，可以帮助了解绘图的效率。

    6. Disable compatibility mode

平板电脑模拟器运行手机程序时，只利用屏幕的中央区域仿真手机的屏幕范围。

7. **Immediately destroy activities**
   设置此项目后，当 Activity 进入 stop 状态时将会立刻结束，下一次要运行需要重新 create。这项设置能够帮助测试 onSaveInstanceState()和 onCreate()方法中的程序代码。

8. **Show CPU Usage**
   设置此项目后，模拟器屏幕上方会以横条的方式显示目前 CPU 的使用状况。最上面的红线代表整体 CPU 的使用率，下面的绿线代表显示这个界面所使用的 CPU 时间。这项功能可以帮助了解程序每一个运行步骤所需的计算量。不过根据测试，这个项目只能够在 Android SDK 1.6 以下（含 1.6）的模拟器上才能设置。

9. **Show background**
   显示程序切换时不属于程序界面的屏幕区域，例如当运行手机程序时，会在平板电脑屏幕的四周显示背景区域（棋盘状底图）。

设置这些程序开发项目后，下一次再启动模拟器时还是有效，因此当不需要使用时必须将它取消。

8-5　AVD 的语言设置、时间设置和上网功能

任何一个 Android 模拟器在建立的时候都内定使用英文接口，如果读者想要变更操作接口的语言，可以从模拟器的 Home screen 中单击 Apps 按钮，然后选择 Settings 项目，再单击其中的 Language & input，最后进入 Language 就可以选择想要使用的接口语言，请读者参考图 8-8（此为平板电脑模拟器的界面，读者可以在手机模拟器上用同样的方式操作）。

图 8-8　设置模拟器所使用的语言

另外 Android 模拟器在建立的时候内定使用格林威治标准时间,但是大陆地区的时间是比格林威治标准时间快了八小时,因此 AVD 的时间和我们计算机的时间并不会一样。如果要将 AVD 的时间调整成和计算机相同,可以从模拟器的 Home screen 中单击 Apps 按钮,然后选择 Settings 项目,再单击其中的 Date & time,然后把右边的 Automatic time zone 项目取消,再单击下面的 Select time zone 就可以进入图 8-9 的时间地区选单,利用计算机的箭头键往下搜寻直到出现 Taipei 项目再将它选取即可。

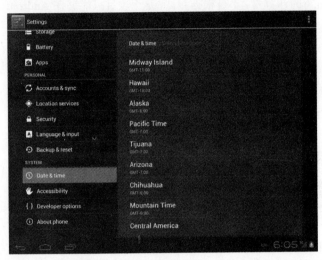

图 8-9 设置模拟器使用的标准时间,此为平板电脑模拟器的界面,读者可以在手机模拟器上用同样的方式操作

最后是有关 Android 模拟器的网络功能,请读者注意必须在启动 AVD 之前先让计算机连上 Internet,例如如果是拨号的 ADSL 上网方式必须先完成 ADSL 拨号的动作,这样启动之后的 AVD 才具备上网功能。

8-6　把实体手机或平板电脑当成模拟器

在开发程序的时候如果手边有 Android 手机或平板电脑,也可以把它当成是模拟器,这样做的好处是运行速度比较快。第一次使用时请依照下列步骤完成设置:

Step 1　在开发程序的 PC 上安装手机或平板电脑原厂提供的 USB 驱动程序。

Step 2　安装好之后将手机或平板电脑以 USB 接口连接到 PC。

Step 3　手机或平板电脑的"设置 > 应用程序"界面启用调试相关功能,例如以 HTC 的手机来说必须勾选"未知的来源"、"开发 > USB 调试中"和"开发 > 允许仿真位置"(测试定位功能时使用)。

Step 4 启动 Eclipse，切换到 DDMS 界面，在左上方的 Devices 窗口中会显示一个特殊名称的模拟器，该模拟器就是实体手机或平板电脑。

Step 5 回到 Eclipse 的 Java 界面开始运行程序项目，该程序就会安装到实体设备中运行。

如果要取得程序的运行界面，可以切换到 Eclipse 的 DDMS 模式，按下左上方 Devices 窗口工具栏中的"照相机"按钮（Screen Capture），就会显示一个窗口让我们进行相关的操作。

Android 版本	1.X	2.X	3.X	4.X
适用性	★	★	★	★

UNIT 9
良好的程序架构是程序开发和维护的重要基础

初学程序设计的人都会以为只要程序可以运行而且运行结果正确就算完工了。如果是学校作业或是专题实现这类小型程序项目是无可厚非,可是如果是要在市面上贩卖的商业程序,事情恐怕就不是这么简单。首先一个商业程序的程序代码规模少则数万行,多则上百万行,这类大型商业程序通常是由多人一起合作完成的。有些人负责操作接口的规划和设计,有些人负责程序代码的编写,像这样要能够做到程序项目开发的分工就必须在程序的结构上做适当的安排。举例来说,如果操作接口的设计和修改都必须牵动到程序代码的改变,那么这二者的分工就会造成运行上的困难。

再者市面上贩卖的商业程序通常会有不同语言的版本,像是中文版、英文版、日文版等,这些不同语言的版本只是在操作界面的消息正文使用不同的语言,程序的运行和功能完全一样,因此如果我们能在程序的结构上把所有的消息正文独立出来,避免掺杂在程序代码中,就能够很方便地修改成各种语言的版本。经过以上的讨论,读者应该能够体会程序架构的重要了吧。

程序开发技术经过数十年的发展与验证后发现,良好的程序架构是把整个程序项目区分成如图9-1 所示的三个部分。

1. Model

    负责运行程序的核心运算和判断逻辑,它通过 View 取得用户输入的数据,然后视需要从数据库查询相关的信息,最后进行计算和判断,再将得到的结果交给 View 来显示。

2. View

    定义程序的用户操作接口。程序操作接口的设计是一门专业的学问,应该使用哪一种接口组件,以及接口组件之间的排列位置和顺序都应该经过审慎的思考和设计,这些工作应该独立出来交由专门负责的研发人员完成。

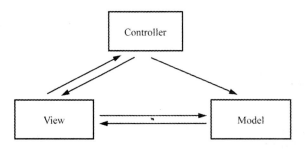

图 9-1　Model、View、Controller 程序架构，简称 MVC

3. Controller

这个部分是负责控制程序的运行流程以及对象之间的互动。它就像是交响乐团的指挥一样，担任整个系统的协调和控制的角色，让程序正常运行。在采用事件驱动的程序架构中，它也负责事件的处理和响应。

以上介绍的程序架构称为 MVC，如果一个程序项目能够依照这样的原则将这 3 个部分作适当切割，就可以很方便地进行分工，而且对于日后程序的维护和改版也能够更容易地进行。如果依照这个 MVC 程序架构重新检查我们的"婚姻建议"范例程序，会发现有下列几点可以改进：

1. findViewById()方法是用来取得宣告在接口布局文件中的接口组件，它是属于 View 的程序代码，而不是 Model 或 Controller。
2. setOnClickListener()是设置接口组件的事件处理程序，它也是属于 View 的程序代码。
3. 有许多字符串直接写在程序代码中，这样会造成不同语言版本移植上的困难。

针对以上问题，我们把程序项目做以下的修改：

1. 新增一个 setupViewComponent()方法负责运行 View 相关的所有程序代码，包括取得接口布局文件中的接口组件和设置接口组件的事件处理程序。
2. 把写在程序代码中的字符串放在 res/values/strings.xml 资源文件中，定义在 strings.xml 资源文件中的字符串在经过编译后会放到资源类 R 中，然后程序代码再从资源类 R 中取得需要的字符串。虽然这样的修改看似化简为繁，但是在开发多国语言的程序版本时，就可以展现它的优点。
3. 在 res/layout/main.xml 接口布局文件中，我们也是把提示字符串直接写在接口组件的属性中，这样也会造成不同语言程序版本移植上的困难，因此我们同样也把这些接口组件上的提示文字定义在 strings.xml 资源文件中，然后再到资源类 R 中取出字符串使用。

要在程序代码中使用资源类 R 中的字符串很简单，只要使用 getString()方法并传入"R.string.字符串名称"即可，读者只要参考下面的范例就可以了解。要在 main.xml 接口布局文件中使用资源类 R 中的字符串也只要使用指令"@string/字符串名称"即可。首先看看修改后的 res/values/

strings.xml 文件，我们新增了 8 个字符串，如以下粗体字的部分：

```xml
<?xml version="1.0" encoding="utf-8"?>
<resources>
    <string name="app_name">婚姻建议</string>
    <string name="promptSex">性别：</string>
    <string name="promptAge">年龄：</string>
    <string name="promptBtnDoSug">婚姻建议</string>
    <string name="sugResult">结果：</string>
    <string name="sugNotHurry">还不急。</string>
    <string name="sugGetMarried">赶快结婚！</string>
    <string name="sugFindCouple">开始找对象。</string>
    <string name="sexMale">男</string>
</resources>
```

接着是修改后的 res/layout/mail.xml 接口布局文件，粗体字的部分是把原来的字符串改成使用 R 类中的字符串：

```xml
<?xml version="1.0" encoding="utf-8"?>
<LinearLayout xmlns:android="http://schemas.android.com/apk/res/android"
    android:orientation="vertical"
    android:layout_width="fill_parent"
    android:layout_height="fill_parent"
    >
<TextView
    android:layout_width="fill_parent"
    android:layout_height="wrap_content"
    android:text="@string/promptSex"
    />
<EditText android:id="@+id/edtSex"
    android:layout_width="fill_parent"
    android:layout_height="wrap_content"
    android:inputType="text"
    android:text=""
    />
<TextView
    android:layout_width="fill_parent"
    android:layout_height="wrap_content"
    android:text="@string/promptAge"
    />
<EditText android:id="@+id/edtAge"
    android:layout_width="fill_parent"
    android:layout_height="wrap_content"
    android:inputType="number"
    android:text=""
    />
<Button android:id="@+id/btnDoSug"
    android:layout_width="fill_parent"
```

```
        android:layout_height="wrap_content"
        android:text="@string/promptBtnDoSug"
        />
    <TextView android:id="@+id/txtResult"
        android:layout_width="fill_parent"
        android:layout_height="wrap_content"
        android:text="@string/sugResult"
        />
</LinearLayout>
```

> **补充说明** 使用"图形操作模式"设计程序接口
>
> 开启 res/layout/main.xml 接口布局文件后在程序编辑窗口下方会有一个 Graphical Layout 标签,如果没有看到这个卷标,请先关闭接口布局文件,再用鼠标右击该文件,然后选择 Open With > Android Layout Editor 就会出现。单击该卷标后就会进入"图形操作模式",我们可以利用"按住拖曳"的方式把各种接口组件加到程序的操作接口,详细的操作方法请读者参考本书光盘中的"Graphical Layout Editor 操作说明.pdf"文档。
> 异常排除:如果进入 Graphical Layout 之后没有出现程序界面,则直接关闭 Eclipse 再重新启动即可。

最后是修改后的程序代码,我们新增了一个 setupViewComponent()方法专门负责运行 View 相关的所有设置。另外在显示判断结果的程序代码中,则是利用 getString()方法从资源类 R 中取得要用的字符串,粗体字的部分代表经过修改的程序代码:

```
package tw.android;

import android.app.Activity;
import android.os.Bundle;
import android.view.View;
import android.widget.*;

public class MarriSug extends Activity {

    private Button btnDoSug;
    private EditText edtSex, edtAge;
    private TextView txtResult;

    /** Called when the activity is first created. */
    @Override
    public void onCreate(Bundle savedInstanceState) {
        super.onCreate(savedInstanceState);
        setContentView(R.layout.main);

        setupViewComponent();
    }

    private void setupViewComponent() {
```

```java
// 从资源类 R 中取得接口组件
btnDoSug = (Button)findViewById(R.id.btnDoSug);
edtSex = (EditText)findViewById(R.id.edtSex);
edtAge = (EditText)findViewById(R.id.edtAge);
txtResult = (TextView)findViewById(R.id.txtResult);

// 设置 button 组件的事件 listener
btnDoSug.setOnClickListener(btnDoSugOnClick);
    }

    private Button.OnClickListener btnDoSugOnClick = new Button.OnClickListener() {
        public void onClick(View v) {
            String strSex = edtSex.getText().toString();
            int iAge = Integer.parseInt(edtAge.getText().toString());

            String strSug = getString(R.string.sugResult);
            if (strSex.equals(getString(R.string.sexMale)))
                if (iAge < 28)
                    strSug += getString(R.string.sugNotHurry);
                else if (iAge > 33)
                    strSug += getString(R.string.sugGetMarried);
                else
                    strSug += getString(R.string.sugFindCouple);
            else
                if (iAge < 25)
                    strSug += getString(R.string.sugNotHurry);
                else if (iAge > 30)
                    strSug += getString(R.string.sugGetMarried);
                else
                    strSug += getString(R.string.sugFindCouple);

            txtResult.setText(strSug);
        }
    };
}
```

在这个单元我们说明了程序架构的重要性以及何谓好的程序架构,同时也学会了如何将它应用在 Android 程序项目的开发上。从第一单元进行到此,读者已经具备编写 Android 应用程序的完整基础,接下来就可以依照自己的需求和兴趣,进一步学习不同的程序设计主题,并根据更多的程序实现经验加强程序设计的能力。

Android 版本	1.X	2.X	3.X	4.X
适用性	★	★		

UNIT 10
升级 Android 手机程序成为平板电脑程序

Android 平板电脑是使用 Android 3.0 以后的版本，由于平板电脑的屏幕尺寸比手机的屏幕尺寸大很多，而且在应用的需求上也有明显的差别，因此 Google 在 Android 3.0 以后的版本中加入许多新的功能，其中最明显的就是操作接口的改变。不过新版的 Android 系统对于手机程序的兼容性非常好，可以在完全不需要修改的情况下运行 Android 手机程序，只是这种运行方式会让原来的手机程序在平板电脑的屏幕上显得"独树一格"，图 10-1 就是在 Android 平板电脑中运行"纯"手机程序的界面，基本上它就是一个放大版的手机程序界面，因此外观和平板电脑程序有明显不同。其实只要我们稍微修改手机程序项目的设置，就可以让手机程序完全融入平板电脑的操作界面中，而且修改后的手机程序也能够在手机上运行，而且完全不会受到影响，这种修改手机程序项目的方法称为"针对 Android 平板电脑作优化"。另外还有一种方法是将手机程序完全升级成 Android 平板电脑程序，这样一来升级后的程序就能够使用新版 Android 平台的所有功能，但是经过修改后的程序就无法在手机上运行，接下来就让我们依序介绍这两种升级手机程序的方法。

补充说明　手机程序在平板电脑的执行模式

除了图 10-1 的放大模式之外（称为 zoom mode），手机程序也可以采用 stretch mode 在平板电脑上运行，不过这需要在程序项目的功能描述文件 AndroidManifest.xml 中加入屏幕模式的设置，详细的做法请参考 UNIT 60 中的说明。

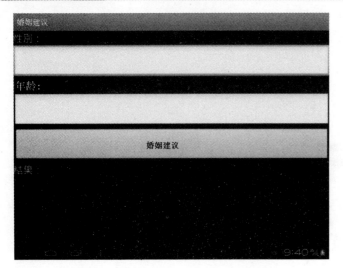

图 10-1　Android 手机程序在平板电脑上的运行界面

10-1　针对 Android 平板电脑作优化

这种修改手机程序项目的方法很简单，短则 5 分钟就可以完成，步骤如下：

Step 1　运行 Eclipse 并开启 Android 平板电脑模拟器，然后将手机程序加载到平板电脑模拟器中并测试功能是否正常，在绝大多数的情况下应该都能正常运行，确定无误后请用模拟器的"回上一页"按钮退出程序。

Step 2　在 Eclipse 中开启手机程序项目的 AndroidManifest.xml 文件，找到其中的<uses-sdk>标签，然后在标签中加上如下粗体字的属性：

```xml
<?xml version="1.0" encoding="utf-8"?>
<manifest …>
    <uses-sdk android:minSdkVersion="3"
              android:targetSdkVersion="11" />
    <application …>
    …
    </application>
</manifest>
```

这个属性值是告知 Android 系统这个程序可以使用 Android 3.0（API 版本号码就是 11）的平板电脑模式运行，注意这里不需要修改 Build Target 属性。

Step 3　将修改后的程序加载到 Android 平板电脑模拟器中进行测试，读者可以发现程序的运行界面已经变成和平板电脑程序一样（如图 10-2 所示），不会再显得格格不入。

图 10-2　修改手机程序项目的 AndroidManifest.xml 文件后的运行结果

Step 4　为了能够充分利用平板电脑增加的屏幕空间,可以在程序项目的 res 文件夹中新增 xlarge 相关的文件夹,并在里头新增大尺寸屏幕设备（也就是平板电脑）专用的资源,像是程序接口布局文件或是高分辨率的图文件,让程序的操作界面更加清楚美观。

使用这种方式修改后的程序项目仍然能够在手机上运行,而且功能完全不受影响。

10-2　将程序升级成为 Android 平板电脑专属程序

使用这种方式修改后的程序项目才能使用 Android 平板电脑的新功能,但是修改后的程序就无法在手机上运行,如果读者确定要采用这种方式,请依照下列步骤操作：

Step 1　运行 Eclipse 开启手机程序项目的 AndroidManifest.xml,找到其中的<uses-sdk>标签,然后修改 android:minSdkVersion 属性如下：

```
<?xml version="1.0" encoding="utf-8"?>
<manifest …>
    <application …>
        …
    </application>
    <uses-sdk android:minSdkVersion="11" />

</manifest>
```

这个属性值是告知 Android 系统这个程序必须使用 Android 3.0（API 版本号码 11）以后的 SDK,因此不可以在旧版的 Android 平台上运行。

Step 2　在 Eclipse 左边的项目检查窗口中用鼠标右击项目名称,然后在弹出的菜单中选择 Properties 就会出现图 10-3 的对话框。先在对话框的左边选择 Android,然后在右边的清单中勾选平板电脑使用的 Android 版本,例如 Android 3.0,再单击 Apply 按钮,最后单击 OK 按钮。在 Eclipse 左边的项目检查窗口中,读者会发现此程序项目所使用的 Android 版本已经改变。

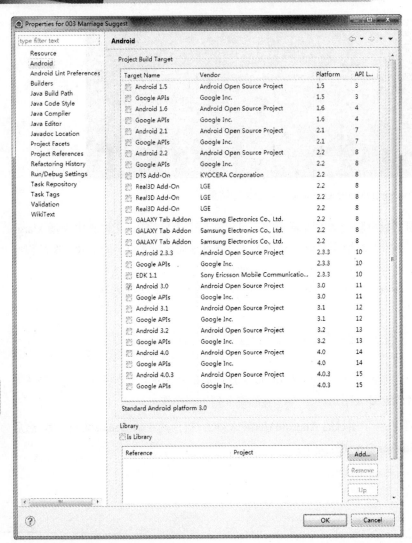

图 10-3 设置程序项目属性的对话框

PART 3 学习使用基本接口组件和布局模式

UNIT 11　学习更多接口组件的属性
UNIT 12　Spinner 下拉式列表组件
UNIT 13　使用 RadioGroup 和 RadioButton 组件建立单选列表
UNIT 14　CheckBox 多选列表和 ScrollView 滚动条
UNIT 15　LinearLayout 接口布局模式
UNIT 16　TableLayout 接口布局模式
UNIT 17　RelativeLayout 接口布局模式
UNIT 18　FrameLayout 接口布局模式和 Tab 标签页

Android 版本	1.X	2.X	3.X	4.X
适用性	★	★	★	★

UNIT 11
学习更多接口组件的属性

　　Android 应用程序是使用按钮、文字输入方块、下拉式菜单等各式各样的接口组件和用户进行互动，因此学习使用各种接口组件是 Android 程序设计的一个重要主题。在前面建立的"婚姻建议"范例程序中，我们已经用过 TextView、EditText 和 Button 三种接口组件，也已经了解 android:id、android:layout_width、android:layout_height 等组件属性的用法，除了这些属性以外，其实还有许多其他属性可以用来改变接口组件的外观、位置、文字颜色、底色等。这个单元就让我们来学习使用更多的组件属性，以便设计出更清楚、更容易使用的程序操作接口。

　　为了能够控制接口组件的外观、位置等特性，Android SDK 制定了非常多的接口组件属性。读者可以利用前面学过的 Android SDK 说明文件来查询所有的组件属性和相关说明。只是这些琳琅满目的属性往往令人有不知从何下手的感觉，因此为了学习上的方便，笔者特别从中挑选出一些常用的属性，另外也把我们之前学过的几个属性一起整理列出，如表 11-1 所示。其中有关长度和大小的属性，像是 textSize、margin 相关和 padding 相关的属性，Android 提供表 11-2 中的 6 种长度单位可以使用。

表 11-1　常用的接口组件属性

属性名称	设置值	使用说明
android:id	组件的名称	设置该接口组件的名称
android:layout_width android:layout_height	fill_parent match_parent wrap_content	设置组件的宽和高，fill_parent 是旧的属性值
android:text	组件中的文字	显示在组件中的文字
android:inputType	Text number date time …	输入的数据类型

续表

属性名称	设置值	使用说明
android:background	颜色(6 个 16 进位数字，例如 FF0000) 储存在 drawable 文件夹中的图形	设置组件的底色或底图，颜色以#开头后面接着 6 个 16 进位的数字，最前面 2 个数字是红色，中间 2 个数字是绿色，最后面 2 个数字是蓝色
android:textSize	数值和长度单位	设置文字大小，总共有 px, pt, dp, sp, in, mm 等 6 种长度单位可以使用，请参考表 11-2 的说明
android:textColor	颜色	设置文字颜色，颜色以#开头后面接着 6 个 16 进位的数字，最前面 2 个数字是红色，中间 2 个数字是绿色，最后面 2 个数字是蓝色
android:password	true false	将输入的文字用暗码显示防止被他人窥视，新版的 Android 程序改成用 inputType 属性控制
android:autoLink	web email phone map all	自动侦测字符串中的超链接数据
android:hint	组件中的提示文字	当 EditText 组件中没有输入任何数据时所显示的字符串
android:layout_margin	数值和长度单位	设置组件四周的间隔距离
android:layout_marginLeft android:layout_marginRight android:layout_marginTop android:layout_marginBottom	数值和长度单位	个别指定组件四周的间隔距离
android:padding	数值和长度单位	设置组件内部的文字和边的距离
android:paddingLeft android:paddingRight android:paddingTop android:paddingBottom	数值和长度单位	个别指定组件内部的文字和边的距离
android:gravity	center_hotizontal center_vertical center	组件中的对象的对齐方式
android:layout_gravity	center_hotizontal center_vertical center	组件相对于它的外框的对齐方式
android:layout_weight	数值	让组件使用固定比例的宽度

表 11-2　接口组件的属性可以使用的长度单位

单位名称	全名	说明
px	pixel	屏幕像素
pt	point	传统印刷使用的字体大小单位，1pt = 1/72 英寸
dp	density-independent point	对应到 160 dpi 屏幕的像素个数，Android 程序应该使用这个单位来设置组件的大小和空间距离
sp	scale-independent point	对应到 160 dpi 屏幕的字体大小，Android 程序应该使用这个单位来设置字体的大小
mm	millimeter	毫米
in	inch	英寸

根据 Android SDK 的建议，如果是设置接口组件的位置、大小和边界距离等相关属性应该使用 dp 长度单位。如果是设置字体大小的相关属性，则应该使用 sp 长度单位。接下来我们利用一些实际范例让读者了解使用这些属性之后的效果。

11-1　match_parent 和 wrap_content 的差别

在 android: layout_width 和 android:layout_height 属性中可以设置 fill_parent、match_parent 或是 wrap_content，fill_parent 和 match_parent 的效果是一样的，两者都是填满组件所在的外框，但是从 Android 2.2 开始应该使用 match_parent 而不要再用 fill_parent。至于 wrap_content 则是依照组件中的文字长度或高度来决定组件的宽或高，请参考以下的程序代码和图 11-1 的范例就可以了解。

图 11-1　使用 match_parent 和 wrap_content 属性值设置接口组件的大小

```
<EditText
    android:layout_width="match_parent"
    android:layout_height="wrap_content"
    android:text="EditText1"
/>
<EditText
    android:layout_width="wrap_content"
    android:layout_height="match_parent"
    android:text="EditText2"
/>
```

11-2　android:inputType 属性的效果

android:inputType 属性可以用来限制输入的字符种类，如果设置成 text 就能够输入任何字符，如果设置成 number 就只能输入数字，当设置成 date 时可以输入数字和斜线 "/" 字符，当设置成 time 时则可以输入数字和分号 ":" 字符以及 pam 等 3 个英文字母，请参考以下范例和图 11-2。

```
<EditText
    android:layout_width="match_parent"
    android:layout_height="wrap_content"
    android:text=""
    android:inputType="text"
/>
<EditText
    android:layout_width="match_parent"
    android:layout_height="wrap_content"
    android:text=""
    android:inputType="number"
/>
<EditText
    android:layout_width="match_parent"
    android:layout_height="wrap_content"
    android:text=""
    android:inputType="date"
/>
<EditText
    android:layout_width="match_parent"
    android:layout_height="wrap_content"
    android:text=""
    android:inputType="time"
/>
```

图 11-2　android:inputType 属性的使用范例

11-3 控制文字大小、颜色和底色

在这个范例中我们使用 android:textSize、android:textColor、android:background 来改变文字的大小、颜色，以及组件的底色，图 11-3 是程序运行的界面。

```
<EditText
    android:layout_width="match_parent"
    android:layout_height="wrap_content"
    android:text="默认的文字大小"
/>
<EditText
    android:layout_width="match_parent"
    android:layout_height="wrap_content"
    android:text="10sp 文字"
    android:textSize="10sp"
/>
<EditText
    android:layout_width="match_parent"
    android:layout_height="wrap_content"
    android:text="20sp 绿色文字"
    android:textSize="20sp"
    android:textColor="#00FF00"
/>
<EditText
    android:layout_width="match_parent"
    android:layout_height="wrap_content"
    android:text="30sp 绿色文字，黑色底色"
    android:textSize="30sp"
    android:textColor="#00FF00"
    android:background="#000000"
/>
```

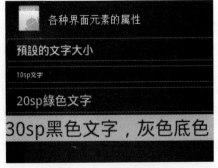

图 11-3 设定文字的大小、颜色和接口组件的底色

11-4　控制组件四周的间隔距离以及组件内部的文字和边的距离

这个范例我们使用 margin 相关属性来增加组件和外框之间的距离，以及使用 padding 相关属性来增加组件内的文字和组件边框的距离，图 11-4 是程序运行的界面。

```
<EditText
    android:layout_width="match_parent"
    android:layout_height="wrap_content"
    android:text="预设的间隔"
/>
<EditText
    android:layout_width="match_parent"
    android:layout_height="wrap_content"
    android:text="设置 padding=20dp"
    android:padding="20dp"
/>
<EditText
    android:layout_width="match_parent"
    android:layout_height="wrap_content"
    android:text="再设置 margin=20dp"
    android:padding="20dp"
    android:layout_margin="20dp"
/>
<EditText
    android:layout_width="match_parent"
    android:layout_height="wrap_content"
    android:text="只设置左右 margin=30dp"
    android:layout_marginLeft="30dp"
    android:layout_marginRight="30dp"
/>
```

图 11-4　margin 和 padding 相关属性的使用范例

根据以上的范例，读者应该可以感受到接口组件属性的妙用，后续在编写 Android 应用程序时，如果需要调整接口组件的大小、位置、颜色，或是其他外观特性，可以利用以上属性，或是查询 Android SDK 说明文件来找出其他适合的属性。

Android 版本	1.X	2.X	3.X	4.X
适用性	★	★	★	★

UNIT 12
Spinner 下拉式菜单组件

当程序需要用户输入数据时,除了让用户自己打字之外,还有一种比较体贴的设计就是列出一组选项让用户挑选,这样就可以避免打字的麻烦。对于手机和平板电脑的应用程序来说,打字是非常不方便的操作方式,因此如果程序可以为用户预先建立好选项列表,就应该避免让用户自行输入数据。这种利用选项列表的操作接口可以使用 Spinner 下拉式菜单元件来完成。

建立 Spinner 下拉式菜单元件的过程比前面使用过的接口组件要复杂一些,我们必须完成下列几件事:

1. 建立选项列表,选项列表中包含许多项目名称,这些项目名称是用数组的方式代表。
2. 把选项列表设置给一个 Spinner 接口组件。
3. 设置 Spinner 组件的菜单显示格式。
4. 设置 Spinner 组件的 OnItemSelectedListener() 事件处理程序,当用户单击某个项目之后,程序必须取得该项目所对应的数据。

建立选项列表有两种方式,一种是直接将选项列表以数组的方式宣告在程序中。这种方式比较简单,但是如果选项的名称会随着不同地区的语言作调整,就不适合使用这种方式,而必须采用第二种方法,就是把选项列表建立在项目的 res/values/strings.xml 文件中,再让程序从项目的资源类 R 中取得选项列表数组。接下来让我们先介绍第一种方式,并以"婚姻建议"程序为范例,将其中的性别输入改成使用 Spinner 下拉式菜单。

12-1 建立 Spinner 下拉式菜单的第一种方法

建立 Spinner 下拉式菜单的过程比较复杂,为了帮助读者了解,我们把整个流程区分成下列几个步骤:

Step 1 在项目界面布局文件 res/layout/main.xml 中建立一个 Spinner 接口组件如以下范例：

```
<Spinner android:id="@+id/spnSex"
    android:layout_width="match_parent"
    android:layout_height="wrap_content"
    android:drawSelectorOnTop="true"
    android:prompt="@string/spnSexPrompt"
    android:spinnerMode="dialog"/>
```

我们设置这个 Spinner 组件的名称叫做 spnSex，android:prompt 属性是设置显示在选项列表上方的说明文字，这里我们使用一个定义在 res/values/strings.xml 资源文件中的字符串。android:spinnerMode 属性则是设置选项列表的显示方式，dialog 的方式是以对话框的型态出现，另一种 dropdown 的方式则是将选项列表列于 Spinner 组件的下方。

接下来的步骤都在程序文件中完成。

Step 2 取得前一个步骤建立的 Spinner 组件。

```
Spinner spnSex = (Spinner)findViewById(R.id.spnSex);
```

Step 3 把要显示的选项列表表示成一个 String 型态的数组，例如：

```
String[] sSexList = new String[] {"男", "女"};
```

Step 4 建立一个 ArrayAdapter 类的对象，将前一个步骤的菜单数组输入该对象并指定使用 Spinner 格式。ArrayAdapter 是一个泛型类（generic class），也就是说它可以处理各种型态的对象。这里我们指定使用 String 型态的对象，因此应使用 ArrayAdapter<String>。

```
ArrayAdapter<String> adapSexList = new ArrayAdapter<String>(
        this, android.R.layout.simple_spinner_item, sSexList);
adapSexList.setDropDownViewResource(
        android.R.layout.simple_spinner_dropdown_item);
```

其中的 android.R.layout.simple_spinner_item 和 android.R.layout.simple_spinner_dropdown_item 都是 Android SDK 本身提供的 Spinner 菜单元格式。另外我们也可以自行使用 xml 文件来定义自己的菜单元格式，后面的实现范例有完整的做法。

Step 5 将上一个步骤建立的 ArrayAdapter 对象设置给前面建立的 Spinner 组件。

```
spnSex.setAdapter(adapSexList);
```

以上几个步骤都是属于 MVC 架构的 View，所以这些程序代码应该放在我们自己建立的 setupViewComponent()方法中。

Step 6 建立 Spinner 的 OnItemSelectedListener 并完成其中的 onItemSelected()和 onNothingSelected() 两个方法，程序代码如下：

```
private Spinner.OnItemSelectedListener spnSexItemSelLis =
    new Spinner.OnItemSelectedListener () {
        public void onItemSelected(AdapterView parent,
                                    View v,
                                    int position,
                                    long id) {
            sSex = parent.getSelectedItem().toString();
        }
        public void onNothingSelected(AdapterView parent) {
        }
    };
```

这两个方法是运行用户单击项目后的数据设置动作。

补充说明 Eclipse 操作技巧

当要在程序代码中建立一个对象，例如以上范例中的 spnSexItemSelLis 对象时，先输入宣告对象的语法，也就是以上范例中的粗体字部分，把里面的方法先空下来，这时候在这一段程序代码会出现红色波浪底线标示语法错误，将鼠标光标移到红色波浪底在线，就会弹出一个说明窗口建议修正方法，单击其中建议的修正项目后，就会自动修正程序代码的错误。利用这种方式可以让程序代码编辑器自动帮我们加入这两个方法，然后再自行输入里头的程序代码。不过这种方式有一个缺点，就是插入程序代码的自变量名称有些是用 arg 命名看不出自变量的功用。遇到这种情况时，建议读者查询 Android SDK 的说明文件（参考 UNIT 5 的介绍），找到该类的说明，然后复制其中的自变量名称。

Step 7 最后就是把上一个步骤建立的 OnItemSelectedListener 对象设置成为 Spinner 组件的事件处理程序。这个动作也是属于 MVC 架构中的 View，因此同样也是写在 setupViewComponent() 方法中。

完成以上所有步骤之后，运行程序就会看到原来性别的输入字段变成下拉式菜单，如图 12-1 所示，当使用者单击性别输入字段的时候就会弹出一个选项列表。在这个范例中我们利用上一个单元介绍的组件属性重新调整接口组件的位置和文字的大小以提升操作的便利性。以下我们列出完整的接口布局文件和程序代码，为了方便原来的程序代码对照，修改的部分使用粗体字代表。

图 12-1 将婚姻建议程序的性别输入字段改成下拉式菜单

■ 界面布局文件：

```xml
<?xml version="1.0" encoding="utf-8"?>
<LinearLayout xmlns:android="http://schemas.android.com/apk/res/android"
    android:orientation="vertical"
    android:layout_width="match_parent"
    android:layout_height="match_parent"
    android:gravity="center_horizontal">
<TextView
    android:layout_width="200dp"
    android:layout_height="wrap_content"
    android:textSize="20sp"
    android:text="@string/promptSex"/>
<Spinner android:id="@+id/spnSex"
    android:layout_width="200dp"
    android:layout_height="wrap_content"
    android:textSize="20sp"
    android:drawSelectorOnTop="true"
    android:prompt="@string/spnSexPrompt"
    android:spinnerMode="dialog"/>
<TextView
    android:layout_width="200dp"
    android:layout_height="wrap_content"
    android:textSize="20sp"
    android:text="@string/promptAge"/>
<EditText android:id="@+id/edtAge"
    android:layout_width="200dp"
    android:layout_height="wrap_content"
    android:textSize="20sp"
    android:inputType="number"
    android:text=""/>
<Button android:id="@+id/btnDoSug"
    android:layout_width="200dp"
    android:layout_height="wrap_content"
    android:textSize="20sp"
    android:text="@string/promptBtnDoSug"/>
<TextView android:id="@+id/txtResult"
    android:layout_width="200dp"
    android:layout_height="wrap_content"
    android:textSize="20sp"
    android:text="@string/sugResult"/>
</LinearLayout>
```

■ 程序代码：

package ...

import ...

```java
public class MarriSug extends Activity {

    private Button btnDoSug;
    private Spinner spnSex
    private EditText edtAge;
    private TextView txtResult;

    private String sSex;

    /** Called when the activity is first created. */
    @Override
    public void onCreate(Bundle savedInstanceState) {
        super.onCreate(savedInstanceState);
        setContentView(R.layout.main);

        setupViewComponent();
    }

    private void setupViewComponent() {
        // 从资源类 R 中取得接口组件
        btnDoSug = (Button)findViewById(R.id.btnDoSug);
        spnSex = (Spinner)findViewById(R.id.spnSex);
        edtAge = (EditText)findViewById(R.id.edtAge);
        txtResult = (TextView)findViewById(R.id.txtResult);

        String[] sSexList = new String[] {"男", "女"};
        ArrayAdapter<String> adapSexList = new ArrayAdapter<String>(
                this, android.R.layout.simple_spinner_item, sSexList);
        adapSexList.setDropDownViewResource(R.layout.spinner_layout);
        spnSex.setAdapter(adapSexList);

        spnSex.setOnItemSelectedListener(spnSexItemSelLis);

        // 设置 button 组件的事件 listener
        btnDoSug.setOnClickListener(btnDoSugOnClick);
    }

    private Spinner.OnItemSelectedListener spnSexItemSelLis =
        new Spinner.OnItemSelectedListener () {
            public void onItemSelected(AdapterView parent,
                                       View v,
                                       int position,
                                       long id) {
                sSex = parent.getSelectedItem().toString();
            }
            public void onNothingSelected(AdapterView parent) {
            }
        };
```

```
private Button.OnClickListener btnDoSugOnClick = new Button.OnClickListener() {
    public void onClick(View v) {
        int iAge = Integer.parseInt(edtAge.getText().toString());

        String strSug = getString(R.string.sugResult);
        if (sSex.equals(getString(R.string.sexMale)))
            if (iAge < 28)
                strSug += getString(R.string.sugNotHurry);
            else if (iAge > 33)
                strSug += getString(R.string.sugGetMarried);
            else
                strSug += getString(R.string.sugFindCouple);
        else
            if (iAge < 25)
                strSug += getString(R.string.sugNotHurry);
            else if (iAge > 30)
                strSug += getString(R.string.sugGetMarried);
            else
                strSug += getString(R.string.sugFindCouple);

        txtResult.setText(strSug);
    }
};
```

在前面步骤四中我们提到可以用 xml 文件定义自己的菜单元格式，以上的程序代码中用底线标示的部分就是使用我们自己项目中的菜单元格式定义文件 res/layout/R.layout.spinner_layout.xml，该档的内容如下，其中只有一个 TextView 组件，也就是说每一个选项都会用该 TextView 组件的格式来显示。

```xml
<?xml version="1.0" encoding="utf-8"?>
<TextView xmlns:android="http://schemas.android.com/apk/res/android"
    android:layout_width="match_parent"
    android:layout_height="wrap_content"
    android:textSize="20sp"
/>
```

12-2　建立 Spinner 下拉式菜单的第二种方法

在第一种方法中，选项列表是以数组的型态直接写在程序代码中，如果将来程序要移植到不同的语言而必须改变项目的名称时，就必须到程序代码中寻找该数组再加以修改，这样对于程序项目的维护是非常不利的。比较好的方法是把和语言相关的字符串定义在程序项目的资源文件中，将来如果要把程序项目移植到其他语言，只要修改项目的资源文件即可，这样可以有效提高程序的维护效率。为了达到这个目的,我们必须修改第一个方法中的第 3 和第 4 步骤。

Step 3 把选项列表数组以 <string-array> 标签的格式宣告在 res/values/strings.xml 文件中如下:

```xml
<?xml version="1.0" encoding="utf-8"?>
<resources>

    <string ...</string>
    …
    <string-array name="spnSexList">
        <item>男</item>
        <item>女</item>
    </string-array>
</resources>
```

Step 4 使用 ArrayAdapter 类的 createFromResource() 方法从项目的资源类 R 中取出项目列表数组,并建立一个 ArrayAdapter 对象,但是该 ArrayAdapter 对象的处理型态必须改成 CharSequence,如以下范例:

```java
ArrayAdapter<CharSequence> adapSexList = ArrayAdapter.createFromResource(
            this, R.array.spnSexList,
            android.R.layout.simple_spinner_item);
```

我们同样可以使用自行定义的 xml 文件来建立自己的菜单元格式。

其他步骤都和第一个方法相同。

这两个方法的运行结果完全一样,因此以下仅列出程序代码的部分,为了和原来的程序代码比较,有关 Spinner 组件的部分全部使用粗体字标示。

```java
package tw.android;

import android.app.Activity;
import android.os.Bundle;
import android.view.View;
import android.widget.*;

public class MarriSug extends Activity {

    private Button btnDoSug;
    private Spinner spnSex;
    private EditText edtAge;
    private TextView txtResult;

    private String sSex;

    /** Called when the activity is first created. */
    @Override
    public void onCreate(Bundle savedInstanceState) {
```

```java
        super.onCreate(savedInstanceState);
        setContentView(R.layout.main);

        setupViewComponent();
    }

    private void setupViewComponent() {
        // 从资源类 R 中取得接口组件
        btnDoSug = (Button)findViewById(R.id.btnDoSug);
        spnSex = (Spinner)findViewById(R.id.spnSex);
        edtAge = (EditText)findViewById(R.id.edtAge);
        txtResult = (TextView)findViewById(R.id.txtResult);

        ArrayAdapter<CharSequence> adapSexList = ArrayAdapter.createFromResource(
                this, R.array.spnSexList, R.layout.spinner_layout);
        spnSex.setAdapter(adapSexList);

        spnSex.setOnItemSelectedListener(spnSexItemSelLis);

        // 设置 button 组件的事件 listener
        btnDoSug.setOnClickListener(btnDoSugOnClick);
    }

    private Spinner.OnItemSelectedListener spnSexItemSelLis =
        new Spinner.OnItemSelectedListener () {
            public void onItemSelected(AdapterView parent,
                                       View v,
                                       int position,
                                       long id) {
                sSex = parent.getSelectedItem().toString();
            }
            public void onNothingSelected(AdapterView parent) {
            }
    };

    private Button.OnClickListener btnDoSugOnClick = new Button.OnClickListener() {
        public void onClick(View v) {
            int iAge = Integer.parseInt(edtAge.getText().toString());

            String strSug = getString(R.string.sugResult);
            if (sSex.equals(getString(R.string.sexMale)))
                if (iAge < 28)
                    strSug += getString(R.string.sugNotHurry);
                else if (iAge > 33)
                    strSug += getString(R.string.sugGetMarried);
                else
                    strSug += getString(R.string.sugFindCouple);
```

```
            else
                if (iAge < 25)
                    strSug += getString(R.string.sugNotHurry);
                else if (iAge > 30)
                    strSug += getString(R.string.sugGetMarried);
                else
                    strSug += getString(R.string.sugFindCouple);

                txtResult.setText(strSug);
            }
        };
    }
```

UNIT 13

使用 RadioGroup 和 RadioButton 组件建立单选清单

Spinner 接口组件是在用户单击之后才会显示选项列表,这种方式比较节省操作接口的空间,但是缺点是要先做一个单击的动作。除了这种下拉式菜单之外,还有另外一种也是属于单选式的接口组件,那就是 RadioGroup 和 RadioButton。请读者参考图 13-1,它是由多个 RadioButton 项目所组成,这组 RadioButton 项目包含在一个 RadioGroup 组件中,用户只能从这些 RadioButton 项目中单击其中一个。

图 13-1　RadioGroup 和 RadioButton

RadioGroup 和 RadioButton 接口组件的使用方式比 Spinner 组件简单,只需要下列 2 个步骤:

Step 1 在 res/layout 文件夹下的接口布局文件中利用 RadioGroup 标签和 RadioButton 标签建立好选项列表如下:

```
<RadioGroup android:id="@+id/radGSex"
    android:layout_width="match_parent"
    android:layout_height="wrap_content"
    android:orientation="vertical"
    android:checkedButton="@+id/radMale">
    <RadioButton android:id="@+id/radMale"
      android:text="男生"/>
    <RadioButton android:id="@+id/radFemale"
      android:text="女生"/>
</RadioGroup>
```

我们必须为 RadioGroup 组件和全部的 RadioButton 组件设置好 id 名称，因为在程序代码中将会用这些 id 名称来控制这些组件。此外还有几个新出现的组件属性，读者从它们的名称应该就可以了解它们的用途。建立好选项列表之后这组列表就可以正常运行，读者如果在接口描述文件中插入以上的程序代码就会看到图 13-1 的效果，我们可以任意单击其中一个项目，单击之后另一个项目的单击状态就会自动被清除。

Step 2 RadioGroup 选项列表的操作都会搭配一个 Button 组件，当按下该 Button 之后，程序才会读取用户单击的项目。要知道用户选择的项目只要调用 RadioGroup 组件的 getCheckedRadioButtonId()方法即可，它会传回目前被用户选取的项目 id 名称，我们只要使用 switch…case…的语法就可以依照被选取的项目来运行对应的程序代码。

```
int iCheckedRadBtn = radGSex.getCheckedRadioButtonId();

switch (iCheckedRadBtn)
{
case R.id.radMale:
    // 选择这个选项所运行的程序代码
    …
case R.id.radFemale:
    // 选择这个选项所运行的程序代码
    …
}
```

根据以上 2 个步骤就可以建立一组 RadioGroup 单选清单，但是如果要把它用到我们的"婚姻建议"程序，还需要用到其他控制 RadioGroup 组件的技巧。

13-1 将"婚姻建议"程序改成使用 RadioGroup 菜单

请读者参考图 13-2 的程序运行界面，当用户在性别选项中单击男生时，下方的年龄选项是以 28 和 33 岁为分界。但是如果使用者单击女生，则下方的年龄选项会变成以 25 和 30 岁为分界。也就是说年龄选项中的文字必须随着单击的性别而改变。要完成这样的功能需要多了解一些 RadioGroup 的运行方式。首先当用户单击 RadioGroup 组件中的 RadioButton 时，Android 系统会发出一个 OnCheckedChange 的事件（我们在前面的单元曾经解释过窗口接口程序都是以"事件驱动"的方式运行），所以我们可以在性别 RadioGroup 组件的 OnCheckedChange 事件的 listener 中完成改变年龄选项文字的动作。详细的作法是我们在程序中建立一个 RadioGroup 组件的 OnCheckedChangeListener 对象，该对象中就是改变年龄选项文字的程序代码，然后把它设置给性别的 RadioGroup 组件。以下是完整的字符串资源文件、接口布局文件和程序文件，粗体字的部分代表经过修改的程序代码。

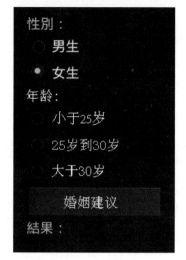

图 13-2 "婚姻建议"程序改用 RadioGroup 菜单的运行界面

- res/values/strings.xml 字符串资源文件：

```
<?xml version="1.0" encoding="utf-8"?>
<resources>
    <string name="app_name">婚姻建议</string>
    <string name="promptSex">性别：</string>
    <string name="promptAge">年龄：</string>
    <string name="promptBtnDoSug">婚姻建议</string>
    <string name="sugResult">结果：</string>
    <string name="sugNotHurry">还不急。</string>
    <string name="sugGetMarried">赶快结婚！</string>
    <string name="sugFindCouple">开始找对象。</string>
    <string name="male">男生</string>
    <string name="female">女生</string>
    <string name="maleAgeRng1">小于 28 岁</string>
    <string name="maleAgeRng2">28~33 岁</string>
    <string name="maleAgeRng3">大于 33 岁</string>
    <string name="femaleAgeRng1">小于 25 岁</string>
    <string name="femaleAgeRng2">25~30 岁</string>
    <string name="femaleAgeRng3">大于 30 岁</string>
</resources>
```

- res/layout/main.xml 界面布局文件：

```
<?xml version="1.0" encoding="utf-8"?>
<LinearLayout xmlns:android="http://schemas.android.com/apk/res/android"
    android:orientation="vertical"
    android:layout_width="match_parent"
    android:layout_height="match_parent"
    android:gravity="center_horizontal">
<TextView
```

```xml
        android:layout_width="200dp"
        android:layout_height="wrap_content"
        android:textSize="20sp"
        android:text="@string/promptSex"/>
    <RadioGroup android:id="@+id/radGSex"
        android:layout_width="200dp"
        android:layout_height="wrap_content"
        android:orientation="vertical"
        android:checkedButton="@+id/radMale">
        <RadioButton android:id="@+id/radMale"
            android:textSize="20sp"
            android:text="@string/male"/>
        <RadioButton android:id="@+id/radFemale"
            android:textSize="20sp"
            android:text="@string/female"/>
    </RadioGroup>
    <TextView
        android:layout_width="200dp"
        android:layout_height="wrap_content"
        android:textSize="20sp"
        android:text="@string/promptAge"/>
    <RadioGroup android:id="@+id/radGAge"
        android:layout_width="200dp"
        android:layout_height="wrap_content"
        android:orientation="vertical"
        android:checkedButton="@+id/radAgeRng1">
        <RadioButton android:id="@+id/radBtnAgeRng1"
            android:textSize="20sp"
            android:text="@string/maleAgeRng1"/>
        <RadioButton android:id="@+id/radBtnAgeRng2"
            android:textSize="20sp"
            android:text="@string/maleAgeRng2"/>
        <RadioButton android:id="@+id/radBtnAgeRng3"
            android:textSize="20sp"
            android:text="@string/maleAgeRng3"/>
    </RadioGroup>
    <Button android:id="@+id/btnDoSug"
        android:layout_width="200dp"
        android:layout_height="wrap_content"
        android:textSize="20sp"
        android:text="@string/promptBtnDoSug"/>
    <TextView android:id="@+id/txtResult"
        android:layout_width="200dp"
        android:layout_height="wrap_content"
        android:textSize="20sp"
        android:text="@string/sugResult"/>
</LinearLayout>
```

■ 程序文件：

```java
package ...

import ...

public class Main extends Activity {

    private Button btnDoSug;
    private RadioGroup radGSex, radGAge;
    private RadioButton radBtnAgeRng1, radBtnAgeRng2, radBtnAgeRng3;
    private TextView txtResult;

    /** Called when the activity is first created. */
    @Override
    public void onCreate(Bundle savedInstanceState) {
        super.onCreate(savedInstanceState);
        setContentView(R.layout.main);

        setupViewComponent();
    }

    private void setupViewComponent() {
        // 从资源类 R 中取得接口组件
        btnDoSug = (Button)findViewById(R.id.btnDoSug);
        radGSex = (RadioGroup)findViewById(R.id.radGSex);
        radGAge = (RadioGroup)findViewById(R.id.radGAge);
        radBtnAgeRng1 = (RadioButton)findViewById(R.id.radBtnAgeRng1);
        radBtnAgeRng2 = (RadioButton)findViewById(R.id.radBtnAgeRng2);
        radBtnAgeRng3 = (RadioButton)findViewById(R.id.radBtnAgeRng3);
        txtResult = (TextView)findViewById(R.id.txtResult);

        // 设置事件 listener
        btnDoSug.setOnClickListener(btnDoSugOnClick);
        radGSex.setOnCheckedChangeListener(radGSexOnCheChanLis);
    }

    private RadioGroup.OnCheckedChangeListener radGSexOnCheChanLis = new RadioGroup.OnCheckedChangeListener () {
        public void onCheckedChanged(RadioGroup group, int checkedId)
        {
            if (checkedId == R.id.radMale) {
                radBtnAgeRng1.setText(getString(R.string.maleAgeRng1));
                radBtnAgeRng2.setText(getString(R.string.maleAgeRng2));
                radBtnAgeRng3.setText(getString(R.string.maleAgeRng3));
            }
            else if (checkedId == R.id.radFemale) {
                radBtnAgeRng1.setText(getString(R.string.femaleAgeRng1));
```

```java
                radBtnAgeRng2.setText(getString(R.string.femaleAgeRng2));
                radBtnAgeRng3.setText(getString(R.string.femaleAgeRng3));
            }
        }
    };

    private Button.OnClickListener btnDoSugOnClick = new Button.OnClickListener() {
        public void onClick(View v) {
            int iCheckedRadBtn = radGSex.getCheckedRadioButtonId();

            String strSug = getString(R.string.sugResult);
            switch (iCheckedRadBtn)
            {
            case R.id.radMale:
                switch (radGAge.getCheckedRadioButtonId())
                {
                case R.id.radBtnAgeRng1:
                    strSug += getString(R.string.sugNotHurry);
                    break;
                case R.id.radBtnAgeRng3:
                    strSug += getString(R.string.sugGetMarried);
                    break;
                default:
                    strSug += getString(R.string.sugFindCouple);
                }
                break;
            case R.id.radFemale:
                switch (radGAge.getCheckedRadioButtonId())
                {
                case R.id.radBtnAgeRng1:
                    strSug += getString(R.string.sugNotHurry);
                    break;
                case R.id.radBtnAgeRng3:
                    strSug += getString(R.string.sugGetMarried);
                    break;
                default:
                    strSug += getString(R.string.sugFindCouple);
                }
                break;
            }

            txtResult.setText(strSug);
        }
    };
}
```

以上的程序代码我们依序解说如下:

1. 在字符串资源文件中我们定义好所有 RadioButton 的选项文字。
2. 在接口布局文件中我们定义了性别和年龄这两个 RadioGroup 和它们内部的 RadioButton 选项，并且每一个组件都指定 id 名称以便在程序中进行控制。另外我们也设置了一些组件的属性（文字大小、组件宽度和水平置中对齐），让程序的界面更加清楚美观。
3. 在程序代码中新增 RadioGroup 和 RadioButton 对象以便储存接口布局文件中建立的接口组件。这些对象会在程序代码的 setupViewComponent() 方法中完成设置。
4. setupViewComponent() 方法的最后一行是把我们在程序代码后面建立的 RadioGroup.OnCheckedChangeListener 对象设置给性别的 RadioGroup 对象，因此当用户改变性别选项时，年龄选项中的文字就会随着改变。建立 RadioGroup.OnCheckedChangeListener 对象可以利用我们在前一个单元建立 Spinner.OnItemSelectedListener 对象时介绍过的操作技巧。
5. 最后当用户按下按钮时，程序代码先取得性别选项，然后再取得年龄选项，最后显示判断结果。

图 13-3 是完整的程序运行界面。我们将"婚姻建议"程序从最原始的文字输入数据方式，修改成使用 Spinner 下拉式菜单的操作接口，现在进一步改成使用 RadioButton 菜单。如果读者仔细体会一下程序的操作流程，应该可以了解用户接口的改变对于程序操作的便利性所带来的影响。对手持式设备的应用程序来说，单击是最方便的操作方式，而且如果界面空间允许，应该尽可能把选项全部列出，让使用者可以一目了然。这个最新版的"婚姻建议"程序应该是界面最清楚且操作最方便的版本。

图 13-3 使用 RadioButton 菜单的"婚姻建议"程序

Android 版本	1.X	2.X	3.X	4.X
适用性	★	★	★	★

UNIT 14
CheckBox 多选清单和 ScrollView 滚动条

如果程序需要提供用户可以复选的选项列表，就必须使用 CheckBox 接口组件。这种可以复选的清单通常包含比较多的数据项，因此有可能会超出屏幕的范围，这时候我们可以搭配使用 ScrollView 滚动条，因此这个单元就让我们一起学习 CheckBox 和 ScrollView 接口组件的用法。

假设程序需要提供一组兴趣相关的选项供用户选取，而且允许使用者可以同时选取多个，这时候我们可以在项目的 res/layout 文件夹下的接口布局文件中加入 CheckBox 组件，每一个 CheckBox 组件都对应到一个兴趣项目，如以下范例：

```xml
<LinearLayout
    …
    >
<CheckBox android:id="@+id/chbItem1"
    android:layout_width="match_parent"
    android:layout_height="wrap_content"
    android:text="项目 1 "
    />
<CheckBox android:id="@+id/chbItem2"
    android:layout_width="match_parent"
    android:layout_height="wrap_content"
    android:text="项目 2 "
    />
…
<Button android:id="@+id/btnSelOK"
    android:layout_width="match_parent"
    android:layout_height="wrap_content"
```

```
        android:text="确定"
        />
</LinearLayout>
```

我们必须为每一个 CheckBox 组件都设置一个 id 名称，因为程序必须检查每一个 CheckBox 组件是否被用户选取。另外在所有 CheckBox 选项的最后也要加上一个按钮，当用户完成选项的勾选之后按下该按钮，程序就会利用 CheckBox 组件的 isChecked()方法来检查每一个 CheckBox 的选取状态。运行上面的接口布局文件后我们可以看到如图 14-1 的界面。

如果程序包含的选项太多以致超出屏幕的范围，可以在接口布局文件的<LinearLayout>标签前面加上<ScrollView>标签，也就是用<ScrollView>标签将<LinearLayout>标签包起来，如以下范例：

```
<ScrollView    xmlns:android="http://schemas.android.com/apk/res/android"
    android:layout_width="match_parent"
    android:layout_height="match_parent"
    >
<LinearLayout
    …
</LinearLayout>
```

完成接口布局文件的设计之后，接下来的工作就是在程序中加上判断 CheckBox 组件勾选状态的程序代码。当用户按下"确定"按钮之后，程序必须逐一检查所有的 CheckBox 组件（利用每个 CheckBox 对象的 isChecked()方法），并记录用户选取的项目，最后我们的范例程序会把被勾选的项目显示在按扭下方的 TextView 组件中。以下是完整的字符串资源文件、接口布局文件和程序代码，我们把程序中用到的所有字符串全部定义在 res/values/strings.xml 文件中，另外在接口布局文件中利用设置文字大小、组件宽度和水平置中对齐等属性将选项的字体放大并显示在屏幕的中央以方便用户浏览和操作，图 14-2 是程序运行的界面。

图 14-1　CheckBox 复选清单

图 14-2　"兴趣选项"程序的运行界面

- 字符串资源文件 res/values/strings.xml：

```xml
<?xml version="1.0" encoding="utf-8"?>
<resources>
    <string name="hello">你好，这个程序的主要类名称叫做 Main</string>
    <string name="app_name">选择兴趣      </string>
    <string name="music">音乐         </string>
    <string name="sing">唱歌          </string>
    <string name="dance">跳舞         </string>
    <string name="travel">旅行        </string>
    <string name="reading">阅读       </string>
    <string name="writing">写作       </string>
    <string name="climbing">爬山      </string>
    <string name="swim">游泳          </string>
    <string name="exercise">运动      </string>
    <string name="fitness">健身       </string>
    <string name="photo">摄影         </string>
    <string name="eating">美食        </string>
    <string name="painting">绘画      </string>
    <string name="promptSelOK">确定       </string>
    <string name="hobList">您的兴趣：</string>
</resources>
```

- 界面布局文件 res/layout/main.xml：

```xml
<?xml version="1.0" encoding="utf-8"?>
<ScrollView   xmlns:android="http://schemas.android.com/apk/res/android"
    android:layout_width="match_parent"
    android:layout_height="match_parent"
    >
<LinearLayout
    android:orientation="vertical"
    android:layout_width="match_parent"
    android:layout_height="match_parent"
    android:gravity="center_horizontal"
    >
<CheckBox android:id="@+id/chbMusic"
    android:layout_width="200dp"
    android:layout_height="wrap_content"
    android:textSize="30sp"
    android:text="@string/music"
    />
<CheckBox android:id="@+id/chbSing"
    android:layout_width="200dp"
    android:layout_height="wrap_content"
    android:textSize="30sp"
    android:text="@string/sing"
    />
```

```xml
<CheckBox android:id="@+id/chbDance"
    android:layout_width="200dp"
    android:layout_height="wrap_content"
    android:textSize="30sp"
    android:text="@string/dance"
    />
<CheckBox android:id="@+id/chbTravel"
    android:layout_width="200dp"
    android:layout_height="wrap_content"
    android:textSize="30sp"
    android:text="@string/travel"
    />
<CheckBox android:id="@+id/chbReading"
    android:layout_width="200dp"
    android:layout_height="wrap_content"
    android:textSize="30sp"
    android:text="@string/reading"
    />
<CheckBox android:id="@+id/chbWriting"
    android:layout_width="200dp"
    android:layout_height="wrap_content"
    android:textSize="30sp"
    android:text="@string/writing"
    />
<CheckBox android:id="@+id/chbClimbing"
    android:layout_width="200dp"
    android:layout_height="wrap_content"
    android:textSize="30sp"
    android:text="@string/climbing"
    />
<CheckBox android:id="@+id/chbSwim"
    android:layout_width="200dp"
    android:layout_height="wrap_content"
    android:textSize="30sp"
    android:text="@string/swim"
    />
<CheckBox android:id="@+id/chbExercise"
    android:layout_width="200dp"
    android:layout_height="wrap_content"
    android:textSize="30sp"
    android:text="@string/exercise"
    />
<CheckBox android:id="@+id/chbFitness"
    android:layout_width="200dp"
    android:layout_height="wrap_content"
    android:textSize="30sp"
    android:text="@string/fitness"
    />
```

```xml
<CheckBox android:id="@+id/chbPhoto"
    android:layout_width="200dp"
    android:layout_height="wrap_content"
    android:textSize="30sp"
    android:text="@string/photo"
    />
<CheckBox android:id="@+id/chbEating"
    android:layout_width="200dp"
    android:layout_height="wrap_content"
    android:textSize="30sp"
    android:text="@string/eating"
    />
<CheckBox android:id="@+id/chbPainting"
    android:layout_width="200dp"
    android:layout_height="wrap_content"
    android:textSize="30sp"
    android:text="@string/painting"
    />
<Button android:id="@+id/btnSelOK"
    android:layout_width="200dp"
    android:layout_height="wrap_content"
    android:text="@string/promptSelOK"
    />
<TextView android:id="@+id/txtHobList"
    android:layout_width="200dp"
    android:layout_height="wrap_content"
    android:text="@string/hobList"
    />
</LinearLayout>
</ScrollView>
```

■ 程序代码：

```java
package …

import …

public class Main extends Activity {

    private CheckBox chbMusic, chbSing, chbDance,
            chbTravel, chbReading, chbWriting,
            chbClimbing, chbSwim, chbExercise,
            chbFitness, chbPhoto, chbEating,
            chbPainting;
    private Button btnSelOK;
    private TextView txtHobList;

    /** Called when the activity is first created. */
```

```java
@Override
public void onCreate(Bundle savedInstanceState) {
    super.onCreate(savedInstanceState);
    setContentView(R.layout.main);

    setupViewComponent();
}

private void setupViewComponent() {
    // 从资源类 R 中取得接口组件
    chbMusic = (CheckBox)findViewById(R.id.chbMusic);
    chbSing = (CheckBox)findViewById(R.id.chbSing);
    chbDance = (CheckBox)findViewById(R.id.chbDance);
    chbTravel = (CheckBox)findViewById(R.id.chbTravel);
    chbReading = (CheckBox)findViewById(R.id.chbReading);
    chbWriting = (CheckBox)findViewById(R.id.chbWriting);
    chbClimbing = (CheckBox)findViewById(R.id.chbClimbing);
    chbSwim = (CheckBox)findViewById(R.id.chbSwim);
    chbExercise = (CheckBox)findViewById(R.id.chbExercise);
    chbFitness = (CheckBox)findViewById(R.id.chbFitness);
    chbPhoto = (CheckBox)findViewById(R.id.chbPhoto);
    chbEating = (CheckBox)findViewById(R.id.chbEating);
    chbPainting = (CheckBox)findViewById(R.id.chbPainting);
    btnSelOK = (Button)findViewById(R.id.btnSelOK);
    txtHobList = (TextView)findViewById(R.id.txtHobList);

    // 设置 button 组件的事件 listener
    btnSelOK.setOnClickListener(btnSelOKOnClick);
}

private Button.OnClickListener btnSelOKOnClick = new Button.OnClickListener() {
    public void onClick(View v) {
        String s = getString(R.string.hobList);

        if (chbMusic.isChecked())
            s += chbMusic.getText().toString();

        if (chbSing.isChecked())
            s += chbSing.getText().toString();

        if (chbDance.isChecked())
            s += chbDance.getText().toString();

        if (chbTravel.isChecked())
            s += chbTravel.getText().toString();

        if (chbReading.isChecked())
            s += chbReading.getText().toString();
```

```java
            if (chbWriting.isChecked())
                s += chbWriting.getText().toString();

            if (chbClimbing.isChecked())
                s += chbClimbing.getText().toString();

            if (chbSwim.isChecked())
                s += chbSwim.getText().toString();

            if (chbExercise.isChecked())
                s += chbExercise.getText().toString();

            if (chbFitness.isChecked())
                s += chbFitness.getText().toString();

            if (chbPhoto.isChecked())
                s += chbPhoto.getText().toString();

            if (chbEating.isChecked())
                s += chbEating.getText().toString();

            if (chbPainting.isChecked())
                s += chbPainting.getText().toString();

            txtHobList.setText(s);
        }
    };
}
```

UNIT 15
LinearLayout 界面编排模式

Android 版本	1.X	2.X	3.X	4.X
适用性	★	★	★	★

在前面已经学过的范例程序中,读者是否注意到在接口布局文件 res/layout/main.xml 中有一个一再出现的<LinearLayout>标签,可是我们一直没有对它加以解释。之前为了把学习的注意力集中在接口组件的使用上,因此刻意将它忽略以免解说太过复杂。现在总算时机成熟了,因此就让我们来一窥它的庐山真面目吧!

LinearLayout 标签是一种接口组件的编排方式,顾名思义它就是依照线性次序由上往下逐一排列接口组件。我们之前所有的范例程序都是采用这种编排方式,如以下范例的运行界面为图 15-1。

图 15-1　LinearLayout 接口编排方式

```
<?xml version="1.0" encoding="utf-8"?>
<LinearLayout xmlns:android="http://schemas.android.com/apk/res/android"
    android:orientation="vertical"
    android:layout_width="match_parent"
    android:layout_height="match_parent">
<TextView
    android:layout_width="match_parent"
    android:layout_height="wrap_content"
    android:text="姓名: "/>
<EditText
    android:layout_width="match_parent"
    android:layout_height="wrap_content"
    android:text="输入姓名"/>
<Button
    android:layout_width="wrap_content"
    android:layout_height="wrap_content"
```

```
        android:text="确定"/>
</LinearLayout>
```

以上程序代码的运行结果和我们之前的所有范例程序一样,全部的接口组件都是由上往下一个接着一个排列。我们可以把 LinearLayout 标签想成是建立一个外框,里头要放入接口组件,这个外框的大小是由 android:layout_width 和 android:layout_height 这两个属性决定。如果把属性的值设置为"match_parent",代表要填满它所在的外框也就是屏幕。如果把它的值设置为"wrap_content",则代表只要满足它内部所包含的接口组件即可。还有一个 android:orientation 属性是决定它里面的接口组件的排列方向是水平还是垂直排列,当然我们也可以把它改成水平排列如下:

```
<?xml version="1.0" encoding="utf-8"?>
<LinearLayout xmlns:android="http://schemas.android.com/apk/res/android"
    android:orientation="horizontal"
    android:layout_width="match_parent"
    android:layout_height="match_parent">
<TextView
    android:layout_width="wrap_content"
    android:layout_height="wrap_content"
    android:text="姓名: "/>
<EditText
    android:layout_width="wrap_content"
    android:layout_height="wrap_content"
    android:text="输入姓名"/>
<Button
    android:layout_width="wrap_content"
    android:layout_height="wrap_content"
    android:text="确定"/>
</LinearLayout>
```

以上粗体字的部分是经过修改的程序代码,除了把排列方向改成"horizontal"以外,接口组件的 android:layout_width 属性都被换成"wrap_content",因为如果还是使用"match_parent",那么第一个组件就会占满整个屏幕宽度,后面的组件就看不到了。以上程序代码的运行界面如图 15-2 所示。

图 15-2 将 LinearLayout 的 orientation 属性设置成 horizontal 的结果

既然 LinearLayout 标签可以想成是建立一个外框,那么是不是可以在一个外框中放入另一个框框呢?答案是肯定的,例如以下的范例,我们在 LinearLayout 标签中加入另一个 LinearLayout 标签。

```
<?xml version="1.0" encoding="utf-8"?>
<LinearLayout xmlns:android="http://schemas.android.com/apk/res/android"
    android:orientation="vertical"
    android:layout_width="match_parent"
    android:layout_height="match_parent">
```

```
<LinearLayout
    android:orientation="horizontal"
    android:layout_width="match_parent"
    android:layout_height="wrap_content"
    android:background="#C0C0C0">
<TextView
    android:layout_width="wrap_content"
    android:layout_height="wrap_content"
    android:textColor="#000000"
    android:text="姓名："/>
<EditText
    android:layout_width="match_parent"
    android:layout_height="wrap_content"
    android:textColor="#000000"
    android:text="输入姓名"/>
</LinearLayout>
<Button
    android:layout_width="wrap_content"
    android:layout_height="wrap_content"
    android:text="确定"/>
</LinearLayout>
```

第一个 LinearLayout 是垂直排列，里面包含的 LinearLayout 则是水平排列，而且其中包含一个 TextView 和一个 EditText 接口组件。请读者注意它们的宽和高的设置值，另外我们也设置里面的 LinearLayout 的底色为灰色。最下面的那一个按钮则是属于第一个 LinearLayout。另外还要注意只需要在第一个 LinearLayout 标签中指定 xmlns:android 属性即可。以上接口布局文件运行后的界面如图 15-3 所示，这个范例告诉我们可以视需要将部分的接口组件用另一个 LinearLayout 标签集合起来，然后设置不一样的属性像是排列方式或是底色。

我们将以上学到的 LinearLayout 使用技巧应用到之前的"婚姻建议"范例程序，把它的操作界面改成如图 15-4 所示。要调整程序操作接口只需要更动 res/layout 文件夹下的接口布局文件即可，完全不需要改变任何程序代码。这里我们用到一个控制接口组件对齐方式的属性 android:gravity 让按钮对齐屏幕的中央，以下是完整的接口布局文件。

图 15-3　使用二层 LinearLayout 的结果　　图 15-4　使用多层 LinearLayout 的"婚姻建议"程序操作接口

```
<?xml version="1.0" encoding="utf-8"?>
<LinearLayout xmlns:android="http://schemas.android.com/apk/res/android"
```

```xml
        android:orientation="vertical"
        android:layout_width="match_parent"
        android:layout_height="match_parent"
        android:gravity="center_horizontal">
    <LinearLayout
        android:orientation="horizontal"
        android:layout_width="200dp"
        android:layout_height="wrap_content">
    <TextView
        android:layout_width="wrap_content"
        android:layout_height="wrap_content"
        android:textSize="20sp"
        android:text="@string/promptSex"/>
    <Spinner android:id="@+id/spnSex"
        android:layout_width="match_parent"
        android:layout_height="wrap_content"
        android:textSize="20sp"
        android:drawSelectorOnTop="true"
        android:prompt="@string/spnSexPrompt"
        android:spinnerMode="dialog"/>
    </LinearLayout>
    <LinearLayout
        android:orientation="horizontal"
        android:layout_width="200dp"
        android:layout_height="wrap_content">
    <TextView
        android:layout_width="wrap_content"
        android:layout_height="wrap_content"
        android:textSize="20sp"
        android:text="@string/promptAge"/>
    <EditText android:id="@+id/edtAge"
        android:layout_width="match_parent"
        android:layout_height="wrap_content"
        android:textSize="20sp"
        android:inputType="number"
        android:text=""/>
    </LinearLayout>
    <Button android:id="@+id/btnDoSug"
        android:layout_width="200dp"
        android:layout_height="wrap_content"
        android:textSize="20sp"
        android:text="@string/promptBtnDoSug"/>
    <TextView android:id="@+id/txtResult"
        android:layout_width="200dp"
        android:layout_height="wrap_content"
        android:textSize="20sp"
        android:text="@string/sugResult"/>
</LinearLayout>
```

Android 版本	1.X	2.X	3.X	4.X
适用性	★	★	★	★

UNIT 16
TableLayout 接口编排模式

这个单元让我们来学习另一种接口组件的编排方式，它叫做 TableLayout。顾名思义，TableLayout 就是把接口组件依照表格的方式排列，也就是由上往下一列接着一列，而且每一个字段都上下对齐。我们以下面的界面布局文件为例，最外层是一个 TableLayout 标签，其中的每一列再用 TableRow 标签将属于这一列的接口组件包裹起来。

```xml
<?xml version="1.0" encoding="utf-8"?>
<TableLayout xmlns:android="http://schemas.android.com/apk/res/android"
    android:layout_width="400dp"
    android:layout_height="match_parent"
    android:layout_gravity="center_horizontal"
    >
<TableRow>
    <TextView android:text="姓名："/>
    <TextView android:text="性别："/>
    <TextView android:text="生日："/>
</TableRow>
<TableRow>
    <EditText android:text="输入姓名"/>
    <EditText android:text="输入性别"/>
    <EditText android:text="输入生日"/>
</TableRow>
<Button android:text="确定"/>
</TableLayout>
```

以上接口布局文件运行后的程序界面如图 16-1 所示。

图 16-1　TableLayout 接口编排方式

在使用 TableLayout 和 TableRow 标签时要注意以下几点：

1. 包裹在 TableRow 标签中的接口组件，它的 android:layout_width 和 android:layout_height 属性都没有作用。也就是说不管把它们的值设置为 match_parent 或是 wrap_content，都一律使用 wrap_content 模式，因此可以直接把这两个属性省略，如以上范例。
2. 由于 TableRow 标签可能重复出现多次，为了方便了解整个 TableLayout 的结构，可以将每一个 TableRow 标签中的接口组件内缩对齐，如此每一列包含哪些接口组件就可以一目了然。
3. 像是 android:background 设置底色的属性、android:gravity 设置对齐方式的属性和 android:layout_margin 设置组件的距离等类似功能的属性都可以用于 TableLayout 和 TableRow 标签中。例如可以在上面范例的第一个 TableRow 标签中加入 android:gravity="center_horizontal"就可以将第一列的接口组件做水平置中对齐，如图 16-2 所示。

图 16-2　使用 android:gravity 属性让同一列的接口组件水平置中

4. 如果 TableLayout 标签中的接口组件没有被包含在 TableRow 标签中，则该组件会自成一列，如上面范例的 Button 组件。
5. 如果要让 TableRow 标签中的组件依比例使用整个 Table 的宽度，例如让上面范例的三个字段自动等分 Table 的宽度，可以借助 android:layout_weight 属性，它会将同一列组件所有的 weight 值加总后，再依照每一个组件的 weight 值的比例计算所占的宽度，如下面范例以及它的运行界面如图 16-3 所示。

```
<?xml version="1.0" encoding="utf-8"?>
<TableLayout xmlns:android="http://schemas.android.com/apk/res/android"
    android:layout_width="400dp"
```

```
        android:layout_height="match_parent"
        android:layout_gravity="center_horizontal"
        >
    <TableRow>
        <TextView android:text="姓名："
            android:layout_weight="1"/>
        <TextView android:text="性别："
            android:layout_weight="1"/>
        <TextView android:text="生日："
            android:layout_weight="1"/>
    </TableRow>
    <TableRow>
        <EditText android:text="输入姓名"
            android:layout_weight="1"/>
        <EditText android:text="输入性别"
            android:layout_weight="1"/>
        <EditText android:text="输入生日"
            android:layout_weight="1"/>
    </TableRow>
    <Button android:text="确定"/>
</TableLayout>
```

图 16-3　使用 android:layout_weight 属性依比例决定组件的宽度

6. TableLayout 标签中的每一个 TableRow 内的组件都和上一个 TableRow 中的组件依序对齐，无法错开。如果读者希望做到如图 16-4 所示的程序界面，可以在 TableLayout 标签中再增加一个 TableLayout 标签，这样就可以让不同列的字段错开，如以下范例：

图 16-4　利用不同的 TableLayout 标签让不同列的字段可以错开

```
<?xml version="1.0" encoding="utf-8"?>
<TableLayout xmlns:android="http://schemas.android.com/apk/res/android"
    android:layout_width="400dp"
    android:layout_height="match_parent"
    android:layout_gravity="center_horizontal"
    >
<TableLayout
    android:layout_width="match_parent"
    android:layout_height="wrap_content"
    >
<TableRow>
    <TextView android:text="姓名:"/>
    <TextView android:text="性别:"/>
    <TextView android:text="生日:"/>
</TableRow>
<TableRow>
    <EditText android:text="输入姓名"/>
    <EditText android:text="输入性别"/>
    <EditText android:text="输入生日"/>
</TableRow>
</TableLayout>
<TableRow>
    <TextView android:text="电话:"/>
    <TextView android:text="地址:"/>
</TableRow>
<TableRow>
    <EditText android:text="请输入电话"/>
    <EditText android:text="请输入地址"/>
</TableRow>
<Button android:text="确定"/>
</TableLayout>
```

最后我们利用 TableLayout 接口编排方式将前面的"婚姻建议"程序的操作接口稍微改变一下，图 16-5 是修改后的程序界面，以下是修改后的接口布局文件，粗体字是有改动的部分。

```
<?xml version="1.0" encoding="utf-8"?>
<TableLayout xmlns:android="http://schemas.android.com/apk/res/android"
    android:layout_width="200dp"
    android:layout_height="match_parent"
    android:layout_gravity="center_horizontal"
    >
<TableRow>
    <TextView
        android:layout_width="wrap_content"
        android:layout_height="wrap_content"
        android:textSize="20sp"
        android:text="@string/promptSex"
```

图 16-5 使用 TableLayout 界面编排方式的"婚姻建议"程序

```
            />
        <RadioGroup android:id="@+id/radGSex"
            android:layout_width="200dp"
            android:layout_height="wrap_content"
            android:orientation="vertical"
            android:checkedButton="@+id/radMale">
            <RadioButton android:id="@+id/radMale"
                android:textSize="20sp"
                android:text="@string/male"
                />
            <RadioButton android:id="@+id/radFemale"
                android:textSize="20sp"
                android:text="@string/female"
                />
        </RadioGroup>
    </TableRow>
    <TableRow>
        <TextView
            android:layout_width="wrap_content"
            android:layout_height="wrap_content"
            android:textSize="20sp"
            android:text="@string/promptAge"
            />
        <RadioGroup android:id="@+id/radGAge"
            android:layout_width="200dp"
            android:layout_height="wrap_content"
            android:orientation="vertical"
            android:checkedButton="@+id/radAgeRng1">
            <RadioButton android:id="@+id/radBtnAgeRng1"
                android:textSize="20sp"
                android:text="@string/maleAgeRng1"
```

```xml
            />
            <RadioButton android:id="@+id/radBtnAgeRng2"
                android:textSize="20sp"
                android:text="@string/maleAgeRng2"
            />
            <RadioButton android:id="@+id/radBtnAgeRng3"
                android:textSize="20sp"
                android:text="@string/maleAgeRng3"
            />
        </RadioGroup>
    </TableRow>
    <Button android:id="@+id/btnDoSug"
        android:layout_width="200dp"
        android:layout_height="wrap_content"
        android:textSize="20sp"
        android:text="@string/promptBtnDoSug"
    />
    <TextView android:id="@+id/txtResult"
        android:layout_width="200dp"
        android:layout_height="wrap_content"
        android:textSize="20sp"
        android:text="@string/sugResult"
    />
</TableLayout>
```

Android 版本	1.X	2.X	3.X	4.X
适用性	★	★	★	★

UNIT 17
RelativeLayout 布局

RelativeLayout 是所谓的相对布局，也就是说接口组件之间是利用指定相对的位置关系来决定它们的排列方式。既然要指定接口组件之间的相对位置，为了能够识别每一个组件，我们必须为每一个组件指定 id 名称。让我们先来看一个简单的范例。

```
<?xml version="1.0" encoding="utf-8"?>
<RelativeLayout xmlns:android="http://schemas.android.com/apk/res/android"
    android:layout_width="400dp"
    android:layout_height="match_parent"
    android:layout_gravity="center_horizontal"
    >
<TextView android:id="@+id/txt1"
    android:layout_width="wrap_content"
    android:layout_height="wrap_content"
    android:text="txt1"/>
<TextView android:id="@+id/txt2"
    android:layout_width="wrap_content"
    android:layout_height="wrap_content"
    android:text="txt2"
    android:layout_toRightOf="@id/txt1"/>
<EditText android:id="@+id/edt1"
    android:layout_width="wrap_content"
    android:layout_height="wrap_content"
    android:text="edt1"
    android:layout_below="@id/txt1"/>
<EditText android:id="@+id/edt2"
    android:layout_width="wrap_content"
```

```
            android:layout_height="wrap_content"
            android:text="edt2"
            android:layout_toRightOf="@id/edt1"/>
    <Button android:id="@+id/btn1"
            android:layout_width="wrap_content"
            android:layout_height="wrap_content"
            android:text="btn1"
            android:layout_below="@id/edt1"/>
</RelativeLayout>
```

请读者注意粗体字的程序代码，首先我们指定使用 RelativeLayout 标签，然后建立一个 TextView 组件叫做 txt1，下一个 TextView 组件是 txt2，并指定它是在 txt1 的右边（使用 android:layout_toRightOf 属性），然后是第 3 个组件叫做 edt1 并指定它在 txt1 的下方（使用 android:layout_below 属性），第 4 个组件是 edt2，它是在 edt1 的右边，最后一个是 Button 组件叫做 btn1，它是在 edt1 的下方，这个接口布局文件的运行结果如图 17-1 所示。

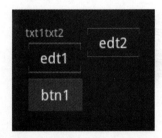

图 17-1　RelativeLayout 布局

读者可以看出这个结果不太能够令人满意，其中最大的问题是两个 EditText 组件出现一高一低没有对齐的情况，这是因为上面的范例只有指定组件之间的相对位置而没有指定如何对齐，也就是说在 RelativeLayout 布局下，必须同时指定组件的相对位置以及如何对齐。指定对齐方式是利用 align 相关的属性，如果我们在 edt2 组件的标签中加上如下的属性设置就会得到如图 17-2 的结果。

    android:layout_alignTop="@id/edt1"

图 17-2　RelativeLayout 布局加上 align 属性的效果

我们将设置相对位置和对齐方式的相关属性列出，如表 17-1 所示。

表 17-1 设置相对位置和对齐方式的属性

属性类	属性名称	属性值	说明
指定相对位置	layout_toLeftOf	"id/某一个接口组件 id 名称"	将接口组件置于指定接口组件的左边
	layout_toRightOf	"id/某一个接口组件 id 名称"	将接口组件置于指定接口组件的右边
	layout_above	"id/某一个接口组件 id 名称"	将接口组件置于指定接口组件的上方
	layout_below	"id/某一个接口组件 id 名称"	将接口组件置于指定接口组件的下方
指定对齐方式	layout_alignLeft	"id/某一个接口组件 id 名称"	将接口组件和指定接口组件的左边对齐
	layout_alignRight	"id/某一个接口组件 id 名称"	将接口组件和指定接口组件的右边对齐
	layout_alignTop	"id/某一个接口组件 id 名称"	将接口组件和指定接口组件的上缘对齐
	layout_alignBottom	"id/某一个接口组件 id 名称"	将接口组件和指定接口组件的下缘对齐
	layout_alignBaseLine	"id/某一个接口组件 id 名称"	将接口组件和指定接口组件的中心线对齐
	layout_alignParentLeft	"true" 或 "false"	将接口组件对齐它所在的外框左边
	layout_alignParentRight	"true" 或 "false"	将接口组件对齐它所在的外框右边
	layout_alignParentTop	"true" 或 "false"	将接口组件对齐它所在的外框上缘
	layout_alignParentBottom	"true" 或 "false"	将接口组件对齐它所在的外框下缘

在指定接口组件的相对位置和对齐方式时必须遵守下列规定：

1. 指定的接口组件 id 必须是在前面已经定义过，也就是说不可以使用定义在后面的接口组件。举例来说，下面的范例我们指定 txt1 要放在 edt1 的右边，可是 edt1 是在后面才定义，这是不合法的。

```
<?xml version="1.0" encoding="utf-8"?>
<RelativeLayout xmlns:android="http://schemas.android.com/apk/res/android"
    android:layout_width="match_parent"
    android:layout_height="match_parent"
    >
<TextView android:id="@+id/txt1"
    android:layout_width="wrap_content"
    android:layout_height="wrap_content"
    android:text="txt1"
    android:layout_toRightOf="@id/edt1"/>
.
.
.
<EditText android:id="@+id/edt1"
    android:layout_width="wrap_content"
    android:layout_height="wrap_content"
    android:text="edt1"
    android:layout_below="@id/txt1"/>
.
.
.
</RelativeLayout>
```

2. 最先定义的接口组件会放在屏幕的左上方,如果后面接口组件的位置是在第一个接口组件的左边或是上方,那么就会看不到这个接口组件。例如以下范例将看不到 edt1 组件。

```
<?xml version="1.0" encoding="utf-8"?>
<RelativeLayout xmlns:android="http://schemas.android.com/apk/res/android"
    android:layout_width="match_parent"
    android:layout_height="match_parent"
    >
<TextView android:id="@+id/txt1"
    android:layout_width="wrap_content"
    android:layout_height="wrap_content"
    android:text="txt1"/>
    .
    .
    .
<EditText android:id="@+id/edt1"
    android:layout_width="wrap_content"
    android:layout_height="wrap_content"
    android:text="edt1"
    android:layout_above="@id/txt1"/>
    .
    .
    .
</RelativeLayout>
```

这个问题有两种解决方法,首先可以利用 android:layout_center 相关属性把第一个定义的组件往屏幕的右下方移动。第二个解决方法是调整接口组件定义的次序,例如把范例中的 edt1 组件放到前面定义,然后指定 txt1 组件置于 edt1 的下方。

最后我们利用以上学到的 RelativeLayout 编排技巧,将图 17-1 的接口组件做最适当的编排,得到如图 17-3 所示的结果。读者可以看到我们将 txt1 和 edt1 的左边对齐,txt2 和 edt2 的左边也对齐。要达到这样的结果必须调整接口组件的定义次序才能符合上述二个规定,修改后的接口布局文件如下:

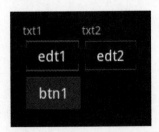

图 17-3 将图 17-1 的接口组件作适当编排后的结果

```
<?xml version="1.0" encoding="utf-8"?>
<RelativeLayout xmlns:android="http://schemas.android.com/apk/res/android"
    android:layout_width="400dp"
    android:layout_height="match_parent"
```

```
            android:layout_gravity="center_horizontal"
            >
<TextView android:id="@+id/txt1"
            android:layout_width="wrap_content"
            android:layout_height="wrap_content"
            android:text="txt1"/>
<EditText android:id="@+id/edt1"
            android:layout_width="wrap_content"
            android:layout_height="wrap_content"
            android:text="edt1"
            android:layout_below="@id/txt1"/>
<EditText android:id="@+id/edt2"
            android:layout_width="wrap_content"
            android:layout_height="wrap_content"
            android:text="edt2"
            android:layout_toRightOf="@id/edt1"
            android:layout_alignTop="@id/edt1"/>
<TextView android:id="@+id/txt2"
            android:layout_width="wrap_content"
            android:layout_height="wrap_content"
            android:text="txt2"
            android:layout_above="@id/edt2"
            android:layout_alignLeft="@id/edt2"/>
<Button android:id="@+id/btn1"
            android:layout_width="wrap_content"
            android:layout_height="wrap_content"
            android:text="btn1"
            android:layout_below="@id/edt1"/>
</RelativeLayout>
```

接下来就让我们利用 RelativeLayout 接口布局来做个简单的"计算机猜拳游戏"程序，要完成这个计算机游戏必须考虑下列几件事：

1．玩家可以任意选择出剪刀、石头或是布，我们必须设计一个简单清楚的接口让玩家操作。

2．计算机必须能够自行决定出拳，而且没有规则性，否则玩家便有可能破解。

3．游戏进行过程中的界面必须美观、简单且易于了解。

接下来就让我们一步一步依序说明如何完成这个计算机游戏程序。

首先关于玩家出拳的操作方式，在前面的单元中我们已经学过 Spinner 下拉式菜单和 RadioGroup 单选按钮。这两种接口组件都可以提供多选一的功能，但是如果仔细考虑，使用这两种接口组件在用户单击之后，还要再按下一个按钮才会开始运行程序。可是如果我们直接将剪刀、石头和布各做成一个独立的按钮，那么用户只要直接按下其中一个按钮就完成出拳的动作，在操作上更简单明确。

至于要让计算机不规则地自动出拳就需要使用"随机数"的功能，所谓"随机数"是一群出现次序不规则的数，也就是随机产生的数字。因为随机数不规则的特性，我们无法根据观察过去的结果来预测未来（像是乐透彩）。随机数经常用来模拟一些随机的结果，例如游戏常用的丢骰子。在 Java 程序中可以调用 Math.random()方法随机产生一个介于 0 和 1 之间的小数（可能是 0，但是永远小于 1），我们可以利用乘法把这个小数放大再转换成整数型态，例如：

```
int iRan = (int)(Math.random()*6 + 1);
```

以上的公式会使 iRan 的值永远是 1～6 之间的一个整数，这不就是我们掷一颗骰子的结果吗！在猜拳游戏中，我们可以用 1、2、3 分别代表剪刀、石头和布，因此我们需要的是 1～3 之间的整数，那么只要将上面公式中的 6 改成 3 即可。

最后我们必须考虑整个游戏界面的设计，包括如何排列使用者出拳的三个按钮，如何显示计算机出拳的结果，以及如何显示输赢的判断等。为了让游戏界面富有变化，我们可以使用 RelativeLayout 布局，请读者参考图 17-4 的游戏操作界面，其中的字体大小、颜色、接口组件排列的顺序和位置都经过仔细的调整以符合清楚、美观的原则。这个游戏界面的接口布局文件如下，其中使用到许多前面已经学过的组件属性，读者可以借此机会复习一下。

图 17-4　"计算机猜拳游戏"的操作界面

```xml
<?xml version="1.0" encoding="utf-8"?>
<RelativeLayout xmlns:android="http://schemas.android.com/apk/res/android"
    android:layout_width="400dp"
    android:layout_height="match_parent"
    android:layout_gravity="center_horizontal"
    >
<TextView android:id="@+id/txtTitle"
    android:layout_width="wrap_content"
    android:layout_height="wrap_content"
    android:text="@string/promptTitle"
    android:textSize="40sp"
    android:textColor="#FF00FF"
```

```xml
        android:textStyle="bold"
        android:layout_centerHorizontal="true"
        android:paddingLeft="20dp"
        android:paddingRight="20dp"
        android:layout_marginTop="20dp"
        android:layout_marginBottom="20dp"
        />
    <TextView android:id="@+id/txtCom"
        android:layout_width="wrap_content"
        android:layout_height="wrap_content"
        android:text="@string/promptComPlay"
        android:layout_below="@id/txtTitle"
        android:layout_alignLeft="@id/txtTitle"
        android:textSize="20sp"
        android:layout_marginBottom="20dp"
        />
    <TextView android:id="@+id/txtMyPlay"
        android:layout_width="wrap_content"
        android:layout_height="wrap_content"
        android:text="@string/promptMyPlay"
        android:layout_below="@id/txtTitle"
        android:layout_alignRight="@id/txtTitle"
        android:textSize="20sp"
        android:layout_marginBottom="20dp"
        />
    <Button android:id="@+id/btnScissors"
        android:layout_width="wrap_content"
        android:layout_height="wrap_content"
        android:text="@string/playScissors"
        android:layout_below="@id/txtMyPlay"
        android:layout_alignLeft="@id/txtMyPlay"
        android:textSize="20sp"
        android:paddingLeft="15dp"
        android:paddingRight="15dp"
        />
    <TextView android:id="@+id/txtComPlay"
        android:layout_width="wrap_content"
        android:layout_height="wrap_content"
        android:text=""
        android:layout_below="@id/btnScissors"
        android:layout_alignLeft="@id/txtCom"
        android:textSize="30sp"
        android:textColor="#FF00FF"
        />
    <Button android:id="@+id/btnStone"
        android:layout_width="wrap_content"
        android:layout_height="wrap_content"
        android:text="@string/playStone"
```

```xml
        android:layout_below="@id/btnScissors"
        android:layout_alignLeft="@id/btnScissors"
        android:textSize="20sp"
        android:paddingLeft="15dp"
        android:paddingRight="15dp"
        />
    <Button android:id="@+id/btnNet"
        android:layout_width="wrap_content"
        android:layout_height="wrap_content"
        android:text="@string/playNet"
        android:layout_below="@id/btnStone"
        android:layout_alignLeft="@id/btnStone"
        android:textSize="20sp"
        android:paddingLeft="25dp"
        android:paddingRight="25dp"
        />
    <TextView android:id="@+id/txtResult"
        android:layout_width="wrap_content"
        android:layout_height="wrap_content"
        android:text="@string/result"
        android:layout_below="@id/btnNet"
        android:layout_alignLeft="@id/txtCom"
        android:textSize="20sp"
        android:textColor="#0FFFFF"
        android:layout_marginTop="20dp"
        />
</RelativeLayout>
```

接下来我们列出完整的字符串资源文件和程序代码，在程序代码中有关接口的设置全部都在 setupViewComponent()方法中完成。另一个重点是当使用者按下剪刀、石头或是布其中一个按钮后，程序会先以随机数决定计算机出拳，接下来就是显示计算机所出的拳并决定胜负，最后显示输赢的判定结果。程序代码中的重点部分特别用粗体字代表请读者留意。根据这个范例程序，读者应该能够更熟悉 Android 程序接口的设计技巧。

- res/values/strings.xml 文件：

```xml
<?xml version="1.0" encoding="utf-8"?>
<resources>
    <string name="app_name">计算机猜拳游戏</string>
    <string name="promptComPlay">计算机出拳：</string>
    <string name="promptMyPlay">玩家出拳：</string>
    <string name="playScissors">剪刀</string>
    <string name="playStone">石头</string>
    <string name="playNet">布</string>
    <string name="playerWin">恭喜，你赢了！</string>
    <string name="playerLose">很可惜，你输了！</string>
    <string name="playerDraw">双方平手！</string>
```

```xml
        <string name="promptTitle">和计算机猜拳</string>
        <string name="result">判定输赢：</string>
</resources>
```

- 程序文件：

```java
package tw.android;

import android.app.Activity;
import android.os.Bundle;
import android.view.View;
import android.widget.*;

public class Main extends Activity {

    private TextView txtComPlay, txtResult;
    private Button btnScissors;
    private Button btnStone;
    private Button btnNet;

    /** Called when the activity is first created. */
    @Override
    public void onCreate(Bundle savedInstanceState) {
        super.onCreate(savedInstanceState);
        setContentView(R.layout.main);

        setupViewComponent();
    }
    private void setupViewComponent() {
        txtComPlay = (TextView)findViewById(R.id.txtComPlay);
        txtResult = (TextView)findViewById(R.id.txtResult);
        btnScissors = (Button)findViewById(R.id.btnScissors);
        btnStone = (Button)findViewById(R.id.btnStone);
        btnNet = (Button)findViewById(R.id.btnNet);

        btnScissors.setOnClickListener(btnScissorsLin);
        btnStone.setOnClickListener(btnStoneLin);
        btnNet.setOnClickListener(btnNetLin);
    }

    private Button.OnClickListener btnScissorsLin = new Button.OnClickListener() {
        public void onClick(View v) {
            // 决定计算机出拳.
            int iComPlay = (int)(Math.random()*3 + 1);

            // 1 - 剪刀, 2 - 石头, 3 - 布.
            if (iComPlay == 1) {
                txtComPlay.setText(R.string.playScissors);
```

```java
            txtResult.setText(getString(R.string.result) +
                        getString(R.string.playerDraw));
        }
        else if (iComPlay == 2) {
            txtResult.setText(getString(R.string.result) +
            txtResult.setText(getString(R.string.result) +
                        getString(R.string.playerLose));
        }
        else {
            txtComPlay.setText(R.string.playNet);
            txtResult.setText(getString(R.string.result) +
                        getString(R.string.playerWin));
        }
    }
};

private Button.OnClickListener btnStoneLin = new Button.OnClickListener() {
    public void onClick(View v) {
        // 决定计算机出拳.
        int iComPlay = (int)(Math.random()*3 + 1);

        // 1 - 剪刀, 2 - 石头, 3 - 布.
        if (iComPlay == 1) {
            txtComPlay.setText(R.string.playScissors);
            txtResult.setText(getString(R.string.result) +
                        getString(R.string.playerWin));
        }
        else if (iComPlay == 2) {
            txtComPlay.setText(R.string.playStone);
            txtResult.setText(getString(R.string.result) +
                        getString(R.string.playerDraw));
        }
        else {
            txtComPlay.setText(R.string.playNet);
            txtResult.setText(getString(R.string.result) +
                        getString(R.string.playerLose));
        }
    }
};

private Button.OnClickListener btnNetLin = new Button.OnClickListener() {
    public void onClick(View v) {
        // 决定计算机出拳.
        int iComPlay = (int)(Math.random()*3 + 1);

        // 1 - 剪刀, 2 - 石头, 3 - 布.
        if (iComPlay == 1) {
            txtComPlay.setText(R.string.playScissors);
```

```
                txtResult.setText(getString(R.string.result) +
                        getString(R.string.playerLose));
            }
            else if (iComPlay == 2) {
                txtComPlay.setText(R.string.playStone);
                txtResult.setText(getString(R.string.result) +
                        getString(R.string.playerWin));
            }
            else {
                txtComPlay.setText(R.string.playNet);
                txtResult.setText(getString(R.string.result) +
                        getString(R.string.playerDraw));
            }
        }
    };
}
```

Android 版本	1.X	2.X	3.X	4.X
适用性	★	★	★①	★①

UNIT 18

FrameLayout 布局和 Tab 卷标页

FrameLayout 是比较特殊的接口编排方式，如果使用以下的接口布局文件，会产生图 18-1 的运行界面，原来在接口布局文件中总共建立了 txt、edt 和 btn 三个组件，可是程序的运行结果却显示三个重迭的接口组件，看到这个结果我们不禁要问：究竟 FrameLayout 有何用途？

图 18-1　FrameLayout 组件布局

```
<?xml version="1.0" encoding="utf-8"?>
<FrameLayout xmlns:android="http://schemas.android.com/apk/res/android"
    android:layout_width="match_parent"
    android:layout_height="match_parent"
    >
<TextView android:id="@+id/txt"
    android:layout_width="wrap_content"
    android:layout_height="wrap_content"
    android:text="txt"/>
```

---

① 可以使用，但是建议改用 Action Bar 上的 Tab 标签页（UNIT 52）。

```
<EditText android:id="@+id/edt"
    android:layout_width="wrap_content"
    android:layout_height="wrap_content"
    android:text="edt"/>
<Button android:id="@+id/btn"
    android:layout_width="wrap_content"
    android:layout_height="wrap_content"
    android:text="btn"/>
</FrameLayout>
```

通常 FrameLayout 接口布局是用在建立 Tab 标签页接口的时候，Tab 标签页接口在微软 Windows 操作系统中经常出现，如果读者开启 IE 网页浏览器中的菜单"工具>因特网选项"，就会看到 Tab 标签页的范例（如图 18-2 所示），它像是多层次的文件夹，我们可以单击上方的标签来检查不同的内容。此外 Tab 标签页的标签也不一定是在上方，在某些软件中 Tab 标签是出现在下面，像是在 Eclipse 中开启 res/layout/main.xml 文件后，在文件编辑窗口的下面会有两个 Tab 标签可以切换不同的检查模式，接着我们就来介绍如何建立 Android 程序的 Tab 标签页。

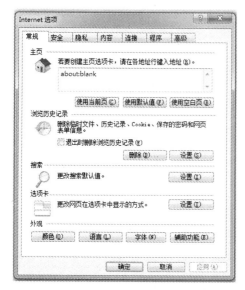

图 18-2 微软 Windows 操作系统的 Tab 标签页

18-1 建立 Tab 标签页的步骤

要完成 Tab 标签页接口需要一个 TabHost 组件，它负责管控整个 Tab 标签页的架构。除了 TabHost 组件之外还需要用到 2 个组件：TabWidget 和 FrameLayout。TabWidget 组件负责控制所有的 Tab 标签，FrameLayout 组件则负责显示每一个 Tab 标签页中的接口组件。

建立 Tab 标签页接口有 2 种方式，一种是把每一个 Tab page 内的接口组件都一齐写在同一个接口布局文件中，但是用不同的<LinearLayout>标签（或是<RelativeLayout>或<TableLayout>）把不同 Tab page 内的接口组件区隔开来。另一种方式是把每一个 Tab page 内的接口组件都写成一个独立的接口布局文件，再利用 Intent 的方式把它加载到对应的 Tab page 中，这种方式需要用到 Intent 对象，因此留待后面的单元再进一步说明，在这个单元中先来学习第一种方式。我们把使用 Tab 标签页接口的方法整理成下列步骤：

Step 1 在项目的 res/layout 文件夹下的接口布局文件中使用<TabHost>、<TabWidget>、<FrameLayout>和<LinearLayout>（或是<RelativeLayout>或<TableLayout>）标签依照下列范例的格式，建立每一个 Tab page 的接口组件。

```xml
<TabHost xmlns:android="http://schemas.android.com/apk/res/android"
    android:id="@+id/tabHost"
    android:layout_width="match_parent"
    android:layout_height="match_parent"
    >
<TabWidget android:id="@android:id/tabs"
    android:layout_width="wrap_content"
    android:layout_height="wrap_content"
    android:gravity="center_horizontal"
    />
    <FrameLayout android:id="@android:id/tabcontent"
        android:layout_width="match_parent"
        android:layout_height="match_parent"
        >
    <LinearLayout android:id="@+id/tab1"
        android:orientation="vertical"
        android:layout_width="match_parent"
        android:layout_height="wrap_content"
        android:paddingTop="70dp"
        >
        第一个 Tab page 的接口组件
        …
    </LinearLayout>
    <LinearLayout android:id="@+id/tab2"
        android:orientation="vertical"
        android:layout_width="match_parent"
        android:layout_height="wrap_content"
        android:paddingTop="70dp"
        >
        第二个 Tab page 的接口组件
        …
    </LinearLayout>
    </FrameLayout>
</TabHost>
```

建立以上的接口布局文件时要注意以下几点:

1. <TabWidget>标签的 id 一定要设置成@android:id/tabs,不可自行更改。如果想让 Tab 标签置中对齐,可以加上 android:gravity 属性,如上述范例。
2. <FrameLayout>标签的 id 一定要设置成@android:id/tabcontent,不可自行更改。
3. <TabHost>标签的 id 可以自行决定(注意,主程序类一样是 extends Activity,不需要改成 extends TabActivity)。
4. 每一个<LinearLayout>(或是<RelativeLayout>或<TableLayout>)标签都要设置 android:paddingTop,也就是把接口组件往下移,以避免和 Tab page 的标题重迭。

接下来的步骤请在程序文件中完成:

Step 2 取得接口布局文件中的 TabHost 组件并完成初始化。

```
TabHost tabHost = (TabHost)findViewById(R.id.tabHost);
tabHost.setup();
```

Step 3 建立第 1 个 Tab page 并设置它的标题、图示(可省略)以及它的接口组件。

```
TabSpec spec=tabHost.newTabSpec("tab1");
spec.setContent(R.id.tab1);
spec.setIndicator("显示在标签上的文字",
getResources().getDrawable(android.R.drawable.ic_lock_idle_alarm));
```

补充说明　Android 4 版本的改变

原来在 Tab page 的标题可以加上小图标,可是在 Android 4 中运行时小图标会隐藏,只显示标题文字。

Step 4 把第 1 个 Tab page 加入 TabHost 对象中。

```
tabHost.addTab(spec);
```

Step 5 重复步骤 3 和 4 加入其他 Tab page。

Step 6 指定程序启动时第一个显示的 Tab page。

```
tabHost.setCurrentTab(0);
```

0 代表第 1 个 Tabe page,1 代表第 2 个 Tabe page,依此类推。

Step 7 如果想要改变 Tab page 标题的字体大小,可以利用以下程序代码:

```
TabWidget tabWidget = (TabWidget)tabHost.findViewById(android.R.id.tabs);
View tabView = tabWidget.getChildTabViewAt(0);      // 0 代表第 1 个
                                                     // Tabe page
TextView tab = (TextView)tabView.findViewById(android.R.id.title);
tab.setTextSize(20);
```

如果要设置第 2 个 Tabe page 标题的字体大小，就把 0 改成 1，依此类推。

18-2 范例程序

接下来我们用一个实际范例来示范 Tab 标签页接口的实现过程，这个范例程序是把 UNIT 16 的"婚姻建议"程序和 UNIT 17 的"计算机猜拳游戏"程序结合起来，放到不同的 Tab page 中。首先请读者建立一个新的程序项目，然后根据前一小节第 1 个步骤中的范例编辑接口布局文件，再把"婚姻建议"程序项目中的接口布局文件里头的接口组件直接复制，贴到其中标示"第一个 Tab page 的接口组件"的位置，由于 UNIT 16 的"婚姻建议"程序是使用<TableLayout>，所以请把<LinearLayout>换成<TableLayout>。另外"计算机猜拳游戏"程序的接口组件则复制贴到其中的"第二个 Tab page 的接口组件"的位置，并且把<LinearLayout>换成<RelativeLayout>，然后从"婚姻建议"程序和"计算机猜拳游戏"程序的程序文件中复制所有接口组件的定义和设置的程序代码并加入建立 Tab 的相关程序代码如下粗体字的部分。由于原来这二个程序项目中都有一个叫做 txtResult 的接口组件，所以复制后会有组件名称重复的错误，请读者将这二个同名的组件在名称的最后加上 1 和 2 的编号以便区分，其他相关程序代码也依此原则修改即可。另外定义在字符串资源文件中的字符串也要复制到新的程序项目的字符串资源文件，完成后的程序运行界面如图 18-3 所示。

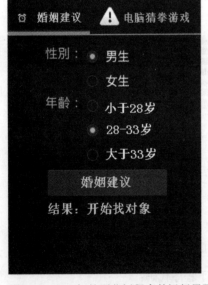

图 18-3　Tab 标签页范例程序的运行界面

package ...

import ...

```java
public class Main extends Activity {

    // 复制"婚姻建议"程序和"计算机猜拳游戏"程序的接口对象定义
    ...

        /** Called when the activity is first created. */
        @Override
        public void onCreate(Bundle savedInstanceState) {
            super.onCreate(savedInstanceState);
            setContentView(R.layout.main);

            setupViewComponent();
        }

        private void setupViewComponent() {
            // 从资源类 R 中取得接口组件并设置属性
            TabHost tabHost = (TabHost)findViewById(R.id.tabHost);
            tabHost.setup();

            TabSpec spec=tabHost.newTabSpec("tab1");
            spec.setContent(R.id.tab1);
            spec.setIndicator("婚姻建议",
                    getResources().getDrawable(android.R.drawable.ic_lock_idle_alarm));
            tabHost.addTab(spec);

            spec=tabHost.newTabSpec("tab2");
            spec.setIndicator("计算机猜拳游戏",
                    getResources().getDrawable(android.R.drawable.ic_dialog_alert));
            spec.setContent(R.id.tab2);
            tabHost.addTab(spec);

            tabHost.setCurrentTab(0);

                // 设置 Tab 标签的字体大小
                TabWidget tabWidget = (TabWidget)tabHost.findViewById(android.R.id.tabs);
                View tabView = tabWidget.getChildTabViewAt(0);
                TextView tab = (TextView)tabView.findViewById(android.R.id.title);
                tab.setTextSize(20);
                tabView = tabWidget.getChildTabViewAt(1);
                tab = (TextView)tabView.findViewById(android.R.id.title);
                tab.setTextSize(20);

                // 复制"婚姻建议"程序和"计算机猜拳游戏"程序的 setupViewComponent()方法内
                   的全部程序代码
                ...
        }
```

```
// 复制"婚姻建议"程序和"计算机猜拳游戏"程序的全部的事件处理程序
   …
}
```

当读者完成这个程序项目后会发现程序代码变得有点冗长，因为我们把两个项目的程序代码放在同一个程序文件中。从程序维护的观点来看，这种方式并不是很好的作法，这也是本单元介绍的建立 Tab 标签页方法的一个缺点，但是它的优点是接口组件只需要放在同一个接口布局文件即可，不需要建立多个文件，因此如果程序的操作接口不会太过复杂，而且程序代码也不会太长，则适合使用本单元介绍的方法。可是如果程序中会用到许多接口组件，而且程序代码又有一定的复杂度，则可以考虑使用后面 UNIT 39 中介绍的方法，利用 Intent 对象的方式来建立 Tab 标签页。这个单元介绍的 Tab 标签页技术可以同时适用在所有的 Android 版本，但如果是针对 Android 3.0 以后所写的程序，建议改用 UNIT 52 所介绍的方法。

PART 4 图像接口组件与动画效果

UNIT 19　ImageButton 和 ImageView 接口组件
UNIT 20　Gallery、GridView 和 ImageSwitcher 接口组件
UNIT 21　使用 Tween 动画效果
UNIT 22　Frame Animation 和 Multi-Thread 游戏程序
UNIT 23　Property Animation 初体验
UNIT 24　Property Animation 加上 Listener 成为动画超人

Android 版本	1.X	2.X	3.X	4.X
适用性	★	★	★	★

UNIT 19
ImageButton 和 ImageView 接口组件

ImageButton 接口组件和前面介绍过的 Button 组件功能完全相同,唯一的差别是 Button 组件上显示的是文字,ImageButton 组件上显示的是图像,因此 ImageButton 让我们可以用比较生动有趣的方式来代表按钮。ImageButton 组件上的图像必须置于项目的 res/drawable 文件夹中,图像文件的格式可以是 png、jpg、gif 或是 bmp,Android 4.0 以上的版本还支持新的文件格式 webp,举例来说,我们可以在项目的 res/layout 文件夹下的接口布局文件中增加一个 ImageButton 组件如下:

```
<ImageButton android:id="@+id/imgBtn"
    android:layout_width="wrap_content"
    android:layout_height="wrap_content"
    android:src="@drawable/图像文件名"
    />
```

每一个 ImageButton 组件都必须指定一个 id 名称以便在程序中使用该 ImageButton,android:src 属性则是决定它上面显示的图像,我们必须在程序代码中建立并设置 ImageButton 组件的 OnClickListener 对象,就像是之前使用 Button 组件时一样,这样当用户按下按钮之后就会运行 OnClickListener 对象中的程序代码。

ImageButton 的目的是让按钮的外观比较富有变化,如果只是要显示图像供用户观看就必须使用 ImageView 组件。ImageView 组件同样是使用 android:src 属性设置要显示的图像文件,请参考以下范例:

```
<ImageView android:id="@+id/imgView"
    android:layout_width="wrap_content"
```

```
android:layout_height="wrap_content"
android:src="@drawable/图像文件名"
/>
```

如果在程序运行的过程中想要改变 ImageView 组件上所显示的图像，可以调用它的 setImageResource()方法，该方法可以从项目的资源类 R 中加载指定的图像文件。

> **补充说明** Eclipse 操作技巧
>
> 资源类 R 中的图像文件都是对应到程序项目的 res/drawable 文件夹下的文件，如果使用 Windows 文件总管在该文件夹中新增或是删除文件，必须先在 Eclipse 左边的项目检查窗口中选定该程序项目，然后再选择主菜单中的 File>Refresh 来更新资源类 R 的内容。

接下来就让我们利用 ImageButton 和 ImageView 接口组件，帮之前完成的"计算机猜拳游戏"程序进行操作接口的改良，主要的改变是把操作接口中的"剪刀"、"石头"、和"布"等文字换成用图像代表，这样不但可以增加程序界面的美观，而且也比较方便让使用者判断输赢的结果，图 19-1 是程序的运行界面，以下是修改后的接口布局文件和程序文件，粗体字代表经过修改的部分。和原来的程序代码相较之下修改的部分并不多，在接口布局文件中就是把原来的 Button 改成使用 ImageButton 并设置适当的图像文件，另外也将计算机出拳的结果换成使用 ImageView 组件显示，在程序文件中也只是更动和接口显示相关的部分，程序的运算逻辑并没有改变。

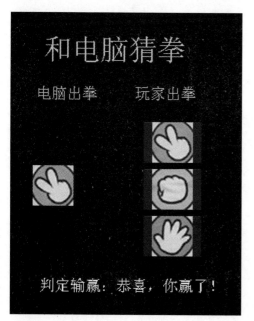

图 19-1 使用 ImageButton 和 ImageView 接口组件的"计算机猜拳游戏"操作界面

- 界面布局文件：

```xml
<?xml version="1.0" encoding="utf-8"?>
<RelativeLayout xmlns:android="http://schemas.android.com/apk/res/android"
    android:layout_width="400dp"
    android:layout_height="match_parent"
    android:layout_gravity="center_horizontal"
    >
<TextView android:id="@+id/txtTitle"
    android:layout_width="wrap_content"
    android:layout_height="wrap_content"
    android:text="@string/promptTitle"
    android:textSize="40sp"
    android:textColor="#FF00FF"
    android:textStyle="bold"
    android:layout_centerHorizontal="true"
    android:paddingLeft="20dp"
    android:paddingRight="20dp"
    android:layout_marginTop="20dp"
    android:layout_marginBottom="20dp"
    />
<TextView android:id="@+id/txtCom"
    android:layout_width="wrap_content"
    android:layout_height="wrap_content"
    android:text="@string/promptComPlay"
    android:layout_below="@id/txtTitle"
    android:layout_alignLeft="@id/txtTitle"
    android:textSize="20sp"
    android:layout_marginBottom="20dp"
    />
<TextView android:id="@+id/txtMyPlay"
    android:layout_width="wrap_content"
    android:layout_height="wrap_content"
    android:text="@string/promptMyPlay"
    android:layout_below="@id/txtTitle"
    android:layout_alignRight="@id/txtTitle"
    android:textSize="20sp"
    android:layout_marginBottom="20dp"
    />
<ImageButton android:id="@+id/btnScissors"
    android:layout_width="wrap_content"
    android:layout_height="wrap_content"
    android:src="@drawable/scissors"
    android:layout_below="@id/txtMyPlay"
    android:layout_alignLeft="@id/txtMyPlay"
    android:paddingLeft="15dp"
    android:paddingRight="15dp"
    />
```

```xml
<ImageView android:id="@+id/imgComPlay"
    android:layout_width="wrap_content"
    android:layout_height="wrap_content"
    android:layout_below="@id/btnScissors"
    android:layout_alignLeft="@id/txtCom"
/>
<ImageButton android:id="@+id/btnStone"
    android:layout_width="wrap_content"
    android:layout_height="wrap_content"
    android:src="@drawable/stone"
    android:layout_below="@id/btnScissors"
    android:layout_alignLeft="@id/btnScissors"
    android:paddingLeft="15dp"
    android:paddingRight="15dp"
/>
<ImageButton android:id="@+id/btnNet"
    android:layout_width="wrap_content"
    android:layout_height="wrap_content"
    android:src="@drawable/net"
    android:layout_below="@id/btnStone"
    android:layout_alignLeft="@id/btnStone"
    android:paddingLeft="15dp"
    android:paddingRight="15dp"
/>
<TextView android:id="@+id/txtResult"
    android:layout_width="wrap_content"
    android:layout_height="wrap_content"
    android:text="@string/result"
    android:layout_below="@id/btnNet"
    android:layout_alignLeft="@id/txtCom"
    android:textSize="20sp"
    android:textColor="#0FFFFF"
    android:layout_marginTop="20dp"
/>
</RelativeLayout>
```

■ 程序文件：

```java
package …

import …

public class Main extends Activity {

    private TextView txtResult;
    private ImageView imgComPlay;
    private ImageButton btnScissors,
            btnStone,
```

```java
                        btnNet;

    /** Called when the activity is first created. */
    @Override
    public void onCreate(Bundle savedInstanceState) {
        super.onCreate(savedInstanceState);
        setContentView(R.layout.main);

        setupViewComponent();
    }

    private void setupViewComponent() {
            imgComPlay = (ImageView)findViewById(R.id.imgComPlay);
            txtResult = (TextView)findViewById(R.id.txtResult);
            btnScissors = (ImageButton)findViewById(R.id.btnScissors);
            btnStone = (ImageButton)findViewById(R.id.btnStone);
            btnNet = (ImageButton)findViewById(R.id.btnNet);

            btnScissors.sctOnClickListener(btnScissorsLin);
            btnStone.setOnClickListener(btnStoneLin);
            btnNet.setOnClickListener(btnNetLin);
    }

    private Button.OnClickListener btnScissorsLin = new Button.OnClickListener() {
        public void onClick(View v) {
            // 决定计算机出拳.
            int iComPlay = (int)(Math.random()*3 + 1);

            // 1 － 剪刀, 2 － 石头, 3 － 布.
            if (iComPlay == 1) {
                imgComPlay.setImageResource(R.drawable.scissors);
                txtResult.setText(getString(R.string.result) +
                            getString(R.string.playerDraw));
            }
            else if (iComPlay == 2) {
                imgComPlay.setImageResource(R.drawable.stone);
                txtResult.setText(getString(R.string.result) +
                            getString(R.string.playerLose));
            }
            else {
                imgComPlay.setImageResource(R.drawable.net);
                txtResult.setText(getString(R.string.result) +
                            getString(R.string.playerWin));
            }
        }
    };

    private Button.OnClickListener btnStoneLin = new Button.OnClickListener() {
```

```
        public void onClick(View v) {
            // 修改部分同 btnScissorsLin
        }
    };

    private Button.OnClickListener btnNetLin = new Button.OnClickListener() {
        public void onClick(View v) {
            // 修改部分同 btnScissorsLin
        }
    };
}
```

Android 版本	1.X	2.X	3.X	4.X
适用性	★	★	★	★

UNIT 20

Gallery、GridView 和 ImageSwitcher 接口组件

现在图像已经是一种很常见的数据，几乎每一中新型手机和平板电脑都已经具备照相的功能，为了能够方便观看图像，图像浏览程序手机和平板电脑的必备工具，因此这个单元就让我们来学习如何实现一个图像浏览程序，这个程序的名称就叫做"图像画廊"。

要完成"图像画廊"程序需要使用和图像相关的接口组件，除了前一个单元介绍的 ImageView 之外，还有 ImageSwitcher、Gallery 和 GridView 都可以用来显示图像。Gallery 组件是提供图像缩图的浏览功能，它可以将图像缩图排成一列让使用者检查并单击，同时也会随着使用者的操作，自动调整图像缩图的位置。图 20-1 上方像是一列底片的部分就是 Gallery 接口组件。GridView 接口组件则是把整个界面切割成许多小格子，每一个格子都显示一张图像缩图，因此和 Gallery 组件相较之下，一次可以检查比较多的图像缩图，如图 20-2 所示。

除了浏览图像缩图的接口组件之外，还需要一个可以显示原来图像的组件。上一个单元学过的 ImageView 虽然可以用来显示图像，但是如果使用以下即将介绍的 ImageSwitcher 组件可以做到更生动有趣的效果。ImageSwitcher 组件，顾名思义就是一个图像切换器，它里头其实就是 ImageView 对象，只是当所显示的图像改变时，它会播放图像切换的动画效果，像是淡入淡出（fade in/fade out）。图 20-1 中央较大的图像就是 ImageSwitcher 组件。本单元的"图像画廊"程序的操作方式就是用户利用屏幕上方的 Gallery 组件浏览图像缩图并单击想要观看的图像，单击后 Gallery 组件会自动将被单击的缩图调整到中央的位置，然后在下方的 ImageSwitcher 组件中显示该缩图的原始图像，在显示的过程中会播放图像切换的动画效果，接着让我们先来学习 Gallery 组件、GridView 组件和 ImageSwitcher 组件的用法。

图像接口组件与动画效果　PART 4

图 20-1　"图像画廊"程序的操作界面

图 20-2　GridView 接口组件

20-1　Gallery 组件和 GridView 组件的使用方法

Gallery 组件和 GridView 组件的用法很类似，我们先介绍 Gallery 组件的使用方式，使用 Gallery 接口组件需要完成以下步骤：

Step 1　在项目的 res/layout 文件夹下的接口布局文件中建立一个 Gallery 组件标签并加以命名和设置属性如下，其中的 android:spacing 属性是设置图像缩图之间的距离。

```
<Gallery android:id="@+id/gal"
    android:layout_width="match_parent"
    android:layout_height="wrap_content"
    android:spacing="10dp"
/>
```

接下来的步骤都在程序文件中进行，首先我们必须新增一个衍生自 BaseAdapter 的类，我们可以将它取名为 ImageAdapter，这个类的功能是管理图像缩图数组，并提供给 Gallery 组件使用。

Step 2　在 Eclipse 左边的项目检查窗口中单击程序项目的 src/(组件名称)。

Step 3　在该组件名称上右击，在弹出的菜单中选择 New>Class，就会看到图 20-3 的对话框。

Step 4　在对话框中的 Name 字段输入 ImageAdapter（也就是我们想要的类名称），然后单击在 Superclass 字段右边的 Browse 按钮，在出现的对话框的最上面的字段输入 BaseAdapter，下方列表就会显示 BaseAdapter 类，用鼠标双击该类就会自动回到原来的对话框，并完成填入 Superclass 字段得到如图 20-3 所示的结果。

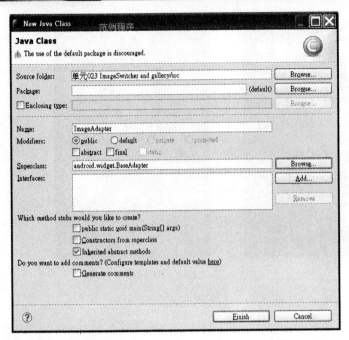

图 20-3 新增类对话框

Step 5 单击 Finish 按钮完成新增类的步骤,Eclipse 会在程序代码窗口中自动开启 ImageAdapter.java 程序文件让用户编辑,请读者输入以下程序代码:

```
package …

import android.content.Context;
import android.view.*;
import android.widget.*;
import android.widget.Gallery.LayoutParams;

public class ImageAdapter extends BaseAdapter {

    private Context cont;
    private Integer[] imgArr;

    public ImageAdapter(Context c) {
        cont = c;
    }

    public void setImageArray(Integer[] imgArr) {
        this.imgArr = imgArr;
    }

    @Override
```

```java
        public int getCount() {
            // TODO Auto-generated method stub
            return imgArr.length;
        }

        @Override
        public Object getItem(int position) {
            // TODO Auto-generated method stub
            return position;
        }

        @Override
        public long getItemId(int position) {
            // TODO Auto-generated method stub
            return position;
        }

        @Override
        public View getView(int position, View convertView, ViewGroup parent) {
            // TODO Auto-generated method stub
            ImageView v = new ImageView(cont);

            v.setImageResource(imgArr[position]);
            v.setAdjustViewBounds(true);
            v.setLayoutParams(new Gallery.LayoutParams(
                    LayoutParams.WRAP_CONTENT, LayoutParams.WRAP_CONTENT));

            return v;
        }

}
```

在 ImageAdapter 类中有两个属性，其中的 cont 属性是用来储存对象的运行环境，imgArr 属性则是用来储存缩图数组，缩图数组中的每一个缩图都是储存它的资源 id 编号，setImageArray() 方法就是用来设置缩图数组，其他的方法包括 getCount()、getItem()、getItemId()、getView() 都是给 Android 系统调用使用。在 getView() 方法中我们先建立一个 ImageView 对象，然后把要显示的缩图设置给它，再设置好显示相关的参数，最后传回该 ImageView 对象。

> **补充说明 Android 程序的 callback 方法**
>
> 上述程序代码范例中的 getCount()、getItem() 等用来提供给 Android 系统调用的方法称为 callback 方法。在程序运行的过程中，经常需要利用这些 callback 方法和 Android 系统进行互动，也就是说 Android 系统会在需要的时候运行程序所提供的 callback 方法，我们必须在这些 callback 方法中完成该做的事。

接下来是编辑主程序文件 Main.java。

Step 6 在主类的程序代码中调用 findViewById()方法取得定义在接口布局文件中的 Gallery 组件，并且建立一个 ImageAdapter 型态的对象，再将该对象设置给 Gallery 组件。

```
// 建立一个 ImageAdapter 型态的对象
ImageAdapter imgAdap = new ImageAdapter(this);
…(设置 ImageAdapter 对象)

Gallery gal = (Gallery) findViewById(R.id.gal);
gal.setAdapter(imgAdap);

// 设置 Gallery 对象的 OnItemSelectedListener，请参考下一个步骤的说明
gal.setOnItemSelectedListener(adaViewItemSelLis);
```

Step 7 建立一个 AdapterView.OnItemSelectedListener 对象并设置给上一个步骤的 Gallery 对象。AdapterView.OnItemSelectedListener 对象的功能是当用户在 Gallery 组件上单击某一个图像缩图时，把该缩图所对应的原始图像显示在 ImageSwitcher 组件中。

```
private AdapterView.OnItemSelectedListener adaViewItemSelLis =
    new AdapterView.OnItemSelectedListener () {
    public void onItemSelected(AdapterView parent,
                               View v,
                               int position,
                               long id) {
        imgSwi.setImageResource(imgArr[position]);
    }
    public void onNothingSelected(AdapterView parent) {
    }
};
```

读者可以利用之前介绍过的程序代码编辑技巧，先完成对象定义的语法，也就是以上程序代码中粗体字的部分，再根据程序编辑窗口的语法修正提示功能，让程序编辑器自动加入需要的方法，然后再自行输入其中的程序代码。

GridView 组件的使用方式和 Gallery 组件很类似，也是要自行建立一个继承自 BaseAdapter 的类（就是上述的 ImageAdapter 类），然后把它设置给 GridView 组件，这个 ImageAdapter 类就是负责做出每一张图像缩图供 GridView 显示，我们将 GridView 的使用步骤说明如下：

Step 1 在项目的 res/layout 文件夹下的接口布局文件中建立一个 GridView 组件标签并完成命名和设置属性如下：

```
<GridView android:id="@+id/grdView"
    android:layout_width="match_parent"
    android:layout_height="match_parent"
    android:numColumns="auto_fit"
    android:verticalSpacing="12dp"
    android:horizontalSpacing="12dp"
```

```
        android:columnWidth="80dp"
        android:stretchMode="columnWidth"
        android:gravity="center"
    />
```

Step 2 在程序代码中调用 findViewById()方法取得上述的 GridView 组件。

Step 3 建立一个衍生自 BaseAdapter 的新类，例如我们可以将它取名为 ImageAdapter，这个类的功能是管理图像缩图数组。ImageAdapter 类中的 getView()方法是负责将指定的图像缩图放到 ImageView 对象中，以供 GridView 组件显示使用。ImageAdapter 类中的方法是提供给 Android 系统调用，我们的程序中不会调用这些方法，新增完 ImageAdapter 类后请在它的程序文件中输入下列的程序代码：

```
package …

import …

public class ImageAdapter extends BaseAdapter {

    private Context cont;
    private Integer[] imgArr;

    public ImageAdapter(Context c) {
        cont = c;
    }

    public void setImageArray(Integer[] imgArr) {
        this.imgArr = imgArr;
    }

    @Override
    public int getCount() {
        // TODO Auto-generated method stub
        return imgArr.length;
    }

    @Override
    public Object getItem(int position) {
        // TODO Auto-generated method stub
        return null;
    }

    @Override
    public long getItemId(int position) {
        // TODO Auto-generated method stub
        return 0;
    }
```

```java
        @Override
        public View getView(int position, View convertView, ViewGroup parent) {
            // TODO Auto-generated method stub
            ImageView v;

            if (convertView == null) {
                v = new ImageView(cont);
                v.setLayoutParams(new GridView.LayoutParams(90, 90));
                v.setScaleType(ImageView.ScaleType.CENTER_CROP);
                v.setPadding(5, 5, 5, 5);
            }
            else
                v = (ImageView) convertView;

            v.setImageResource(imgArr[position]);

            return v;
        }
    }
```

这一段程序代码和前面介绍 Gallery 的 ImageAdapter 类的程序代码非常类似，主要的差别是在 getView()方法，这里会用到 convertView 自变量，如果该自变量是 null，我们必须建立一个新的 ImageView 对象并设置好相关属性。如果它不是 null，我们就把它转换成 ImageView 对象，再把图像缩图设置给它即可。

Step 4 在主类的程序代码中建立一个 ImageAdapter 类的对象，然后调用它的 setImageArray() 方法传入图像缩图数组，最后再把这个 ImageAdapter 对象设置给步骤二的 GridView 组件如下：

```java
private Integer[] thumbImgArr = {
            R.drawable.img01, R.drawable.img02, R.drawable.img03,
            R.drawable.img04, R.drawable.img05, R.drawable.img06};

GridView grdView = (GridView)findViewById(R.id.grdView);

ImageAdapter imgAdap = new ImageAdapter(this);
imgAdap.setImageArray(thumbImgArr);
grdView.setAdapter(imgAdap);
```

20-2　ImageSwitcher 组件的使用方法

ImageSwitcher 组件的目的是用来显示图像，它和 ImageView 组件的差异是 ImageSwitcher 组件可以设置图像消失和出现时的效果，使用 ImageSwitcher 组件的步骤如下：

Step 1 在项目的 res/layout 文件夹下的接口布局文件中建立一个 ImageSwitcher 标签并设置 id 名称和属性如下：

```
<ImageSwitcher android:id="@+id/imgSwi"
    android:layout_width="match_parent"
    android:layout_height="wrap_content"
/>
```

Step 2 在程序项目的主类程序文件（也就是建立项目时自动产生的那一个程序文件）中让类实现 ViewFactory 接口，如下粗体字的部分：

public class 主程序类名称 extends Activity implements ViewFactory {

ViewFactory 接口中定义了一个 makeView()方法，ImageSwitcher 对象需要这个方法来建立内部的 ImageView 对象。makeView()方法是由 Android 系统调用，我们的程序不会自己调用这个方法。让主类实现 ViewFactory 接口的 makeView()方法后，就可以把主类设置给 ImageSwitcher 对象。makeView()方法的程序代码如下，首先是建立一个 ImageView 对象，然后设置相关参数，最后传回该 ImageView 对象。

```
public View makeView() {
    ImageView v = new ImageView(this);
    v.setBackgroundColor(0xFF000000);
    v.setScaleType(ImageView.ScaleType.FIT_CENTER);
    v.setLayoutParams(new ImageSwitcher.LayoutParams
        (LayoutParams.MATCH_PARENT,
        LayoutParams.MATCH_PARENT));
    return v;
}
```

Step 3 在步骤二的程序文件中调用 findViewById()方法取得步骤一的 ImageSwitcher 组件，接着调用 ImageSwitcher 对象的 setFactory()方法把主类传入 ImageSwitcher 中（使用 this），然后调用 ImageSwitcher 对象的 setInAnimation()和 setOutAnimation()方法指定图像切换时的动画效果，请读者参考以下程序代码，其中我们使用 Android 系统内建的淡入淡出（fade in/fade out）效果。

```
ImageSwitcher imgSwi = (ImageSwitcher) findViewById(R.id.imgSwi);
imgSwi.setFactory(this);
imgSwi.setInAnimation(AnimationUtils.loadAnimation(this,
        android.R.anim.fade_in));
imgSwi.setOutAnimation(AnimationUtils.loadAnimation(this,
        android.R.anim.fade_out));
```

20-3 完成"图像画廊"程序

和前面单元的范例程序相较之下,建立"图像画廊"程序需要比较多的步骤和程序代码,而且运行流程也比较复杂,因为 ImageSwitcher 组件和 Gallery 组件都需要建立额外的 callback 方法供 Android 系统调用使用,或许读者刚开始会觉得有些复杂,但是相信只要仔细阅读以上的说明,再自己动手实现一次程序应该就能够了解。以下再补充三点说明:

1. 为了让图像显示在屏幕的中央,在接口布局文件中我们使用二层 LinearLayout 的方式以及 android:layout_weight 属性,来设置 Gallery 和 ImageSwitcher 的位置。
2. 程序显示的图像文件是储存在 imgArr 数组中,该数组是 Integer 型态,因为它只需记录图像文件在资源类 R 中的 id 编号即可。图像缩图则是储存在 thumbImgArr 数组,该数组同样是 Integer 型态,因为它也是记录资源类 R 中的 id 编号。
3. 设置 ImageSwitcher 组件和 Gallery 组件的相关程序代码都放在我们自己建立的 setupViewComponent()方法中。

最后我们列出完整的接口布局文件和主类程序文件供读者参考,ImageAdapter 程序文件已经列于前面的步骤说明中。

- 界面布局文件:

```xml
<?xml version="1.0" encoding="utf-8"?>
<LinearLayout xmlns:android="http://schemas.android.com/apk/res/android"
    android:orientation="horizontal"
    android:layout_width="match_parent"
    android:layout_height="match_parent"
    >
    <LinearLayout
        android:orientation="vertical"
        android:layout_width="0dp"
        android:layout_height="match_parent"
        android:layout_weight="1"
    />
    <LinearLayout
        android:orientation="vertical"
        android:layout_width="0dp"
        android:layout_height="match_parent"
        android:layout_weight="1"
        >
        <Gallery android:id="@+id/gal"
            android:layout_width="match_parent"
            android:layout_height="wrap_content"
            android:spacing="10dp"
        />
```

```xml
<ImageSwitcher android:id="@+id/imgSwi"
    android:layout_width="match_parent"
    android:layout_height="wrap_content"
    android:layout_marginTop="20dp"
    />
</LinearLayout>
<LinearLayout
    android:orientation="vertical"
    android:layout_width="0dp"
    android:layout_height="match_parent"
    android:layout_weight="1"
    />
</LinearLayout>
```

- 主类程序文件：

```java
package …

import …

public class ImageGallery extends Activity implements ViewFactory {

    private ImageSwitcher imgSwi;
    private Gallery gal;

    private Integer[] imgArr = {
            R.drawable.img01, R.drawable.img02, R.drawable.img03,
            R.drawable.img04, R.drawable.img05, R.drawable.img06,
            R.drawable.img07, R.drawable.img08};

    private Integer[] thumbImgArr = {
            R.drawable.img01th, R.drawable.img02th, R.drawable.img03th,
            R.drawable.img04th, R.drawable.img05th, R.drawable.img06th,
            R.drawable.img07th, R.drawable.img08th};

    /** Called when the activity is first created. */
    @Override
    public void onCreate(Bundle savedInstanceState) {
        super.onCreate(savedInstanceState);
        setContentView(R.layout.main);

        setupViewComponent();
    }

    private void setupViewComponent() {
        // 从资源类 R 中取得接口组件并设置相关属性
        imgSwi = (ImageSwitcher) findViewById(R.id.imgSwi);
        imgSwi.setFactory(this); // 必须 implements ViewSwitcher.ViewFactory
```

```java
        imgSwi.setInAnimation(AnimationUtils.loadAnimation(this,
                android.R.anim.fade_in));
        imgSwi.setOutAnimation(AnimationUtils.loadAnimation(this,
                android.R.anim.fade_out));

        ImageAdapter imgAdap = new ImageAdapter(this);
        imgAdap.setImageArray(thumbImgArr);

        gal = (Gallery) findViewById(R.id.gal);
        gal.setAdapter(imgAdap);
        gal.setOnItemSelectedListener(adaViewItemSelLis);
    }

    public View makeView() {
        ImageView v = new ImageView(this);
        v.setBackgroundColor(0xFF000000);
        v.setScaleType(ImageView.ScaleType.FIT_CENTER);
        v.setLayoutParams(new ImageSwitcher.LayoutParams(LayoutParams.MATCH_PARENT,
                LayoutParams.MATCH_PARENT));
        return v;
    }

    private AdapterView.OnItemSelectedListener adaViewItemSelLis =
        new AdapterView.OnItemSelectedListener () {
        public void onItemSelected(AdapterView parent,
                                    View v,
                                    int position,
                                    long id) {
            imgSwi.setImageResource(imgArr[position]);
        }
        public void onNothingSelected(AdapterView parent) {
        }
    };
}
```

Android 版本	1.X	2.X	3.X	4.X
适用性	★	★	★	★

UNIT 21
使用 Tween 动画效果

在前一个单元的 ImageSwitcher 接口组件中，我们第一次体验了 Android 程序的动画效果，只要简单地使用 ImageSwitcher 的 setInAnimation()和 setOutAnimation()这两个方法，就可以完成图像切换时的淡入淡出效果，让程序的操作界面更加生动有趣。在这个单元中，我们将进一步学习如何在 Android 程序中自行建立动画。

Android 程序可以对显示在屏幕上的对象做出二种类型的动画效果：Tween animation 和 Frame animation，这两种动画都叫做 View animation。所谓 Tween animation 是根据指定动画开始和结束时的对象属性，例如位置、Alpha 值（透明度）、大小、角度等，以及动画播放的时间长度，Android 系统就会自动产生动画播放过程中的所有界面。Frame animation 则类似卡通动画的制作过程，我们必须指定每一个帧所播放的图像文件和时间长度，Android 系统再依照我们的设置播放动画。

不管是 Tween animation 或 Frame animation Android 都有两种建立动画的方法，一种是在项目的 res 文件夹下建立动画资源文件（xml 文件格式），该动画资源会自动加入项目的资源类 R 中，程序再从资源类中加载动画来使用。第二种方式是直接在程序代码中建立动画对象并设置相关属性。

Android 提供 4 种类型的 Tween 动画效果：

1. Alpha
   根据改变图像的透明度来达成动画效果，当图像的 alpha 值是 1 时代表图像完全不透明，此时是最清楚的状态。当图像的 alpha 值由 1 减到 0 时，图像变得越来越透明，也就是越来越不清楚直到看不见（alpha 值为 0）。

2. Scale
   根据改变图像的大小来达成动画效果，图像的 scale 值也是用 0~1 来代表，0 代表完全看不到，1 代表原来图像的大小。scale 值可以在 x 和 y 两个方向独立设置，x 方向是图像的宽，y 方向是图像的高。

3. Translate
根据改变图像的位置来达成动画效果，图像的位置是根据 x 和 y 方向上的位移量来决定。

4. Rotate
根据改变图像的旋转角度来达成动画效果。

接着就让我们来学习如何建立动画资源文件。

21-1 建立动画资源文件

建立动画资源文件的过程包含下列 7 个步骤：

Step 1 在 Eclipse 左边的项目检查窗口中打开项目下的 res 文件夹。

Step 2 在 res 文件夹上右击，再从弹出的快捷菜单中选择 New>Folder，就会出现图 21-1 所示的对话框。

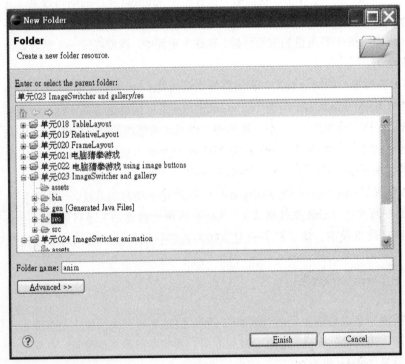

图 21-1 在程序项目的 res 文件夹中新增 anim 文件夹

Step 3 在对话框中的 Folder name 字段输入 anim，注意要使用小写英文字母，然后按下 Finish 按钮。

图像接口组件与动画效果　PART 4

补充说明

在 Android 程序项目中，res 文件夹里头的子文件夹名称和文件名都只能使用小写英文字母、底线字符或数字，不可以使用大写英文字母。

Step 4　在 Eclipse 左边的项目检查窗口中会出现新增的 anim 文件夹，在该文件夹上右击，再从弹出的快捷菜单中选择选择 New>File，就会出现图 21-2 所示的对话框。

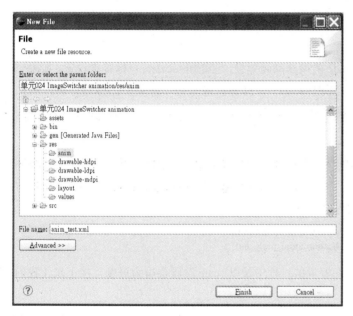

图 21-2　新增动画资源文件的对话框

Step 5　在对话框中的 File name 字段输入动画资源文件的名称再按下 Finish 按钮，注意文件名只能用小写的英文字母、数字、或是底线字符，例如 anim_scale_out.xml。在 anim 文件夹中的每一个文件都是一个动画效果，每一个动画效果都可以是单独的 Alpha、Scale、Translate 或是 Rotate 类型。也可以是将一种以上的动画类型结合起来，例如 Scale 加上 Rotate。在本单元的范例程序中我们将把自行建立的动画设置给 ImageSwitcher 对象，由于在 ImageSwitcher 对象中会用到图像消失和出现两种动画效果，所以我们在动画资源文件的最后加上 in 和 out 来区别，in 代表用在图像出现时的动画，out 代表用在图像消失时的动画。

Step 6　用鼠标左键双击上述的动画资源文件将它开启，然后开始建立动画并设置相关属性，例如我们将下列的程序代码写入上一个步骤建立的 anim_scale_out.xml 动画资源文件，我们将在下一个小节中详细介绍动画建立时所使用的属性和设置值：

153

```xml
<?xml version="1.0" encoding="utf-8"?>
<set xmlns:android="http://schemas.android.com/apk/res/android">
    <scale android:interpolator="@android:anim/linear_interpolator"
           android:fromXScale="0.0"
           android:toXScale="1.0"
           android:fromYScale="0.0"
           android:toYScale="1.0"
           android:pivotX="50%"
           android:pivotY="50%"
           android:startOffset="3000"
           android:duration="3000"
    />
</set>
```

在这个范例中我们建立一个 scale 类型的动画效果，它是用<scale…/>标签来设置。如果想建立其他类型的动画效果，只要把<scale…/>标签换成其他动画类型的标签即可，例如<translate…/>。如果想要结合两种以上的动画类型，则是依序加上不同动画类型的标签，后续我们会有实际范例。

Step 7 完成以上动画资源文件后，就能够在程序代码中加载使用，例如我们将上一个单元"图像画廊"程序中的 imgSwi 对象，改成使用我们自行建立的 anim_scale_out 动画效果，程序代码只需要修改以下粗体字的部分即可。

imgSwi.setOutAnimation(AnimationUtils.loadAnimation(thisCont,
                       R.anim.anim_scale_out));

以上是建立动画资源文件的步骤，在解说的过程中是以 scale 动画类型为范例，接下来让我们学习如何建立各种不同类型的动画。

21-2 建立各种类型的动画

为了让读者方便查询建立动画所使用的属性和设置值，我们将 4 种动画类型的相关属性整理成下列表格：

表 21-1　四种 Tween 动画类型和它们的相关属性

动画类型	属性名称	属性值	说明
Alpha	android:interpolator	"@android:anim/linear_interpolator" "@android:anim/accelerate_interpolator" "@android:anim/decelerate_interpolator" "@android:anim/accelerate_decelerate_interpolator"	设置动画过程中变化的快慢 第一个值是一样快 第二个值是越来越快 第三个值是越来越慢 第四个值是中间快前后慢
	android:fromAlpha	0～1	动画开始时的图像 alpha 值

续表

动画类型	属性名称	属性值	说明
Alpha	android:toAlpha	0～1	动画结束时的图像 alpha 值
	android:startOffset	整数值	启动动画后要等多久才真正开始运移动画，以毫秒为单位
	android:duration	整数值	动画持续时间，以毫秒为单位
Scale	android:interpolator	同前面说明	同前面说明
	android:fromXScal	0～	动画开始时图像的 x 方向大小比例，1 以上的值代表放大
	android:toXScale	0～	动画结束时图像的 x 方向大小比例，1 以上的值代表放大
	android:fromYScale	0～	动画开始时图像的 y 方向大小比例，1 以上的值代表放大
	android:toYScale	0～	动画结束时图像的 y 方向大小比例，1 以上的值代表放大
	android:pivotX	0～1	动画开始时图像的 x 坐标，0 代表最左边，1 代表最右边
	android:pivotY	0～1	动画开始时图像的 y 坐标，0 代表上缘，1 代表下缘
	android:startOffset	同前面说明	同前面说明
	android:duration	同前面说明	同前面说明
Translate	android:interpolator	同前面说明	同前面说明
	android:fromXDelta	整数值	动画开始时图像的 x 坐标的位移量
	android:toXDelta	整数值	动画结束时图像的 x 坐标的位移量
	android:fromYDelta	整数值	动画开始时图像的 y 坐标的位移量
	android:toYDelta	整数值	动画结束时图像的 y 坐标的位移量
	android:startOffset	同前面说明	同前面说明
	android:duration	同前面说明	同前面说明
Rotate	android:interpolator	同前面说明	同前面说明
	android:fromDegrees	整数值	动画开始时图像的角度
	android:toDegrees	整数值	动画结束时图像的角度
	android:pivotX	同前面说明	同前面说明
	android:pivotY	同前面说明	同前面说明
	android:startOffset	同前面说明	同前面说明
	android:duration	同前面说明	同前面说明

以下是 6 个动画资源文件的范例，文件名依序为 anim_alpha_in.xml、anim_alpha_out.xml、anim_scale_rotate_in.xml、anim_scale_rotate_out.xml、anim_trans_in.xml、anim_trans_out.xml。这些动画资源文件就是使用表 21-1 中的动画属性所建立，读者可以一边阅读一边对照表 21-1 中的说明

就可以了解，这些动画资源文件将被用在"图像画廊"程序中，让图像切换时可以显示更多样化的动画效果。

- anim_alpha_in.xml 文件：

```xml
<?xml version="1.0" encoding="utf-8"?>
<set xmlns:android="http://schemas.android.com/apk/res/android">
    <alpha android:interpolator="@android:anim/linear_interpolator"
        android:fromAlpha="0.0"
        android:toAlpha="1.0"
        android:duration="3000"/>
</set>
```

- anim_alpha_out.xml 文件：

```xml
<?xml version="1.0" encoding="utf-8"?>
<set xmlns:android="http://schemas.android.com/apk/res/android">
    <alpha android:interpolator="@android:anim/linear_interpolator"
        android:fromAlpha="1.0"
        android:toAlpha="0.0"
        android:duration="3000"/>
</set>
```

- anim_scale_rotate_in.xml 文件：

```xml
<?xml version="1.0" encoding="utf-8"?>
<set xmlns:android="http://schemas.android.com/apk/res/android">
    <scale android:interpolator="@android:anim/linear_interpolator"
        android:fromXScale="0.0"
        android:toXScale="1.0"
        android:fromYScale="0.0"
        android:toYScale="1.0"
        android:pivotX="50%"
        android:pivotY="50%"
        android:startOffset="3000"
        android:duration="3000"/>
    <rotate
        android:interpolator="@android:anim/accelerate_decelerate_interpolator"
        android:fromDegrees="0"
        android:toDegrees="360"
        android:pivotX="50%"
        android:pivotY="50%"
        android:startOffset="3000"
        android:duration="3000"/>
</set>
```

- anim_scale_rotate_out.xml 文件：

```xml
<?xml version="1.0" encoding="utf-8"?>
<set xmlns:android="http://schemas.android.com/apk/res/android">
```

```xml
<scale android:interpolator="@android:anim/linear_interpolator"
    android:fromXScale="1.0"
    android:toXScale="0.0"
    android:fromYScale="1.0"
    android:toYScale="0.0"
    android:pivotX="50%"
    android:pivotY="50%"
    android:duration="3000"/>
<rotate
    android:interpolator="@android:anim/accelerate_decelerate_interpolator"
    android:fromDegrees="0"
    android:toDegrees="360"
    android:pivotX="50%"
    android:pivotY="50%"
    android:duration="3000"/>
</set>
```

- anim_trans_in.xml 文件：

```xml
<?xml version="1.0" encoding="utf-8"?>
<set xmlns:android="http://schemas.android.com/apk/res/android">
    <translate android:interpolator="@android:anim/linear_interpolator"
        android:fromXDelta="0"
        android:toXDelta="0"
        android:fromYDelta="-300"
        android:toYDelta="0"
        android:duration="3000"
    />
</set>
```

- anim_trans_out.xml 文件：

```xml
<?xml version="1.0" encoding="utf-8"?>
<set xmlns:android="http://schemas.android.com/apk/res/android">
    <translate android:interpolator="@android:anim/linear_interpolator"
        android:fromXDelta="0"
        android:toXDelta="0"
        android:fromYDelta="0"
        android:toYDelta="300"
        android:duration="3000" />
</set>
```

21-3 使用随机动画的"图像画廊"程序

在前一个小节中总共建立了 3 组动画，每一组都有 in/out 两个动画资源文件，这些动画将被用在"图像画廊"程序中，当用户在"图像画廊"程序上方的 Gallery 组件中单击一个图像缩图时，程序会随机选择一组动画效果来完成切换图像的动作。这里会再次用到前面单元介绍过的随机数功能，程序代码的部分必须把原来在 setupViewComponent()方法中，设置 imgSwi 组件的 In/OutAnimation 的程序代码，搬到 adaViewItemSelLis 对象的 onItemSelected()方法内，因为每一次用户单击一个图像缩图后，我们必须重新选择动画效果，而这个动画效果是用随机数来决定。修改后的程序代码如下，粗体字代表经过修改的程序代码。另外我们增加一个属性 thisCont 用来储存程序的运行环境，它会在 adaViewItemSelLis 对象的 onItemSelected()方法内使用，程序运行的界面如图 21-3 所示。

图 21-3　"图像画廊"程序使用随机动画的效果

```
package tw.android;

import …

public class Main extends Activity implements ViewFactory {
    …
    private Main thisCont = this;
        …

    private AdapterView.OnItemSelectedListener adaViewItemSelLis =
        new AdapterView.OnItemSelectedListener () {
        public void onItemSelected(AdapterView parent,
                        View v,
                        int position,
                        long id) {
            switch ((int)(Math.random()*3 + 1)) {
            case 1:
                    imgSwi.setInAnimation(AnimationUtils.loadAnimation
                        (thisCont, R.anim.anim_alpha_in));
                    imgSwi.setOutAnimation(AnimationUtils.loadAnimation
                        (thisCont, R.anim.anim_alpha_out));
                    break;
            case 2:
                    imgSwi.setInAnimation(AnimationUtils.loadAnimation
                        (thisCont, R.anim.anim_trans_in));
                    imgSwi.setOutAnimation(AnimationUtils.loadAnimation
                        (thisCont, R.anim.anim_trans_out));
                    break;
            case 3:
                    imgSwi.setInAnimation(AnimationUtils.loadAnimation
```

```
                              (thisCont,    R.anim.anim_scale_rotate_in));
                   imgSwi.setOutAnimation(AnimationUtils.loadAnimation
                              (thisCont, R.anim.anim_scale_rotate_out));
                   break;
            }

            imgSwi.setImageResource(imgArr[position]);
        }
        public void onNothingSelected(AdapterView parent) {
        }
    };
}
```

21-4　在程序代码中建立动画效果

除了使用动画资源文件来建立动画效果之外，我们也可以利用程序代码来建立动画，在程序中我们可以建立 AlphaAnimation、ScaleAnimation、TranslateAnimation 和 RotateAnimation 四种动画对象，它们分别对应到动画资源文件的<alpha…/>、<scale…/>、<translate…/>和<rotate…/>四种动画类型标签，而且这些动画对象所使用的参数和它们对应的动画类型标签中的属性也是有清楚的对应关系，以 ScaleAnimation 对象为例，它的建构参数如下：

```
ScaleAnimation(float fromX,
               float toX,
               float fromY,
               float toY,
               int pivotXType,
               float pivotXValue,
               int pivotYType,
               float pivotYValue)
```

读者可以在表 21-1 中找到这些参数名称所对应的属性，这些动画对象的建构参数和它们对应的属性的功能完全相同。除此之外在动画对象的建构参数中多了 pivotXType 和 pivotYType，它们的作用是指定后面的 pivotXValue 和 pivotYValue 参数的参考基准为何，如果设置为 Animation.RELATIVE_TO_SELF 代表是以自己为基准，这也是动画标签中的属性的作法。细心的读者会发现动画对象的建构参数中没有 interpolator、startOffset 和 duration，这些属性是使用动画对象的方法来设置，例如我们可以利用下列的程序代码，建立和前面 anim_trans_out.xml 动画资源文件完全一样的动画效果。读者也可以尝试写出其他动画效果的程序代码，如果要查询动画对象的详细参数说明，可以利用 UNIT 5 介绍的方法查询 Android SDK 说明文件。

```
TranslateAnimation anim_trans_out = new TranslateAnimation(Animation.RELATIVE_TO_SELF, 0,
                                        Animation.RELATIVE_TO_SELF, 0,
                                        Animation.RELATIVE_TO_SELF, 0,
                                        Animation.RELATIVE_TO_SELF, 300);
```

```
anim_trans_out.setInterpolator(new LinearInterpolator());
anim_trans_out.anim_rotate_in.setDuration(3000);
imgSwi.setInAnimation(anim_trans_in);
```

最后一个问题是如何结合多个动画对象来产生混合的动画效果,像是前面 anim_scale_rotate_in.xml 动画资源文件要如何使用程序代码来完成呢?答案是利用 AnimationSet 对象,它可以加入多个单一效果的动画对象,例如我们可以利用下列的程序代码得到和前面 anim_scale_rotate_in.xml 动画资源文件完全一样的动画效果,读者可以使用这一段程序代码取代原来加载动画资源的程序代码并得到完全相同的结果。

```
ScaleAnimation anim_scale_in = new ScaleAnimation(0.0f, 1.0f, 0.0f, 1.0f,
                    Animation.RELATIVE_TO_SELF, 0.5f, Animation.RELATIVE_TO_SELF, 0.5f);
anim_scale_in.setInterpolator(new LinearInterpolator());
anim_scale_in.setStartOffset(3000);
anim_scale_in.setDuration(3000);

RotateAnimation anim_rotate_in = new RotateAnimation(0, 360,
                    Animation.RELATIVE_TO_SELF, 0.5f, Animation.RELATIVE_TO_SELF, 0.5f);
anim_rotate_in.setInterpolator(new LinearInterpolator());
anim_rotate_in.setStartOffset(3000);
anim_rotate_in.setDuration(3000);

AnimationSet anim_set = new AnimationSet(false);
anim_set.addAnimation(anim_scale_in);
anim_set.addAnimation(anim_rotate_in);

imgSwi.setInAnimation(anim_set);
```

21-5 应该使用动画资源文件还是在程序代码中建立动画对象

读者学完了两种建立动画的方法心中可能会有一个疑问:既生瑜何生亮?为什么要有两种建立动画的方法?到底该用哪一种比较好?这个问题的答案在于程序架构的考虑。前面我们介绍过 MVC 程序架构,它的精神是希望把程序的接口设计和程序代码的运算逻辑分开,以方便程序的分工和日后的维护。那么动画效果应该属于哪一类?是接口设计还是核心运算?就我们的"图像画廊"程序而言,动画效果应该是属于接口设计的部分,因此应该把它从程序代码中独立出来写成动画资源文件比较好,日后如果想要修改动画效果,只要修改项目的 res/anim 文件夹下的动画资源文件即可,不需要改变程序代码。但是如果其他程序把动画效果当成是核心的运算逻辑,那么就可以考虑采用在程序代码中建立动画对象的方法。

Android 版本	1.X	2.X	3.X	4.X
适用性	★	★	★	★

UNIT 22

Frame animation 和 Multi-Thread 游戏程序

除了 Tween animation 类型的动画效果以外，Android 程序也可以建立 Frame animation 类型的动画。Frame animation 动画的建立过程就像制作卡通影片一样，我们必须指定每一个界面使用的图像文件和停留时间的长短，当开始播放动画的时候，就会依照我们的设置依序显示指定的图像。有两种方法可以建立 Frame animation，以下我们依次介绍。

22-1 建立 Frame animation 的两种方法

我们先介绍使用 xml 动画资源文件建立 Frame animation 的方法。Frame animation 动画资源文件中用到的标签和属性比 Tween animation 简单得多，以下是一个完整的 Frame animation 动画资源文件：

```
<?xml version="1.0" encoding="utf-8"?>
<animation-list xmlns:android="http://schemas.android.com/apk/res/android"
    android:oneshot="false">
    <item android:drawable="@drawable/image01" android:duration="100" />
    <item android:drawable="@drawable/image02" android:duration="100" />
    <item android:drawable="@drawable/image03" android:duration="100" />
</animation-list>
```

<animation-list…>标签代表这是一个 Frame animation，android:oneshot 属性是用来控制动画是否要重复播放，true 代表只要从头到尾播放一次，false 代表播放完毕之后还要再从头播放。<item…>

标签则是用来设置每一帧所使用的图像文件和播放的时间长度。以上面的范例来说，第一个影格是显示 res/drawable 文件夹中的 image01 图像文件，播放的时间长度是 0.1 秒（android:duration 属性的值是以千分之一秒为单位）。Frame animation 动画资源文件必须放在程序项目的 res/drawable 文件夹中，然后再利用以下的程序代码加载到程序中使用：

```
Resources res = getResources();
AnimationDrawable animDraw = (AnimationDrawable)res.getDrawable(R.drawable.anim_drawable);
```

首先我们调用 getResources()方法取得资源对象，再利用资源对象的 getDrawable()方法取得 Frame animation 动画资源文件（假设文件名为 anim_drawable.xml）。

除了使用动画资源文件来建立 Frame animation 以外，如同我们在前一个单元介绍的 Tween animation 一样，也可以利用程序代码的方式建立 Frame animation，例如以下程序可以产生和上面的动画资源文件完全一样的 Frame animation：

```
AnimationDrawable animDraw = new AnimationDrawable();
animDraw.setOneShot(false);

Resources res = getResources();
animDraw.addFrame(res.getDrawable(R.drawable.image01), 100); // 100 是 duration
animDraw.addFrame(res.getDrawable(R.drawable.image02), 100);
animDraw.addFrame(res.getDrawable(R.drawable.image03), 100);
```

这一段程序代码并不复杂，读者从每一个方法的名称就可以了解它的功能，唯一要注意的是 Frame animation 是用 AnimationDrawable 对象来代表。最后把建立好的 Frame animation 设置给程序接口的 ImageView 组件就可以开始播放动画，我们将以上讨论的流程整理成下列步骤：

Step 1 在接口布局文件中建立一个 ImageView 组件。

Step 2 在程序代码中取得接口布局文件中的 ImageView 组件。

Step 3 从程序项目资源中取得 Frame animation，或是利用程序代码建立 Frame animation。

Step 4 运行 ImageView 对象的 setImageDrawable()方法或是 setBackgroundDrawable()方法把 Frame animation 设置给 ImageView 组件。

Step 5 开始播放动画。

以下是程序代码范例，其中 imgView 是在接口布局文件中建立的 ImageView 组件：

```
ImageView animImgView = (ImageView)findViewById(R.id.imgView);

Resources res = getResources();
AnimationDrawable animDraw = (AnimationDrawable)res.getDrawable(R.drawable.anim_drawable);
animImgView.setImageDrawable(animDraw);
animDraw.start();
```

根据以上的说明，要让程序播放 Frame animation 并不难，但是这个单元我们要利用 Frame

animation 完成一个"掷骰子游戏"程序，这个程序必须一边播放动画一边运行计时的工作，这就需要使用 Multi-Thread 的程序架构。

22-2　Multi-Thread"掷骰子游戏"程序

"掷骰子游戏"程序的运行界面如图 22-1 所示，当使用者按下"掷骰子"按钮后，上方的骰子图片会开始播放点数不断跳动的动画，5 秒之后动画自动停止并以随机数的方式得到最后的点数。要完成这个程序首先必须考虑的问题是：如何在播放掷骰子动画的同时运行计时的动作？最直接的作法是在启动动画之后立刻进入一个循环不断地检查系统时间，等 5 秒之后再停止动画并随机决定骰子最后的点数。如果读者以这种方式实现将会发现在循环运行期间不会出现掷骰子动画，等循环结束后才会出现最后的点数，发生这种情况的原因在于 Android 系统只有在主程序（main thread）处于闲置的情况下才会更新界面，当 main thread 忙于运行程序代码时，程序界面会暂停更新，为了解决这个问题，我们必须使用 Multi-Thread 程序架构，也就是说在启动骰子动画之后，运行另一个 thread（称为 background thread）来负责计时的工作，等 5 秒钟之后再停止动画并产生最后点数。

图 22-1　掷骰子动画游戏程序的界面

> **补充说明**　Android 程序的 main thread 和 background thread
>
> 当 Android 程序开始运行时，所建立的 thread 称为 main thread，main thread 也叫做 UI thread 因为程序的所有接口组件都属于 main thread。除了 main thread 之外，其他后来产生的 thread 都叫做 background thread 或 worker thread。

经过以上讨论我们已经解决播放掷骰子动画和计时必须同时进行的问题,接下来还有一个要考虑的情况是当计时完成后,如何显示最后骰子的点数,这个工作将牵涉 background thread 和 main thread 之间的信息沟通。

22-3　使用 Handler 对象传送信息

　　为什么显示骰子点数的问题会和信息传送扯上关系?我们前面得到的结论是让程序的 main thread 负责播放动画,然后启动另一个 background thread 来运行计时的工作,等时间一到再用随机数的方式得到最后的点数,并更新程序界面的骰子图像,因此最简单的作法是让 background thread 在计时完毕后就直接随机产生骰子点数并将它显示在程序界面上。可惜的是这个方法行不通,因为程序界面中的所有接口组件都是属于 main thread,基于程序运行时的安全考虑,Android 系统不允许 background thread 取用 main thread 的接口组件,所以 background thread 无法更新程序界面的骰子图像,解决方法是让 background thread 送给 main thread 一个信息(message)通知计时完成,再由 main thread 运行产生骰子点数和显示骰子图像的工作。

　　要让 background thread 传送信息给 main thread 必须借助 Handler 对象,我们在主程序类(也就是 main thread 的程序代码)中建立一个 Handler 对象,于是 background thread 就可以利用这个 Handler 对象将信息放到 main thread 的 message queue 中,再由 Android 系统通知 main thread 处理该信息。我们将最后得到的"掷骰子游戏"程序的运行架构用图 22-2 代表。

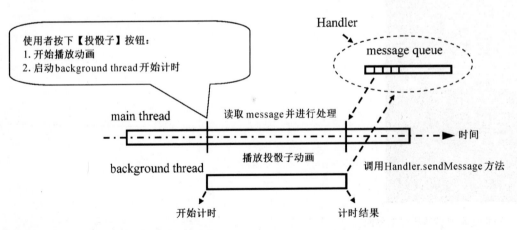

图 22-2　"掷骰子游戏"程序的运行架构

22-4　实现"掷骰子游戏"程序

　　我们把完成"掷骰子游戏"程序的过程整理成以下步骤:

Step 1 运行 Eclipse 新增一个 Android 程序项目，项目的属性设置请依照之前的惯例即可。

Step 2 在 Eclipse 左边的项目检查窗口中展开此项目的 res/ layout 文件夹，开启其中的接口布局文件 main.xml，读者可以在程序代码编辑窗口下方，单击最右边的 Tab 标签页切换到文字编辑模式，然后依序加入一个 ImageView 组件、一个 TextView 组件和一个 Button 组件，并设置它们的 id 和外观属性如下：

```xml
<?xml version="1.0" encoding="utf-8"?>
<LinearLayout xmlns:android="http://schemas.android.com/apk/res/android"
    android:orientation="vertical"
    android:layout_width="match_parent"
    android:layout_height="match_parent"
    android:gravity="center_horizontal"
    >
<ImageView android:id="@+id/imgRollingDice"
    android:layout_width="150dp"
    android:layout_height="150dp"
    />
<TextView android:id="@+id/txtDiceResult"
    android:layout_width="150dp"
    android:layout_height="wrap_content"
    android:text="@string/diceResult"
    android:textSize="20sp"
    android:layout_marginTop="20dp"
    />
<Button android:id="@+id/btnRollDice"
    android:layout_width="wrap_content"
    android:layout_height="wrap_content"
    android:text="@string/btnRollDice"
    android:textSize="20sp"
    android:layout_marginTop="20dp"
    />
</LinearLayout>
```

Step 3 请读者准备 6 个不同点数的骰子图像文件，或是使用本书光盘中的范例程序所附的图像文件，再利用 Windows 文件管理器将骰子图像文件复制到此程序项目文件夹中的 res/drawable-hdpi 子文件夹。

Step 4 在 Eclipse 左边的项目检查窗口中展开 res/drawable-hdpi 文件夹，用鼠标右击 drawable-hdpi 文件夹，再选择 New>File 就会出现如图 22-3 所示的对话框，然后在对话框中输入 Frame animation 动画文件的文件名，例如 anim_drawable.xml，注意文件名只能用小写的英文字母、数字或是底线字符，且扩展名必须是 xml。完成后结束对话框，该文件就会自动开启在程序代码编辑窗口中。在程序代码编辑窗口下方有 2 个 Tab 标签页可以切换编辑模式，请读者选择右边的文本模式然后输入以下程序代码，在这个动画文件中我们用不规则的方式播放 6 个不同点数的骰子图像（以 diceXX 命名），每一张图像的播放时间是 0.1 秒，且设置重复播放：

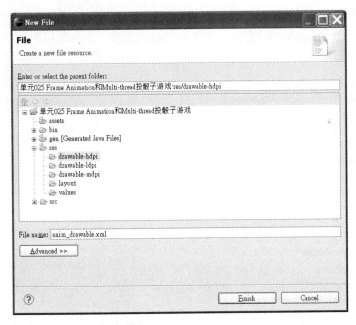

图 22-3　新增文件对话框

```xml
<?xml version="1.0" encoding="utf-8"?>
<animation-list xmlns:android="http://schemas.android.com/apk/res/android"
    android:oneshot="false">
    <item android:drawable="@drawable/dice03" android:duration="100" />
    <item android:drawable="@drawable/dice02" android:duration="100" />
    <item android:drawable="@drawable/dice05" android:duration="100" />
    <item android:drawable="@drawable/dice01" android:duration="100" />
    <item android:drawable="@drawable/dice06" android:duration="100" />
    <item android:drawable="@drawable/dice04" android:duration="100" />
</animation-list>
```

Step 5 在 Eclipse 左边的项目检查窗口中展开 "src/ (程序组件名称)" 文件夹，开启其中的程序文件，在这个程序文件中我们必须完成以下工作：

1. 建立一个 setupViewComponent() 方法，这个方法负责设置好程序中需要用到的接口对象，以及 "掷骰子" 按钮的 OnClickListener，然后在 onCreate() 中运行 setupViewComponent() 方法。

2. 建立一个 Handler 对象并完成其中的 handleMessage() 方法，该方法是当有信息传送到 main thread 的 message queue 时会自动运行，根据前面的讨论，我们要在这个方法中以随机的方式产生骰子最后的点数，并更新程序界面的骰子图像。

3. 在"掷骰子"按钮的 OnClickListener 中,我们先从程序的资源中加载动画,然后设置给 ImageView 对象并开始播放,接着建立一个 Thread 对象并将它启动运行,这个 Thread 对象会先进入 sleep 状态 5 秒钟,然后便会停止动画,再运行 Handler 对象的 sendMessage()方法传送信息给 main thread,于是 main thread 便会运行上面第 2 项的 handleMessage()方法。

以下是完成后的程序文件,请读者特别注意粗体标示的程序代码,另外补充说明一点,在"掷骰子"按钮的 OnClickListener 中加载的动画对象 animDraw,因为要在另一个 Thread 中使用,所以必须定义成 final。

```
package …

import …

public class Main extends Activity {

    private ImageView mImgRollingDice;
    private TextView mTxtDiceResult;
    private Button mBtnRollDice;

    private Handler handler=new Handler() {

        public void handleMessage(Message msg) {
            super.handleMessage(msg);

            int iRand = (int)(Math.random()*6 + 1);

            String s = getString(R.string.diceResult);
            mTxtDiceResult.setText(s + iRand);
            switch (iRand) {
            case 1:
                mImgRollingDice.setImageResource(R.drawable.dice01);
                break;
            case 2:
                mImgRollingDice.setImageResource(R.drawable.dice02);
                break;
            case 3:
                mImgRollingDice.setImageResource(R.drawable.dice03);
                break;
            case 4:
                mImgRollingDice.setImageResource(R.drawable.dice04);
                break;
            case 5:
                mImgRollingDice.setImageResource(R.drawable.dice05);
                break;
            case 6:
```

```java
                    mImgRollingDice.setImageResource(R.drawable.dice06);
                    break;
            }
        }
    };

    /** Called when the activity is first created. */
    @Override
    public void onCreate(Bundle savedInstanceState) {
        super.onCreate(savedInstanceState);
        setContentView(R.layout.main);

        setupViewComponent();
    }

    private void setupViewComponent() {
        mImgRollingDice = (ImageView)findViewById(R.id.imgRollingDice);
        mTxtDiceResult = (TextView)findViewById(R.id.txtDiceResult);
        mBtnRollDice = (Button)findViewById(R.id.btnRollDice);

        mBtnRollDice.setOnClickListener(btnRollDiceOnClickLis);
    }

    private Button.OnClickListener btnRollDiceOnClickLis = new Button.OnClickListener() {
        public void onClick(View v) {

            String s = getString(R.string.diceResult);
            mTxtDiceResult.setText(s);

            // 从程序资源中取得动画文件，设置给 ImageView 对象，然后开始播放
            Resources res = getResources();
            final AnimationDrawable animDraw =
                    (AnimationDrawable)res.getDrawable
                    (R.drawable.anim_drawable);
            mImgRollingDice.setImageDrawable(animDraw);
            animDraw.start();

            // 启动 background thread 进行计时
            new Thread(new Runnable() {

                @Override
                public void run() {
                    // TODO Auto-generated method stub
                    try {
                        Thread.sleep(5000);
                    } catch (Exception e) {
                        // TODO Auto-generated catch block
                        e.printStackTrace();
```

```
                }
                animDraw.stop();
                handler.sendMessage(handler.obtainMessage());
            }
        }
        ).start();
    }
};
}
```

完成以上步骤之后便可以运行程序，读者将会看到如图 22-1 所示的运行界面。由于程序是使用定义在动画资源文件中的动画，因此每一次掷骰子都会看到同样的骰子变换顺序，只有最后出现的点数会不一样，如果想要让骰子点数的变化顺序每一次都不相同，可以使用本单元第一小节中介绍的方法，在使用者按下"掷骰子"按钮后，利用随机数和程序代码的方式来建立 Frame animation，这样程序就会在每一次掷骰子的时候，播放点数变化顺序不一样的骰子动画。

Android 版本	1.X	2.X	3.X	4.X
适用性			★	★

UNIT 23

Property animation 初体验

从学过的 Tween animation 和 Frame animation 已经能够体会 Android 的动画功能，但其实 Tween animation 的用途并不局限于只能用在图像，许多接口组件也都能使用 Tween animation，例如我们在程序项目的 res/anim 文件夹中建立一个名为 rotate.xml 的动画资源文件，它的内容如下：

```xml
<?xml version="1.0" encoding="utf-8"?>
<set xmlns:android="http://schemas.android.com/apk/res/android">
  <rotate
        android:interpolator="@android:anim/accelerate_decelerate_interpolator"
        android:fromDegrees="0"
        android:toDegrees="360"
        android:pivotX="50%"
        android:pivotY="50%"
        android:duration="3000"/>
</set>
```

然后在接口布局文件中建立 Button 组件并取名为 btn，接着在程序文件中取得该 Button 组件，并设置给一个同名的 Button 对象，最后从程序资源中加载 rotate 动画，并调用 Button 对象的 startAnimation()方法将动画套用在 Button 对象，如下列程序代码，于是就会看到程序界面的按钮开始转动，如图 23-1 所示。

```java
Button btn = (Button)findViewById(R.id.btn);
Animation anim = AnimationUtils.loadAnimation(Main.this, R.anim.rotate);
btn.startAnimation(anim);
```

图 23-1　使用 Tween animation 让按钮转动

我们可以依照上述的方法建立各种类型的 Tween animation，再套用到各种接口组件让程序的操作界面"动"起来，看到这里相信读者对于 Android 的动画功能一定更加佩服，但是如果再仔细想一想，例如以上面的旋转按钮范例来说，我们可不可以让按钮的背景颜色也能随着时间改变，或是让按钮中的文字大小随着时间变化，甚至在按钮转动的过程中，当到达特定的角度时能够运行某个特定的工作。很可惜对于 Tween animation 来说，并无法做到这样的功能，但这并不代表 Android 程序无法达成这种功能，事实上运用比较复杂的程序技术，例如 multi-thread 架构，就可以做到。但是从 Android 3.0 版开始，加入了另一种称为 Property animation 的动画技术，这项新的动画技术大大地拓展了动画的应用范围，它让前面 Tween animation 无法做到的动画效果变成可能，接下来就让我们开始介绍这个功能更强大的 Property animation。

23-1　Property animation 的基本用法

为了让读者了解 Property animation 和 Tween animation 的差别，让我们先来谈一谈构成动画的几个基本要素：

1. 动画的主角——物体
   也就是要动起来的对象，可能是一张图像、一个按钮、一个字符串……
2. 时间
   就是动画从开始到结束的时间长短。
3. 物体状态的变化
   就是让用户看到物体状态不断的变化，例如位置、大小、颜色、角度等。

以 Tween animation 来说，它的 xml 动画资源文件中只包含了时间和状态变化两个要素，第一个要素，也就是动画的主角，是在程序代码中才会决定，例如在上一小节的旋转按钮范例中，我们调用 Button 对象的 startAnimation() 将动画套用到 Button，这时候该 Button 对象就成为动画的主角。

Property animation 则使用和 Tween animation 不同的方式来建立动画，它不使用 xml 动画文件，而是直接在程序代码中利用 Animator 对象设置三个动画要素，然后便可以开始播放动画，以旋转按钮的范例来说，以下的 Property animation 程序代码可以达到和 Tween animation 完全一样的效果。

```
ObjectAnimator animBtnRotate = ObjectAnimator.ofFloat(btn, "rotation", 0, 360);
animBtnRotate.setDuration(3000);
animBtnRotate.start();
```

第一行程序代码是使用 ObjectAnimator 的 ofFloat()方法建立一个 ObjectAnimator 型态的动画对象，使用 ofFloat()方法的原因是我们所要改变的物体状态，也就是旋转角度，是用浮点数的代表，如果物体的状态是以整数型态代表，则换成使用 ofInt()。ofFloat()方法的第一个参数是指定动画的主角，也就是要改变状态的对象，第二个参数是指定所要改变的状态，也就是属性，这个属性名称和在接口布局文件中使用的属性名称相同，第三和第四个参数是指定状态的起始值和结束值。第二行程序代码是调用 ObjectAnimator 对象的 setDuration()方法设置动画播放的时间长度，第三行程序代码则是调用 start()方法开始运移动画。

看完以上范例读者会发现使用 Property animation 的方法建立动画似乎比使用 Tween animation 的方式简单，的确如此，而且 Property animation 的功能更强大，它几乎可以让程序中所有的对象动起来，甚至包括没有显示在程序界面的对象，不过在这里我们先卖个关子，等下一个单元再介绍 Property animation 的进阶用法。现在让我们先回到前面的 Property animation 范例，除了可以调用 setDuration()设置动画运行的时间长短之外，还可以使用以下的方法设置其他动画的属性：

1. setRepeatCount()
   设置动画回放的次数，如果想要连续播放不要停可以设置为 ObjectAnimator.INFINITE。

2. setRepeatMode()
   设置动画回放的方式，一种是从头播放（ObjectAnimator.RESTART），一种是反向从最后往前拨（ObjectAnimator.REVERSE）。

3. setStartDelay()
   当运行 start()之后，要延迟多少时间（以千分之一秒为单位）才开始播放动画。

4. setInterpolator()
   设置物体状态的变化率和时间的关系，例如固定的变化率（linear interpolator），或是变化率会随着时间变快（accelerate interpolator），Android 提供许多不同的 interpolator 模式让我们选用，请读者参考表 23-1。

表 23-1　各种 interpolator 模式

Interpolator 名称	说明
LinearInterpolator	播放动画时，物体状态的变化速度为固定
AccelerateInterpolator	播放动画时，物体状态的变化速度越来越快
DecelerateInterpolator	播放动画时，物体状态的变化速度越来越慢
AccelerateDecelerateInterpolator	播放动画时，物体状态的变化速度从慢变快，再从快变慢，也就是在中间的时候变化最快

续表

Interpolator 名称	说明
AnticipateInterpolator	播放动画时,物体状态的变化会先往反方向,再开始依照设置的方式改变,就像是选手起跑前会先往后退,再往前冲刺
OvershootInterpolator	动画播放到最后的时候会冲过头,然后再回到终点
AnticipateOvershootInterpolator	结合 AnticipateInterpolator 和 OvershootInterpolator 两种模式的特点
BounceInterpolator	动画播放到终点的时候会反弹,就像是皮球掉到地板一样,反弹几次后才会停下来
CycleInterpolator	可以设置动画重复播放的次数
TimeInterpolator	可以让我们自行设置物体状态变化的方式

接下来就让我们实际建立一个程序项目来测试 Property animation 的效果吧!

23-2 范例程序

我们将建立一个套用在 TextView 组件的动画程序,程序的运行界面如图 23-2 所示,其中有三个按钮分别用来展示旋转、改变 Alpha 值和移动位置三种不同型态的动画效果,请读者依照下列步骤操作以完成此程序项目:

图 23-2　Property animation 范例程序的执行画面

Step 1 运行 Eclipse 新增一个 Android 程序项目,项目的所有设置依照惯例即可。

Step 2 开启程序项目的 res/layout 文件夹中的接口布局文件,编辑其中的内容如下:

```
<?xml version="1.0" encoding="utf-8"?>
<LinearLayout xmlns:android="http://schemas.android.com/apk/res/android"
```

```xml
    android:id="@+id/linLay"
    android:orientation="vertical"
    android:layout_width="match_parent"
    android:layout_height="match_parent"
    android:gravity="center_horizontal">
<LinearLayout
    android:orientation="horizontal"
    android:layout_width="match_parent"
    android:layout_height="wrap_content"
    android:gravity="center_horizontal">
<Button android:id="@+id/btnRotateAnim"
    android:layout_width="wrap_content"
    android:layout_height="wrap_content"
    android:text="旋转文字"/>
<Button android:id="@+id/btnAlphaAnim"
    android:layout_width="wrap_content"
    android:layout_height="wrap_content"
    android:text="透明文字"/>
<Button android:id="@+id/btnFallAnim"
    android:layout_width="wrap_content"
    android:layout_height="wrap_content"
    android:text="文字往下掉"
    />
</LinearLayout>
<LinearLayout
    android:orientation="vertical"
    android:layout_width="match_parent"
    android:layout_height="match_parent"
    android:gravity="center">
<TextView android:id="@+id/txt"
    android:layout_width="wrap_content"
    android:layout_height="wrap_content"
    android:text="@string/hello"/>
</LinearLayout>
</LinearLayout>
```

我们利用两层 LinearLayout 的架构（最外层的 LinearLayout 里面包含两个 LinearLayout）先让三个 Button 做水平排列，然后让 TextView 位于 Button 的下方并且水平垂直置中对齐（使用 android:gravity 属性）。另外我们用到一个 id 名称叫做 hello 的字符串，它定义在 res/values/strings.xml 资源文件中如下：

```xml
<?xml version="1.0" encoding="utf-8"?>
<resources>
    <string name="hello">示范 Property Animation</string>
    <string name="app_name">Property Animation 范例程序</string>
</resources>
```

Step 3 开启程序项目的"src/(组件名称)"文件夹中的程序文件，然后将其中的程序代码编辑如下：

```
package …

import …

public class Main extends Activity {

    private LinearLayout mLinLay;
    private TextView mTxt;
    private Button mBtnFallAnim,
            mBtnAlphaAnim,
            mBtnRotateAnim;
    private float y, yEnd;
    private boolean mIsFallingFirst = true;

    /** Called when the activity is first created. */
    @Override
    public void onCreate(Bundle savedInstanceState) {
        super.onCreate(savedInstanceState);
        setContentView(R.layout.main);

        setupViewComponent();
    }

    private void setupViewComponent() {
        mLinLay = (LinearLayout)findViewById(R.id.linLay);
        mTxt = (TextView)findViewById(R.id.txt);
        mBtnFallAnim = (Button)findViewById(R.id.btnFallAnim);
        mBtnAlphaAnim = (Button)findViewById(R.id.btnAlphaAnim);
        mBtnRotateAnim = (Button)findViewById(R.id.btnRotateAnim);

        mBtnFallAnim.setOnClickListener(btnFallAnimOnClickLis);
        mBtnAlphaAnim.setOnClickListener(btnAlphaAnimOnClickLis);
        mBtnRotateAnim.setOnClickListener(btnRotateAnimOnClickLis);
    }

    private Button.OnClickListener btnRotateAnimOnClickLis = new
                            Button.OnClickListener() {
        public void onClick(View v) {
            ObjectAnimator animTxtRotate =
                ObjectAnimator.ofFloat(mTxt, "rotation", 0, 360);
            animTxtRotate.setDuration(3000);
            animTxtRotate.setRepeatCount(1);
            animTxtRotate.setRepeatMode
              (ObjectAnimator.REVERSE);
```

```
                    animTxtRotate.setInterpolator
                     (new   AccelerateDecelerateInterpolator());
                    animTxtRotate.start();
            }
        };

        private Button.OnClickListener btnAlphaAnimOnClickLis = new
                                Button.OnClickListener() {
            public void onClick(View v) {
                    ObjectAnimator animTxtAlpha =
                            ObjectAnimator.ofFloat(mTxt, "alpha", 1, 0);
                    animTxtAlpha.setDuration(2000);
                    animTxtAlpha.setRepeatCount(1);
                    animTxtAlpha.setRepeatMode(ObjectAnimator.REVERSE);
                    animTxtAlpha.setInterpolator(new   LinearInterpolator());
                    animTxtAlpha.start();
            }
        };

        private Button.OnClickListener btnFallAnimOnClickLis = new
                                Button.OnClickListener() {
            public void onClick(View v) {
                    if (mIsFallingFirst) {
                        // 计算掉落的 y 坐标
                        y = mTxt.getY();
                        yEnd = mLinLay.getHeight() - mTxt.getHeight();

                        mIsFallingFirst = false;
                    }

                    ObjectAnimator animTxtFalling =
                            ObjectAnimator.ofFloat(mTxt, "y", y, yEnd);
                    animTxtFalling.setDuration(2000);
                    animTxtFalling.setRepeatCount(ObjectAnimator.INFINITE);
                    animTxtFalling.setInterpolator(new   BounceInterpolator());
                    animTxtFalling.start();
            }
        };
    }
```

依照前面的惯例，我们自行建立一个 setupViewComponent() 方法负责设置好所有的见面对象以及它们的事件 Listener。在三个按钮的 OnClickListener 中，我们利用上一个小节介绍的 Property animation 让 TextView 对象（对象名称叫做 mTxt）完成旋转、改变透明度和往下掉落的动画效果。

读者完成此程序项目后便可以运行程序测试动画效果，例如当按下"旋转文字"按钮时，便会看到界面中央的字符串开始旋转，如图 23-3 所示。

图 23-3　按下"旋转文字"按钮时看到的字符串旋转效果

UNIT 24
Property Animation 加上 Listener 成为动画超人

Android 版本	1.X	2.X	3.X	4.X
适用性			★	★

前一个单元我们只是让 Property Animation 小试身手，在这个单元我们将见识它的真正威力，首先登场的是 AnimatorSet 对象，它的功能就像是动画的指挥家，专门负责安排动画的播放顺序，接下来我们将介绍动画事件的 Listener，以及 ValueAnimator 的用法。

24-1 使用 AnimatorSet

在上一个单元中我们已经学会使用 ObjectAnimator 对象建立单一动画，可是如果我们已经建立好多个 ObjectAnimator 动画，要如何让它们依序播放，甚至是同时播放呢？答案是使用 AnimatorSet 对象。使用 AnimatorSet 的方式很简单，它提供以下方法来设置动画的播放顺序：

1. play(动画对象)
   设置要播放的动画对象，设置好之后必须调用 start()方法才会开始播放。但其实这个方法的重点不是在设置播放的动画，而是它会传回一个 AnimatorSet.Builder 对象，这个对象可以让我们进一步设置动画对象之间的播放顺序，它提供以下的方法：
   a. before(动画对象)
      把指定的动画对象放在后面播放。
   b. after(动画对象)
      把指定的动画对象放在前面播放。
   c. with(动画对象)

同时播放指定的动画对象。

读者光看以上的说明一定觉得似懂非懂，但是只要配合以下的范例就会恍然大悟，不过要先提醒读者，这里我们使用的是 Java 程序常见的"匿名对象"语法格式，也就是说当调用一个方法后，如果它传回一个对象，我们可以直接在该方法后面继续调用它的传回对象的方法，请参考以下范例就可明白，假设 A、B、C 是三个已经建立好的 ObjectAnimator 动画对象：

```
AnimatorSet animSet = new AnimatorSet();
animSet.play(A).before(B);
animSet.play(B).before(C);
animSet.start();
```

则播放动画的顺序为 ABC，或者我们也可以写成：

```
animSet.play(B).after(A);
animSet.play(B).before(C);
```

同样也是设置 ABC 的播放顺序，也就是说 before()和 after()方法可以随意搭配使用。至于 with()请参考以下范例：

```
animSet.play(B).after(A);
animSet.play(B).with(C);
```

会先播放 A，然后再同时播放 B 和 C，利用 before()、after()和 with()这三个方法，我们就可以任意设置动画的播放顺序。

> **补充说明　Java 程序常用的匿名对象语法**
>
> Java 程序经常使用类似以下的程序代码：
> object.method1(…).method2(…).method3(…);
> 这其实是利用匿名对象的方式让程序代码更简洁的一种写法，它的意思是先运行 object.method1(…)，它会传回一个对象，接着再调用该对象的 method2(…)，然后又传回另一个对象，最后再调用该对象的 method3(…)。

2. playSequentially(动画对象, 动画对象, …)
   依照列出的动画对象顺序播放。

3. playTogether(动画对象, 动画对象, …)
   同时播放列出的动画对象。

4. start()
   开始播放动画。

以上一个单元建立的三个动画 animTxtRotate、animTxtAlpha 和 animTxtFalling 为例，如果要让它们依序播放可以使用下列程序代码：

```
animSet.play(animTxtRotate).before(animTxtAlpha);
animSet.play(animTxtAlpha).before(animTxtFalling);
```

或是直接使用 playSequentially()方法写成：

```
animSet.playSequentially(animTxtRotate, animTxtAlph, animTxtFalling);
```

或者我们可以先同时播放前两个动画，再播放最后一个动画如下：

```
animSet.play(animTxtRotate).with(animTxtAlpha);
animSet.play(animTxtFalling).after(animTxtAlpha);
```

24-2 加上动画事件 listener

如果从"事件"的观点来看播放动画这件事，从动画开始播放到结束的过程中会经历许多事件，包含开始（start）、界面更新（update）、结束（end）、重复（repeat）和取消（cancel）。有些时候我们希望当动画正在播放时，在某个时间点或是某些情况下运行特定的工作，例如当动画结束或是使用者取消动画时，让对象回到原来的状态，或是当动画对象移动到某个位置或角度，就运行特定的工作。要得知这些动画事件，可以利用 ObjectAnimator 的 addListener()和 addUpdateListener()方法加上事件 listener（就像是帮 Button 对象加上 OnClickListener 一样），这样当发生所指定的事件时，就会运行其中的程序代码，以下我们分别说明 addListener()和 addUpdateListener()的用法：

1. addListener()

    addListener()的第一种用法如下：

    ```java
    ObjectAnimator.addListener(new AnimatorListener(){
        @Override
        public void onAnimationCancel(Animator animation) {
            // TODO Auto-generated method stub

        }

        @Override
        public void onAnimationEnd(Animator animation) {
            // TODO Auto-generated method stub

        }

        @Override
        public void onAnimationRepeat(Animator animation) {
            // TODO Auto-generated method stub

        }

        @Override
    ```

```
        public void onAnimationStart(Animator animation) {
            // TODO Auto-generated method stub

        }});
```

这里提醒读者可以利用之前学过的程序代码编辑技巧，先输入"animScreenBackColor.addListener(new AnimatorListener(){});"，再根据程序编辑器的语法修正提示功能自动加入以上四个方法。

现在回到上面的程序代码，AnimatorListener 是一个接口，规定我们要完成四个方法，这四个方法就是当动画开始播放到结束的过程中所经历的 start、end、repeat 和 cancel 四种事件，只要将程序代码写在这四个方法中，它们就会在该时间点运行。但是有时候我们不一定会用到全部四种事件，因此还有一种比较简短的格式如下：

ObjectAnimator.addListener(new **AnimatorListenerAdapter**(){});

也就是换成使用 AnimatorListenerAdapter 抽象类，这样我们就可以只写出需要的事件即可，例如以下范例：

```
ObjectAnimator.addListener(new AnimatorListenerAdapter(){
    @Override
    public void onAnimationEnd(Animator animation) {
        // TODO Auto-generated method stub
        super.onAnimationEnd(animation);

        // 以下加上自己的程序代码
}});
```

这里再提示另一个编辑程序代码的技巧，请读者先输入"ObjectAnimator 对象.addListener(new AnimatorListenerAdapter(){});"，然后在"AnimatorListenerAdapter"单字上右击，就会出现图 24-1 的对话框，在对话框的左边会列出 AnimatorListenerAdapter 类中可以加上的方法，读者可以发现这些方法和前面 AnimatorListener 界面所使用的方法一样，但是我们可以只挑选自己需要的即可，这样可以减少程序代码的复杂度。

2. addUpdateListener()

这个方法的使用方式就简单多了，它是利用 AnimatorUpdateListener 接口加入 update 事件的 listener 如下，在 onAnimationUpdate()方法中的程序代码是当每一次动画界面要更新时会被运行。

```
ObjectAnimator 物件.addUpdateListener(new AnimatorUpdateListener(){
    @Override
    public void onAnimationUpdate(ValueAnimator animation) {
        // TODO Auto-generated method stub
```

```
            // 以下加上自己的程序代码
        }
    });
```

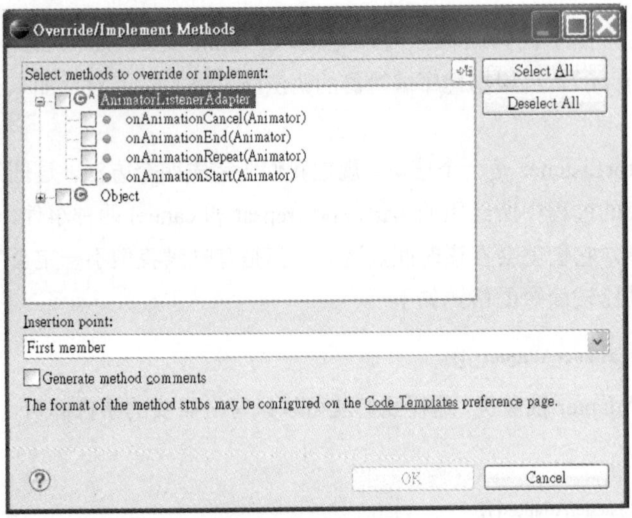

图 24-1 新增"预设类方法"的对话框

利用以上介绍的两个动画事件 listener，我们就可以在动画运行的过程中做出许多变化，这个单元的实现程序将会有完整的范例，但是现在我们还要再介绍另一个 Property animation 也会用到的动画对象 ValueAnimator。

24-3　ValueAnimator

在建立 ObjectAnimator 对象时我们会指定动画套用的对象和它的属性，可是在少数的情况下可能不适合使用这种方式，像是要改变物体或是背景的颜色。由于计算机是用一个整数来代表颜色，其中高字节代表红色强度，中字节代表绿色强度，低字节代表蓝色强度，如果我们想要根据改变红色强度来显示颜色动画，那么颜色整数值的变化就不会是连续的，因为我们只能改变高字节的部分，中字节和低字节的值都必须固定。这种情况就无法使用 ObjectAnimator 直接建立动画，而必须改成使用 ValueAnimator，配合动画事件 listener 来达成。建立 ValueAnimator 对象时我们只需要设置动画参数的起始值和结束值，然后如同上一单元的介绍，设置动画的 duration、interpolator 和 repeat 模式，请参考以下范例：

```
ValueAnimator animVal = ValueAnimator.ofInt(0, 100);
animVal.setDuration(3000);
animVal.setInterpolator(new LinearInterpolator());
animVal.start();
animVal.addUpdateListener(new AnimatorUpdateListener(){
```

```
@Override
public void onAnimationUpdate(ValueAnimator animation) {
    // TODO Auto-generated method stub

    // 显示每一个动画界面，我们可以取得目前界面的参数数值如下
    int val = (Integer)animation.getAnimatedValue();
}
});
```

也就是说 ValueAnimator 只是帮我们计算每一个动画界面所对应的参数值，这个参数值会根据 animation 自变量传进来，然后我们的程序代码便根据这个参数值，自行决定动画物体的真正属性值，并将它绘制在程序界面上。

24-4　范例程序

我们接续上一单元的程序项目，继续加上"放大文字"、"左右移动文字"和"改变背景颜色"三种动画效果，如图 24-2 所示，请读者依照以下步骤操作：

图 24-2　新增三个动画效果后的程序界面

Step 1　开启程序项目的 res/layout 文件夹中的接口布局文件，加上三个 Button 组件如下：

```
<?xml version="1.0" encoding="utf-8"?>
<LinearLayout …
    >
<LinearLayout
    …
    >
…
<Button android:id="@+id/btnScaleAnim"
```

```xml
        android:layout_width="wrap_content"
        android:layout_height="wrap_content"
        android:text="放大文字"
        />
<Button android:id="@+id/btnMoveAnim"
        android:layout_width="wrap_content"
        android:layout_height="wrap_content"
        android:text="文字左右移动"
        />
<Button android:id="@+id/btnBackColorAnim"
        android:layout_width="wrap_content"
        android:layout_height="wrap_content"
        android:text="改变背景颜色"
        />
</LinearLayout>
<LinearLayout
    …
    >
…
</LinearLayout>
</LinearLayout>
```

Step 2 开启程序项目的"src/(组件名称)"文件夹中的程序文件,在setupViewComponent()方法中设置好新增的三个按钮对象和它们的OnClickListener,如下粗体字标示的部分:

```java
package …

import …

public class Main extends Activity {

    private LinearLayout mLinLay;
    private TextView mTxt;
    private Button mBtnFallAnim,
                   mBtnAlphaAnim,
                   mBtnRotateAnim,
                   mBtnMoveAnim,
                   mBtnScaleAnim,
                   mBtnBackColorAnim;

    …

    private void setupViewComponent() {
        mLinLay = (LinearLayout)findViewById(R.id.linLay);
        mTxt = (TextView)findViewById(R.id.txt);
        mBtnFallAnim = (Button)findViewById(R.id.btnFallAnim);
        mBtnAlphaAnim = (Button)findViewById(R.id.btnAlphaAnim);
        mBtnRotateAnim = (Button)findViewById(R.id.btnRotateAnim);
```

```java
        mBtnMoveAnim = (Button)findViewById(R.id.btnMoveAnim);
        mBtnScaleAnim = (Button)findViewById(R.id.btnScaleAnim);
        mBtnBackColorAnim = (Button)findViewById(R.id.btnBackColorAnim);

        mBtnFallAnim.setOnClickListener(btnFallAnimOnClickLis);
        mBtnAlphaAnim.setOnClickListener(btnAlphaAnimOnClickLis);
        mBtnRotateAnim.setOnClickListener(btnRotateAnimOnClickLis);
        mBtnMoveAnim.setOnClickListener(btnMoveAnimOnClickLis);
        mBtnScaleAnim.setOnClickListener(btnScaleAnimOnClickLis);
        mBtnBackColorAnim.setOnClickListener(btnBackColorAnimOnClickLis);
    }

    ...

    private Button.OnClickListener btnScaleAnimOnClickLis = new
                        Button.OnClickListener() {
        public void onClick(View v) {
            ValueAnimator animTxtScale = ValueAnimator.ofInt(0, 50);
                animTxtScale.setDuration(4000);
                animTxtScale.setRepeatCount(1);
                animTxtScale.setRepeatMode(ObjectAnimator.REVERSE);
                animTxtScale.setInterpolator(new LinearInterpolator());
                animTxtScale.addUpdateListener(new AnimatorUpdateListener(){
                    @Override
                    public void onAnimationUpdate(ValueAnimator animation) {
                    // TODO Auto-generated method stub
                    int val = (Integer)animation.getAnimatedValue();
                        mTxt.setTextSize(TypedValue.COMPLEX_UNIT_SP, 15+val);
                    }
            });
            animTxtScale.start();
        }
    };

    private Button.OnClickListener btnMoveAnimOnClickLis = new
                        Button.OnClickListener() {
        public void onClick(View v) {
                float x, xEnd1, xEnd2;

                x = mTxt.getX();
                xEnd1 = 0;
                xEnd2 = mLinLay.getWidth() - mTxt.getWidth();

                ObjectAnimator animTxtMove1 =
                    ObjectAnimator.ofFloat(mTxt, "x", x, xEnd1);
                animTxtMove1.setDuration(2000);
                animTxtMove1.setInterpolator(new
                        AccelerateDecelerateInterpolator());
```

```java
                ObjectAnimator animTxtMove2 =
                ObjectAnimator.ofFloat(mTxt, "x", xEnd1, xEnd2);
                animTxtMove2.setDuration(3000);
                animTxtMove2.setInterpolator(new
                AccelerateDecelerateInterpolator());

                ObjectAnimator animTxtMove3 =
                    ObjectAnimator.ofFloat(mTxt, "x", xEnd2, x);
                animTxtMove3.setDuration(2000);
                animTxtMove3.setInterpolator(new
                AccelerateDecelerateInterpolator());

                AnimatorSet animTxtMove = new AnimatorSet();
                animTxtMove.playSequentially(animTxtMove1,
                animTxtMove2, animTxtMove3);
                animTxtMove.start();
            }
        };

        private Button.OnClickListener btnBackColorAnimOnClickLis = new
                        Button.OnClickListener() {
            public void onClick(View v) {
                int iBackColorRedVal, iBackColorRedEnd;
                final int iBackColor =
                        ((ColorDrawable)(mLinLay.getBackground())).
                        getColor();
                iBackColorRedVal = (iBackColor & 0x00FF0000) >> 16;

                if (iBackColorRedVal > 127)
                        iBackColorRedEnd = 0;
                else
                        iBackColorRedEnd = 255;

                ValueAnimator animScreenBackColor =
                        ValueAnimator.ofInt(iBackColorRedVal,
                        iBackColorRedEnd);
                animScreenBackColor.setDuration(3000);
                animScreenBackColor.setInterpolator(new
                LinearInterpolator());
                animScreenBackColor.start();
                animScreenBackColor.addUpdateListener(new
                AnimatorUpdateListener(){
                    @Override
                    public void onAnimationUpdate(ValueAnimator
                    animation) {
                        // TODO Auto-generated method stub
                        int val = (Integer)animation.getAnimatedValue();
                        mLinLay.setBackgroundColor(iBackColor &
```

```
                    0xFF00FFFF | val << 16);
                }
            });
        }
    };
}
```

"放大文字"的过程是利用上一小节介绍的 ValueAnimator，因为文字大小包括数值和单位两个部分，所以必须根据调用 setTextSize()方法来完成。我们利用 ValueAnimator 控制文字大小改变的范围、duration、repeat 和 interpolator 模式，然后设置 update listener，再利用 setTextSize()来改变文字的大小。

"左右移动文字"的方式是先建立三个动画对象，再利用 AnimatorSet 依序播放这三段动画。"改变背景颜色"则是如同上一小节的讨论只改变红色的强度，因此我们利用 bit 模式的运算符，也就是&和|，来取得和设置整数变量中的特定字节，然后利用 ValueAnimator 加上设置 update listener 来完成变换背景颜色的动画效果。

完成以上修改后就可以运行程序并测试新的动画效果，图 24-3 是运行"放大文字"的界面，另外读者还可以试看看连续按下多个按钮会得到什么结果。

图 24-3 运行"放大文字"的界面

PART 5　Fragment 与进阶接口组件

UNIT 25　使用 Fragment 让程序接口一分为多
UNIT 26　动态 Fragment 让程序成为变形金刚
UNIT 27　Fragment 的进阶用法
UNIT 28　Fragment 和 Activity 之间的 Callback 机制
UNIT 29　ListView 和 ExpandableListView
UNIT 30　AutoCompleteTextView 自动完成文字输入
UNIT 31　SeekBar 和 RatingBar 接口组件

Android 版本	1.X	2.X	3.X	4.X
适用性			★	★

UNIT 25

使用 Fragment 让程序界面一分为多

在设计手机或是平板电脑的应用程序时,经常面临的一个问题就是不同设备的屏幕尺寸大小不一致。对于小型移动设备像是手机来说,程序界面能够显示的接口组件个数不多,因此如果需要用到比较多的操作组件,就要将操作接口分成多个界面依序显示,但是如果程序是运行在屏幕比较大的移动设备像是平板电脑,就可以一次显示全部的接口组件,因此最好的方式是提供一个接口设计机制,让程序能够依照运行环境自动调整操作接口的显示方式。

Android 系统针对以上的问题从 3.0 版开始新增一个名为 Fragment 的类,利用这个新类我们可以将程序的界面分割成数个区域,这些不同区域的程序界面可以各自显示或隐藏,以适应不同屏幕尺寸的设备,这种 Fragment 类型的程序接口具有以下特性:

1. 程序的运行界面可以由多个 Fragment 组成。
2. 每一个 Fragment 都有各自独立的运行状态,并且接收各自的处理事件。
3. 在程序运行的过程中,Fragment 可以动态加入和移除。

图 25-1 是平板电脑模拟器中的 Settings 系统设置程序,它就是使用 Fragment 型态的操作接口,左边的 Fragment 负责显示选项类,右边的 Fragment 则显示类中的项目。

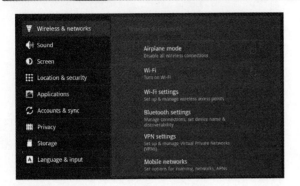

图 25-1　使用 Fragment 类型的程序接口

25-1　使用 Fragment 的步骤

要在程序中使用 Fragment 需要完成以下几个步骤：

Step 1　在程序项目中新增一个继承自 Fragment 的新类。

Step 2　在步骤一的新类中加上需要处理的状态转换方法，例如：

1. onCreate()

   当 Fragment 刚被建立时会运行这个方法，例如我们可以在这个方法中完成变量的初始设置。

2. onCreateView()

   当 Fragment 将要显示在屏幕上时会运行这个方法，我们必须在这个方法中设置好 Fragment 所使用的接口。

3. onPause()

   当 Fragment 要从屏幕上消失时会运行这个方法，我们可以在这个方法中储存用户的操作状态和资料，以便下次 Fragment 重新显示在屏幕上时，用户可以继续之前的工作。

除了以上介绍的状态转换方法之外，还有许多其他不同的方法，读者可以依照 UNIT 5 中的说明查询 SDK 说明文件以取得更多相关数据。

Step 3　建立 Fragment 使用的界面布局文件，Fragment 的操作接口同样是由程序项目中的 res/layout 文件夹下的接口布局文件来定义，只是我们必须在 Fragment 的 onCreateView() 方法中进行接口的设置而不是在 onCreate() 中进行，请读者参考以下范例：

```
public class MyFragment extends Fragment {

    @Override
    public View onCreateView(LayoutInflater inflater, ViewGroup container,
```

```
            Bundle savedInstanceState) {
        // TODO Auto-generated method stub
        return inflater.inflate(R.layout.fragment_layout,
            container, false);
    }

}
```

我们利用 inflater 对象的 inflate()方法取得 res/layout/fragment_layout.xml 接口布局文件，并将最后的结果传回给系统，这样就完成了 Fragment 的接口设置。

Step 4 把前面建立好的 Fragment 类加入程序中，我们可以利用<fragment>标签，在主程序的接口布局文件中完成加入 Fragment 的动作如以下范例：

```xml
<?xml version="1.0" encoding="utf-8"?>
<LinearLayout xmlns:android="http://schemas.android.com/apk/res/android"
    android:orientation="horizontal"
    android:layout_width="match_parent"
    android:layout_height="match_parent"
    >
    <fragment android:id="@+id/frag"
        android:name="tw.android.MyFragment"
        android:layout_width="0dp"
        android:layout_height="wrap_content"
        android:layout_weight="1"
        />
</LinearLayout>
```

使用<fragment>标签时要注意以下几点：

1. fragment 的开头字母必须小写。
2. 每一个<fragment>标签都要设置 android:id 属性。
3. <fragment>标签的 android:name 属性是指定所使用的 Fragment 类，且必须加上完整的组件路径名称。
4. 在<fragment>标签中可以使用 android:layout_weight 属性，以设置比例的方式控制每一个 Fragment 所占的屏幕宽度，此时 android:layout_width 属性必须设置为"0dp"。

25-2 为 Fragment 加上外框并重设大小和位置

Fragment 本身并无法做出像图 25-1 右边的外框效果，如果要在 Fragment 外围加上框线，必须配合使用<FrameLayout>标签，也就是说在接口布局文件中用一个<FrameLayout>标签把需要加上外框的 Fragment 包覆起来，如以下范例：

```xml
<?xml version="1.0" encoding="utf-8"?>
<LinearLayout xmlns:android="http://schemas.android.com/apk/res/android"
```

```
        android:orientation="horizontal"
        android:layout_width="match_parent"
        android:layout_height="match_parent"
        >
<FrameLayout android:layout_weight="1"
    android:layout_width="0dp"
    android:layout_height="wrap_content"
    android:background="?android:attr/detailsElementBackground"
    >
<fragment android:id="@+id/frag"
        android:name="tw.android.MyFragment"
        android:layout_width="match_parent"
        android:layout_height="wrap_content"
        />
</FrameLayout>
</LinearLayout>
```

而且必须指定 android:background 属性为"?android:attr/detailsElementBackground"，并且把 android:layout_weight 和 android:layout_width 属性的设置复制到<FrameLayout>标签中，最后把 fragment 的 android:layout_width 属性设置成"match_parent"。如果要控制 Fragment 在屏幕上显示的大小和位置，例如让它出现在屏幕的中央并且在四周留下空间以增加美观和操作的便利性，可以设置 padding 相关属性，接下来我们用一个实际的程序项目来示范使用 Fragment 的完整过程。

25-3　范例程序

本单元的范例程序是将前面单元完成的"计算机猜拳游戏"程序加上局数统计的功能，并且利用 Fragment 来显示。程序中将建立两个 Fragment，一个用来当成游戏的主界面，一个则是用来显示局数统计数字，程序的运行界面如图 25-2 所示，以下我们逐步列出完成此程序项目的详细过程。

图 25-2　使用 Fragment 接口的"计算机猜拳游戏"程序

Step 1 运行 Eclipse 新增一个 Android 程序项目，项目的设置值依照我们之前的惯例即可。

Step 2 建立游戏程序主界面的 Fragment 的接口布局文件。请读者在 Eclipse 左边的项目检查窗口中单击"此程序项目名称/res/layout"，然后用鼠标右击 layout 文件夹，再选择 New>File 就会出现图 25-3 的对话框，然后在对话框中输入 game.xml（也就是接口布局文件文件名），注意档名只能用小写的英文字母、数字或是底线字符，而且扩展名必须是 xml。完成后结束对话框，该文件就会自动开启在程序代码编辑窗口中。在程序代码编辑窗口下方有两个 Tab 标签页可以切换编辑模式，请读者选择右边的文本模式并从 UNIT 17 的"计算机猜拳游戏"程序项目中复制 res/layout/main.xml 接口布局文件的全部内容到此文件中。

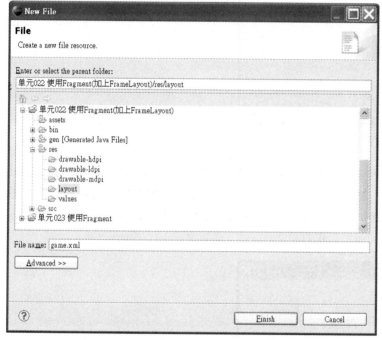

图 25-3 新增接口布局文件对话框

Step 3 建立一个继承自 Fragment 的新类，这个新类将用来显示游戏程序的主界面。请读者在 Eclipse 左边的项目检查窗口中展开此程序项目的"src/(组件路径名称)"文件夹，然后在"组件路径名称"上右击，在弹出的菜单中选择 New>Class，就会看到图 25-4 的对话框。在对话框中的 Name 字段输入 GameFragment（也就是类名称），然后按下在 Superclass 字段右边的 Browse 按钮，在出现的对话框的最上面的字段输入 Fragment，下方列表就会显示 Fragment 类，用鼠标双击该类就会自动回到原来的对话框，并完成填入 Superclass 字段，最后单击 Finish 按钮就会在 Eclipse 中央的程序代码窗口中就会出现 GameFragment.java 程序文件的编辑界面。

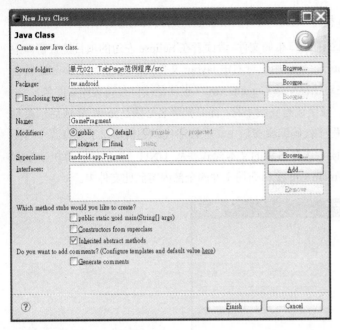

图 25-4　新增 Class 对话框

Step 4 请在程序代码编辑窗口中右击，在出现的快捷菜单中选择 Source > Override/Implement Methods…，就会出现图 25-5 的对话框。在对话框左边的清单中会列出 Fragment 类中还没有使用的方法，请读者勾选其中的 onActivityCreated()和 onCreateView()两个方法，然后单击 OK 按钮，这样就完成加入我们需要的方法，然后将程序代码编辑如下：

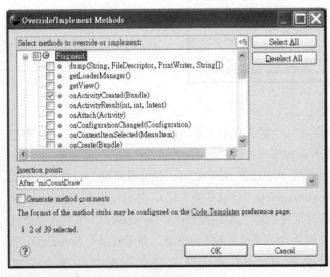

图 25-5　选择加入类方法的对话框

> **补充说明** **Eclipse 操作技巧**
>
> 学习 Android 程序的过程中我们会遇到越来越多的 class，每一个 class 都有许多内定的方法，我们必须视程序的需要 override 其中的某些方法，像是本单元范例所使用的 Fragment 类中的 onActivityCreated()和 onCreateView()。当要 override 覆盖方法时，先将文字输入光标移到要插入方法的位置，再右击选择快捷菜单中的 Source > Override/Implement Methods 就会显示方法列表对话框。如果程序中有不同对象的程序代码，改变文字输入光标的位置可以叫出不同的方法清单对话框。

```
package …

import …

public class GameFragment extends Fragment {

    private Button mBtnScissors,
                   mBtnStone,
                   mBtnNet;
    private TextView mTxtComPlay,
                     mTxtResult;
    private TextView mEdtCountSet,
                     mEdtCountPlayerWin,
                     mEdtCountComWin,
                     mEdtCountDraw;
    private int miCountSet = 0,
                miCountPlayerWin = 0,
                miCountComWin = 0,
                miCountDraw = 0;

    @Override
    public View onCreateView(LayoutInflater inflater, ViewGroup container,
            Bundle savedInstanceState) {
        // TODO Auto-generated method stub
        return inflater.inflate(R.layout.game, container, false);
    }

    @Override
    public void onActivityCreated(Bundle savedInstanceState) {
        // TODO Auto-generated method stub
        super.onActivityCreated(savedInstanceState);

        setupViewComponent();
    }

    private void setupViewComponent() {
mTxtComPlay = (TextView)getView().findViewById(R.id.txtComPlay);
```

```java
        mTxtResult = (TextView)getView().findViewById(R.id.txtResult);
        mBtnScissors = (Button)getView().findViewById(R.id.btnScissors);
        mBtnStone = (Button)getView().findViewById(R.id.btnStone);
        mBtnNet = (Button)getView().findViewById(R.id.btnNet);

        mEdtCountSet = (EditText)getActivity().findViewById(R.id.edtCountSet);
        mEdtCountPlayerWin = (EditText)getActivity().findViewById(R.id.edtCountPlayerWin);
        mEdtCountComWin = (EditText)getActivity().findViewById(R.id.edtCountComWin);
        mEdtCountDraw = (EditText)getActivity().findViewById(R.id.edtCountDraw);

        mBtnScissors.setOnClickListener(btnScissorsLin);
        mBtnStone.setOnClickListener(btnStoneLin);
        mBtnNet.setOnClickListener(btnNetLin);
    }

    private Button.OnClickListener btnScissorsLin = new Button.
OnClickListener() {
        public void onClick(View v) {
            // Decide computer play.
            int iComPlay = (int)(Math.random()*3 + 1);

            miCountSet++;
            mEdtCountSet.setText(new Integer(miCountSet).
toString());

            // 1 - scissors, 2 - stone, 3 - net.
            if (iComPlay == 1) {
                mTxtComPlay.setText(R.string.playScissors);
                mTxtResult.setText(getString(R.string.result) +
                                    getString(R.string.playerDraw));
                miCountDraw++;
                mEdtCountDraw.setText(new Integer(miCountDraw).
toString());
            }
            else if (iComPlay == 2) {
                mTxtComPlay.setText(R.string.playStone);
                mTxtResult.setText(getString(R.string.result) +
                                    getString(R.string.playerLose));
                miCountComWin++;
                mEdtCountComWin.setText
(new Integer(miCountComWin).toString());
            }
            else {
                mTxtComPlay.setText(R.string.playNet);
                mTxtResult.setText(getString(R.string.result) +
                                    getString(R.string.playerWin));
                miCountPlayerWin++;
                mEdtCountPlayerWin.setText(new
```

```
                    Integer(miCountPlayerWin).toString());
            }
        }
    };

    private Button.OnClickListener btnStoneLin = new Button.
    OnClickListener() {
        public void onClick(View v) {
            // 类似前面 btnScissorsLin 对象中的程序代码，只需要修
            改输赢的判断规则
                ...
        }
    };

    private Button.OnClickListener btnNetLin = new Button.
    OnClickListener() {
        public void onClick(View v) {
            // 类似前面 btnScissorsLin 对象中的程序代码，只需要修改输赢的
            判断规则
                ...
        };
}
```

在 onCreateView()方法中我们调用 inflater 对象的 inflate()方法取得 res/layout/game.xml 接口布局文件，并将最后的结果传回给系统。在 onActivityCreated()方法中我们调用自定义的 setupViewComponent()方法设置好所有的接口组件，有关设置接口组件的程序代码都和原来"计算机猜拳游戏"程序类似，除了在取得 Fragment 自己的接口组件时要加上 getView()方法，和在取得其他 Fragment 的接口组件时要加上 getActivity()方法（如以上用粗体标示的程序代码），因此读者可以复制原来的程序代码再加以修改。另外为了统计游戏的局数和输赢的结果，在程序中增加了四个变量（以底线标示）以及相关的程序代码。

Step 5 请读者仿照步骤二的方法新增一个接口布局文件 res/layout/game_result.xml，它将用来显示局数统计数据，该文件的内容如下：

```xml
<?xml version="1.0" encoding="utf-8"?>
<LinearLayout xmlns:android="http://schemas.android.com/apk/res/android"
    android:orientation="vertical"
    android:layout_width="match_parent"
    android:layout_height="match_parent"
    >
<TextView
    android:layout_width="match_parent"
    android:layout_height="wrap_content"
    android:text="@string/promptCountSet"
    />
<EditText android:id="@+id/edtCountSet"
    android:layout_width="match_parent"
```

```xml
        android:layout_height="wrap_content"
        android:editable="false"
        android:text=""
        />
    <TextView
        android:layout_width="match_parent"
        android:layout_height="wrap_content"
        android:text="@string/promptPlayerWin"
        />
    <EditText android:id="@+id/edtCountPlayerWin"
        android:layout_width="match_parent"
        android:layout_height="wrap_content"
        android:editable="false"
        android:text=""
        />
    <TextView
        android:layout_width="match_parent"
        android:layout_height="wrap_content"
        android:text="@string/promptComWin"
        />
    <EditText android:id="@+id/edtCountComWin"
        android:layout_width="match_parent"
        android:layout_height="wrap_content"
        android:editable="false"
        android:text=""
        />
    <TextView
        android:layout_width="match_parent"
        android:layout_height="wrap_content"
        android:text="@string/promptDraw"
        />
    <EditText android:id="@+id/edtCountDraw"
        android:layout_width="match_parent"
        android:layout_height="wrap_content"
        android:editable="false"
        android:text=""
        />
</LinearLayout>
```

Step 6 仿照步骤三和步骤四新增一个继承自 Fragment 的新类，此类的名称叫做 GameResultFragment，它的程序代码如下：

```java
package …

import …

public class GameResultFragment extends Fragment {

    @Override
    public View onCreateView(LayoutInflater inflater, ViewGroup container,
```

```
        Bundle savedInstanceState) {
    // TODO Auto-generated method stub
    return inflater.inflate(R.layout.game_result, container, false);
}
```

}

其中只有一个 onCreateView() 方法用来设置显示的接口，其他有关显示局数统计数据的程序代码都在游戏主程序的 Fragment 中完成。

Step 7 在 Eclipse 左边的项目检查窗口中找到此程序项目的主要接口描述文件 res/layout/main.xml，用鼠标双击将它开启编辑成如下内容：

```xml
<?xml version="1.0" encoding="utf-8"?>
<LinearLayout xmlns:android="http://schemas.android.com/apk/res/android"
    android:orientation="horizontal"
    android:layout_width="match_parent"
    android:layout_height="match_parent"
    android:paddingLeft="100dp"
    android:paddingRight="100dp"
    android:paddingTop="50dp"
    >
<fragment android:id="@+id/fragGame"
    android:name="tw.android.GameFragment"
    android:layout_width="0dp"
    android:layout_height="match_parent"
    android:layout_weight="1"
    />
<FrameLayout android:layout_weight="1"
    android:layout_width="0dp"
android:layout_height="wrap_content"
    android:background="?android:attr/detailsElementBackground"
    >
<fragment android:id="@+id/fragGameResult"
    android:name="tw.android.GameResultFragment"
    android:layout_width="match_parent"
    android:layout_height="wrap_content"
    />
</FrameLayout>
</LinearLayout>
```

其中我们使用 <FrameLayout> 标签在右边 Fragment 的周围加上外框，并且利用 padding 相关属性调整程序界面的位置以增加操作的便利性。

在这个单元中我们第一次学到 Fragment 界面的用法，为了不要让问题太过复杂，我们先以静态的方式来使用 Fragment，在下一个单元我们将进一步学习如何在程序运行的过程中动态加入和移除 Fragment。

Android 版本	1.X	2.X	3.X	4.X
适用性			★	★

UNIT 26

动态 Fragment 让程序成为变形金刚

在前一个单元我们介绍了 Fragment 的几个特性，其中比较特别的是在程序运行过程中可以动态加入或是移除 Fragment，或是用一个新的 Fragment 取代原来的 Fragment。利用这个功能，我们可以让程序的操作接口依照用户的喜好，或是运行环境的限制做出各种变化。想要在程序运行的过程中动态控制 Fragment 需要使用 FragmentManager，接下来就让我们来介绍本单元的主角——FragmentManager。

26-1 Fragment 的总管——FragmentManager

要在程序中改变 Fragment 必须借助 FragmentManager 对象，FragmentManager 对象提供以下三个控制 Fragment 的方法：

1. add()
   把 Fragment 加入程序的操作接口。

2. remove()
   从程序的操作接口中移除指定的 Fragment。

3. replace()
   用新的 Fragment 取代目前程序界面中的 Fragment。

改变程序操作界面的 Fragment 的整个过程称为一个 Transaction（中文的意思是交易），

Transaction 的特性是整个过程必须从头到尾全部完成才可以，不能够只运行其中一部份，如果不能够从头到为全部完成，就要回到原来未运行 Transaction 的状态。在使用 FragmentManager 控制 Fragment 的过程中必须依序运行以下几个步骤：

Step 1 调用 getFragmentManager()方法取得 FragmentManager：

> FragmentManager fragMgr = getFragmentManager();

Step 2 调用 FragmentManager 的 beginTransaction()方法建立一个 FragmentTransaction 开始记录我们要对 Fragment 进行的操作：

> FragmentTransaction fragTran = fragMgr.beginTransaction();

Step 3 利用 FragmentTransaction 的 add()、remove()和 replace()等方法控制 Fragment，请读者参考以下的程序代码范例：

> MyFragment myFrag = new MyFragment();
> MyFragment2 myFrag2 = new MyFragment2();
>
> **fragTran.add(R.id.frameLayout, myFrag, "My fragment"); fragTran.replace(R.id.frameLayout, myFrag, "My fragment 2");**
>
> MyFragment2 frag = (MyFragment2)fragMgr.findFragmentByTag("My fragment 2");
> **fragTran.remove(frag);**

在调用 add()方法之前，我们必须先建立好要加入程序界面的 Fragment 对象。调用 add()方法时必须指定要将 Fragment 加入哪一个接口组件中，另外我们也可以帮这个新加入的 Fragment 取一个名称（称为 Tag，就是 add()方法的第三个参数），这个 Tag 可以让我们在后续的运行过程中找到这个新加入的 Fragment。

replace()方法的功能是用新的 Fragment 取代指定接口组件中的 Fragment，如果指定接口组件中没有 Fragment，则效果和 add()方法相同。remove()方法是从接口组件中移除指定的 Fragment。

Step 4 调用 FragmentTransaction 的 commit()方法送出以上建立的 Fragment 处理流程，系统会根据这个处理流程来更新 Fragment。

> fragTran.commit();

我们可以利用之前介绍过的"匿名对象"语法，将上述加入 Fragment 对象的程序代码简化成一行如下：

> getFragmentManager().beginTransaction().add(R.id.frameLayout, myFrag, "My fragment").commit();

26-2　范例程序

在上一个单元中我们使用两个 Fragment 来完成"计算机猜拳游戏"程序的操作接口，其中一个 Fragment 用来显示用户的操作组件，另一个则是用来显示局数统计数据，这两个 Fragment 都是固定出现在程序界面中，无法将它们隐藏。在这个单元中我们将把局数统计数据的 Fragment 改成可以动态显示和隐藏，并且新增第二种局数统计数据的显示格式，让用户可以依喜好自由选择，图 26-1 是新的程序运行界面，以下是完成这个程序项目的详细步骤：

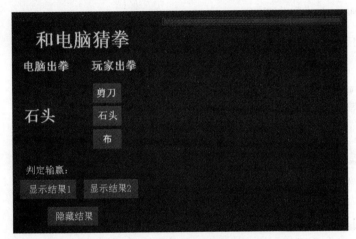

图 26-1　使用动态 Fragment 的"计算机猜拳游戏"程序

Step 1　利用 Windows 文件管理器复制前一个单元的程序项目文件夹，复制后可以变更文件夹的名称，然后运行 Eclipse 利用主菜单的 File>Import 功能加载复制的程序项目。

Step 2　依照前一个单元中的方法，在新加载的程序项目的 res/layout 文件夹中新增一个接口布局文件，我们可以将它取名为 game_result2.xml，并将它开启编辑成如下内容，读者可以从原来的 game_result.xml 文件中复制程序代码到这个新文件再进行修改，这个新的接口布局文件是使用 TableLayout 的方式显示局数统计数据。

```xml
<?xml version="1.0" encoding="utf-8"?>
<TableLayout xmlns:android="http://schemas.android.com/apk/res/android"
    android:layout_width="match_parent"
    android:layout_height="match_parent"
    android:layout_gravity="center_horizontal"
    >
<TableRow>
    <TextView android:text="@string/promptCountSet"
        android:textSize="20sp"
        android:textColor="#0FFFFF"
```

```xml
            />
        <EditText android:id="@+id/edtCountSet"
            android:editable="false"
            android:text=""
            android:layout_weight="1"
            />
</TableRow>
<TableRow>
        <TextView android:text="@string/promptPlayerWin"
            android:textSize="20sp"
            android:textColor="#0FFFFF"
            />
        <EditText android:id="@+id/edtCountPlayerWin"
            android:editable="false"
            android:text=""
            android:layout_weight="1"
            />
</TableRow>
<TableRow>
        <TextView android:text="@string/promptComWin"
            android:textSize="20sp"
            android:textColor="#0FFFFF"
            />
        <EditText android:id="@+id/edtCountComWin"
            android:editable="false"
            android:text=""
            android:layout_weight="1"
            />
</TableRow>
<TableRow>
        <TextView android:text="@string/promptDraw"
            android:textSize="20sp"
            android:textColor="#0FFFFF"
            />
        <EditText android:id="@+id/edtCountDraw"
            android:editable="false"
            android:text=""
            android:layout_weight="1"
            />
</TableRow>
</TableLayout>
```

Step 3 仿照之前的操作方法新增一个继承自 Fragment 的新类，我们可以将它取名为 GameResultFragment2，然后将它开启并编辑如下，这个新类是使用步骤二所建立的接口布局文件。

```java
package ...

import ...

public class GameResultFragment2 extends Fragment {

    @Override
    public View onCreateView(LayoutInflater inflater,
                    ViewGroup container,
                    Bundle savedInstanceState) {
        // TODO Auto-generated method stub
        return inflater.inflate(R.layout.game_result2,
            container, false);
    }

    @Override
    public void onResume() {
        // TODO Auto-generated method stub
        super.onResume();

        GameFragment frag =
                (GameFragment)getFragmentManager().
                    findFragmentById(R.id.fragGame);
        frag.mEdtCountSet = (EditText)getActivity().
            findViewById(R.id.edtCountSet);
        frag.mEdtCountPlayerWin =
            (EditText)getActivity().findViewById
            (R.id.edtCountPlayerWin);
        frag.mEdtCountComWin =
            (EditText)getActivity().findViewById
            (R.id.edtCountComWin);
        frag.mEdtCountDraw = (EditText)getActivity().
            findViewById(R.id.edtCountDraw);
    }
}
```

以上的程序代码中除了 onCreateView()方法之外，还多了一个 onResume()方法（在程序代码中插入新方法的操作步骤请参考前面单元中的"补充说明"），这个方法中的程序代码是设置用来显示局数统计数据的接口对象，这些接口对象是定义在 GameFragment 类中，并且必须等使用者开启局数统计的 Fragment 之后才能使用（未开启局数统计 Fragment 之前这些接口对象并不存在，因此也无法使用）。

当用户建立一个 GameResultFragment2 对象，并将它加入程序界面时，onCreateView()方法会先运行以完成设置界面的工作，当 GameResultFragment2 对象即将显示在屏幕上开始运行时会运行 onResume()方法，因此我们在 onResume()方法中设置好所有的接口对象。这里我们使用 FragmentManager 的 findFragmentById()方法取得程序中的 GameFragment

对象，再将其中定义的接口对象设置成对应的接口组件。

Step 4 在 Eclipse 左边的程序项目检查窗口中，找到此程序项目的 src/(组件路径名称)/GameResultFragment.java 文件，用鼠标双击将它开启，并依照步骤三的方法新增一个 onResume()方法如下，这一段程序代码的功能如同步骤三中的解释，也就是说 GameResultFragment 和 GameResultFragment2 类除了接口不同之外，内部的功能完全相同：

```
package …

import …

public class GameResultFragment extends Fragment {

    @Override
    public View onCreateView(LayoutInflater inflater,
                    ViewGroup container,
                    Bundle savedInstanceState) {
        …
    }

    @Override
    public void onResume() {
        // TODO Auto-generated method stub
        super.onResume();

        GameFragment frag =
                (GameFragment)getFragmentManager().
                findFragmentById(R.id.fragGame);
        frag.mEdtCountSet = (EditText)getActivity().
        findViewById(R.id.edtCountSet);
        frag.mEdtCountPlayerWin =
                (EditText)getActivity().findViewById
                (R.id.edtCountPlayerWin);
        frag.mEdtCountComWin =
                (EditText)getActivity().findViewById
                (R.id.edtCountComWin);
        frag.mEdtCountDraw = (EditText)getActivity().
        findViewById(R.id.edtCountDraw);
    }
}
```

Step 5 仿照步骤四的方法找到 res/layout/main.xml 文件，然后删除在<FrameLayout>标签中的<fragment>标签，并且在<FrameLayout>标签的属性中新增一个 android:id 如下粗体字的部分，因为我们将在程序代码中依照使用者的操作，把不同的 Fragment 放进这个

FrameLayout 组件中。

```xml
<?xml version="1.0" encoding="utf-8"?>
<LinearLayout xmlns:android="http://schemas.android.com/apk/res/android"
    ...
    >
<fragment …
    />
<FrameLayout android:id="@+id/frameLayout"
    android:layout_weight="1"
    android:layout_width="0px"
    android:layout_height="wrap_content"
    android:background="?android:attr/detailsElementBackground"
    />
</LinearLayout>
```

Step 6 开启 res/layout/game.xml 文件（游戏主界面操作接口文件），我们在文件的最后加入三个按钮，用户可以利用这些按钮启动两种局数统计资料界面或是隐藏局数统计资料界面。这里我们用到许多控制接口组件位置和外观的属性，以增加程序界面的美观和操作的便利性。

```xml
<?xml version="1.0" encoding="utf-8"?>
<RelativeLayout xmlns:android="http://schemas.android.com/apk/res/android"
    android:layout_width="400dp"
    android:layout_height="match_parent"
    android:layout_gravity="center_horizontal"
    >
...
<Button android:id="@+id/btnShowResult1"
    android:layout_width="wrap_content"
    android:layout_height="wrap_content"
    android:text="@string/btnShowResult1"
    android:layout_below="@id/txtResult"
    android:layout_alignLeft="@id/txtCom"
    android:textSize="20sp"
    android:layout_marginTop="10dp"
    />
<Button android:id="@+id/btnShowResult2"
    android:layout_width="wrap_content"
    android:layout_height="wrap_content"
    android:text="@string/btnShowResult2"
    android:layout_below="@id/txtResult"
    android:layout_alignLeft="@id/btnNet"
    android:textSize="20sp"
    android:layout_marginTop="10dp"
    />
<Button android:id="@+id/btnHiddenResult"
```

```
        android:layout_width="wrap_content"
        android:layout_height="wrap_content"
        android:text="@string/btnHiddenResult"
        android:layout_below="@id/btnShowResult1"
        android:layout_centerInParent="true"
        android:textSize="20sp"
        android:layout_marginTop="10dp"
        />
</RelativeLayout>
```

Step 7 开启 src/(组件路径名称)/GameFragment.java 文件，这是游戏主程序文件，我们在里头加入前一个步骤中新增的三个按钮的对应对象和 OnClickListener，如以下粗体字的部分：

```
package …

import …

public class GameFragment extends Fragment {

    private Button mBtnScissors,
                   mBtnStone,
                   mBtnNet,
                   mBtnShowResult1,
                   mBtnShowResult2,
                   mBtnHiddenResult;
    private boolean mbShowResult = false;

    private final static String TAG_FRAGMENT_RESULT_1 = "Result 1",
                                TAG_FRAGMENT_RESULT_2 = "Result 2";

    …


    @Override
    public View onCreateView(LayoutInflater inflater,
                             ViewGroup container,
                             Bundle savedInstanceState) {
        …
    }

    @Override
    public void onActivityCreated(Bundle savedInstanceState) {
        …
    }

    private void setupViewComponent() {
        …
        mBtnShowResult1 = (Button)getView().findViewById
```

```java
            (R.id.btnShowResult1);
        mBtnShowResult2 = (Button)getView().findViewById
            (R.id.btnShowResult2);
        mBtnHiddenResult = (Button)getView().findViewById
            (R.id.btnHiddenResult);

        mBtnShowResult1.setOnClickListener(btnShowResult1Lin);
        mBtnShowResult2.setOnClickListener(btnShowResult2Lin);
        mBtnHiddenResult.setOnClickListener(btnHiddenResultLin);
    }

    private Button.OnClickListener btnShowResult1Lin = new Button.OnClickListener() {
        public void onClick(View v) {
            GameResultFragment fragGameResult = new
                GameResultFragment();
            FragmentTransaction fragTran = getFragmentManager().
                beginTransaction();
            fragTran.replace(R.id.frameLayout, fragGameResult,
                    TAG_FRAGMENT_RESULT_1);
            fragTran.commit();

            mbShowResult = true;
        }
    };

    private Button.OnClickListener btnShowResult2Lin = new Button.OnClickListener() {
        public void onClick(View v) {
                GameResultFragment2 fragGameResult2 = new
                    GameResultFragment2();
            FragmentTransaction fragTran = getFragmentManager().
            beginTransaction();
            fragTran.replace(R.id.frameLayout, fragGameResult2,
                    TAG_FRAGMENT_RESULT_2);
            fragTran.commit();

            mbShowResult = true;
        }
    };

    private Button.OnClickListener btnHiddenResultLin = new
    Button.OnClickListener() {
        public void onClick(View v) {
            mbShowResult = false;

            FragmentManager fragMgr = getFragmentManager();
```

```java
            GameResultFragment fragGameResult = (GameResultFragment)
                        fragMgr.findFragmentByTag(TAG_FRAGMENT_RESULT_1);
            if (null != fragGameResult) {
                FragmentTransaction fragTran = fragMgr.beginTransaction();
                fragTran.remove(fragGameResult);
                fragTran.commit();

                return;
            }

            GameResultFragment2 fragGameResult2 = (GameResultFragment2)
                        fragMgr.findFragmentByTag(TAG_FRAGMENT_RESULT_2);
            if (null != fragGameResult2) {
                FragmentTransaction fragTran = fragMgr.beginTransaction();
                fragTran.remove(fragGameResult2);
                fragTran.commit();

                return;
            }
        }
    };

    private Button.OnClickListener btnScissorsLin = new Button.
OnClickListener() {
        public void onClick(View v) {
            // Decide computer play.
            int iComPlay = (int)(Math.random()*3 + 1);

            miCountSet++;
//          mEdtCountSet.setText(new Integer(miCountSet).toString());

            // 1 - scissors, 2 - stone, 3 - net.
            if (iComPlay == 1) {
            mTxtComPlay.setText(R.string.playScissors);
            mTxtResult.setText(getString(R.string.result) +
                            getString(R.string.playerDraw));
            miCountDraw++;
//          mEdtCountDraw.setText(new Integer(miCountDraw).
            toString());
            }
            else if (iComPlay == 2) {
            mTxtComPlay.setText(R.string.playStone);
            mTxtResult.setText(getString(R.string.result) +
                            getString(R.string.playerLose));
            miCountComWin++;
//          mEdtCountComWin.setText(new Integer(miCountComWin).
            toString());
            }
```

```java
                else {
                    mTxtComPlay.setText(R.string.playNet);
                    mTxtResult.setText(getString(R.string.result) +
                                    getString(R.string.playerWin));
                    miCountPlayerWin++;
//                  mEdtCountPlayerWin.setText(new
//                              Integer(miCountPlayerWin).toString());
                }

                if (mbShowResult) {
                    mEdtCountSet.setText(new Integer(miCountSet).toString());
                    mEdtCountDraw.setText(new Integer(miCountDraw).toString());
                    mEdtCountComWin.setText(new Integer(miCountComWin).toString());
                    mEdtCountPlayerWin.setText(
                                    new Integer(miCountPlayerWin).toString());
                }
            }
        };

        private Button.OnClickListener btnStoneLin = new Button.OnClickListener() {
            public void onClick(View v) {
                // 参考前面 btnScissorsLin 对象的程序代码修改
                …
            }
        };

        private Button.OnClickListener btnNetLin = new Button.OnClickListener() {
            public void onClick(View v) {
                // 参考前面 btnScissorsLin 对象的程序代码修改
                …
            }
        };
    }
```

此外为了记录程序是否开启局数统计的显示界面，我们新增一个 mbShowResult 变量，并在三个按钮的 OnClickListener 中对它进行设置。另外在这些按钮的 OnClickListener 中就是依照上一个小节介绍的 Fragment 操作方法分别显示或隐藏局数统计界面。最后我们在"剪刀"、"石头"和"布"的按钮的 OnClickListener 中将显示局数统计数据的程序代码集中放到最后，并当 mbShowResult 变数的值是 true 的时候再显示，请读者特别注意以粗体字标示的程序代码。

完成程序项目的修改后，我们可以将它启动运行，并测试两个显示不同局数统计界面的按钮，读者将会看到如图 26-2 和图 26-3 的结果，如果要取消局数统计界面，可以按下"隐藏结果"按钮。

这个程序项目示范了动态控制 Fragment 的基本方法，但是还有一些美中不足的地方，例如当没有显示局数统计界面时，程序右边会留下一个空的 FrameLayout，另外 Android 系统还针对 Fragment 的操作提供回复（back stack）的功能，这些进阶的 Fragment 操作技巧我们留待下一个单元再继续介绍。

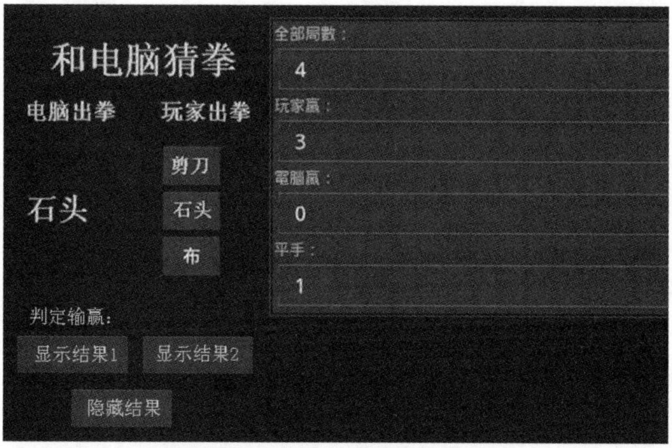

图 26-2　第一种显示局数统计数据的界面

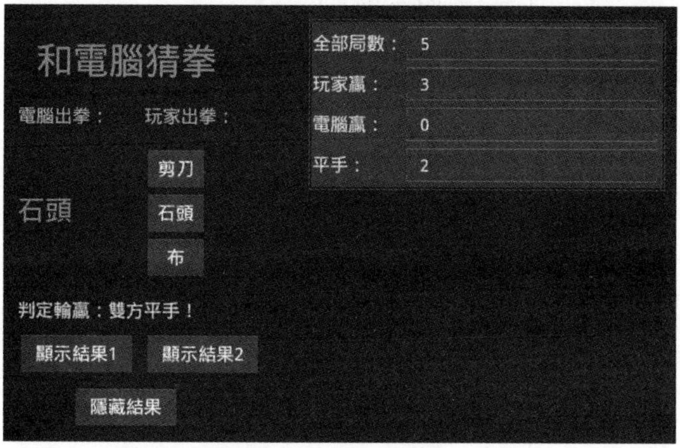

图 26-3　第二种显示局数统计数据的界面

Android 版本	1.X	2.X	3.X	4.X
适用性			★	★

UNIT 27
Fragment 的进阶用法

在使用 Fragment 的时候，如果能够配合运用控制接口组件的技巧以及更多 Fragment 的功能，就能够设计出界面更美观、操作更方便的程序接口。为了能够更完整地运用 Fragment，我们必须了解 Fragment 在运行过程中的各种状态转变阶段，请读者参考图 27-1，它是 Fragment 在运行过程中的状态转换过程（为了避免太过复杂，我们先省略因使用者切换 Activity 或 Fragment 所导致的状态转变）。在"计算机猜拳游戏"程序中我们已经使用了 onCreateView()、onActivityCreated() 和 onResume() 这三个状态转换方法，让负责游戏程序主界面的 GameFragment，以及两个动态加入的 GameResultFragment 和 GameResultFragment2 能够正确地运行，接下来要改善的是显示在程序右边的 FrameLayout，我们希望在没有显示局数统计数据时将它隐藏。

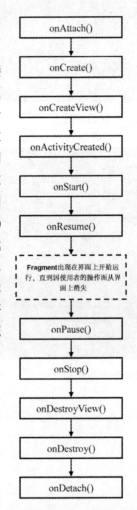

图 27-1 Fragment 从开始到结束所经历的状态变化

27-1 控制 FrameLayout 的显示和隐藏

我们希望程序刚开始运行的时候先不要显示界面右边的 FrameLayout，等到使用者启动局数统计界面时才让 FrameLayout 出现。设置 FrameLayout 的隐藏和显示只要利用 Visibility 属性，我们先在程序的接口描述文件中将这个属性设置为 gone，让 FrameLayout 先从屏幕上消失。接下来的问题是如何在程序运行的时候，让 FramemLayout 和 GameResultFragment 以及 GameResultFragment2 这两个接口同步出现和消失？这其实也不难，我们再回头检查一下图 27-1，当 GameResultFragment 以及 GameResultFragment2 即将出现在屏幕上的时候会运行 onResume()，当它们即将从屏幕上消失的时候会运行 onPause()，因此只要我们在 onResume() 方法中让 FramemLayout 出现，在 onPause() 方法中让 FramemLayout 消失就可以达到目的，详细的实现过程如下：

Step 1 运行 Eclipse 开启前一个单元的"计算机猜拳游戏"程序项目的接口布局文件 res/layout/main.xml，在<FrameLayout>标签中加入以下粗体字的属性：

```xml
<?xml version="1.0" encoding="utf-8"?>
<LinearLayout xmlns:android="http://schemas.android.com/apk/res/android"
    … >
<fragment android:id="@+id/fragGame"
    … />
<FrameLayout android:id="@+id/frameLayout"
    …
    android:visibility="gone"/>
</LinearLayout>
```

Step 2 开启程序文件 src/(组件路径名称)/GameResultFragment.java，在程序代码编辑窗口中右击，然从在出现的快选菜单中选择 Source > Override/Implement Methods…，在对话框左边的清单中勾选 onPause()再按下 OK 按钮，然后加入以下粗体字的程序代码：

```
package …

import …

public class GameResultFragment extends Fragment {

    @Override
    public View onCreateView(LayoutInflater inflater,
                    ViewGroup container,
                    Bundle savedInstanceState) {
        …
    }

    @Override
```

```java
        public void onResume() {
            // TODO Auto-generated method stub
            super.onResume();
            …(原来的程序代码)
            getActivity().findViewById(R.id.frameLayout).
            setVisibility(View.VISIBLE);
        }

        @Override
        public void onPause() {
            // TODO Auto-generated method stub
            super.onPause();
            getActivity().findViewById(R.id.frameLayout).
            setVisibility(View.GONE);
        }
    }
```

我们利用 getActivity() 方法取得 Fragment 所属的 Activity，再利用 Activity 的 findViewById() 取得 FrameLayout，然后设置它的 Visibility 属性，当设置为 View.VISIBLE 时会让 FrameLayout 出现在屏幕上，如果设置为 View. GONE 则会让 FrameLayout 隐藏，这里我们使用"匿名对象"的语法格式以缩短程序代码。

Step 3 仿照步骤二的方法修改程序文件 src/(此程序项目的组件名称) /GameResultFragment2.java 如下：

```java
package …

import …

public class GameResultFragment2 extends Fragment {

    @Override
    public View onCreateView(LayoutInflater inflater,
                    ViewGroup container,
                    Bundle savedInstanceState) {
        …
    }

    @Override
    public void onResume() {
        // TODO Auto-generated method stub
        super.onResume();
        …(原来的程序代码)
        getActivity().findViewById(R.id.frameLayout).
        setVisibility(View.VISIBLE);
    }
```

```
        @Override
        public void onPause() {
            // TODO Auto-generated method stub
            super.onPause();
            getActivity().findViewById(R.id.frameLayout).
            setVisibility(View.GONE);
        }
    }
```

完成以上修改之后运行程序就会发现原来在程序右边的 FrameLayout 已经消失, 如图 27-2 所示, 如果用户单击显示局数统计的按钮, FrameLayout 就会自动出现, 如果单击"隐藏结果"按钮, FrameLayout 就会再度消失。最后我们还要让程序能够记住每一次使用者对于 Fragment 所做的改变, 这样当使用按下平板电脑屏幕左下方的回复按钮时, 就可以回到前一个状态。

图 27-2　程序启动后原来显示在右边的 FrameLayout 已经消失

27-2　使用 Fragment 的 Back Stack 功能和动画效果

所谓 Back Stack, 就是指当使用者在操作某一个程序时又启动另一个程序, 在这种情况下, Android 系统会先暂停目前的程序, 然后切换到新的程序, 当用户按下"回上一页"按钮时, 就会回到前一个程序。对于程序和程序之间的切换, Android 系统会自动使用 Back Stack 功能, 但是对于同一个程序内的 Fragment 之间的切换, Android 系统并不会自动记录 Back Stack, 而是要由程序设计者自行处理。如果只是单纯记录 Fragment 切换的 Back Stack 并不难, 只要使用

FragmentTransaction 对象提供的 addToBackStack()方法即可（必须在 commit()方法之前调用）。但是使用 Back Stack 时还有一个需要考虑的问题是 Fragment 对象的生命周期（Lifecycle）会因此改变，这有可能影响到程序中对于 Fragment 对象的控管机制。

  当没有将 Fragment 加入 Back Stack 时，如果 Fragment 被移出界面（也就是从屏幕上消失）就会被系统删除，可是如果将 Fragment 加入 Back Stack 中，当 Fragment 被移出界面时并不会被删除而是处于停止状态，以便使用者按下回复键的时候能够重新显示在屏幕上。依照这个机制重新检查我们的"计算机猜拳游戏"程序，当用户按下"显示结果"的按钮时会加入一个新的 Fragment 对象并赋予它一个 Tag，只是这个 Tag 在下一次用户按下同一个按钮时又会使用相同的名称（也就是 TAG_FRAGMENT_RESULT_1 或 TAG_FRAGMENT_RESULT_2），因此造成程序中会有名称重复的 Fragment（如果没有使用 Back Stack，则从界面消失的 Fragment 会被系统删除，所以没有名称重复的问题）。为了让程序支持 Back Stack 功能，我们需要修改两个部分：

1. 将 Fragment 对象的 Tag 名称改成使用流水号，也就是依序从 1 开始往后编号，让每一个 Fragment 都有不同的名称。
2. 原来使用 mbShowResult 属性控制程序是否需要显示局数统计数据，该属性是在程序的"显示结果"和"隐藏结果"按钮中进行设置，这种方式当用户换成用系统的回复键进行切换时就会出现问题，因此我们改成检查 FrameLayout 组件的显示或隐藏状态，来决定是否需要显示局数统计资料。

  另外，FragmentTransaction 对象还提供 setTransition()方法可以设置 Fragment 切换的动画效果，它的使用方式很简单，请读者直接参考后面的程序范例就可以了解。

  以上的修改都是针对 GameFragment.java 程序文件，不会更动到其他程序文件和接口布局文件，以下我们列出需要更动的程序代码（以粗体字标示），程序的操作方式和运行界面都和前一个单元相同，只是新增对于系统"回上一页"按键的支持，读者在完成程序代码的修改之后可以进行测试就能够了解。

```
package ...

import ...

public class GameFragment extends Fragment {

    private Button mBtnScissors,
                   mBtnStone,
                   mBtnNet,
                   mBtnShowResult1,
                   mBtnShowResult2,
                   mBtnHiddenResult;
//  public boolean mbShowResult = false;
    private TextView mTxtComPlay,
                     mTxtResult;
```

```java
        public TextView mEdtCountSet,
                  mEdtCountPlayerWin,
                  mEdtCountComWin,
                  mEdtCountDraw;
    private int miCountSet = 0,
              miCountPlayerWin = 0,
              miCountComWin = 0,
              miCountDraw = 0;

    private final static String TAG = "Result";
    private int mTagCount = 0;

    @Override
    public View onCreateView(LayoutInflater inflater, ViewGroup container,
                Bundle savedInstanceState) {
        ...
    }

    @Override
    public void onActivityCreated(Bundle savedInstanceState) {
        // TODO Auto-generated method stub
        super.onActivityCreated(savedInstanceState);

        setupViewComponent();
    }

    private void setupViewComponent() {
        ...
    }

    private Button.OnClickListener btnShowResult1Lin = new Button.OnClickListener() {
            public void onClick(View v) {
                GameResultFragment fragGameResult = new GameResultFragment();
                FragmentTransaction fragTran = getFragmentManager().beginTransaction();
                mTagCount++;
                String sFragTag = TAG + new Integer(mTagCount).toString();
                fragTran.replace(R.id.frameLayout, fragGameResult, sFragTag);
                fragTran.setTransition(FragmentTransaction.
                    TRANSIT_FRAGMENT_FADE);
                fragTran.addToBackStack(null);
                fragTran.commit();

//              mbShowResult = true;
            }
    };

    private Button.OnClickListener btnShowResult2Lin = new Button.
```

```java
            OnClickListener() {
                public void onClick(View v) {
                    GameResultFragment2 fragGameResult2 = new GameResultFragment2();
                    FragmentTransaction fragTran = getFragmentManager().
                    beginTransaction();
                    mTagCount++;
                    String sFragTag = TAG + new Integer(mTagCount).toString();
                    fragTran.replace(R.id.frameLayout, fragGameResult2, sFragTag);
                    fragTran.setTransition(FragmentTransaction.TRANSIT_FRAGMENT_FADE);
                    fragTran.addToBackStack(null);
                    fragTran.commit();
//                  mbShowResult = true;
                }
            };

            private Button.OnClickListener btnHiddenResultLin = new Button.
            OnClickListener() {
                public void onClick(View v) {
//                  mbShowResult = false;

                    FragmentManager fragMgr = getFragmentManager();
                    String sFragTag = TAG + new Integer(mTagCount).toString();
                    Fragment fragGameResult = (Fragment)fragMgr.findFragmentByTag
                      (sFragTag);
                    FragmentTransaction fragTran = fragMgr.beginTransaction();
                    fragTran.remove(fragGameResult);
                    fragTran.setTransition(FragmentTransaction.TRANSIT_FRAGMENT_FADE);
                    fragTran.addToBackStack(null);
                    fragTran.commit();
                }
            };

            private Button.OnClickListener btnScissorsLin = new Button.
            OnClickListener() {
                public void onClick(View v) {
                    // Decide computer play.
                    int iComPlay = (int)(Math.random()*3 + 1);

                    miCountSet++;

                    // 1 - scissors, 2 - stone, 3 - net.
                    if (iComPlay == 1) {
                        mTxtComPlay.setText(R.string.playScissors);
                        mTxtResult.setText(getString(R.string.result) +
                                    getString(R.string.playerDraw));
                        miCountDraw++;
                    }else if (iComPlay == 2) {
```

```java
                    mTxtComPlay.setText(R.string.playStone);
                    mTxtResult.setText(getString(R.string.result) +
                            getString(R.string.playerLose));
                    miCountComWin++;
                }else {
                    mTxtComPlay.setText(R.string.playNet);
                    mTxtResult.setText(getString(R.string.result) +
                            getString(R.string.playerWin));
                    miCountPlayerWin++;
                }

                if (getActivity().findViewById(R.id.frameLayout).isShown()) {
                    mEdtCountSet.setText(new Integer(miCountSet).
                    toString());
                        mEdtCountDraw.setText(new Integer(miCountDraw) .
                    toString());
                        mEdtCountComWin.setText(new Integer(miCountComWin).
                    toString());
                        mEdtCountPlayerWin.setText(new Integer
                    (miCountPlayerWin).toString());
                }
            }
        };

        private Button.OnClickListener btnStoneLin = new Button.OnClickListener() {
            public void onClick(View v) {
                // 参考前面 btnScissorsLin 对象的程序代码修改
                ...
            }
        };

        private Button.OnClickListener btnNetLin = new Button.OnClickListener() {
            public void onClick(View v) {
                // 参考前面 btnScissorsLin 对象的程序代码修改
                ...
            }
        };
    }
```

Android 版本	1.X	2.X	3.X	4.X
适用性			★	★

UNIT 28

Fragment 和 Activity 之间的 callback 机制

截至目前为止我们已经用过许多 callback 方法，这些 callback 方法是由我们的程序提供给 Android 系统调用。在这个单元我们将建立由我们自己程序使用的 callback 机制。其实 callback 的概念很简单，就是两个函数或是两个对象为了要共同完成一件工作，其中被动的一方（以下称为 A）必须提供一个特定的函式或方法让另一方（以下称为 B）调用，而 B 方是担任主动的角色，也就是工作的运行是由它来控制，B 方在过程中会视需要调用 A 方提供的 callback 函式，最后 A 和 B 一起合作将工作完成。使用 callback 的原因通常是基于程序架构的考虑，为了让比较大型的程序项目有良好的系统架构，我们必须将程序代码适当的切割成许多模块或是类，而 callback 就是让模块和类能够协同运行的一种机制。在这个单元中我们将从程序架构的观点重新检查"计算机猜拳游戏"程序，并运用 callback 机制加以改良。

28-1 检查"计算机猜拳游戏"程序架构

首先我们将"计算机猜拳游戏"程序的架构以图 28-1 来代表，图中列出了程序中的三个主要工作：

1. 累计游戏的局数统计资料。
2. 控制 Main 中的游戏局数界面的显示和隐藏。
3. 更新 GameResultFragment 或 GameResultFragment2 的游戏局数显示。

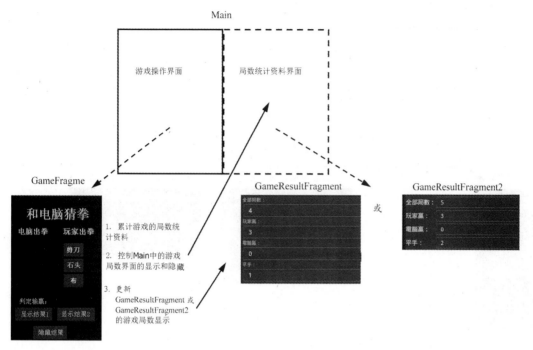

图 28-1 "计算机猜拳游戏"程序的架构

从图 28-1 中可以看出以上三个工作都是由 GameFragment 完成,其他类并没有负担任何工作,从程序架构的观点来考虑,这并不是很好的作法,原因如下:

1. 游戏局数界面是属于 Main 的组件,因此控制显示和隐藏的工作应该由 Main 负责,如果需要显示或隐藏游戏局数界面,GameFragment 应该向 Main 提出要求再由 Main 来运行。
2. 显示游戏局数数据的接口是属于 GameResultFragment 和 GameResultFragment2,GameFragment 应该只是负责提供数据数据,当 GameResultFragment 或 GameResultFragment2 收到数据后再将它们显示出来。
3. 就架构上来说,GameFragment、GameResultFragment 和 GameResultFragment2 都是属于 Main 的组件,因此 Main 应该担任 Controller 的角色,GameFragment 应该将数据传给 Main,然后 Main 再视情况将数据传给 GameResultFragment 或 GameResultFragment2。

上述考虑的出发点是希望程序中每一个对象的层级,应该和它所运行的工作一致,不应该有越俎代庖的情况,因此基于以上讨论,原来的程序运行流程应该修改成图 28-2 的方式。

比较图 28-1 和图 28-2 的运行流程,读者会发现修改后的版本反而变得比较复杂,虽然如此,可是从程序的运行架构来说,修改后的运行逻辑比较清楚,因为每个对象都扮演其应有的角色,如果将来程序需要增加功能,可以比较容易完成。如果用原来的运行架构,所有功能的程序代码将会集中于 GameFragment 而变得难以维护。

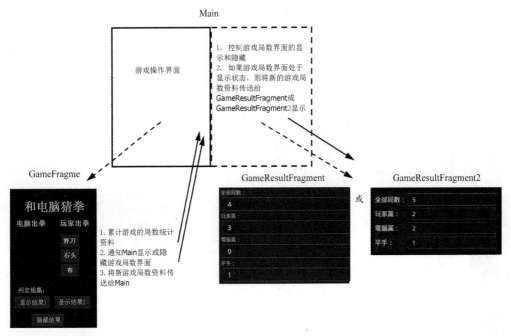

图 28-2 改良后的"计算机猜拳游戏"程序的架构

28-2 实现 Fragment 和 Activity 之间的 callback 机制

要让 GameFragment 能够将请求传送给 Main，我们可以使用 callback 的方式来达成，也就是在 Main 类中针对 GameFragment 的需要实现 callback 方法，再将该 callback 方法传给 GameFragment。建立这样的 callback 机制需要以下步骤：

Step 1 在 GameFragment 类中定义一个 Interface，其中包含所有需要由 Main 类提供的 callback 方法，例如：

```
package …

import …

public class GameFragment extends Fragment {

    // Main Activity 必须实现以下接口中的 callback 方法
    public interface CallbackInterface {
        public void method1(…);
        public void method2(…);
        …
    };
```

…
}

Step 2 让 Main 类实现步骤一所建立的 CallbackInterface。

```
package …

import …

public class Main extends Activity implements GameFragment.
CallbackInterface {

    public void method1(…) {
        …
    }

    public void method2(…) {
        …
    }
}
```

Step 3 将 Main 类中的 callback 方法传给 GameFragment，这个步骤可以在 GameFragment 的 onAttach()状态转换方法中完成，因为在 Android 系统调用 onAttach()方法的时候会传入程序的 Activity，也就是 Main 对象，我们在 GameFragment 类中定义一个 CallbackInterface 的对象，然后把系统传入的 Activity 转型成为 CallbackInterface 对象（因为我们在步骤二中已经让 Main 实现 CallbackInterface）并存入该 CallbackInterface 对象，这样在 GameFragment 类中就可以根据此 CallbackInterface 对象调用 Main 中的 callback 方法。

```
package …

import …

public class GameFragment extends Fragment {

    // Main Activity 必须实现以下接口中的 callback 方法
    public interface CallbackInterface {
        public void method1(…);
        public void method2(…);
        …
    };

    private CallbackInterface mCallback;

    public void onAttach(Activity activity) {
        // TODO Auto-generated method stub
        super.onAttach(activity);
```

```
            try {
                mCallback = (CallbackInterface) activity;
            } catch (ClassCastException e) {
                throw new ClassCastException(activity.toString() +
                        "must implement GameFragment.
                        CallbackInterface.");
            }
        }
        ...
    }
```

28-3 范例程序

接下来就让我们依照图 28-2 的架构修改"计算机猜拳游戏"程序，读者可以先复制前一个单元的程序项目文件夹，复制后可以变更文件夹名称，再运行 Eclipse 主菜单的 File > Import 加载该复制的程序项目进行修改。除了加入前一个小节所讨论的 callback 机制之外，还要针对相关的运行流程进行下列修改：

1. 由于 GameFragment 必须向 Main 提出显示或隐藏游戏局数界面的要求，为了区分所要运行的功能，我们在 GameFragment 中定义了一个 enum GameResultType 型别。
2. 当用户按下"显示结果"或"隐藏结果"按钮时，直接调用 callback 方法由 Main 负责处理。在用户按下任何一个出拳按钮后，同样调用 callback 方法将目前的游戏局数统计资料传给 Main，再由 Main 决定是否需要显示。
3. 原来在 GameFragment 中记录 GameResultFragment 和 GameResultFragment2 对象的流水号换成在 Main 中运行。
4. 在 GameResultFragment 和 GameResultFragment2 类中定义自己的接口对象并且在 onResume() 方法中设置好这些接口对象，除此之外也要在 onResume()方法中将 Main 对象内的局数统计界面对象（定义成 public）设置为目前这个对象。
5. 在 GameResultFragment 和 GameResultFragment2 类中新增一个 updateGameResult()方法供 Main 调用以更新局数统计资料。
6. 在 Main 类中定义记录 GameResultFragment 和 GameResultFragment2 对象流水号的属性（TAG 和 mTagCount），以及记录目前使用的局数统计界面对象（定义成 public）。
7. 在 updateGameResult()这个 callback 方法中，先检查目前是否显示局数统计数据界面，如果是，再根据局数统计资料界面的类型（GameResultFragment 或 GameResultFragment2）调用它的 updateGameResult()方法完成显示局数统计数据的工作。
8. 在 enableGameResult()这个 callback 方法中，则根据指定的局数统计界面类型，建立 GameResultFragment 或 GameResultFragment2 对象并设置它们的流水号，如果是要移除局数统计界面，则移除目前的对象。

在这个单元的最后我们列出修改后的程序代码，粗体字代表有更动的部分，没有修改的程序代码则视情况省略，读者完成修改之后可以启动程序并进行测试，操作接口完全没有改变，只是内部的运行方式已经不同。

其实如果针对这个"计算机猜拳游戏"程序来说，不使用 callback 也可以让 GameFragment 调用 Main 的方法，我们可以在 GameFragment 中调用 getActivity() 来取得 Main 对象，然后就可以调用 Main 的方法。这种方式和 callback 机制的差异在于利用 Interface 定义的 callback 带有强迫性，也就是说如果要使用 GameFragment，就要依照它的规定实现 callback 方法，例如当 GameFragment 是由别人所编写的程序代码时，程序作者就可以根据定义 Callback Interface 的方式要求其他使用者遵照它的规定来使用。如果是完全由自己开发的程序项目，就可以选择不使用 callback。

■ GameFragment.java 程序文件：

```java
package …

import …

public class GameFragment extends Fragment {

    // 所属的 Activity 必须实现以下接口中的 callback 方法
    public interface CallbackInterface {
        public void updateGameResult(int iCountSet,
                                     int iCountPlayerWin,
                                     int iCountComWin,
                                     int iCountDraw);
        public void enableGameResult(GameResultType type);
    };

    enum GameResultType {
        TYPE_1, TYPE_2, TURN_OFF;
    }

    private CallbackInterface mCallback;

    …(和原来程序代码相同)

/*      换成由 GameResultFragment 和 GameResultFragment2 自行控制
    public TextView mEdtCountSet,
                    mEdtCountPlayerWin,
                    mEdtCountComWin,
                    mEdtCountDraw;
*/
    …
//    private final static String TAG = "Result";
```

```java
//      private int mTagCount = 0;

        @Override
        public View onCreateView(LayoutInflater inflater, ViewGroup container,
                Bundle savedInstanceState) {
            …(和原来程序代码相同)
        }

        @Override
        public void onActivityCreated(Bundle savedInstanceState) {
            // TODO Auto-generated method stub
            …(和原来程序代码相同)
        }

        @Override
        public void onAttach(Activity activity) {
            // TODO Auto-generated method stub
            super.onAttach(activity);

            try {
                mCallback = (CallbackInterface) activity;
            } catch (ClassCastException e) {
                throw new ClassCastException(activity.toString() +
                        "must implement GameFragment.CallbackInterface.");
            }
        }

        private void setupViewComponent() {
            …(和原来程序代码相同)
        }

        private Button.OnClickListener btnShowResult1Lin = new Button.OnClickListener() {
            public void onClick(View v) {
                mCallback.enableGameResult(GameResultType.TYPE_1);
            }
        };

        private Button.OnClickListener btnShowResult2Lin = new Button.OnClickListener() {
            public void onClick(View v) {
                mCallback.enableGameResult(GameResultType.TYPE_2);
            }
        };

        private Button.OnClickListener btnHiddenResultLin = new Button.OnClickListener() {
            public void onClick(View v) {
                mCallback.enableGameResult(GameResultType.TURN_OFF);
            }
        };
```

```java
private Button.OnClickListener btnScissorsLin = new Button.OnClickListener() {
    public void onClick(View v) {
        // Decide computer play.
        int iComPlay = (int)(Math.random()*3 + 1);

        miCountSet++;

        // 1 - scissors, 2 - stone, 3 - net.
        if (iComPlay == 1) {
            mTxtComPlay.setText(R.string.playScissors);
            mTxtResult.setText(getString(R.string.result) +
                        getString(R.string.playerDraw));
            miCountDraw++;
        }
        else if (iComPlay == 2) {
            mTxtComPlay.setText(R.string.playStone);
            mTxtResult.setText(getString(R.string.result) +
                        getString(R.string.playerLose));
            miCountComWin++;
        }
        else {
            mTxtComPlay.setText(R.string.playNet);
            mTxtResult.setText(getString(R.string.result) +
                        getString(R.string.playerWin));
            miCountPlayerWin++;
        }

        mCallback.updateGameResult(miCountSet, miCountPlayerWin,
                miCountComWin, miCountDraw);
    }
};

private Button.OnClickListener btnStoneLin = new Button.OnClickListener() {
    public void onClick(View v) {
        // 參考前面 btnScissorsLin 對象的程序代碼修改
        …
    }
};

private Button.OnClickListener btnNetLin = new Button.OnClickListener() {
    public void onClick(View v) {
        // 參考前面 btnScissorsLin 對象的程序代碼修改
        …
    }
};
}
```

■ GameResultFragment.java 程序文件：

```java
package …

import …

public class GameResultFragment extends Fragment {

    private TextView mEdtCountSet,
                     mEdtCountPlayerWin,
                     mEdtCountComWin,
                     mEdtCountDraw;

    @Override
    public View onCreateView(LayoutInflater inflater, ViewGroup container,
            Bundle savedInstanceState) {
        // TODO Auto-generated method stub
        return inflater.inflate(R.layout.game_result, container, false);
    }

    @Override
    public void onResume() {
        // TODO Auto-generated method stub
        super.onResume();

        mEdtCountSet = (EditText)getActivity().findViewById(R.id.edtCountSet);
        mEdtCountPlayerWin = (EditText)getActivity().findViewById
            (R.id.edtCountPlayerWin);
        mEdtCountComWin = (EditText)getActivity().findViewById(R.id.edtCountComWin);
        mEdtCountDraw = (EditText)getActivity().findViewById(R.id.edtCountDraw);

        getActivity().findViewById(R.id.frameLayout).setVisibility
            (View.VISIBLE);
        ((Main)getActivity()).mGameResultType = GameResultType.TYPE_1;
        ((Main)getActivity()).fragResult = this;
    }

    @Override
    public void onPause() {
        // TODO Auto-generated method stub
        super.onPause();

        getActivity().findViewById(R.id.frameLayout).setVisibility(View.GONE);
    }

    public void updateGameResult(int iCountSet,
                                 int iCountPlayerWin,
                                 int iCountComWin,
```

```java
                            int iCountDraw) {
        mEdtCountSet.setText(new Integer(iCountSet).toString());
        mEdtCountDraw.setText(new Integer(iCountDraw).toString());
        mEdtCountComWin.setText(new Integer(iCountComWin).toString());
        mEdtCountPlayerWin.setText(new Integer(iCountPlayerWin).toString());
    }
}
```

- GameResultFragment2.java 程序文件：

```java
package …

import …

public class GameResultFragment2 extends Fragment {

    private TextView mEdtCountSet,
                     mEdtCountPlayerWin,
                     mEdtCountComWin,
                     mEdtCountDraw;

    @Override
    public View onCreateView(LayoutInflater inflater, ViewGroup container,
            Bundle savedInstanceState) {
        // TODO Auto-generated method stub
        return inflater.inflate(R.layout.game_result2, container, false);
    }

    @Override
    public void onResume() {
        // TODO Auto-generated method stub
        super.onResume();

        mEdtCountSet = (EditText)getActivity().findViewById(R.id.edtCountSet);
        mEdtCountPlayerWin = (EditText)getActivity().findViewById
                (R.id.edtCountPlayerWin);
        mEdtCountComWin = (EditText)getActivity().findViewById(R.id.edtCountComWin);
        mEdtCountDraw = (EditText)getActivity().findViewById(R.id.edtCountDraw);

        getActivity().findViewById(R.id.frameLayout).setVisibility(View.VISIBLE);
        ((Main)getActivity()).mGameResultType = GameResultType.TYPE_2;
        ((Main)getActivity()).fragResult = this;
    }

    @Override
    public void onPause() {
        // TODO Auto-generated method stub
        super.onPause();

        getActivity().findViewById(R.id.frameLayout).setVisibility(View.GONE);
```

```java
    }

    public void updateGameResult(int iCountSet,
                                 int iCountPlayerWin,
                                 int iCountComWin,
                                 int iCountDraw) {
        mEdtCountSet.setText(new Integer(iCountSet).toString());
        mEdtCountDraw.setText(new Integer(iCountDraw).toString());
        mEdtCountComWin.setText(new Integer(iCountComWin).toString());
        mEdtCountPlayerWin.setText(new Integer(iCountPlayerWin).toString());
    }
}
```

- **Main.java 程序文件：**

```java
package …

import …

public class Main extends Activity implements GameFragment.CallbackInterface {

    private final static String TAG = "Result";
    private int mTagCount = 0;
    public GameFragment.GameResultType mGameResultType;
    public Fragment fragResult;

    /** Called when the activity is first created. */
    @Override
    public void onCreate(Bundle savedInstanceState) {
        super.onCreate(savedInstanceState);
        setContentView(R.layout.main);
    }

    @Override
    public void updateGameResult(int iCountSet, int iCountPlayerWin,
            int iCountComWin, int iCountDraw) {
        // TODO Auto-generated method stub

        if (findViewById(R.id.frameLayout).isShown()) {
            switch (mGameResultType) {
            case TYPE_1:
                ((GameResultFragment)fragResult).updateGameResult
                    (iCountSet, iCountPlayerWin, iCountComWin, iCountDraw);
                break;
            case TYPE_2:
                ((GameResultFragment2)fragResult).updateGameResult
                    (iCountSet, iCountPlayerWin, iCountComWin, iCountDraw);
                break;
            }
        }
```

```java
    }

    @Override
    public void enableGameResult(GameResultType type) {
        // TODO Auto-generated method stub

        FragmentTransaction fragTran;
        String sFragTag;

        switch (type) {
        case TYPE_1:
            GameResultFragment frag = new GameResultFragment();
            fragTran = getFragmentManager().beginTransaction();
            mTagCount++;
            sFragTag = TAG + new Integer(mTagCount).toString();
            fragTran.replace(R.id.frameLayout, frag, sFragTag);
            fragTran.setTransition(FragmentTransaction.TRANSIT_
            FRAGMENT_FADE);
            fragTran.addToBackStack(null);
            fragTran.commit();
            break;
        case TYPE_2:
            GameResultFragment2 frag2 = new GameResultFragment2();
            fragTran = getFragmentManager().beginTransaction();
            mTagCount++;
            sFragTag = TAG + new Integer(mTagCount).toString();
            fragTran.replace(R.id.frameLayout, frag2, sFragTag);
            fragTran.setTransition(FragmentTransaction.TRANSIT_
            FRAGMENT_FADE);
            fragTran.addToBackStack(null);
            fragTran.commit();
            break;
        case TURN_OFF:
            FragmentManager fragMgr = getFragmentManager();
            sFragTag = TAG + new Integer(mTagCount).toString();
            Fragment fragGameResult = (Fragment)fragMgr.findFragmentByTag
            (sFragTag);
            fragTran = fragMgr.beginTransaction();
            fragTran.remove(fragGameResult);
            fragTran.setTransition(FragmentTransaction.TRANSIT_
            FRAGMENT_FADE);
            fragTran.addToBackStack(null);
            fragTran.commit();
            break;
        }
    }
}
```

Android 版本	1.X	2.X	3.X	4.X
适用性	★	★	★	★

UNIT 29

ListView 和 ExpandableListView

ListView 是一种列表形式的操作接口，如果用户单击平板电脑模拟器首页右上方的 Apps 按钮，然后选择应用程序界面中的 Settings 就会看到图 29-1 的操作接口，这个操作接口的左边就是 ListView。ListView 本身是一个接口组件，我们可以在程序的接口布局文件中加上 ListView 标签，然后利用前面学过的方法，在程序代码中取得该 ListView 组件并对它进行控制。只是这种方式比较麻烦，为了简化 ListView 的操作，我们可以另外采用一种专门用于 ListView 的 Activity 类，它的名称叫做 ListActivity，它提供比较简便的方式来完成清单的建立和控制。

图 29-1 此操作界面的左边就是 ListView 菜单

29-1 使用 ListActivity 建立 ListView 菜单

以下我们直接以范例程序来说明 ListActivity 类的使用步骤：

Step 1 请读者依照之前的方法建立一个新的程序项目。

Step 2 开启"src/(组件路径名称)"文件夹中的程序文件,把它所继承的基础类从 Activity 改成 ListActivity,如以下粗体字的部分:

```
package …

import …

public class Main extends ListActivity {
```

Step 3 开启 res/layout/main.xml 接口布局文件并编辑如下:

```
<?xml version="1.0" encoding="utf-8"?>
<LinearLayout xmlns:android="http://schemas.android.com/apk/res/android"
    android:orientation="horizontal"
    android:layout_width="match_parent"
    android:layout_height="match_parent"
    >
<LinearLayout
    android:orientation="vertical"
    android:layout_width="0dp"
    android:layout_height="match_parent"
    android:layout_weight="1"
    />
<LinearLayout
    android:orientation="vertical"
    android:layout_width="0dp"
    android:layout_height="match_parent"
    android:layout_weight="1"
    >
<TextView android:id="@+id/txtResult"
    android:layout_width="match_parent"
    android:layout_height="wrap_content"
    android:text=""
    />
<ListView android:id="@id/android:list"
        android:layout_width="match_parent"
        android:layout_height="match_parent"
        android:layout_weight="1"
        android:drawSelectorOnTop="false"
        />
</LinearLayout>
<LinearLayout
    android:orientation="vertical"
    android:layout_width="0dp"
    android:layout_height="match_parent"
    android:layout_weight="1"
```

```
        />
</LinearLayout>
```

在最外层的 LinearLayout 中我们建立了三个水平排列的 LinearLayout，并利用 android:layout_weight 属性（注意 android:layout_width 必须设置为"0dp"），设置每一个 LinearLayout 所占的宽度比例，让程序的操作接口置于屏幕中央。第一个 TextView 组件是用来显示用户单击的项目名称。第二个 ListView 组件是用来显示选项列表，请读者特别注意，它的 id 一定要和以上的范例相同不可以改变，因为 ListActivity 固定使用这个 id。

> **补充说明** 让程序的操作画面同时适用手机和平板电脑的技巧
>
> 本单元的范例程序利用二层 LinearLayout 配合设置 android:layout_weight 属性，让操作界面置于屏幕中央，这种方法适用在大屏幕的平板电脑，但是如果是在手机上运行，就直接使用一个 LinearLayout 即可。为了让程序能够同时适用不同屏幕尺寸的设备，可以参考 UNIT 58 中介绍的资源文件夹命名技术，分别设计应用于不同屏幕大小的接口布局文件。

Step 4 在程序代码中需要再一次用到 ArrayAdapter 这个泛型类，我们把要显示的选项列表存入一个 ArrayAdapter 型态的对象，然后调用 setListAdapter()方法将这个对象设置给 ListView。选项列表可以用数组的型态定义在程序代码中，或是定义在 res/values/strings.xml 文件中再从程序代码中加载使用。从日后程序代码的维护来考虑以后者为佳，因此以下的范例是使用第二种方法。

- res/values/strings.xml 文件：

```xml
<?xml version="1.0" encoding="utf-8"?>
<resources>
    …
    <string-array name="weekday">
        <item>星期日</item>
        <item>星期一</item>
        <item>星期二</item>
        <item>星期三</item>
        <item>星期四</item>
        <item>星期五</item>
        <item>星期六</item>
    </string-array>
</resources>
```

- 程序代码：

```
ArrayAdapter<CharSequence> adapWeekday = ArrayAdapter.createFromResource(
                    this, R.array.weekday, android.R.layout.simple_list_item_1);
setListAdapter(adapWeekday);
```

从程序代码中加载数组资源时，我们指定使用 Android SDK 提供的内置格式

android.R.layout.simple_list_item_1。

Step 5 在程序中必须建立一个 AdapterView.OnItemClickListener 的对象，当用户单击 ListView 中的某个项目后，便会运行该对象中的程序代码，以下范例是把用户单击的项目名称显示在程序界面的 TextView 元件中：

```
AdapterView.OnItemClickListener listviewOnItemClkLis = new
AdapterView.OnItemClickListener() {
        public void onItemClick(AdapterView<?> parent, View view,
            int position, long id) {
            // 在 TextView 组件中显示用户单击的项目名称
            mTxtResult.setText(((TextView) view).getText());
        }
};
```

Step 6 最后的工作是取得接口布局文件中定义的 ListView 组件，然后把步骤五建立的对象利用 setOnItemClickListener()方法设置给它。

```
ListView listview = getListView();
listview.setOnItemClickListener(listviewOnItemClkLis);
```

以下是完整的程序代码，启动程序并单击其中一个项目就会看到程序界面的上方显示所单击的项目名称，如图 29-2 所示。

图 29-2 范例程序的运行界面

```
package …

import …
```

```java
public class Main extends ListActivity {

    private TextView mTxtResult;

    /** Called when the activity is first created. */
    @Override
    public void onCreate(Bundle savedInstanceState) {
        super.onCreate(savedInstanceState);
        setContentView(R.layout.main);

        setupViewComponent();
    }

    private void setupViewComponent() {
        mTxtResult = (TextView)findViewById(R.id.txtResult);

        ArrayAdapter<CharSequence> adapWeekday = ArrayAdapter.createFromResource(
                this, R.array.weekday, android.R.layout.simple_list_item_1);
        setListAdapter(adapWeekday);

        ListView listview = getListView();
        listview.setTextFilterEnabled(true);

        listview.setOnItemClickListener(listviewOnItemClkLis);
    }

    AdapterView.OnItemClickListener listviewOnItemClkLis = new
    AdapterView.OnItemClickListener() {
        public void onItemClick(AdapterView<?> parent, View view,
            int position, long id) {
            mTxtResult.setText(((TextView) view).getText());
        }
    };
}
```

29-2 帮 ListView 加上小图标

在前一个小节中我们已经学会建立 ListView 菜单的方法，如果读者比较图 29-2 和图 29-1 的 ListView 界面，可以发现图 29-1 中每一个选项前面都有一个小图标，这种包含图标的菜单看起来比纯文本的菜单更生动有趣，也能够提升程序的质感，但是要如何完成呢？

要在每一个选项的前面加上小图标必须改变选项的显示格式，在前一个小节的范例程序中，决定每一个选项显示格式的关键是在 ArrayAdapter 这个类。我们把要显示的选项列表数组从程序项目的资源类 R 中读入，然后指定使用 Android SDK 的内定格式 android.R.layout.simple_list_item_1。

如果要换成具有小图标的格式，则每一个项目都变成要有小图标和文字两个部分，这样的格式就要由我们自己设置。设置显示格式的方法就是在 res/layout 中新增一个接口布局文件 list_item.xml，然后把它的程序代码编辑如下，这个格式就是一个 ImageView 组件后面跟着一个 TextView 组件，而且是水平排列，这就是我们想要的 ListView 项目的显示格式。

```xml
<?xml version="1.0" encoding="utf-8"?>
<LinearLayout xmlns:android="http://schemas.android.com/apk/res/android"
    android:orientation="horizontal"
    android:layout_width="match_parent"
    android:layout_height="match_parent"
    android:paddingTop="5dp"
    android:paddingBottom="5dp"
    >
    <ImageView android:id="@+id/imgView"
        android:layout_width="wrap_content"
        android:layout_height="wrap_content"
        />
    <TextView android:id="@+id/txtView"
        android:layout_width="wrap_content"
        android:layout_height="wrap_content"
        android:textSize="20sp"
        android:layout_gravity="center_vertical"
        />
</LinearLayout>
```

接下来在程序文件中利用 ArrayList 对象储存每一个项目的数据，由于每一个项目都有图标和文字两个部分，因此我们用 Map 这个泛型类把每一个项目的图标和文字集合起来，并且指定对应的接口组件 id 如下：

```java
mList = new ArrayList<Map<String,Object>>();

for (int i = 0; i < listFromResource.length; i++) {
    Map<String, Object> item = new HashMap<String, Object>();
    item.put("imgView", android.R.drawable.ic_menu_my_calendar);
    item.put("txtView", listFromResource[i]);
    mList.add(item);
}
```

接着再建立一个 SimpleAdapter 对象，把上述建立好的菜单数组存入该对象并指定使用我们前面建立的接口布局文件 list_item.xml，另外也要指定格式文件中的接口组件 id，最后调用 setListAdapter() 方法把 SimpleAdapter 对象设置给 ListView。

```java
SimpleAdapter adapter = new SimpleAdapter(this, mList,
                R.layout.list_item,
                new String[] { "imgView", "txtView" },
```

```
            new int[] { R.id.imgView ,R.id.txtView });

setListAdapter(adapter);
```

建立和设置 AdapterView.OnItemClickListener 对象的方法和前一个小节类似，以下是完成后的程序代码，图 29-3 是程序的运行界面。

图 29-3　加上小图标的 ListView 菜单

```
package …

import …

public class Main extends ListActivity {

    private TextView mTxtResult;
    List<Map<String, Object>> mList;

    /** Called when the activity is first created. */
    @Override
    public void onCreate(Bundle savedInstanceState) {
        super.onCreate(savedInstanceState);
        setContentView(R.layout.main);

        setupViewComponent();
    }

    private void setupViewComponent() {
        mTxtResult = (TextView)findViewById(R.id.txtResult);

        String[] listFromResource = getResources().getStringArray(R.array.weekday);

        mList = new ArrayList<Map<String,Object>>();

        for (int i = 0; i < listFromResource.length; i++) {
            Map<String, Object> item = new HashMap<String, Object>();
            item.put("imgView", android.R.drawable.ic_menu_my_calendar);
            item.put("txtView", listFromResource[i]);
```

```
            mList.add(item);
        }

        SimpleAdapter adapter = new SimpleAdapter(this, mList,
            R.layout.list_item,
            new String[] { "imgView", "txtView" },
            new int[] { R.id.imgView ,R.id.txtView });

        setListAdapter(adapter);

        ListView listview = getListView();
        listview.setTextFilterEnabled(true);
        listview.setOnItemClickListener(listviewOnItemClkLis);
    }

    AdapterView.OnItemClickListener listviewOnItemClkLis = new
    AdapterView.OnItemClickListener() {
        public void onItemClick(AdapterView<?> parent, View view,
            int position, long id) {
            String s =((TextView)view.findViewById(R.id.txtView)).
            getText().toString();
            mTxtResult.setText(s);
        }
    };
}
```

29-3 ExpandableListView 二层式选项列表

　　ListView 选项列表只有一层，也就是说使用者看到的就是可以选择的项目。如果先把这些选项分成几个群组，然后群组下面才是真正可以选择的项目，这样就变成二层式的选项列表，这种型态的列表可以使用 ExpandableListView 接口组件来显示，如图 29-4 所示。使用者操作时必须先单击组名展开该群组的选项，然后再从中选取一个项目。

　　ExpandableListActivity 是针对 ExpandableListView 组件的操作所设计的类，它可以简化 ExpandableListView 组件的使用步骤。在使用这种二层式可展开选项列表时，比较麻烦的是选项列表的建立。因为第一层是组名，它是用一个一维数组来储存，第二层则是选项名称，必须用二维数组来储存，第一列是第一个群组的选项，第二列是第二个群组的选项，依此类推。而且每一个组名或是选项名称还可以使用两个说

图 29-4　ExpandableListView 范例

明字符串，第一个字符串是群组或项目名称，第二个字符串是说明文字，这必须利用 Map 泛型类来完成。

首先我们在开启 res/layout/main.xml 接口布局文件中加入一个 ExpandableListView 组件，如以下范例：

```xml
<?xml version="1.0" encoding="utf-8"?>
<LinearLayout xmlns:android="http://schemas.android.com/apk/res/android"
    android:orientation="horizontal"
    android:layout_width="match_parent"
    android:layout_height="match_parent"
    >
<LinearLayout
    android:orientation="vertical"
    android:layout_width="0dp"
    android:layout_height="match_parent"
    android:layout_weight="1"
    />
<LinearLayout
    android:orientation="vertical"
    android:layout_width="0dp"
    android:layout_height="match_parent"
    android:layout_weight="1"
    >
<TextView android:id="@+id/txtResult"
    android:layout_width="match_parent"
    android:layout_height="wrap_content"
    android:text=""
    />
<ExpandableListView android:id="@id/android:list"
        android:layout_width="match_parent"
        android:layout_height="match_parent"
        android:layout_weight="1"
        android:drawSelectorOnTop="false"
        />
</LinearLayout>
<LinearLayout
    android:orientation="vertical"
    android:layout_width="0dp"
    android:layout_height="match_parent"
    android:layout_weight="1"
    />
</LinearLayout>
```

我们使用三个水平排列的 LinearLayout，并配合 android:layout_weight 属性（注意 android:layout_width 必须设置为"0dp"），来设置每一个 LinearLayout 所占的宽度比例让程序的操作接口置于屏幕中央。第一个 TextView 组件是用来显示用户所单击的项目位置和 id 编号，第二个

ExpandableListView 组件就是用来显示选项列表，请读者特别注意，它的 id 一定要和以上的范例相同不可以改变，因为 ExpandableListActivity 对象会固定使用这个 id。

在程序代码中我们使用 List 和 Map 这两个泛型类。List 类是用数组的形式来储存数据，我们在 List 对象中储存 Map 型态的对象，每一个 Map 的对象都包含两个字符串，一个是群组或项目的名称，另一个是说明文字，相关程序代码如下：

```java
mTxtResult = (TextView)findViewById(R.id.txtResult);

List<Map<String, String>> groupList = new ArrayList<Map<String, String>>();
List<List<Map<String, String>>> childList2D = new ArrayList<List<Map<String, String>>>();

for (int i = 0; i < 5; i++) {
    Map<String, String> group = new HashMap<String, String>();
    group.put(ITEM_NAME, "选项组" + i);
    group.put(ITEM_SUBNAME, "说明" + i);
    groupList.add(group);

    List<Map<String, String>> childList = new ArrayList<Map<String, String>>();
    for (int j = 0; j < 2; j++) {
        Map<String, String> child = new HashMap<String, String>();
        child.put(ITEM_NAME, "选项" + i + j);
        child.put(ITEM_SUBNAME, "说明" + i + j);
        childList.add(child);
    }
    childList2D.add(childList);
}

// 设置我们的 ExpandableListAdapter
mExpaListAdap = new SimpleExpandableListAdapter(
    this,
    groupList,
    android.R.layout.simple_expandable_list_item_2,
    new String[] {ITEM_NAME, ITEM_SUBNAME},
    new int[] {android.R.id.text1, android.R.id.text2},
    childList2D,
    android.R.layout.simple_expandable_list_item_2,
    new String[] {ITEM_NAME, ITEM_SUBNAME},
    new int[] {android.R.id.text1, android.R.id.text2}
);

setListAdapter(mExpaListAdap);
```

最后一个步骤是当用户单击列表中的项目时，程序必须取得该项目的位置。这个工作可以根据 ExpandableListActivity 类的 onChildClick() 方法来完成。图 29-5 是程序的运行界面，在本书的光盘中有完整的程序项目源文件供读者参考。

```
public boolean onChildClick(ExpandableListView parent, View v,
    int groupPosition, int childPosition, long id) {
    // TODO Auto-generated method stub
    String s = "选择：群组" + groupPosition + ",  选项" + childPosition + ", ID" + id;
    mTxtResult.setText(s);

    return super.onChildClick(parent, v, groupPosition, childPosition, id);
}
```

图 29-5　ExpandableListView 范例程序的运行界面

Android 版本	1.X	2.X	3.X	4.X
适用性	★	★	★	★

UNIT 30

AutoCompleteTextView 自动完成文字输入

所谓"自动完成文字输入"的功能就像是我们在 Baidu 搜索页输入搜寻字符串时,只要输入前几个字,下方就会自动出现候选字供我们选择,我们可以直接用鼠标单击想要的字符串,免去重复打字的麻烦,请读者参考图 30-1 就能了解。Android 应用程序也可以做到相同的功能,提供这种功能的接口组件叫做 AutoCompleteTextView,它的使用步骤如下:

图 30-1 Baidu 搜索页的"自动完成文字输入"功能

Step 1 在 res/layout 文件夹下的接口布局文件中建立一个 AutoCompleteTextView 组件如下,这个

AutoCompleteTextView 组件需要指定一个 id 名称，因为在程序代码中必须对它进行设置。

```
<AutoCompleteTextView android:id="@+id/autCompTextView"
    android:layout_width="match_parent"
    android:layout_height="wrap_content"
/>
```

Step 2 在程序代码中建立一个 ArrayAdapter 类的对象。ArrayAdapter 是一个泛型类，在前面介绍 Spinner 组件的单元中我们曾经使用过，这一次要再度利用它来设置"自动完成文字功能"的候选字符串，ArrayAdapter 对象是当成数据输入给 AutoCompleteTextView 的接口，请读者参考以下代码：

```
AutoCompleteTextView autCompTextView =
(AutoCompleteTextView)findViewById(R.id.autCompTextView);

String[] sArr = new String[] {"候选字符串 1", "候选字符串 2", "候选字符串 3"};

ArrayAdapter<String> adapAutoCompText = new ArrayAdapter<String>(
        this, android.R.layout.simple_dropdown_item_1line, sArr);

autCompTextView.setAdapter(adapAutoCompText);
```

完成以上 2 个步骤之后，AutoCompleteTextView 组件就可以正常运行了。

以上的说明是用固定的候选字符串为例，也就是说我们是用定义在程序代码中的固定数组来提供自动完成的候选字符串。但是在实际应用时，候选字符串通常会动态改变，例如当现在输入了一个新的字符串之后，下一次要在同样的字段输入数据时，这个新的字符串就会自动成为候选字符串。要动态改变候选字符串的关键在于 ArrayAdapter 这个对象，除了可以利用固定的数组来设置 ArrayAdapter 对象的内容之外，也可以根据调用 ArrayAdapter 对象的 add()方法把新的数据加入 ArrayAdapter 对象中，另外也可以使用 clear()方法清除其中的数据，在这个单元我们要完成一个可以动态加入候选字的 AutoCompleteTextView 组件，它的运行界面如图 30-2 所示。用户可以在上方的字段输入任何字符串，然后按下"加入自动完成文字"按钮，程序就会记住该字符串，用户可以重复以上动作加入任意字符串。当用户在输入字符串的时候，系统会自动检查是否有类似的候选字符串存在，如果有就会出现在输入文字的下方，如图 30-3 所示。

以下列出完整的字符串资源文件、接口布局文件和程序代码，请读者特别留意粗体字的部分，关于程序代码我们补充说明如下：

1. adapAutoCompText 是 ArrayAdapter 类的对象，它被定义在方法的外面成为公用对象，因为它在多个方法中都会使用。
2. 在 setupViewComponent()方法中我们设置 adapAutoCompText 对象是空的，也就是在建构式中没有指定任何数组，并且把这个 adapAutoCompText 对象设置给 autCompTextView 对象（它就是程序中的 AutoCompleteTextView 组件）。

图 30-2　AutoCompleteTextView 范例程序的运行界面　　图 30-3　自动完成文字输入的效果

3. 当用户输入字符串后按下"加入自动完成文字"按钮时，程序先取得输入的字符串，然后调用 add()方法把它加入 adapAutoCompText 对象中，再把输入的字符串清除，以方便用户输入下一个字符串。
4. 当用户按下"清除自动完成文字"按钮时，程序会把 adapAutoCompText 对象中的全部字符串清除。

■　字符串资源文件：

```xml
<?xml version="1.0" encoding="utf-8"?>
<resources>
    <string name="app_name">AutoCompleteTextView 范例程序</string>
    <string name="promptAutoCompleteText">请输入文字</string>
    <string name="promptBtnAddAutoCompleteText">加入自动完成文字</string>
    <string name="promptBtnClrAutoCompleteText">清除自动完成文字</string>
</resources>
```

■　界面布局文件：

```xml
<?xml version="1.0" encoding="utf-8"?>
<LinearLayout xmlns:android="http://schemas.android.com/apk/res/android"
    android:orientation="horizontal"
    android:layout_width="match_parent"
    android:layout_height="match_parent"
    >
<LinearLayout
    android:orientation="vertical"
    android:layout_width="0dp"
    android:layout_height="match_parent"
    android:layout_weight="1"
    />
<LinearLayout
    android:orientation="vertical"
```

```xml
        android:layout_width="0dp"
        android:layout_height="match_parent"
        android:layout_weight="1"
        >
<TextView
        android:layout_width="match_parent"
        android:layout_height="wrap_content"
        android:text="@string/promptAutoCompleteText"
        />
<AutoCompleteTextView android:id="@+id/autCompTextView"
        android:layout_width="match_parent"
        android:layout_height="wrap_content"
        />
<Button android:id="@+id/btnAddAutoCompleteText"
        android:layout_width="match_parent"
        android:layout_height="wrap_content"
        android:text="@string/promptBtnAddAutoCompleteText"
        />
<Button android:id="@+id/btnClrAutoCompleteText"
        android:layout_width="match_parent"
        android:layout_height="wrap_content"
        android:text="@string/promptBtnClrAutoCompleteText"
        />
</LinearLayout>
<LinearLayout
        android:orientation="vertical"
        android:layout_width="0dp"
        android:layout_height="match_parent"
        android:layout_weight="1"
        />
</LinearLayout>
```

■ 程序代码：

```java
package …

import …

public class Main extends Activity {

    private Button btnAddAutoCompleteText,
                   btnClrAutoCompleteText;
    private AutoCompleteTextView autCompTextView;

    private ArrayAdapter<String> adapAutoCompText;

    /** Called when the activity is first created. */
    @Override
```

```java
public void onCreate(Bundle savedInstanceState) {
    super.onCreate(savedInstanceState);
    setContentView(R.layout.main);

    setupViewComponent();
}

private void setupViewComponent() {
    // 从资源类 R 中取得接口组件
    btnAddAutoCompleteText = (Button)findViewById(R.id.btnAddAutoCompleteText);
    btnClrAutoCompleteText = (Button)findViewById(R.id.btnClrAutoCompleteText);
    autCompTextView = (AutoCompleteTextView)findViewById(R.id.autCompTextView);

    adapAutoCompText = new ArrayAdapter<String>(
            this, android.R.layout.simple_dropdown_item_1line);

    autCompTextView.setAdapter(adapAutoCompText);

    // 设置 button 组件的事件 listener
    btnAddAutoCompleteText.setOnClickListener(btnAddAutoCompleteTextOnClickLis);
    btnClrAutoCompleteText.setOnClickListener(btnClrAutoCompleteTextOnClickLis);
}

private Button.OnClickListener btnAddAutoCompleteTextOnClickLis = new
Button.OnClickListener() {
    public void onClick(View v) {
        String s = autCompTextView.getText().toString();
        adapAutoCompText.add(s);
        autCompTextView.setText("");
    }
};

private Button.OnClickListener btnClrAutoCompleteTextOnClickLis = new
Button.OnClickListener() {
    public void onClick(View v) {
        adapAutoCompText.clear();
    }
};
}
```

Android 版本	1.X	2.X	3.X	4.X
适用性	★	★	★	★

UNIT 31

SeekBar 和 RatingBar 接口组件

SeekBar 接口组件的功能类似微软 Word 程序中用来浏览文件的滚动条，滚动条上有一个可以拖曳的控制钮，滚动条的长度代表文件全部的范围，控制钮则代表目前在文件中的位置，图 31-1 是一个 SeekBar 组件的范例。要建立 SeekBar 组件只要在项目的 res/layout 文件夹下的程序接口布局文件中增加一个<SeekBar …/>标签即可，图 31-1 就是由以下的程序代码所产生：

```xml
<SeekBar android:id="@+id/seekBar"
    android:layout_width="match_parent"
    android:layout_height="wrap_content"
    android:max="100"
    android:progress="40"
    />
```

图 31-1　SeekBar 接口组件的范例

SeekBar 组件上的控制钮可由用户拖曳，程序再根据它的位置来调整显示的内容，因此我们必须在程序中建立一个 SeekBar 组件的 OnSeekBarChangeListener 对象，当用户改变了 SeekBar 上的控制扭的位置时，Android 系统会自动运行该事件处理程序。在这个 OnSeekBarChangeListener 对象中我们必须建立以下 3 个方法：

```java
public void onProgressChanged(SeekBar seekBar, int progress, boolean fromUser) {

}
```

```
public void onStartTrackingTouch(SeekBar seekBar) {

}
public void onStopTrackingTouch(SeekBar seekBar) {

}
```

第一个方法 onProgressChanged()是当 SeekBar 上的控制钮位置改变时运行，后面两个方法是当控制钮被按下准备拖曳时和控制钮被放开时运行。一般情况下我们只需要处理 onProgressChanged()这个方法，它输入的自变量包括所操作的 SeekBar 对象和目前控制钮的位置（progress 自变量的值）。

RatingBar 接口组件的外观比较特别，它像是一般杂志上经常用来代表商品评比结果的一排星星。在评比的时候可能会用到一颗完整的星星或是半颗星星，RatingBar 组件也是一样，如图 31-2 所示。建立 RatingBar 组件的方法是在接口布局文件中加上<RatingBar …/>标签，如以下范例：

```
<RatingBar android:id="@+id/ratBar"
    android:layout_width="wrap_content"
    android:layout_height="wrap_content"
    style="?android:attr/ratingBarStyle"
    android:numStars="5"
    android:rating="3.5"
    />
```

图 31-2　RatingBar 接口组件的范例

RatingBar 组件可以利用 style 属性来设置外观，ratingBarStyle 是默认型态，我们也可以设置为 ratingBarStyleSmall 或是 ratingBarStyleIndicator，如果设置为 ratingBarStyleIndicator 代表只是用来显示评分值，用户不能够对它进行变更。android:numStars 属性可以用来设置星星的数目，android:rating 属性则是设置目前的评分值，评分值可以有小数点（SeekBar 组件的 progress 属性不可以有小数点）。其实 RatingBar 组件的评分值也可以用 progress 属性来代表，progress 属性和 rating 属性之间的关系是 progress 属性值刚好是 rating 属性值的两倍，也就是说如果 rating 是 0.5，那么 progress 就是 1，如果 rating 是 1，那么 progress 就是 2，依此类推。程序代码中可以调用 RatingBar 组件的 setRating()方法改变 rating 的值，用户也可以用鼠标按下 RatingBar 组件上的星星来设置 rating 的值，如果程序要取得用户的设置值，就要建立一个 OnRatingBarChangeListener 对象，该对象中需要建立以下方法，方法中的第一个自变量是目前操作的 RatingBar 组件，第二个自变量是使用者设置的评分值。如果想取得 progress 的值，可以调用 getProgress()方法。

```
public void onRatingChanged(RatingBar ratingBar, float rating,
            boolean fromUser) {

}
```

接下来我们用一个范例程序来示范 SeekBar 和 RatingBar 的功能，程序的运行界面如图 31-3 所示。用户可以拖曳 SeekBar 组件上的控制钮，下方的说明文字会实时更新 progress 的值。用户也可以用鼠标单击下方的 RatinBar 组件上的星星，下方的说明文字会显示目前设置的 rating 值和 progress 值。这个程序的字符串资源文件、接口布局文件和程序文件如下。程序代码的部分主要是 SeekBar 和 RatingBar 这两个接口组件的事件处理程序，请读者特别留意粗体字的部分。

图 31-3 　SeekBar 和 RatingBar 范例程序的运行界面

- 字符串资源文件：

```
<?xml version="1.0" encoding="utf-8"?>
<resources>
    <string name="app_name">SeekBar 和 RatingBar 范例程序</string>
    <string name="resultSeekBar">SeekBar 的 Progress 值：</string>
    <string name="result1RatBar">RatingBar 的 Rating 值：</string>
    <string name="result2RatBar">RatingBar 的 Progress 值：</string>
</resources>
```

- 界面布局文件：

```
<?xml version="1.0" encoding="utf-8"?>
<LinearLayout xmlns:android="http://schemas.android.com/apk/res/android"
    android:orientation="horizontal"
    android:layout_width="match_parent"
    android:layout_height="match_parent"
    >
<LinearLayout
    android:orientation="vertical"
    android:layout_width="0dp"
    android:layout_height="match_parent"
    android:layout_weight="1"
    />
```

```xml
<LinearLayout
    android:orientation="vertical"
    android:layout_width="0dp"
    android:layout_height="match_parent"
    android:layout_weight="1"
    >
<SeekBar android:id="@+id/seekBar"
    android:layout_width="match_parent"
    android:layout_height="wrap_content"
    android:max="100"
    android:layout_marginTop="20dp"
    />
<TextView android:id="@+id/txtSeekBar"
    android:layout_width="match_parent"
    android:layout_height="wrap_content"
    android:text="@string/resultSeekBar"
    />
<RatingBar android:id="@+id/ratBar"
    android:layout_width="wrap_content"
    android:layout_height="wrap_content"
    android:numStars="5"
    android:layout_marginTop="20dp"
    />
<TextView android:id="@+id/txt1RatBar"
    android:layout_width="match_parent"
    android:layout_height="wrap_content"
    android:text="@string/result1RatBar"
    />
<TextView android:id="@+id/txt2RatBar"
    android:layout_width="match_parent"
    android:layout_height="wrap_content"
    android:text="@string/result2RatBar"
    />
</LinearLayout>
<LinearLayout
    android:orientation="vertical"
    android:layout_width="0dp"
    android:layout_height="match_parent"
    android:layout_weight="1"
    />
</LinearLayout>
```

■ 程序文件：

package …

import …

```java
public class Main extends Activity {
    private RatingBar mRatBar;
    private SeekBar mSeekBar;
    private TextView mTxtSeekBar,
                    mTxt1RatBar,
                    mTxt2RatBar;

    /** Called when the activity is first created. */
    @Override
    public void onCreate(Bundle savedInstanceState) {
        super.onCreate(savedInstanceState);
        setContentView(R.layout.main);

        setupViewComponent();
    }

    private void setupViewComponent() {
        // 从资源类 R 中取得接口组件
        mRatBar = (RatingBar)findViewById(R.id.ratBar);
        mSeekBar = (SeekBar)findViewById(R.id.seekBar);
        mTxtSeekBar = (TextView)findViewById(R.id.txtSeekBar);
        mTxt1RatBar = (TextView)findViewById(R.id.txt1RatBar);
        mTxt2RatBar = (TextView)findViewById(R.id.txt2RatBar);

        mSeekBar.setOnSeekBarChangeListener(seekBarOnChangeLis);
        mRatBar.setOnRatingBarChangeListener(ratBarOnChangeLis);
    }

    SeekBar.OnSeekBarChangeListener seekBarOnChangeLis =   new
    SeekBar.OnSeekBarChangeListener() {
        public void onProgressChanged(SeekBar seekBar, int progress,
        boolean fromUser) {
            String s = getString(R.string.resultSeekBar);
            mTxtSeekBar.setText(s + Integer.toString(progress));
        }
        public void onStartTrackingTouch(SeekBar seekBar) {

        }
        public void onStopTrackingTouch(SeekBar seekBar) {

        }
    };

    RatingBar.OnRatingBarChangeListener ratBarOnChangeLis =   new
    RatingBar.OnRatingBarChangeListener()              {
        @Override
```

```java
            public void onRatingChanged(RatingBar ratingBar, float rating,
                    boolean fromUser) {
                // TODO Auto-generated method stub
                String s = getString(R.string.result1RatBar);
                mTxt1RatBar.setText(s + Float.toString(rating));
                s = getString(R.string.result2RatBar);
                mTxt2RatBar.setText(s + Integer.toString(mRatBar.
                    getProgress()));
            }
        };
    }
```

PART 6 其他接口组件与对话框

UNIT 32　时间日期接口组件和对话框
UNIT 33　ProgressBar、ProgressDialog 和 Multi-Thread 程序
UNIT 34　AlertDialog 对话框
UNIT 35　Toast 信息框
UNIT 36　自定义 Dialog 对话框

Android 版本	1.X	2.X	3.X	4.X
适用性	★	★	★	★

UNIT 32
时间日期接口组件和对话框

如果程序需要让用户选择日期或时间，或是显示日期和时间来提醒使用者，就可以考虑使用 DatePicker 和 TimePicker 接口组件。这两个接口组件不但提供简单清楚的操作接口，而且也会自动处理日期和时间不规则的数值范围问题，举例来说，一天有 24 个小时，一个小时却有 60 分钟。另外像是每个月的日期天数也不尽相同，因此如果要程序设计者自行处理日期和时间的资料，将会是一个令人头痛的问题，幸好 Android 系统提供了日期和时间组件，让我们可以轻易地完成日期和时间的处理。

32-1　DatePicker 日期接口组件

如果程序需要显示日期，只要在项目的 res/layout 文件夹下的接口布局文件中加入 DatePicker 标签即可，如以下范例：

```
<DatePicker android:id="@+id/datePik"
    android:layout_width="wrap_content"
    android:layout_height="wrap_content"
/>
```

这个 DatePicker 接口组件被指定一个 id 名称，因为在程序代码中需要对它进行操作。DatePicker 组件一开始会自动显示模拟器的日期，如果读者运行程序时发现 DatePicker 接口组件显示的日期不是现在计算机上的日期，这是因为模拟器的日期设置不正确，读者可以参考 UNIT 8 中的说明，调整模拟器的日期时间设置。DatePicker 接口组件的外观如图 32-1 所示，其中的上三角形和下三角形按钮可以分别用来调整年、月、日的数据，另外也可以直接按住年、月、日的数字往上或往下拖曳，或是单击右边月历中的日期，还可以按住月历往上或往下拖曳来浏览不同的月份。

图 32-1　DatePicker 接口组件

程序中如果要取得用户设置的年、月、日，可以调用 DatePicker 组件的 getYear()、getMonth() 和 getDayOfMonth()方法。但是要注意的是 getMonth()方法会传回月份减 1，也就是传回 0 代表 1 月、1 代表 2 月，依此类推。

32-2　TimePicker 时间接口组件

TimePicker 组件和 DatePicker 组件的用法类似，只要在程序接口布局文件中加入 TimePicker 标签即可，如以下范例：

```
<TimePicker android:id="@+id/timePik"
    android:layout_width="wrap_content"
    android:layout_height="wrap_content"
/>
```

TimePicker 接口组件同样被指定一个 id 名称，因为在程序代码中需要对它进行操作。TimePicker 组件一开始会显示平板电脑模拟器的时间，它的外观如图 32-2 所示，其中的上三角形和下三角形按钮可以分别用来调整时和分，程序中如果要取得用户设置的时和分，可以调用 TimePicker 组件的 getCurrentHour()和 getCurrentMinute()方法。

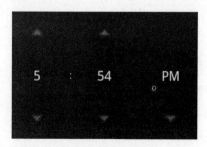

图 32-2　TimePicker 接口组件

32-3 范例程序

DatePicker 和 TimePicker 接口组件的用法并不复杂,以下我们直接用一个程序来示范,这个程序的运行界面如图 32-3 所示,其中包含一个 DatePicker 和一个 TimePicker 接口组件,用户可以利用前面介绍的方法调整日期和时间,然后按下确定按钮,程序会在按钮下方显示用户所设置的日期和时间。此程序项目的字符串资源文件、接口布局文件和程序文件如下,程序代码中主要就是设置"确认"按钮的 OnClickListener 对象,它负责取得用户设置的日期和时间,并显示在按钮下方的 TextView 组件中。

图 32-3 DatePicker 和 TimePicker 接口组件的范例程序

■ 字符串资源文件:

```
<?xml version="1.0" encoding="utf-8"?>
<resources>
    <string name="app_name">DatePicker 和 TimePicker 范例程序</string>
    <string name="promptBtnOK">确定</string>
    <string name="result">您选择的日期和时间是</string>
</resources>
```

■ 界面布局文件:

```
<?xml version="1.0" encoding="utf-8"?>
<LinearLayout xmlns:android="http://schemas.android.com/apk/res/android"
    android:orientation="horizontal"
    android:layout_width="match_parent"
    android:layout_height="match_parent">
<LinearLayout
```

```xml
        android:orientation="vertical"
        android:layout_width="0dp"
        android:layout_height="match_parent"
        android:layout_weight="1"/>
<LinearLayout
        android:orientation="vertical"
        android:layout_width="0dp"
        android:layout_height="match_parent"
        android:layout_weight="1">
<DatePicker android:id="@+id/datePik"
        android:layout_width="wrap_content"
        android:layout_height="wrap_content"/>
<TimePicker android:id="@+id/timePik"
        android:layout_width="wrap_content"
        android:layout_height="wrap_content"/>
<Button android:id="@+id/btnOK"
        android:layout_width="match_parent"
        android:layout_height="40dp"
        android:textSize="15sp"
        android:text="@string/promptBtnOK"/>
<TextView android:id="@+id/txtResult"
        android:layout_width="match_parent"
        android:layout_height="wrap_content"
        android:text="@string/result"/>
</LinearLayout>
<LinearLayout
        android:orientation="vertical"
        android:layout_width="0dp"
        android:layout_height="match_parent"
        android:layout_weight="1"/>
</LinearLayout>
```

■ 程序文件：

```java
package …

import …

public class Main extends Activity {

    private DatePicker mDatePik;
    private TimePicker mTimePik;
    private TextView mTxtResult;
    private Button mBtnOK;

    /** Called when the activity is first created. */
    @Override
    public void onCreate(Bundle savedInstanceState) {
        super.onCreate(savedInstanceState);
```

```
        setContentView(R.layout.main);

        setupViewComponent();
    }

    private void setupViewComponent() {
        // 从资源类 R 中取得接口组件
        mDatePik = (DatePicker)findViewById(R.id.datePik);
        mTimePik = (TimePicker)findViewById(R.id.timePik);
        mTxtResult = (TextView)findViewById(R.id.txtResult);
        mBtnOK = (Button)findViewById(R.id.btnOK);

        mBtnOK.setOnClickListener(btnDoOKOnClick);
    }

    private Button.OnClickListener btnDoOKOnClick = new Button.OnClickListener() {
        public void onClick(View v) {
            String s = getString(R.string.result);
            mTxtResult.setText(s + mDatePik.getYear() + "年" +
                              (mDatePik.getMonth()+1) + "月" +
                              mDatePik.getDayOfMonth() + "日" +
                              mTimePik.getCurrentHour() + "点" +
                              mTimePik.getCurrentMinute() + "分");
        }
    };
}
```

32-4　DatePickerDialog 和 TimePickerDialog 对话框

　　DatePickerDialog 和 TimePickerDialog 的功能与前面介绍的 DatePicker 和 TimePicker 接口组件一样，只不过换成是以对话框的形态出现。DatePickerDialog 对话框本身是一个类，我们只要建立一个它的对象，就会产生一个 DatePickerDialog 对话框。剩下的工作就是设置它的标题、信息、图标等，最后再把它显示在屏幕上，整个过程我们用下列的步骤说明：

Step 1 建立一个 DatePickerDialog 类的对象。建立对象的同时必须指定它的拥有者，以及它的 OnDateSetListener（也就是当使用者按下对话框的确定按钮后所运行的事件处理程序），同时我们要设置好对话框显示的日期，请参考下列程序代码：

```
Calendar now = Calendar.getInstance();

DatePickerDialog datePicDlg = new DatePickerDialog(类名称.this,
                        datePicDlgOnDateSelLis,
                        now.get(Calendar.YEAR),
                        now.get(Calendar.MONTH),
                        now.get(Calendar.DAY_OF_MONTH));
```

我们利用 Calendar 类中的方法取得系统现在的日期，然后建立一个 DatePickerDialog 的对象，并传入建构式所需的参数，其中的 datePicDlgOnDateSelLis 是一个 OnDateSetListener 对象，它的程序代码如下：

```
private DatePickerDialog.OnDateSetListener datePicDlgOnDateSelLis = new DatePickerDialog.OnDateSetListener()
{
    public void onDateSet (DatePicker view, int year,
                  int monthOfYear, int dayOfMonth) {
        // 当用户按下 DatePickerDialog 对话框中的确定按钮后要运行的程序
        …
    }
};
```

也就是说我们把用户按下设置日期的按钮后要运行的程序写在 OnDateSetListener 对象中的 onDateSet()方法内。

Step 2 设置对话框的标题、信息、图标，另外还要将 Cancelable 属性设置为 false，它让用户无法利用屏幕左下方的"回上一页"按钮离开对话框，程序代码如下：

```
datePicDlg.setTitle("选择日期");
datePicDlg.setMessage("请选择适合您的日期");
datePicDlg.setIcon(android.R.drawable.ic_dialog_info);
datePicDlg.setCancelable(false);
```

Step 3 调用 show()方法显示对话框。

TimePickerDialog 和 DatePickerDialog 一样也是一个类，我们同样也是根据建立一个它的对象来产生一个 TimePickerDialog 对话框，然后设置它的标题、信息和图标，最后再把它显示在屏幕上。在本书的光盘中包含一个完整的 DatePickerDialog 和 TimePickerDialog 范例程序，它的运行界面如图 32-4 所示，如果需要读者可以自行参考其中的程序代码。

图 32-4　DatePickerDialog 和 TimePickerDialog 范例程序的运行界面

260

Android 版本	1.X	2.X	3.X	4.X
适用性	★	★	★	★

UNIT 33

ProgressBar、ProgressDialog 和 Multi-Thread 程序

如果程序需要运行比较费时的工作，通常会显示一个进度列让用户了解目前工作完成的百分比，这就是本单元要介绍的 ProgressBar 的功能。ProgressBar 接口组件本身的使用方式很简单，只要在项目的 res/layout 文件夹下的程序接口布局文件中增加一个 ProgressBar 标签就可以显示一个进度列，如以下范例：

```
<ProgressBar android:id="@+id/proBar"
    style="?android:attr/progressBarStyleHorizontal"
    android:layout_width="match_parent"
    android:layout_height="wrap_content"
    android:max="100"
    android:progress="30"
    android:secondaryProgress="50"
/>
```

我们将此 ProgressBar 设置一个 id 名称，以便在程序中更新它显示的进度。另外请读者留意 style 属性的设置格式。在这个范例中我们把 ProgressBar 的最大进度值设置为 100（android:max 属性），目前显示的进度值设为 30（android: progress 属性），还有一个称为第二进度值（secondary progress）的属性，它的效果请读者参考图 33-1，第二进度值是以比较淡的颜色显示，如果读者曾经看过 YouTuBe 网站的影片，当影片播放时在播放器下方的进度列除了显示目前影片的进度之外，还可以看到一条比较淡的颜色跑在影片进度的前面，它是代表目前影片已经完成下载的百分比，这就是第二进度值的功能。

图 33-1　progressBarStyleHorizontal 型态的 ProgressBar 接口组件

除了图 33-1 型态的进度列之外，Android 系统还提供另外 3 种不同类型的进度列，只是这 3 种类型的进度列都是环状的形式，如图 33-2 所示，它们在接口布局文件中的建立方式如下：

```xml
<ProgressBar
    android:layout_width="wrap_content"
    android:layout_height="wrap_content"
/>
<ProgressBar
    style="?android:attr/progressBarStyleLarge"
    android:layout_width="wrap_content"
    android:layout_height="wrap_content"
/>
<ProgressBar
    style="?android:attr/progressBarStyleSmall"
    android:layout_width="wrap_content"
    android:layout_height="wrap_content"
/>
```

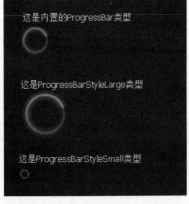

图 33-2　其他 3 种类型的 ProgressBar

这 3 种环状形式的进度列并没有提供百分比的信息，只是代表目前工作正在进行中请使用者等待。通常这种环状形式的进度列是用在工作完成度无法掌握的情况，像是网络正在联机中。

虽然 ProgressBar 的使用方法很简单，但是当需要使用它的时候，通常代表程序正要运行比较费时的工作，为了让即将进行的工作不要阻碍整个系统的运行，ProgressBar 通常需要配合使用 "多任务" 的程序架构。这个 "多任务" 的程序架构必须同时进行下列三件事：首先程序必须不断地更新 ProgressBar 所显示的进度，第二是程序必须持续运行该项工作，第三是程序必须持续对用户的操作作出响应。为了能够完整地呈现 ProgressBar 的应用，我们需要先了解"多任务"程序，也就是 Multi-Thread 程序的架构。

33-1 Multi-Thread 程序

所谓 Multi-Thread 程序就是在目前运行的程序中再产生一个"同时"进行的工作。读者可以回想我们之前的范例程序，它们在运行过程中不论何时都只有一个工作在进行。像是"计算机猜拳游戏"，首先是用户按下出拳按钮，然后计算机再出拳，最后决定胜负。或是在"图像画廊"程序中，刚开始是用户浏览图像缩图，当用户单击一个图像缩图后，再将原始图像显示在屏幕上，这些工作都是依序进行。但是现在我们必须让程序同时运行多项工作。不过我们称它"同时"也不全然正确，因为如果系统只有一个 CPU，它是将多个工作快速地轮流运行，所以感觉上像是多项工作一起进行，其实在任何一个时间点都只有一项工作在运行，但是如果系统中有多个 CPU 核心，则确实会有多个工作同时运行。

"多任务程序"的实现方法就是建立 Thread 对象。Thread 是一个 java 类，只要我们根据它来建立一个自己的 Thread 类，然后把要同时运行的程序代码写入该类的 run() 方法中，最后再产生该 Thread 类的对象并调用它的 start() 方法，就可以让写在 run() 方法中的程序代码和原来启动它的程序代码一起运行如以下范例，这样我们就解决了多项工作要同时运行的问题。接下来是如何更新 ProgressBar，这就牵涉不同 Thread 之间的信息沟通，这项工作需要使用 Handler 对象。

```
public class MyThread extends Thread {

    public void run () {
        // 要和主程序一起运行的程序代码
        …
    }
}
```

33-2 使用 Handler 对象完成 Thread 之间的信息沟通

如果依照直觉的作法，ProgressBar 的更新应该是由上一小节的 MyThread 对象负责，因为 MyThread 对象的任务就是运行一项长时间的工作，它应该根据目前的进度不断地更新 ProgressBar 上显示的进度值。可是这种作法就会遇到我们在 UNIT 22 中讨论过的问题，也就是 ProgressBar 组件的所有权是属于 main thread，Android 系统并不允许其他的 background thread 取用 main thread 的接口组件，如果 background thread 要更新接口组件的状态，必须通知 main thread，再由 main thread 来运行。要完成这件事必须借助 Handler 对象，Handler 对象可以让 background thread 存取 main thread

的消息队列（message queue），要使用 Handler 对象需要下列两个步骤：

Step 1 在主程序类中建立一个 Handler 对象。

Step 2 在 MyThread 类中使用该 Handler 对象的 post()方法把更新 ProgressBar 的工作（包装成 Runnable 对象），放到 main thread 的 message queue 中让 main thread 运行。

33-3 第一版的 Multi-Thread ProgressBar 范例程序

接下来我们根据前面介绍的 ProgressBar 接口组件和 Multi-Thread 程序架构来完成第一个 ProgressBar 范例程序，这个程序使用一个循环不断地读取系统的时间，并根据计算时间差来更新 ProgressBar 上显示的进度值，每隔一秒增加 2%的进度值和 4%的第二进度值，另外我们也同时显示其他 3 种环状形式的 ProgressBar。我们将整个程序的建立过程以下列步骤说明：

Step 1 在项目的 res/layout 文件夹下的接口布局文件中建立 4 个不同类型的 ProgressBar 组件如以下范例：

```xml
<?xml version="1.0" encoding="utf-8"?>
<LinearLayout xmlns:android="http://schemas.android.com/apk/res/android"
    android:orientation="vertical"
    android:layout_width="match_parent"
    android:layout_height="match_parent"
    >
<TextView
    android:layout_width="match_parent"
    android:layout_height="wrap_content"
    android:text="这是内定的 ProgressBar 型态："
    />
<ProgressBar android:id="@+id/proBar1"
    android:layout_width="wrap_content"
    android:layout_height="wrap_content"
    />
<TextView
    android:layout_width="match_parent"
    android:layout_height="wrap_content"
    android:text="这是 progressBarStyleHorizontal 的型态："
    android:layout_marginTop="20dp"
    />
<ProgressBar android:id="@+id/proBar2"
    style="?android:attr/progressBarStyleHorizontal"
    android:layout_width="match_parent"
    android:layout_height="wrap_content"
    android:max="100"
    />
<TextView
```

```
            android:layout_width="match_parent"
            android:layout_height="wrap_content"
            android:text="这是 progressBarStyleLarge 的型态："
            android:layout_marginTop="20dp"
            />
    <ProgressBar android:id="@+id/proBar3"
            style="?android:attr/progressBarStyleLarge"
            android:layout_width="wrap_content"
            android:layout_height="wrap_content"
            />
    <TextView
            android:layout_width="match_parent"
            android:layout_height="wrap_content"
            android:text="这是 progressBarStyleSmall 的型态："
            android:layout_marginTop="20dp"
            />
    <ProgressBar android:id="@+id/proBar4"
            style="?android:attr/progressBarStyleSmall"
            android:layout_width="wrap_content"
            android:layout_height="wrap_content"
            />
</LinearLayout>
```

Step 2 在 Eclipse 左边的项目检查窗口中展开此程序项目的"src/(组件路径名称)"文件夹。在组件路径名称上右击，在弹出的菜单中选择 New>Class 就会出现图 33-3 的对话框。

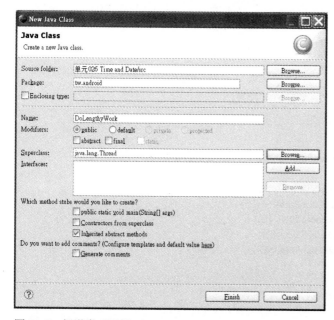

图 33-3 新增类对话框

Step 3 在对话框中的 Name 字段输入我们想要的类名称（例如 DoLengthyWork），然后按下在 Superclass 字段右边的 Browse 按钮就会出现图 33-4 的对话框，在对话框的最上面的字段中输入 Thread，下方列表中就会显示 Thread 类，用鼠标双击该 Thread 类就会自动回到原来的对话框，并完成填入 Superclass 字段。

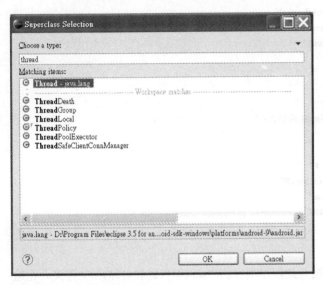

图 33-4 选择继承类的对话框

Step 4 按下 Finish 按钮后，在 Eclipse 中央的程序代码窗口中，就会出现程序文件的编辑界面，请读者输入以下程序代码：

```java
public class DoLengthyWork extends Thread {

    private Handler mHandler;
    private ProgressBar mProBar;

    public void run () {
        Calendar begin = Calendar.getInstance();
        do {
            Calendar now = Calendar.getInstance();
            final int iDiffSec = 60 * (now.get(Calendar.MINUTE) –
                            begin.get(Calendar.MINUTE)) +
                            now.get(Calendar.SECOND) –
                            begin.get(Calendar.SECOND);

            if (iDiffSec * 2 > 100) {
                mHandler.post(new Runnable() {
                    public void run() {
                        mProBar.setProgress(100);
```

```java
                    }
                });
                break;
            }
            mHandler.post(new Runnable() {
                public void run() {
                    mProBar.setProgress(iDiffSec * 2);
                }
            });

            if (iDiffSec * 4 < 100)
                mHandler.post(new Runnable() {
                    public void run() {
                        mProBar.setSecondaryProgress(iDiffSec * 4);
                    }
                });
            else
                mHandler.post(new Runnable() {
                    public void run() {
                        mProBar.setSecondaryProgress(100);
                    }
                });
        } while (true);
    }

    void setProgressBar(ProgressBar proBar) {
        mProBar = proBar;
    }

    void setHandler(Handler h) {
        mHandler = h;
    }
}
```

在这个 DoLengthyWork 类中有 2 个私有对象，分别为 mHandler 和 mProBar。mHandler 对象是用来运行 post 更新 ProgressBar 的工作，mProBar 对象是用来储存要处理的 ProgressBar 对象。这 2 个私有对象都有各自的方法（setHandler()和 setProgressBar()）来设置它们的值。在 run()方法中主要就是读取系统时间的循环,并持续 post 更新 ProgressBar 的工作。这里我们使用 Calendar 对象的 getInstance()方法来读取系统时间，然后根据时间差来更新 mProBar 对象的进度值和第二进度值。

Step 5 在主类的程序代码中建立一个 Handler 对象，并且在 setupViewComponent()方法内建立一个步骤四的 DoLengthyWork 类的对象，并设置好它内部的 2 个对象，然后调用它的 start()

方法开始运行，请读者参考下列程序代码：

```
package …

import …

public class Main extends Activity {

    private Handler mHandler = new Handler();

    /** Called when the activity is first created. */
    @Override
    public void onCreate(Bundle savedInstanceState) {
        super.onCreate(savedInstanceState);
        setContentView(R.layout.main);

        setupViewComponent();
    }

    private void setupViewComponent() {
        // 从资源类 R 中取得接口组件
        final ProgressBar proBar = (ProgressBar)findViewById
        (R.id.proBar2);

        DoLengthyWork work = new DoLengthyWork();
        work.setHandler(mHandler);
        work.setProgressBar(proBar);
        work.start();
    }
}
```

完成以上步骤之后运行程序就可以看到 ProgressBar 的运行界面，如图 33-5 所示。

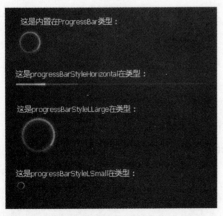

图 33-5　ProgressBar 范例程序的运行界面

33-4 第二版的 Multi-Thread ProgressBar 范例程序

接下来我们介绍另一种 Multi-Thread 程序的作法，这种方式比前一个方法简单。前面的范例是先建立一个 Thread 类，然后再产生一个该类的对象。如果像这样只需要建立一个 Thread 对象的情况，我们可以省略建立 Thread 类的步骤，直接在主类程序代码中利用 new Thread 指令建立 Thread 对象，然后把原来在 run() 方法中的程序代码包装成一个 Runnable 对象传给 Thread 对象。以下是修改后的 setupViewComponent() 方法，其他的程序代码和前一版的程序完全相同。在这一个版本中不需要建立 Thread 类的程序文件（也就是 DoLengthyWork.java）。完成修改后请读者再运行一次程序就可以看到和前面完全一样的结果。

```java
private void setupViewComponent() {
    // 从资源类 R 中取得接口组件
    final ProgressBar proBar = (ProgressBar)findViewById(R.id.proBar2);

    new Thread(new Runnable() {
        public void run() {
            Calendar begin = Calendar.getInstance();
            do {
                Calendar now = Calendar.getInstance();
                final int iDiffSec = 60 * (now.get(Calendar.MINUTE) –
                                    begin.get(Calendar.MINUTE)) +
                                    now.get(Calendar.SECOND) –
                                    begin.get(Calendar.SECOND);

                if (iDiffSec * 2 > 100) {
                    mHandler.post(new Runnable() {
                        public void run() {
                            proBar.setProgress(100);
                        }
                    });

                    break;
                }

                mHandler.post(new Runnable() {
                    public void run() {
                        proBar.setProgress(iDiffSec * 2);
                    }
                });

                if (iDiffSec * 4 < 100)
                    mHandler.post(new Runnable() {
                        public void run() {
```

```
                            proBar.setSecondaryProgress(iDiffSec * 4);
                        }
                    });
                else
                    mHandler.post(new Runnable() {
                        public void run() {
                            proBar.setSecondaryProgress(100);
                        }
                    });
            } while (true);
        }
    }).start();
}
```

33-5　ProgressDialog 对话框

　　ProgressDialog 对话框的功能和前面介绍的 ProgressBar 接口组件的功能相同，都是要告知用户目前程序正在运行一项比较费时的工作，两者的不同点在于 ProgressBar 接口组件是固定出现在程序的操作接口中，而 ProgressDialog 对话框则是需要的时候才会显示，等工作完成后就会自动消失。

　　ProgressDialog 也有两种形式，一种是会显示运行进度的百分比，而且也可以有第二进度值，另一种则是环状的等待循环。ProgressDialog 本身也是一个类，因此要建立一个 ProgressDialog 对话框只需要产生一个 ProgressDialog 类的对象即可，然后再设置它的标题、信息、图标，接着把它显示出来。显示 ProgressDialog 之后程序便开始进入 Multi-Thread 模式，其中一个 backgroud thread 负责运行主要工作，其他的 thread 负责更新 ProgressDialog 的进度值。这些负责更新进度值的程序代码同样是以 post 的方式放到 main thread 的 message queue 中，再由 main thread 来运行。在本书的光盘中包含一个完整的 ProgressDialog 范例程序，它的运行界面如图 33-6 所示，如果需要读者可以自行参考其中的程序代码。

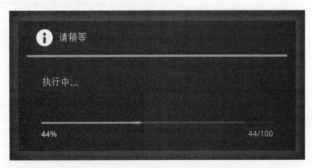

图 33-6　ProgressDialog 范例程序的运行界面

Android 版本	1.X	2.X	3.X	4.X
适用性	★	★	★	★

UNIT 34
AlertDialog 对话框

AlertDialog 对话框的功能是通知使用者某些讯息，这个讯息可以单纯只是一些信息、或是警告讯息或错误讯息，甚至是询问使用者一个问题，然后使用者再以对话框下方的按钮进行响应。AlertDialog 对话框中的按钮数目和按钮上所显示的文字可以由程序进行控制，但是最多只能有 3 个按钮，图 34-1 是一个 AlertDialog 对话框的范例。建立 AlertDialog 对话框有两种方法，一种是利用 AlertDialog.Builder 类别，另一种是利用 AlertDialog 类别，以下我们先介绍如何使用 AlertDialog.Builder 类别。

图 34-1 AlertDialog 对话框

34-1 使用 AlertDialog.Builder 类别建立 AlertDialog 对话框

使用 AlertDialog.Builder 类别建立 AlertDialog 对话框的步骤如下：

Step 1 建立一个 AlertDialog.Builder 类别的对象，建立对象的同时必须指定它的拥有者，请参考下列程序代码范例：

```
AlertDialog.Builder altDlgBldr = new AlertDialog.Builder(类别名称.this);
```

以上的程序代码是建立一个属于指定程序类别的 AlertDialog.Builder 对象，对象名称叫做 altDlgBldr。

Step 2 设定对话框的标题、讯息、图标，另外我们还要将 Cancelable 属性设定为 false，它让使用者无法利用"回上一页"按钮离开对话框，程序代码如下：

```
altDlgBldr.setTitle("AlertDialog");
altDlgBldr.setMessage("AlertDialog 范例");
altDlgBldr.setIcon(android.R.drawable.ic_dialog_info);
altDlgBldr.setCancelable(false);
```

Step 3 加上按钮，AlertDialog 对话框可以视需要加上按钮，当然也可以不加，只是没有按钮的话我们必须把上一个步骤中的 Cancelable 属性设定为 true，这样使用者才可以利用"回上一页"按钮离开对话框。可以加到 AlertDialog 对话框中的按钮有 3 个，分别为 PositiveButton、NegativeButton 和 NeutralButton，它们分别用不同的方法加入，请参考以下的程序代码范例：

```
altDlgBldr.setPositiveButton("是",
        new DialogInterface.OnClickListener() {
            @Override
            public void onClick(DialogInterface dialog, int which) {
                // 按下 PositiveButton 后要执行的程序代码
            }
        }
);
altDlgBldr.setNegativeButton("否",
        new DialogInterface.OnClickListener() {
            @Override
            public void onClick(DialogInterface dialog, int which) {
                // 按下 NegativeButton 后要执行的程序代码
            }
        });
altDlgBldr.setNeutralButton("取消",
        new DialogInterface.OnClickListener() {
            @Override
            public void onClick(DialogInterface dialog, int which) {
                // 按下 NeutralButton 后要执行的程序代码
            }
        });
```

这些加入不同按钮的方法都使用类似的自变量格式，第一个自变量是要显示在按钮上的文字，第二个自变量是设定按下按钮后的事件处理程序，它必须是一个 DialogInterface.OnClickListener 对象，而且其中必须包含一个 OnClick()方法。我们直接将产生对象的程序代码和 OnClick()方法中的程序代码写在 SetXXXButton()方法的自变量中，这也是一种匿名物件的写法。

Step 4 呼叫 show()方法显示对话框。

```
altDlgBldr.show();
```

完成以上步骤之后，就可以在屏幕上显示一个 AlertDialog 对话框，接下来我们介绍第二种建立 AlertDialog 对话框的方法。

34-2 使用 AlertDialog 类别建立 AlertDialog 对话框

读者或许会直觉地认为可以借由建立一个 AlertDialog 的对象来产生对话框，如果真是如此，那又何必有 AlertDialog.Builder 类别的存在。如果读者不相信，可以尝试建立一个 AlertDialog 的对象试看看，相信会出现错误讯息，原因在于 AlertDialog 类别把建构式定义成 protected，所以我们无法直接产生它的对象，解决方法是要用继承的方式，也就是我们要自己新增一个继承自 AlertDialog 的类别，然后在该类别中建立一个建构式呼叫 AlertDialog 类别的建构式，然后我们就可以利用自己的类别产生一个 AlertDialog 的物件。以下的程序代码就是我们自己新增的类别，名称叫做 MyAlertDialog。它的程序代码很简单，第一它是继承 AlertDialog 类别，第二是它只有一个 public 的建构式，而且该建构式就只有呼叫 AlertDialog 类别的建构式，没有其他的程序代码。

```
package …

import …

public class MyAlertDialog extends AlertDialog {
    public MyAlertDialog(Context context) {
        super(context);
        // TODO Auto-generated constructor stub
    }
}
```

建立好这个 MyAlertDialog 类别之后，就可以利用以下的步骤来产生 AlertDialog 对话框：

Step 1 建立一个 MyAlertDialog 类别的对象，建立对象的同时必须指定它的拥有者，请参考下列程序代码范例：

```
MyAlertDialog myAltDlg = new MyAlertDialog(类别名称.this);
```

以上的程序代码是建立一个属于指定类别的 MyAlertDialog 对象，对象名称叫做 myAltDlg。

Step 2 设定对话框的标题、信息、图标，另外我们还要将 Cancelable 属性设定为 false，它让使用者无法利用"回上一页"按钮离开对话框。

```
myAltDlg.setTitle("AlertDialog");
myAltDlg.setMessage("使用 MyAlertDialog 类别产生");
myAltDlg.setIcon(android.R.drawable.ic_dialog_info);
myAltDlg.setCancelable(false);
```

在 MyAlertDialog 类别的程序代码中除了建构式之外，我们并没有定义任何其他方法，因此以上程序代码用到的方法都是继承 AlertDialog 类别而来。

Step 3 加上按钮，加上按钮的方法和 AlertDialog.Builder 类别的方式不一样。AlertDialog 类别是利用自变量来决定要加上 Positive、Negative 或是 Neutral 按钮，请读者参考以下范例就可以了解：

```
myAltDlg.setButton(DialogInterface.BUTTON_POSITIVE, "是", altDlgOnClkPosiBtnLis);
myAltDlg.setButton(DialogInterface.BUTTON_NEGATIVE, "否", altDlgOnClkNegaBtnLis);
myAltDlg.setButton(DialogInterface.BUTTON_NEUTRAL, "取消", altDlgOnClkNeutBtnLis);
```

另外这一次我们是把按钮的 OnClickListener 对象先建立好，再传给建立按钮的方法，建立这些按钮的 OnClickListener 对象时可以利用我们之前介绍的技巧，先输入建立对象的语法，把其中的方法先空下来，再借由语法修正建议让程序编辑窗口自动帮我们加入需要的方法。

```
private  DialogInterface.OnClickListener altDlgOnClkPosiBtnLis = new
DialogInterface.OnClickListener() {
        @Override
        public void onClick(DialogInterface dialog, int which) {
            // 按下 PositiveButton 后要执行的程序代码
        }
};

private  DialogInterface.OnClickListener altDlgOnClkNegaBtnLis = new
DialogInterface.OnClickListener() {
        @Override
        public void onClick(DialogInterface dialog, int which) {
            // 按下 NegativeButton 后要执行的程序代码
        }
};

private  DialogInterface.OnClickListener altDlgOnClkNeutBtnLis = new
DialogInterface.OnClickListener() {
        @Override
        public void onClick(DialogInterface dialog, int which) {
            // 按下 NeutralButton 后要执行的程序代码
        }
};
```

Step 4 呼叫 show() 方法显示对话框。

```
myAltDlg.show();
```

34-3　范例程序

我们在程序的操作画面中加入 2 个按钮，一个按钮是用 AlertDialog 衍生类别的方式建立 AlertDialog 对话框，另一个按钮则使用 AlertDialog.Builder 类别来建立 AlertDialog 对话框。当使用者在 AlertDialog 对话框中按下任何一个按钮之后回到主程序画面，程序会在屏幕上显示使用者按下的按钮，程序的执行画面如图 34-2 所示，程序的接口布局文件和程序代码如下。程序代码的内容主要是在这 2 个按钮的 OnClickListener 对象中，分别使用前面介绍的两种方法建立 AlertDialog 对话框。另外在 AlertDialog 对话框内按钮的 OnClickListener 对象中显示使用者所按下的按钮。

图 34-2　使用 AlertDialog 衍生类别和 AlertDialog.Builder 类别建立 AlertDialog 对话框

- 界面布局文件：

```
<?xml version="1.0" encoding="utf-8"?>
<LinearLayout xmlns:android="http://schemas.android.com/apk/res/android"
    android:orientation="horizontal"
    android:layout_width="match_parent"
    android:layout_height="match_parent">
<LinearLayout
    android:orientation="vertical"
    android:layout_width="0dp"
    android:layout_height="match_parent"
    android:layout_weight="1"/>
<LinearLayout
    android:orientation="vertical"
    android:layout_width="0dp"
    android:layout_height="match_parent"
    android:layout_weight="1">
<Button android:id="@+id/btnAlertDlg"
    android:layout_width="match_parent"
    android:layout_height="wrap_content"
    android:text="AlertDialog"/>
<Button android:id="@+id/btnAlertDlgBld"
    android:layout_width="match_parent"
    android:layout_height="wrap_content"
    android:text="AlertDialogBuilder"/>
<TextView android:id="@+id/txtResult"
    android:layout_width="wrap_content"
```

275

```xml
        android:layout_height="wrap_content"
        android:text=""/>
</LinearLayout>
<LinearLayout
    android:orientation="vertical"
    android:layout_width="0dp"
    android:layout_height="match_parent"
    android:layout_weight="1"/>
</LinearLayout>
```

■ 程序代码:

```java
package …

import …

public class Main extends Activity {

    private Button mBtnAlertDlg,
            mBtnAlertDlgBld;
    private TextView mTxtResult;

    /** Called when the activity is first created. */
    @Override
    public void onCreate(Bundle savedInstanceState) {
        super.onCreate(savedInstanceState);
        setContentView(R.layout.main);
        setupViewComponent();
    }

    private void setupViewComponent() {
        mBtnAlertDlg = (Button)findViewById(R.id.btnAlertDlg);
        mBtnAlertDlgBld = (Button)findViewById(R.id.btnAlertDlgBld);
        mTxtResult = (TextView)findViewById(R.id.txtResult);
        mBtnAlertDlg.setOnClickListener(btnAlertDlgOnClkLis);
        mBtnAlertDlgBld.setOnClickListener(btnAlertDlgBldOnClkLis);
    }

    private Button.OnClickListener btnAlertDlgOnClkLis = new Button.OnClickListener() {
        public void onClick(View v) {
            mTxtResult.setText("");
            MyAlertDialog myAltDlg = new MyAlertDialog(Main.this);
            myAltDlg.setTitle("AlertDialog");
            myAltDlg.setMessage("AlertDialog 的使用时机是要自己建立一个 class 来继承它");
            myAltDlg.setIcon(android.R.drawable.ic_dialog_info);
            myAltDlg.setCancelable(false);
            myAltDlg.setButton(DialogInterface.BUTTON_POSITIVE, "是", altDlgOnClkPosiBtnLis);
            myAltDlg.setButton(DialogInterface.BUTTON_NEGATIVE, "否", altDlgOnClk NegaBtnLis);
```

```java
            myAltDlg.setButton(DialogInterface.BUTTON_NEUTRAL, "取消", altDlgOnClkNeutBtnLis);
            myAltDlg.show();
        }
    };

    private DialogInterface.OnClickListener altDlgOnClkPosiBtnLis = new
DialogInterface.OnClickListener() {
        @Override
        public void onClick(DialogInterface dialog, int which) {
            // TODO Auto-generated method stub
            mTxtResult.setText("你启动了 AlertDialog 而且按下了\"是\"按钮");
        }
    };

    private DialogInterface.OnClickListener altDlgOnClkNegaBtnLis = new
DialogInterface.OnClickListener() {
        @Override
        public void onClick(DialogInterface dialog, int which) {
            // TODO Auto-generated method stub
            mTxtResult.setText("你启动了 AlertDialog 而且按下了\"否\"按钮");
        }
    };

    private DialogInterface.OnClickListener altDlgOnClkNeutBtnLis = new
DialogInterface.OnClickListener() {
        @Override
        public void onClick(DialogInterface dialog, int which) {
            // TODO Auto-generated method stub
            mTxtResult.setText("你启动了 AlertDialog 而且按下了\"取消\"按钮");
        }
    };

    private Button.OnClickListener btnAlertDlgBldOnClkLis = new
Button.OnClickListener() {
        public void onClick(View v) {
            mTxtResult.setText("");
            AlertDialog.Builder altDlgBldr = new AlertDialog.Builder(Main.this);
            altDlgBldr.setTitle("AlertDialog");
            altDlgBldr.setMessage("由 AlertDialog.Builder 产生");
            altDlgBldr.setIcon(android.R.drawable.ic_dialog_info);
            altDlgBldr.setCancelable(false);
            altDlgBldr.setPositiveButton("是",
                new DialogInterface.OnClickListener() {
                    @Override
                    public void onClick(DialogInterface dialog,
                        int which) {
                        // TODO Auto-generated method stub
                        mTxtResult.setText("你启动了
```

```java
                        AlertDialogBuilder 而且按下了\"是\"按钮");
                }
            });
            altDlgBldr.setNegativeButton("否",
                    new DialogInterface.OnClickListener() {
                        @Override
                        public void onClick(DialogInterface
                                                    dialog, int which) {
                            // TODO Auto-generated method stub
                            mTxtResult.setText("你启动了 AlertDialogBuilder 而且按下了\"否\"按钮");
                        }
                    });
            altDlgBldr.setNeutralButton("取消",
                    new DialogInterface.OnClickListener() {
                        @Override
                        public void onClick(DialogInterface
                                                    dialog, int which) {
                            // TODO Auto-generated method stub
                            mTxtResult.setText("你启动了 AlertDialogBuilder 而且按下了\"取消\"按钮");
                        }
                    });
            altDlgBldr.show();
        }
    };
}
```

Android 版本	1.X	2.X	3.X	4.X
适用性	★	★	★	★

UNIT 35
Toast 消息框

Toast 组件的功能和对话框有些类似，但是使用上更简单。使用 Toast 组件的目的只有一个，就是在屏幕上弹出一个信息窗口告知用户某个信息，而且这个信息窗口没有任何按钮，经过几秒钟后就会自动消失。如果使用者不注意，可能会来不及看清楚，所以只有在显示的信息不是很重要的情况下才会使用 Toast 快显信息。要使用 Toast 显示信息时，只需要调用它的 makeText()和 show()方法即可：

```
Toast t = Toast.makeText(类名称.this, R.string.字符串的 id, Toast.LENGTH_LONG 或 Toast.LENGTH_SHORT);
t.show();
```

第一行程序是调用 makeText()方法设置 Toast 的拥有者、要显示的字符串（上面的范例是使用字符串资源文件中的字符串）和信息出现的时间长短。其中要显示的字符串可以是资源类 R 中的字符串，或是在程序中建立的 String 对象，例如：

```
String s = "要显示的信息字符串";
Toast t = Toast.makeText(类名称.this, s, Toast.LENGTH_LONG 或 Toast.LENGTH_SHORT);
t.show();
```

makeText()方法会传回一个 Toast 对象，然后我们调用该对象的 show()方法就可以完成信息的显示。我们可以把以上的程序代码简化如下：

```
Toast.makeText(类名称.this, R.string.字符串的 id, Toast.LENGTH_LONG 或 Toast.LENGTH_SHORT)
.show();
```

它的意思就是运行完 makeText()方法后传回一个对象，再接着调用该对象的 show()方法，图 35-1 是一个 Toast 的实例。

图 35-1　Toast 信息组件

其实使用 AlertDialog 对话框也可以只显示信息而不要加上按钮和标题，如图 35-2 所示。那么 Toast 和 AlertDialog 的使用时机又有何不同呢？根据前面的说明，Toast 组件的特点是经过几秒钟后就会自动消失，使用者如果稍不注意就会遗漏该信息。但是 AlertDialog 对话框并不会自己消失，如果 AlertDialog 对话框没有按钮，用户必须按下模拟器的"回上一页"按钮才能关闭对话框（程序代码中必须将 AlertDialog 对话框的 Cancelable 属性设为 true），因此用户一定会看到该信息。所以说如果用户一定要知道的信息就使用 AlertDialog 对话框，否则就使用 Toast 快显信息即可。接下来我们把 Toast 快显信息应用到前面完成的"计算机猜拳游戏"程序中，当用户出拳后，程序会以 Toast 显示输赢的结果，程序的运行界面如图 35-3 所示，以下我们列出原来程序代码中必须修改的部分（以粗体字标示）。

图 35-2　没有加上按钮的 AlertDialog 对话框

图 35-3　使用 Toast 快显讯息的"计算机猜拳游戏"程序

```
private Button.OnClickListener btnScissorsLin = new Button.OnClickListener() {
    public void onClick(View v) {
        // 决定计算机出拳
        int iComPlay = (int)(Math.random()*3 + 1);
```

```java
            // 1 - 剪刀, 2 - 石头, 3 - 布.
            if (iComPlay == 1) {
                imgComPlay.setImageResource(R.drawable.scissors);
//              txtResult.setText(getString(R.string.result) +
//                              getString(R.string.playerDraw));
                Toast.makeText(Main.this, R.string.playerDraw,
                            Toast.LENGTH_LONG).show();
            }
            else if (iComPlay == 2) {
                imgComPlay.setImageResource(R.drawable.stone);
//              txtResult.setText(getString(R.string.result) +
//                              getString(R.string.playerLose));
                Toast.makeText(Main.this, R.string.playerLose,
                            Toast.LENGTH_LONG).show();
            }
            else {
                imgComPlay.setImageResource(R.drawable.net);
//              txtResult.setText(getString(R.string.result) +
//                              getString(R.string.playerWin));
                Toast.makeText(Main.this, R.string.playerWin,
                            Toast.LENGTH_LONG).show();
            }
        }
    };

    private Button.OnClickListener btnStoneLin = new Button.OnClickListener() {
        public void onClick(View v) {
            // 请以同样方式修改
        }
    };

    private Button.OnClickListener btnNetLin = new Button.OnClickListener() {
        public void onClick(View v) {
            // 请以同样方式修改
        }
    };
```

Android 版本	1.X	2.X	3.X	4.X
适用性	★	★	★	★

UNIT 36
自定义 Dialog 对话框

前面介绍的对话框像是 TimePickerDialog、ProgressDialog 等都是已经具备特定功能的对话框，它们的外观和按钮数目都已经固定，如果读者的程序刚好需要这些对话框，就可以直接使用不需要再自己动手设计，但是如果读者需要其他功能的对话框，那么就需要自己建立。自己设计对话框需要多花一些精神和时间，但是却给我们最大的发挥空间，这个单元就让我们来学习如何建立自己的对话框。

自定义对话框必须使用 Dialog 类，首先必须写好一个对话框专用的接口布局文件，也就是在主程序的接口布局文件 main.xml 之外，再新增一个接口布局文件，其中定义了在对话框中用到的所有接口组件和它们的编排方式。我们前面学过的所有接口组件和布局都可以用在对话框的接口布局文件中。接着在程序代码中建立一个 Dialog 类的对象，然后把上述的对话框接口布局文件加载到该对话框对象中，并设置好标题和其他属性以及事件处理程序，最后把该对话框显示出来，这整个过程我们以下列步骤说明：

Step 1 编写对话框的接口布局文件，该文件必须放在项目的 res/layout 文件夹中，读者可以利用 Eclipse 左边的项目检查窗口，在项目的 layout 文件夹上右击，然后选择 New>File，然后在对话框中输入文件名。注意文件名只能用小写的英文字母、数字或是底线字符，且扩展名必须是 xml，例如 my_dlg.xml。完成后结束对话框，然后在该文件名上用鼠标左键双击把它打开，开始设计对话框操作界面，我们可以使用前面学过的所有接口组件和布局来设计对话框接口布局文件。

Step 2 在程序代码中建立一个 Dialog 类的对象，建立对象的同时必须指定它的拥有者，请参考下列程序代码：

```
Dialaog myDlg = new Dialog(类名称.this);
```

以上的程序代码是建立一个属于指定类的 Dialog 对象，它的名称叫做 myDlg。

Step 3 设置对话框的标题和 Cancelable 属性（让用户无法利用"回上一页"按钮离开对话框），然后把对话框的接口布局文件加载到该对话框物件中：

```
myDlg.setTitle("对话框标题");
myDlg.setCancelable(false);
myDlg.setContentView(R.layout.对话框的接口布局文件文件名);
```

Step 4 如果对话框中的接口组件需要设置事件处理程序，例如 Button，就需要建立相关的 Listener 对象。以下的程序代码是以 Button 的 OnClickListener 为例：

```
private Button.OnClickListener myDlgBtnOKOnClkLis = new Button.OnClickListener() {
    public void onClick(View v) {
        // Button 按下后要运行的程序代码
        …
        myDlg.cancel();
    }
};
```

当运行完按钮中的工作之后，记得最后要调用对话框的 cancel()方法来结束对话框。另外要提醒读者，如果需要取得对话框中的接口组件的数据，例如要知道用户在对话框的 EditText 组件中输入的字符串，必须先调用对话框的 findViewById()方法取得该接口组件，例如：

```
EditText edt = (EditText)myDlg.findViewById(R.id.接口组件的 id);
```

Step 5 取得对话框中需要设置 Listener 的接口组件，然后把建立好的 Listener 对象设置给它，例如：

```
Button btn = (Button)myDlg.findViewById(R.id.接口组件 id);
btn.setOnClickListener(myDlgBtnOKOnClkLis);
```

其中的 myDlgBtnOKOnClkLis 是在步骤 4 所建立的对象。

Step 6 调用 show()方法显示对话框。

```
myDlg.show();
```

以上就是自己建立对话框的完整过程，接下来让我们示范一个很常见的系统登入对话框。程序的主界面会先显示一个"登入系统"按钮，当用户按下该按钮之后，会出现一个登入系统的对话框让用户输入账号名称和密码，并且有"确定登入"和"取消"二个按钮。比较特殊的是当用户输入密码时，屏幕上不会显示用户输入的字符，而是以一个黑点取代以避免被旁人窥视。要做到这样的功能只需要设置该 EditText 组件的 android:inputType 属性即可，当用户完成输入并按下"确定登入"按钮之后，程序界面会显示用户输入的账号和密码。图 36-1 是程序运行的界面，主程序的接口布局文件和对话框的接口布局文件以及完整的程序代码如下。程序代码主要是在"登入系统"按钮的 OnClickListener 对象中,依照前面介绍的步骤建立一个 Dialog 对象并将它显示出来。这个 Dialog 对象中有 2 个按钮,因此程序中分别建立了这 2 个按钮的 OnClickListener 对象,其中包含按下按钮之后要运行的程序代码。

图 36-1　自行建立系统登入对话框的运行界面

- 主程序接口布局文件:

```xml
<?xml version="1.0" encoding="utf-8"?>
<LinearLayout xmlns:android="http://schemas.android.com/apk/res/android"
    android:orientation="horizontal"
    android:layout_width="match_parent"
    android:layout_height="match_parent"
    >
<LinearLayout
    android:orientation="vertical"
    android:layout_width="0dp"
    android:layout_height="match_parent"
    android:layout_weight="1"
    />
<LinearLayout
    android:orientation="vertical"
    android:layout_width="0dp"
    android:layout_height="match_parent"
    android:layout_weight="1"
    >
<Button android:id="@+id/btnLoginDlg"
    android:layout_width="match_parent"
    android:layout_height="wrap_content"
    android:text="登入系统"
    />
<TextView android:id="@+id/txtResult"
    android:layout_width="wrap_content"
    android:layout_height="wrap_content"
    android:text=""
    />
</LinearLayout>
```

```xml
<LinearLayout
    android:orientation="vertical"
    android:layout_width="0dp"
    android:layout_height="match_parent"
    android:layout_weight="1"
    />
</LinearLayout>
```

- "登入系统"对话框接口布局文件：

```xml
<?xml version="1.0" encoding="utf-8"?>
<LinearLayout xmlns:android="http://schemas.android.com/apk/res/android"
    android:orientation="vertical"
    android:layout_width="300dp"
    android:layout_height="wrap_content"
    >
<TextView
    android:layout_width="wrap_content"
    android:layout_height="wrap_content"
    android:text="使用者名称："
    />
<EditText android:id="@+id/edtUserName"
    android:layout_width="match_parent"
    android:layout_height="wrap_content"
    android:text=""
    />
<TextView
    android:layout_width="wrap_content"
    android:layout_height="wrap_content"
    android:text="密码："
    />
<EditText android:id="@+id/edtPassword"
    android:layout_width="match_parent"
    android:layout_height="wrap_content"
    android:inputType="textPassword"
    android:text=""
    />
<LinearLayout
    android:orientation="horizontal"
    android:layout_width="match_parent"
    android:layout_height="wrap_content"
    android:gravity="center"
    >
<Button android:id="@+id/btnOK"
    android:layout_width="160dp"
    android:layout_height="wrap_content"
    android:text="确定登入"
    />
```

```xml
<Button android:id="@+id/btnCancel"
    android:layout_width="80dp"
    android:layout_height="wrap_content"
    android:text="取消"
    />
</LinearLayout>
</LinearLayout>
```

■ 程序代码：

```java
package …

import …

public class Main extends Activity {

    private Button mBtnLoginDlg;
    private TextView mTxtResult;
    private Dialog mLoginDlg;

    /** Called when the activity is first created. */
    @Override
    public void onCreate(Bundle savedInstanceState) {
        super.onCreate(savedInstanceState);
        setContentView(R.layout.main);

        setupViewComponent();
    }

    private void setupViewComponent() {
        mBtnLoginDlg = (Button)findViewById(R.id.btnLoginDlg);
        mTxtResult = (TextView)findViewById(R.id.txtResult);

        mBtnLoginDlg.setOnClickListener(btnLoginDlgOnClkLis);
    }

    private Button.OnClickListener btnLoginDlgOnClkLis = new
    Button.OnClickListener() {
        public void onClick(View v) {
            mTxtResult.setText("");

            mLoginDlg = new Dialog(Main.this);
            mLoginDlg.setTitle("登入系统");
            mLoginDlg.setCancelable(false);
            mLoginDlg.setContentView(R.layout.login_dlg);
            Button loginBtnOK = (Button)mLoginDlg.findViewById
            (R.id.btnOK);
            Button loginBtnCancel = (Button)mLoginDlg.findViewById
```

```
            (R.id.btnCancel);
            loginBtnOK.setOnClickListener(loginDlgBtnOKOnClkLis);
            loginBtnCancel.setOnClickListener(loginDlgBtnCancelOnClkLis);
            mLoginDlg.show();
        }
    };

    private Button.OnClickListener loginDlgBtnOKOnClkLis = new
    Button.OnClickListener() {
        public void onClick(View v) {
            EditText edtUserName = (EditText)mLoginDlg.findViewById
            (R.id.edtUserName);
            EditText edtPassword = (EditText)mLoginDlg.findViewById
            (R.id.edtPassword);

            mTxtResult.setText("你输入的使用者名称:" + edtUserName.
                        getText().toString() + ", 密码: " +
                        edtPassword.getText().toString());
            mLoginDlg.cancel();
        }
    };

    private Button.OnClickListener loginDlgBtnCancelOnClkLis = new
    Button.OnClickListener() {
        public void onClick(View v) {
            mTxtResult.setText("你按下\"取消\"按钮。");
            mLoginDlg.cancel();
        }
    };
}
```

PART 7　Intent、Intent Filter 和传送数据

UNIT 37　项目中的 AndroidManifest.xml 程序功能描述文件
UNIT 38　Intent 粉墨登场
UNIT 39　Tab 标签页接口——使用 Intent 对象
UNIT 40　Intent Filter 让程序也能帮助别人
UNIT 41　让 Intent 对象附带数据
UNIT 42　要求 Activity 返回数据

Android 版本	1.X	2.X	3.X	4.X
适用性	★	★	★	★

UNIT 37

工程中的 AndroidManifest.xml 程序功能描述文件

如果在 Eclipse 左边的项目检查窗口中打开某一个 Android 应用程序项目，就可以看到类似如下的项目内容：

```
Android 程序项目名称
    ├── src 文件夹
    ├── gen 文件夹
    ├── Android X.X 文件夹
    ├── assets 文件夹
    ├── bin 文件夹
    ├── res 文件夹
    ├── AndroidManifest.xml 文件
    ├── default.properties 文件
    └── proguard.cfg 文件
```

其中有些文件夹读者应该已经非常熟悉，像是 src、gen、res，但是其他文件夹或是文件到目前为止我们还没有使用过，因此也还未曾作过说明，这个单元就让我们针对整个 Android 应用程序项目的文件架构做一个完整的介绍。

1. src 文件夹

   用来存放项目中所有的 Java 程序文件，此文件夹下可以再建立数个不同的组件路径文件夹，以方便将程序文件区分成不同的组件分开存放。

2. gen 文件夹

   储存由 Android 程序编译程序产生的项目资源，其中包含一个资源类 R，这些程序项目资源是根据 res 文件夹下的资源文件所产生，我们不能修改 gen 文件夹中的内容。

3. Android X.X 文件夹

   这是程序项目使用的 Android 版本，我们也不能直接修改，但是可以根据开启项目的属性对话框来加以变更。首先在 Eclipse 左边的项目检查窗口中用鼠标右击该应用程序项目，然后在快捷菜单中选择 Properties 选项就会出现图 37-1 的对话框，在对话框左边的列表项目中单击 Android，右边便会出现目前安装的 Android 版本，请勾选其中一个，然后按下 Apply 按钮，再按下 OK 按钮，读者便会发现后面 X.X 的版本号码已经改变。

图 37-1　程序项目的属性对话框

4. assets 文件夹

   这个文件夹的功能和 res 文件夹有些类似，都是用来存放程序中会用到的其他文件资源，例如图像文件，和 res 文件夹不同的是，Android 编译程序不会将存放在 assets 文件夹的文件加入项目的资源类 R 中。程序如果要使用 assets 文件夹中的文件，必须自行指定它们的路径才能使用。

5. bin 文件夹
   储存由 Android 程序编译程序产生的文件，我们不需要修改其中的内容。
6. res 文件夹
   用来存放项目中需要用到的资源，包括字符串、接口布局文件、动画资源文件、图像文件等。这些资源文件会被 Android 编译程序自动加入 gen 文件夹下的资源类 R。
7. AndroidManifest.xml 文件
   这个文件就是本单元即将介绍的主角，其中记录了此程序项目的架构和功能等相关信息，它是程序用来和 Android 系统沟通的重要数据，程序开发者必须自行维护它的内容，以下我们将详细介绍其中的项目。
8. default.properties 文件
   这也是由 Android 程序项目产生器自动建立的文件，我们不能对它进行修改。
9. proguard.cfg 文件
   这个文件我们也不需对它进行修改。

如果打开"计算机猜拳游戏"程序项目中的 AndroidManifest.xml 文件将会看到如下的内容，AndroidManifest.xml 文件的编辑窗口有多种不同的检查模式，请读者在编辑窗口下方的 tab 标签页选择 AndroidManifest.xml 就可以用源文件的模式检查。

```xml
<?xml version="1.0" encoding="utf-8"?>
<manifest xmlns:android="http://schemas.android.com/apk/res/android"
    package="tw.android"
    android:versionCode="1"
    android:versionName="1.0">
  <uses-sdk android:minSdkVersion="11" />

  <application android:icon="@drawable/icon" android:label="@string/app_name">
      <activity android:name=".Main"
            android:label="@string/app_name">
          <intent-filter>
              <action android:name="android.intent.action.MAIN" />
              <category android:name="android.intent.category.LAUNCHER" />
          </intent-filter>
      </activity>
  </application>
</manifest>
```

最外层的<manifest>标签是用来记录项目的相关信息，其中有 3 个属性：
1. package
   指定项目中的程序文件的组件路径。

2. android:versionCode
   记录此项目的版本号码。

3. android:versionName
   记录此项目的版本名称。

＜uses-sdk＞标签是用来记录此程序项目使用的 Android SDK 版本，android:minSdkVersion="11" 代表此程序项目必须在 Android SDK 3.0 以上的环境才能运行（Android SDK 3.0 的版本编号就是 11）。＜application＞标签是用来记录程序的相关信息，它使用两个属性：

1. android:icon
   指定程序运行时显示的小图标。

2. android:label
   程序运行时显示在平板电脑屏幕上方的程序名称。

＜application＞标签中包含一个＜activity＞标签。Activity 是 Android 应用程序的运行单元，截至目前为止，我们的范例程序都只有一个 Activity，虽然在 ProgressBar 的范例程序中有新增一个类，但是该类是 Thread 而不是 Activity。其实在一个 Android 应用程序项目中可以建立多个 Activity，每一个 Activity 都必须在 AndroidManifest.xml 文件中建立一个对应的＜activity＞标签描述该 Activity 的相关信息。接下来就让我们示范如何在一个程序项目中使用多个 Activity。

首先我们建立一个新的 Android 程序项目，然后在程序接口布局文件中建立一个按钮。接下来我们一步一步说明如何建立第二个 Activity，这个新增的 Activity 就是前面完成的"计算机猜拳游戏"程序，当用户按下主程序界面上的按钮后，就会启动这个"计算机猜拳游戏"程序。

Step 1 每个 Activity 都需要一个专用的接口布局文件，请在 Eclipse 左边的项目检查窗口中展开此项目的 res/layout 文件夹，用鼠标右击 layout 文件夹，然后选择 New>File，接着在文件对话框中输入文件名，注意文件名只能用小写英文字母、数字或是底线字符，且扩展名必须是 xml，例如 game_activity.xml，按下 Finish 按钮后该文件会自动开启在程序代码编辑窗口中，请把"计算机猜拳游戏"项目中的接口布局文件的内容全部复制过来。

Step 2 在 Eclipse 左边的项目检查窗口中展开此项目的"src/(组件路径名称)"文件夹，然后用鼠标右击"组件路径名称"文件夹，在弹出的菜单中选择 New>Class。

Step 3 在类对话框中的 Name 字段输入我们想要的类名称（例如 Game），然后按下 Superclass 字段右边的 Browse 按钮，在出现的对话框最上面的字段中输入 Activity，下方列表中就会显示 Activity 类，用鼠标双击该类就会自动回到原来的对话框，并完成填入 Superclass 字段。

Step 4 按下 Finish 按钮后，在 Eclipse 中央的程序代码编辑窗口就会出现该 Activity 程序文件的内容，请读者找出"计算机猜拳游戏"项目中的程序文件，然后将其中的程序代码复制到这个新的 Activity 类中，但是要注意下列几点：

1. 请保留这个 Activity 原来的类名称，不要改变它。
2. 程序代码中出现 Main.this 的地方请将它改成 Game.this（Game 就是我们新增的 Activity 类的名称）。
3. 原来 Activity 类程序代码的第一行 package 的定义也不要更动。import 组件的程序代码在复制之后如果出现错误，也要视情况进行适当地修改。
4. 把这个 Activity 使用的接口布局文件换成步骤一所建立的文件，如下粗体字的部分：

```
...
public void onCreate(Bundle savedInstanceState) {
    super.onCreate(savedInstanceState);
    setContentView(R.layout.game_activity);

    setupViewComponent();
}
```

Step 5 打开此项目的界面布局文件 res/layout/main.xml，在操作接口中加入一个按钮如下：

```
<Button android:id="@+id/btnExecGame"
    android:layout_width="match_parent"
    android:layout_height="wrap_content"
    android:text="运行"计算机猜拳游戏"程序"
/>
```

Step 6 打开此项目的主类程序文件（也就是项目一开始就有的那一个程序文件），依照前面学过的方法新增一个按钮的 OnClickListener 对象，然后在此对象中加入下列粗体部分的程序代码，最后将此 OnClickListener 对象设置给前一个步骤所建立的按钮。

```
public void onClick(View v) {
    Intent it = new Intent();
    it.setClass(Main.this, Game.class);
    startActivity(it);
}
```

看到这一段程序代码读者心中难免有些疑问，因为里头出现了一个陌生的 Intent 对象。Intent 是 Android 系统中具有重要功能的对象，它可以让我们的程序和 Android 系统中的其他程序进行互动，或是让 Android 系统帮忙寻找适当的程序来运行 Intent 对象中所描述的任务。在下一个单元我们将对 Intent 对象作详细的介绍，这里我们先利用它来启动"计算机猜拳游戏"这个 Activity。

Step 7 把原来"计算机猜拳游戏"项目的 res/values/strings.xml 文件中所定义的字符串，复制到此项目的同名文件中。

Step 8 把原来"计算机猜拳游戏"项目的 res/drawable-hdpi 文件夹中的剪刀、石头、布图像文件复制到此项目的同名文件夹中。

以上是完成此程序项目的程序代码和操作接口，最后我们还要编辑此项目的 AndroidManifest.xml 文件，在里头加入这个新 Activity 的信息。

Step 9 打开此项目的 AndroidManifest.xml 文件，加入下列粗体字的部分：

```xml
<?xml version="1.0" encoding="utf-8"?>
<manifest xmlns:android="http://schemas.android.com/apk/res/android"
    package="tw.android"
    android:versionCode="1"
    android:versionName="1.0">
    <uses-sdk android:minSdkVersion="11" />

    <application android:icon="@drawable/icon" android:label="@string/app_name">
        <activity android:name=".Main"
                  android:label="@string/app_name">
            <intent-filter>
                <action android:name="android.intent.action.MAIN" />
                <category android:name="android.intent.category.LAUNCHER" />
            </intent-filter>
        </activity>
        <activity android:name=".Game"
                  android:label="@string/gameTitle">
        </activity>
    </application>
</manifest>
```

新增的程序代码是告诉 Android 系统这个项目中新增加一个 Activity，android:name=".Game" 属性是指定这个 Activity 的路径和名称，以附点开头代表和项目的主要 Activity 放在同一个路径。如果路径不相同，就要写出完整的路径，例如 android:name="tw.android.Game"。android:label 属性是设置 Activity 运行时要显示在屏幕上方的程序标题，上面的范例是使用定义在字符串资源文件中名为 gameTitle 的字符串，因此读者必须在 res/values/strings.xml 文件中定义好该字符串。

> **补充说明** 使用"交互式接口"编辑 AndroidManifest.xml
>
> 开启 AndroidManifest.xml 文件后在程序编辑窗口下方有一排标签，由左至右依序为 Manifest、Application、Permissions…，它们其实就是对应到 AndroidManifest.xml 文件中的标签架构，我们可以利用这些标签页来编辑 AndroidManifest.xml 文件的内容，详细的操作方法请读者参考本书光盘中的 Graphical Manifest Editor 操作说明.pdf 文件。其他像是 menu、drawable 等各种程序项目资源文件也都有对应的 Editor 可以使用，我们可以用鼠标右击文件，然后选择 Open With 就会出现各种 Editor 的清单让我们选择，读者可以自行尝试。

以上的步骤虽然有些繁复，但却也让我们更加了解 Android 应用程序的架构和功能。请读者一一完成所有步骤并运行程序，就会在模拟器的屏幕上看到如图 37-2 所示的界面。当用户按下按钮后，就会启动"计算机猜拳游戏"程序，我们可以利用"回上一页"按钮离开"计算机猜拳游戏"程序，回到原来的主程序界面。

图 37-2　在程序中启动另一个 Activity

Android 版本	1.X	2.X	3.X	4.X
适用性	★	★	★	★

UNIT 38

Intent 粉墨登场

在前一个单元中我们先请 Intent 对象小试身手,帮我们启动程序项目中的另一个 Activity,这个单元就让我们正式介绍 Intent 这个 Android 系统中的最佳男主角。Intent 翻成中文的意思是"意图",说的白话一些就是"我想要…",也就是说目前运行中的 Activity 想请其他的 Activity 或是 Android 系统中的其他程序来完成一件工作,并且把运行权交给对方,然后目前的 Activity 便会进入休息的状态,等到对方完成工作交回运行权后,才会重新回到运行状态。

我们可以把 Intent 对象视为是程序和 Android 系统互动的媒介,当程序需要另一个 Activity 或是程序来完成工作时,便可以建立一个 Intent 对象,然后在该对象中填入相关的数据,最后调用 startActivity()方法将此 Intent 对象传给 Android 系统。Android 系统收到 Intent 对象之后,便会根据其中的信息启动适当的 Activity 或程序来进行处理。

让我们先回顾上一个单元使用 Intent 的程序代码:

```
Intent it = new Intent();
it.setClass(Main.this, Game.class);
startActivity(it);
```

这是使用 Intent 对象最简单的格式,我们利用 setClass()方法直接指定要启动的 Activity 类(就是 Game.class),并填入此 Intent 物件的拥有者(Main.class),接着调用 startActivity()方法将它送给 Android 系统。Intent 对象的另一种使用方式是只记录要处理的数据以及处理方法,例如检查、传送、加入或删除,然后将它送出。Android 系统收到这个 Intent 对象之后,会根据其中的数据类型以及处理方式,从系统的记录列表中挑选一个合适的程序来处理。这种型态的 Intent 对象需要使用一个 Uri 对象,它是用来储存要处理的数据,例如程序中如果需要开启一个网页,可以利用以下的程序代码请 Android 系统运行网页浏览程序:

```
Uri uri = Uri.parse("http://www....");
Intent it = new Intent(Intent.ACTION_VIEW, uri);
startActivity(it);
```

　　Intent 对象可以完成许多类型的工作，像是开启网页、传送信息、传送 email、播放影片、播放音乐、开启图片、拍摄照片、安装程序、移除程序等。以下我们直接以程序来示范它的使用方法，这个范例程序使用 Intent 对象请求外部程序完成开启网页、播放 MP3 和检查图片等 3 项工作。它的操作方式就是利用 3 个按钮分别运行上述 3 项工作。程序的运行界面如图 38-1 所示，读者可以利用"回上一页"按钮回到主程序界面。程序接口布局文件和程序代码列出如下，完成每一种类型的工作都有其固定的程序代码格式，这些程序代码都只有短短数行，请读者直接参考下列程序代码范例就可以了解，其中重要的部分以粗体字标示。

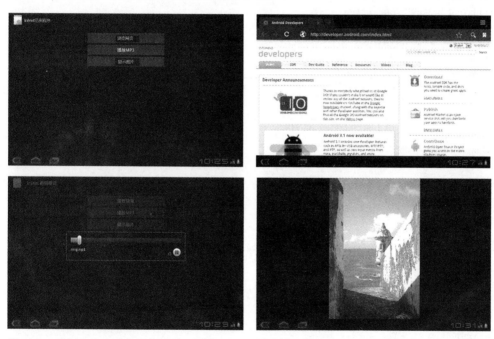

图 38-1　使用 Intent 对象开启网页、播放 MP3 和检查图片的范例

■　界面布局文件：

```
<?xml version="1.0" encoding="utf-8"?>
<LinearLayout xmlns:android="http://schemas.android.com/apk/res/android"
    android:orientation="horizontal"
    android:layout_width="match_parent"
    android:layout_height="match_parent">
<LinearLayout
    android:orientation="vertical"
    android:layout_width="0dp"
```

```xml
        android:layout_height="match_parent"
        android:layout_weight="1"/>
    <LinearLayout
        android:orientation="vertical"
        android:layout_width="0dp"
        android:layout_height="match_parent"
        android:layout_weight="1">
        <Button android:id="@+id/btnBrowseWWW"
            android:layout_width="match_parent"
            android:layout_height="wrap_content"
            android:text="浏览网页"
            android:layout_marginTop="20dp"/>
        <Button android:id="@+id/btnPlayMP3"
            android:layout_width="match_parent"
            android:layout_height="wrap_content"
            android:text="播放 MP3"/>
        <Button android:id="@+id/btnViewImg"
            android:layout_width="match_parent"
            android:layout_height="wrap_content"
            android:text="显示图片"/>
    </LinearLayout>
    <LinearLayout
        android:orientation="vertical"
        android:layout_width="0dp"
        android:layout_height="match_parent"
        android:layout_weight="1"/>
</LinearLayout>
```

■ 程序代码:

```java
package …

import …

public class Main extends Activity {

private Button mBtnBrowseWWW,
            mBtnPlayMP3,
            mBtnViewImg;

    /** Called when the activity is first created. */
    @Override
    public void onCreate(Bundle savedInstanceState) {
        super.onCreate(savedInstanceState);
        setContentView(R.layout.main);

        setupViewComponent();
    }
```

```java
private void setupViewComponent() {
    mBtnBrowseWWW = (Button)findViewById(R.id.btnBrowseWWW);
    mBtnPlayMP3 = (Button)findViewById(R.id.btnPlayMP3);
    mBtnViewImg = (Button)findViewById(R.id.btnViewImg);

    mBtnBrowseWWW.setOnClickListener(btnBrowseWWWOnClickLis);
    mBtnPlayMP3.setOnClickListener(btnPlayMP3OnClickLis);
    mBtnViewImg.setOnClickListener(btnViewImgOnClickLis);
}

private Button.OnClickListener btnBrowseWWWOnClickLis = new
Button.OnClickListener() {
    public void onClick(View v) {
        Uri uri = Uri.parse("http://developer.android.com/");
        Intent it = new Intent(Intent.ACTION_VIEW, uri);
        startActivity(it);
    }
};

private Button.OnClickListener btnPlayMP3OnClickLis = new
Button.OnClickListener() {
    public void onClick(View v) {
        Intent it = new Intent(Intent.ACTION_VIEW);
        File file = new File("/sdcard/song.mp3");
        it.setDataAndType(Uri.fromFile(file), "audio/*");
        startActivity(it);
    }
};

private Button.OnClickListener btnViewImgOnClickLis = new
Button.OnClickListener() {
    public void onClick(View v) {
        Intent it = new Intent(Intent.ACTION_VIEW);
        File file = new File("/sdcard/image.png");
        it.setDataAndType(Uri.fromFile(file), "image/*");
        startActivity(it);
    }
};
}
```

要运行这个程序需要将 MP3 文件和图像文件储存在模拟器的 SD 卡中。把文件上传到模拟器的 SD 卡需要使用 Eclipse 的 DDMS 功能，另外如果想在 SD 卡中建立或删除文件夹则需要进入模拟器的 Linux 操作系统，接下来我们就来介绍相关的操作方法。

38-1　Eclipse 的 DDMS 功能以及模拟器的 Linux 命令行模式

以下是将文件上传到模拟器的 SD 卡的步骤，其中的范例图片是以平板电脑模拟器为例，手机模拟器的操作步骤也完全相同：

Step 1 首先必须建立一个含有 SD 卡的模拟器，请单击 Eclipse 菜单上的"Window>AVD Manager"就会出现图 38-2 的对话框，按下右上方的 New 按钮就会出现图 38-3 的对话框，在 Name 字段中给这个模拟器取一个名称，例如 android_4.0_sd_card，在 Target 字段中选择一个 Android 的版本，然后在下方的字段中指定 SD 卡的容量，例如 50（注意右边的容量单位请选择 MiB），最后按下 Create AVD 按钮。

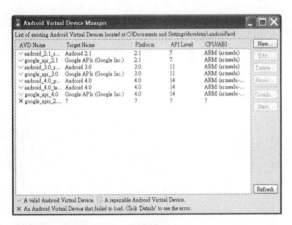

图 38-2　新增含有 SD 卡的模拟器　　　　　图 38-3　设置模拟器的属性

Step 2 启动步骤一建立的模拟器，然后按下模拟器界面上的 Apps 按钮，再从 Apps 列表中选择 Settings，然后单击 Storage。

Step 3 模拟器界面会显示 SD 卡的总容量和可用容量，如图 38-4 所示，如果有问题，可以选择 Unmount SD card，然后单击 Format SD card，完成后再单击 Mount SD card。

图 38-4　模拟器的 SD 卡操作菜单

Intent、Intent Filter 和传送数据　PART 7

Step 4　回到 Eclipse 程序，找到工具栏最右边一个名称叫 Open Perspective 的按钮，当把鼠标光标停在工具栏上的按钮时会弹出一个按钮名称的信息，读者可以利用这个方法找到指定的按钮。单击 Open Perspective 按钮，然后选择其中的 DDMS 项目，Eclipse 的操作界面会随之改变，如图 38-5 所示。在左边的窗口中会显示目前运行中的模拟器名称，请单击含有 SD 卡的模拟器，要注意的是这里的模拟器名称不是我们在步骤一所输入的名称，而是 Android SDK 自定义的模拟器程序名称。

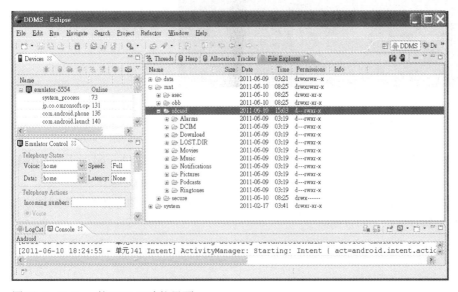

图 38-5　Eclipse 的 DDMS 功能界面

Step 5　在右边窗口上方的标签中单击 File Explorer，然后在下方显示的文件夹结构中展开 mnt/sdcard，并单击 sdcard 文件夹。

Step 6　在 File Explorer 窗口右上方有一个名称叫做 Push a file onto the device 按钮（利用鼠标停留在按钮上的方式显示按钮名称），单击它后会出现文件浏览对话框，选择要上传到模拟器的文件后，再按下右下方的"开启"按钮就会将选择的文件上传至模拟器的 SD 卡。如果所选择的模拟器没有使用 SD 卡，则当文件上传后在下方的 Console 窗口中会显示错误信息。

Step 7　完成上传文件后再利用 Eclipse 工具栏最右边的 Java 按钮回到原来 Eclipse 的界面。

　　以上步骤只能上传文件到模拟器的 SD 卡，或是从 SD 卡中删除文件（利用 Push a file onto the device 按钮右边的那个减号按钮）。如果是要新增或是删除文件夹，就必须使用命令行模式窗口进入模拟器的 Linux 操作系统，请读者依照下列方式操作：

Step 1　从 Windows 的"开始>所有程序>附属应用程序"中运行"命令提示字符"程序。

301

Step 2　将"命令提示字符"程序的运行目录切换到 Eclipse 程序文件夹下的 android-sdk/platform-tools 文件夹（旧版 Android SDK 为 android-sdk-windows/tools 文件夹）。

Step 3　运行指令"adb -s emulator-5554 shell"，其中 emulator-5554 是在 Eclipse 的 DDMS 操作界面中显示的模拟器程序名称，请读者参考图 38-6。

图 38-6　进入模拟器 Linux 操作系统

Step 4　完成之后会显示一个#号提示字符代表我们已经进入模拟器的 Linux 操作系统，接下来运行"cd sdcard"进入 SD 卡磁盘，然后使用 Linux 的操作指令，像是 ls、cd、mkdir、rmdir 等进行文件夹的相关操作，等全部完成后再输入 exit 指令离开 Linux 操作系统。

请读者依照上述方法将程序运行时需要的 song.mp3 文件和 image.png 图像文件上传到模拟器的 SD 卡，然后就可以运行本单元的范例程序。

Android 版本	1.X	2.X	3.X	4.X
适用性	★	★	★	★①

UNIT 39
Tab 标签页接口——使用 Intent 对象

在 UNIT 18 中我们已经介绍过建立 Tab 标签页接口的方法,当时是把不同的标签页中的接口组件放在同一个接口布局文件,而且程序代码也是集中放在主类的程序文件。这种方法的好处是项目中的文件数目不会增加,但是如果 Tab 标签页的接口组件个数比较多,或是程序代码比较复杂,把它们放在同一个文件会造成日后程序维护上的困难。前面我们已经学过 Intent 对象的用法,这个单元就让我们改用 Intent 对象来建立 Tab 标签页。这个新方法其实是基于一个很简单的概念,就是每一个 Tab 标签页都对应到一个独立的 Activity 类,因此不同标签页的接口布局文件和程序代码都是各自独立,如此一来可以减少接口布局文件和程序代码的复杂度,让它们比较容易阅读和维护。这个新的方法还是必须用到 TabHost、TabWidget 和 FrameLayout 这三个组件。以下我们就直接用一个实际的范例来解说,这个范例是把之前完成的 DatePicker 和 TimePicker 程序以及 ProgressBar 程序结合起来,然后放在不同的 Tab 标签页,请读者依照下列步骤操作:

Step 1 建立一个新的 Android 应用程序项目,然后把 res/layout 文件夹下的接口布局文件修改如下:

```
<?xml version="1.0" encoding="utf-8"?>
<TabHost xmlns:android="http://schemas.android.com/apk/res/android"
android:id="@android:id/tabhost"
    android:layout_width="match_parent"
    android:layout_height="match_parent"
```

---

① 可以使用,但是建议改用 Action Bar 上的 Tab 标签页(UNIT 52)。

```xml
>
<TabWidget android:id="@android:id/tabs"
    android:layout_width="match_parent"
    android:layout_height="wrap_content"
    />
<FrameLayout android:id="@android:id/tabcontent"
    android:layout_width="match_parent"
    android:layout_height="match_parent"
    android:paddingTop="70dp"
  />
</TabHost>
```

请读者注意<TabHost>、<TabWidget>和<FrameLayout>这三个组件的 id 名称一定要完全和上面的范例相同，否则程序运行时会发生错误。另外我们在<FrameLayout>组件中设置 android:paddingTop="70dp"以免标签页中的接口组件和标签的标题重迭。

Step 2 依照前面学过的方法在 res/layout 文件夹下新增一个接口布局文件，例如 date_time_picker.xml，然后把之前 DatePicker 和 TimePicker 程序项目中的接口布局文件的内容复制到这个新的接口布局文件中。如果接口布局文件中有使用到字符串资源，也要在新的项目中建立这些字符串资源。

Step 3 依照步骤二的方法新增另一个接口布局文件，例如 prog_bar_demo.xml，然后把之前 ProgressBar 程序项目中的接口布局文件的内容复制到这个新的接口布局文件中。如果接口布局文件中有使用到字符串资源，也要在新的项目中建立这些字符串资源。

Step 4 依照之前面学过的方法，在 src/(组件路径名称)文件夹中新增 2 个 Activity 类。这 2 个类的程序代码请分别复制前面的 DatePicker 和 TimePicker 程序项目以及 ProgressBar 程序项目中的程序代码。但是注意 setContentView()方法中加载的接口布局文件名称必须改成步骤二和步骤三的接口布局文件名称。

Step 5 将主程序类所继承的基础类从 Activity 改成 TabActivity 如下：

```
package …

import …

public class 主类名称 extends TabActivity {
…
```

Step 6 在主类程序文件中加入下列程序代码：

```
TabHost tabHost = getTabHost();

Intent it = new Intent();
it.setClass(Main.this, DateTimePicker.class);
TabSpec spec=tabHost.newTabSpec("tab1");
```

```
spec.setContent(it);
spec.setIndicator("日期和时间",
        getResources().getDrawable(android.R.drawable.ic_lock_idle_alarm));
tabHost.addTab(spec);

it = new Intent();
it.setClass(Main.this, ProgBarDemo.class);
spec=tabHost.newTabSpec("tab2");
spec.setIndicator("ProgressBar",
        getResources().getDrawable(android.R.drawable.ic_dialog_alert));
spec.setContent(it);
tabHost.addTab(spec);

tabHost.setCurrentTab(0);
```

这一段程序代码和之前介绍的建立 Tab 标签页的第一个方法有些类似，二者只有下列三点不同：

1. 建立 TabHost 对象是根据调用 getTabHost()方法而不是 findViewById()方法。
2. 不需要调用 TabHost 对象的 setup()方法。
3. 调用 TabSpec 对象的 setContent()方法时，是传入一个设置好 Activity 类的 Intent 对象。

补充说明　Android 4 版本的改变

原来在 Tab page 的标题可以加上小图标，可是在 Android 4 中运行时小图标会隐藏只显示标题文字。

Step 7 在工程的 AndroidManifest.xml 文件中加入步骤四建立的 2 个 Activity 的信息，如下面程序代码范例的粗体字部分。

```
<?xml version="1.0" encoding="utf-8"?>
<manifest xmlns:android="http://schemas.android.com/apk/res/android"
        …>
    <application …>
        <activity …
                …
        </activity>
        <activity android:name=".DateTimePicker"/>
        <activity android:name=".ProgBarDemo"/>
    </application>
</manifest>
```

Step 8 如果想要改变 Tab page 标题的字体大小，可以利用以下程序代码：

```
TabWidget tabWidget = (TabWidget)tabHost.findViewById(android.R.id.tabs);
View tabView = tabWidget.getChildTabViewAt(0);   // 0 代表第 1 个 Tabe page
```

```
TextView tab = (TextView)tabView.findViewById(android.R.id.title);
tab.setTextSize(20);
```

如果要设置第 2 个 Tabe page 标题的字体大小,就把 0 改成 1,依此类推。

完成上述所有步骤之后运行程序就可以看到如图 39-1 所示的界面。以下列出完整的主类程序代码,其他程序文件中的程序代码都是从 DatePicker 和 TimePicker 程序项目以及 ProgressBar 程序项目复制而来,因此不再赘述。在主类程序文件中我们把建立 Tab 标签页的程序代码写在 setupViewComponent()方法内。这个单元介绍的 Tab 标签页技术可以同时适用在所有的 Android 版本,但如果是针对 Android 3.0 以后所写的程序,建议改用 UNIT 52 所介绍的方法。

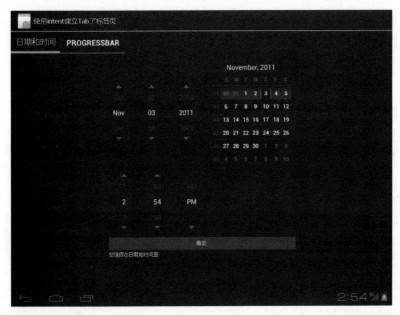

图 39-1 使用 Tab 标签页结合 DatePicker 和 TimePicker 程序以及 ProgressBar 程序

```
package …

import …

public class Main extends TabActivity {
    /** Called when the activity is first created. */
    @Override
    public void onCreate(Bundle savedInstanceState) {
        super.onCreate(savedInstanceState);
        setContentView(R.layout.main);

        setupViewComponent();
    }
```

```java
private void setupViewComponent() {
    TabHost tabHost = getTabHost();

    Intent it = new Intent();
    it.setClass(Main.this, DateTimePicker.class);
    TabSpec spec=tabHost.newTabSpec("tab1");
    spec.setContent(it);
    spec.setIndicator("日期和时间",
            getResources().getDrawable(android.R.drawable.ic_lock_idle_alarm));
    tabHost.addTab(spec);

    it = new Intent();
    it.setClass(Main.this, ProgBarDemo.class);
    spec=tabHost.newTabSpec("tab2");
    spec.setIndicator("ProgressBar",
            getResources().getDrawable(android.R.drawable.ic_dialog_alert));
    spec.setContent(it);
    tabHost.addTab(spec);

    tabHost.setCurrentTab(0);

    // 设置 tab 标签的字体大小
    TabWidget tabWidget = (TabWidget)tabHost.findViewById(android.R.id.tabs);
    View tabView = tabWidget.getChildTabViewAt(0);
    TextView tab = (TextView)tabView.findViewById(android.R.id.title);
    tab.setTextSize(20);
    tabView = tabWidget.getChildTabViewAt(1);
    tab = (TextView)tabView.findViewById(android.R.id.title);
    tab.setTextSize(20);
}
}
```

Android 版本	1.X	2.X	3.X	4.X
适用性	★	★	★	★

UNIT 40

Intent Filter 让程序也能帮助别人

Intent 对象是当程序需要帮手的时候对 Android 系统发出的求助信息，Android 系统收到 Intent 对象之后，会根据其中的描述，启动适合的程序来运行。Intent 对象对于要运行的工作有两种描述方式。第一种是直接指名道姓，指定要运行的 Activity，这种方式称为显式的 Intent（Explicit Intent），就像是 UNIT 37 的范例程序，当用户按下主程序界面中的按钮时，我们在 Intent 对象中指定要启动 Game Activity。第二种方式是在 Intent 对象中记录数据和对数据的操作，这种方式称为隐式的 Intent（Implicit Intent），例如 UNIT 38 的范例程序，我们只把网址、MP3 文件名或图像文件名称写入 Intent 对象，并设置想要的操作方式，例如 ACTION_VIEW。当 Android 系统收到这种 Intent 对象的时候，会自动搜寻可以完成这项工作的程序。如果没有找到任何可以运行的程序，Android 系统就会显示如图 40-1 所示的信息，然后终止原来发出这个 Intent 的程序。如果刚好找到一个可以完成这项工作的程序，Android 系统就会直接启动该程序，如果找到多个程序可以运行这项工作，就会显示一个菜单窗口，请用户挑选要启动的程序，如图 40-2 所示。当 Android 系统要帮隐式的 Intent 挑选合适的程序时，必须使用 Intent Filter 机制，有关 Intent Filter 的信息是记录在程序项目的 AndroidManifest.xml 文件中。

图 40-1　找不到可以运行 Intent 对象的程序时显示的错误信息

图 40-2 找到多个可以运行 Intent 对象的程序时所出现的菜单界面

40-1 设置 AndroidManifest.xml 文件中的 Intent Filter

请读者在 Eclipse 中开启前一个单元范例程序的 AndroidManifest.xml 文件，就会看到如下的程序代码：

```xml
<?xml version="1.0" encoding="utf-8"?>
<manifest xmlns:android="http://schemas.android.com/apk/res/android"
    package="tw.android"
    android:versionCode="1"
    android:versionName="1.0">
    <uses-sdk android:minSdkVersion="11" />

    <application android:icon="@drawable/icon" android:label="@string/app_name">
        <activity android:name=".Main"
                  android:label="@string/app_name">
            <intent-filter>
                <action android:name="android.intent.action.MAIN" />
                <category android:name="android.intent.category.LAUNCHER" />
            </intent-filter>
        </activity>

    </application>
</manifest>
```

有关 AndroidManifest.xml 文件的架构我们已经在 UNIT 37 中作过介绍，只剩下<intent-filter>标签还没有解释，现在就请读者注意以上程序代码中粗体字的部分。<intent-filter>标签是包含在<activity>…</activity>的标签中，它是告诉 Android 系统这个 Activity 的功能。这个范例中使用<action … />和<category …/>这两个标签，这一组标签的设置值是告诉 Android 系统，这个 Activity 是程序项目运行的入口，也就是启动程序项目时，第一个运行的 Activity。

如果程序项目中新增了其他的 Activity，每一个新的 Activity 都要在程序项目的 AndroidManifest.xml 文件中描述它的相关信息，我们可以为 Activity 加上<intent-filter>标签，然后在里面使用下列三种标签描述它的功能：

1. <action .../>标签
   描述这个 Activity 可以运行的操作类型，例如 VIEW、EDIT、DIAL、SEND 等。

2. <data .../>标签
   描述这个 Activity 可以处理的数据或是数据类型。比较常见的是指定数据类型，例如我们可以利用属性 android:mimeType 指定要处理图像或是影片数据：image/png、image/jpg、image/*、video/*等（后续会以范例说明），其中*代表所有格式都可以接受，或是利用属性 android:scheme 指定数据类型，例如 http、tel、file 等。

3. <category .../>标签
   指定这个 Activity 所属的类，如果是程序项目启动时要运行的 Activity，就必须设置为 LAUNCHER 类，如下：

   `<category android:name="android.intent.category.LAUNCHER" />`

   如果是其他的 Activity，通常都设置为 DEFAULT 类。

   `<category android:name="android.intent.category.DEFAULT" />`

接着我们看以下的范例：

```
<activity android:name=".MyImageAct"
          android:label="@string/title_MyImageAct">
    <intent-filter>
        <action android:name="android.intent.action.VIEW" />
        <action android:name="android.intent.action.EDIT" />
        <category android:name="android.intent.category.DEFAULT" />
        <data android:mimeType="image/*" />
    </intent-filter>
    <intent-filter>
        <action android:name="android.intent.action.VIEW" />
        <category android:name="android.intent.category.DEFAULT" />
        <data android:scheme="http" />
    </intent-filter>
</activity>
```

这一段程序代码告诉 Android 系统在此程序项目中有一个 Activity 名称叫做 MyImageAct，它的内部包含两组<intent-filter>标签，这是告诉 Android 系统它可以对两种不同数据类型的 Intent 对象提供服务，而服务的内容是由<intent-filter>标签内的描述来决定，以第一组<intent-filter>标签为例：

1. <intent-filter>标签内的第一行和第二行是说这个 Activity 可以用来检查（VIEW）和编辑（EDIT）某种型态的数据。

2. <intent-filter>标签内的第三行是说这个 Activity 是属于 DEFAULT 类，也就是可以为一般的 Intent 对象提供服务。

3. <intent-filter>标签内的第四行是说这个Activity处理的数据型态是图像,而且可以接受所有图像格式。

第二组<intent-filter>标签的内容和以上的说明类似,只是在<data>的描述中换成使用android:scheme属性指定数据的类型。编辑好Activity的Intent Filter数据后,接下来的问题是当程序送出一个Intent对象之后,Android系统如何找出可以处理这个Intent对象的程序?

40-2　Android系统检查Intent和Intent Filter的规则

当Android系统收到一个Intent对象之后,它会把全部有设置<intent-filter>标签的程序依照下列规则进行比对。

1. 检查action项目

   Activity在<intent-filter>标签中设置的action项目必须含有Intent对象中指定要运行的action项目。

2. 检查category项目

   Activity在<intent-filter>标签中设置的category必须和Intent对象指定的category相同。如果Intent对象中没有指定category,则视为DEFAULT。

3. 检查data项目

   data项目的比对方式比较复杂。Android系统会从Intent所附带的数据中尝试抽取出type、scheme、authority和path四个部分(有些部分可能是空的),然后再和Activity的<intent-filter>标签中的<data>设置进行比对看有没有相符。

符合以上三项条件的Activity都会被挑选出来,列入接受此Intent对象的列表。然后如同我们在前面的说明,Android系统会根据列表中的Activity数目采取不同的后续处理方式。

40-3　程序接收到Intent对象的工作

当Android系统决定接收Intent对象的Activity之后,该Activity就会被启动,然后运行它的onCreate()方法。我们在onCreate()方法中完成以下工作:

Step 1 调用getIntent()方法取得Android系统传入的Intent对象。

```
Intent it = getIntent();
```

Step 2 利用Intent对象的方法取得data、action、scheme、category等资料,并根据数据类型和指定的操作方式处理资料。

```
String sAct = it.getAction();
String sScheme = it.getScheme();
```

```
        if (sScheme.equals("http")) {
            // 运行开启网页的程序代码
            …
        }
        else if (sScheme.equals("file")) {
            if (sAct.equals("android.intent.action.VIEW")) {
                // 运行检查文件的程序代码
                …
            }
            else if (sAct.equals("android.intent.action.EDIT")) {
                // 运行编辑文件的程序代码
                …
            }
        }
```

以上就是 Activity、Intent 对象和 Intent Filter 三者之间的运行机制和使用方法，讨论起来好像有点复杂，但是实现并不困难。

40-4　范例程序

我们使用一个程序项目来示范实现的过程，这个范例是把前一个单元的程序项目略作修改，原来的程序是示范如何使用 Intent 对象启动网页浏览器、MP3 播放器和图像检查程序，这个单元我们要在该项目中新增一个具有 Intent Filter 功能的 Activity，它可以接收浏览网页、检查和编辑图像等工作的 Intent 对象，请读者依下列步骤操作：

Step 1 使用 Windows 文件管理器复制前一个单元的程序项目文件夹，复制之后可以变更复制文件夹的名称。

Step 2 开启 Eclipse，使用主菜单 File > Import 加载复制的项目。

Step 3 在程序项目中新增一个继承 Activity 类的新类，我们可以将它取名为 MyImageAct（新增类的步骤已经操作过许多次，如果需要可以参考前面单元的说明），另外我们还要为这个 MyImageAct 类建立一个接口布局文件 my_image_act.xml（置于 res/layout 文件夹中），它的接口很简单，只有一个 TextView 组件如下：

```xml
<?xml version="1.0" encoding="utf-8"?>
<LinearLayout xmlns:android="http://schemas.android.com/apk/res/android"
    android:orientation="vertical"
    android:layout_width="match_parent"
    android:layout_height="match_parent"
    android:gravity="center_horizontal"
    >
<TextView android:id="@+id/txtResult"
    android:layout_width="300dp"
    android:layout_height="wrap_content"
```

```
    android:text=""
    />
</LinearLayout>
```

Step 4 这个新增的 MyImageAct 类将被设置对 image 类型的数据提供 VIEW 和 EDIT 以及对网址数据运行 VIEW，只是在实际程序代码中我们不会真的实现这些功能，只会将收到的资料和所指定的操作显示在屏幕上，以下是 MyImageAct 类的完整程序代码：

```
package …

import …

public class MyImageAct extends Activity {

    private TextView mTxtResult;

    /** Called when the activity is first created. */
    @Override
    public void onCreate(Bundle savedInstanceState) {
        super.onCreate(savedInstanceState);
        setContentView(R.layout.my_image_act);

        setupViewComponent();
        showResult();
    }

    private void setupViewComponent() {
        mTxtResult = (TextView)findViewById(R.id.txtResult);
    }

    private void showResult() {
        Intent it = getIntent();
        String sAct = it.getAction();
        String sScheme = it.getScheme();
        if (sScheme.equals("http")) {
            String s = "接收到的 Intent 对象要求\"开启网页\"" +
                       it.getData().toString();
            mTxtResult.setText(s);
        }
        else if (sScheme.equals("file")) {
            if (sAct.equals("android.intent.action.VIEW")) {
                String s = "接收到的 Intent 对象要求\"检查\"" +
                           it.getData().toString();
                mTxtResult.setText(s);
            }
            else if (sAct.equals("android.intent.action.EDIT")) {
```

```
            String s = "接收到的 Intent 对象要求\"编辑\"" +
                    it.getData().toString();
            mTxtResult.setText(s);
        }
    }
  }
}
```

我们把取得 Intent 对象和附带数据的程序代码,以及运行数据处理的程序代码全部写在 showResult()方法中,然后在 onCreate()方法内调用 showResult()方法。

Step 5 在程序项目中的 AndroidManifest.xml 文件中加上 MyImageAct 的描述,包括它的 Intent Filter,请读者留意以下程序代码的粗体字部分,其中我们使用了定义在字符串资源文件中名为 title_MyImageAct 的字符串,因此读者必须在此程序项目的 res/values/strings.xml 文件中新增该字符串的定义。

```xml
<?xml ...>
<manifest ...>
    <application ...">
        <activity android:name=".Main" ...>
            ...
        </activity>
        <activity android:name=".MyImageAct"
                android:label="@string/title_MyImageAct">
            <intent-filter>
                <action android:name="android.intent.action.EDIT" />
                <action android:name="android.intent.action.VIEW" />
                <category android:name="android.intent.category.DEFAULT" />
                <data android:mimeType="image/*" />
            </intent-filter>
            <intent-filter>
                <action android:name="android.intent.action.VIEW" />
                <category android:name="android.intent.category.DEFAULT" />
                <data android:scheme="http" />
            </intent-filter>
        </activity>
    </application>
</manifest>
```

Step 6 主程序的接口布局文件和程序代码请修改如下,我们将原来播放 MP3 的按钮换成编辑图片的按钮以便示范不同操作的处理方法,程序的运行界面如图 40-3 所示,图 40-4 是分别按下三个按钮后的运行结果(按下第一个按钮后必须选择启动我们的程序)。

图 40-3　程序的运行界面　　　　　　　　图 40-4　按下程序界面的三个按钮后的运行结果

■　界面布局文件：

```xml
<?xml version="1.0" encoding="utf-8"?>
<LinearLayout xmlns:android="http://schemas.android.com/apk/res/android"
    android:orientation="vertical"
    android:layout_width="match_parent"
    android:layout_height="match_parent"
    android:gravity="center_horizontal"
    >
<Button android:id="@+id/btnBrowseWWW"
    android:layout_width="wrap_content"
    android:layout_height="wrap_content"
    android:text="浏览网页"
    android:layout_marginTop="20dp"
    android:paddingLeft="50dp"
    android:paddingRight="50dp"
    />
<Button android:id="@+id/btnEditImg"
    android:layout_width="wrap_content"
    android:layout_height="wrap_content"
    android:text="编辑图片"
    android:paddingLeft="50dp"
    android:paddingRight="50dp"
    />
<Button android:id="@+id/btnViewImg"
    android:layout_width="wrap_content"
    android:layout_height="wrap_content"
    android:text="检查图片"
    android:paddingLeft="50dp"
    android:paddingRight="50dp"
    />
</LinearLayout>
```

■ 程序代码：

```java
package …

import …

public class Main extends Activity {

    private Button mBtnBrowseWWW,
                   mBtnEditImg,
                   mBtnViewImg;

    /** Called when the activity is first created. */
    @Override
    public void onCreate(Bundle savedInstanceState) {
        super.onCreate(savedInstanceState);
        setContentView(R.layout.main);

        setupViewComponent();
    }

    private void setupViewComponent() {
        mBtnBrowseWWW = (Button)findViewById(R.id.btnBrowseWWW);
        mBtnEditImg = (Button)findViewById(R.id.btnEditImg);
        mBtnViewImg = (Button)findViewById(R.id.btnViewImg);

        mBtnBrowseWWW.setOnClickListener(btnBrowseWWWOnClickLis);
        mBtnEditImg.setOnClickListener(btnEditImgOnClickLis);
        mBtnViewImg.setOnClickListener(btnViewImgOnClickLis);
    }

    private Button.OnClickListener btnBrowseWWWOnClickLis = new Button.OnClickListener() {
        public void onClick(View v) {
            Uri uri = Uri.parse("http://developer.android.com/");
            Intent it = new Intent(Intent.ACTION_VIEW, uri);
            startActivity(it);
        }
    };

    private Button.OnClickListener btnEditImgOnClickLis = new Button.OnClickListener() {
        public void onClick(View v) {
            Intent it = new Intent(Intent.ACTION_EDIT);
            File file = new File("/sdcard/image.png");
            it.setDataAndType(Uri.fromFile(file), "image/*");
            startActivity(it);
        }
```

```
    };

    private Button.OnClickListener btnViewImgOnClickLis = new
Button.OnClickListener() {
        public void onClick(View v) {
            Intent it = new Intent(Intent.ACTION_VIEW);
            File file = new File("/sdcard/image.png");
            it.setDataAndType(Uri.fromFile(file), "image/*");
            startActivity(it);
        }
    };
}
```

Android 版本	1.X	2.X	3.X	4.X
适用性	★	★	★	★

UNIT 41

让 Intent 对象附带数据

Intent 对象可以用来启动其他的 Activity，或是调用其他程序来完成特定的工作。有些时候，调用的程序可能需要将一些数据传给对方，这时候可以把这些资料先储存在一个 Bundle 对象中，然后把这个 Bundle 对象传给 Intent 对象，于是 Bundle 对象中的数据就会随着 Intent 对象传给对方，对方收到 Intent 对象之后再取出其中的数据，这整个流程我们分成两个部分来说明。

41-1 传送数据的 Activity 需要完成的工作

传送数据的 Activity 必须完成以下步骤：

Step 1 建立一个 Intent 对象，然后调用 setClass()方法设置拥有者和要启动的 Activity，通常 Intent 对象的拥有者就是建立它的 Activity。

```
Intent it = new Intent();
it.setClass(类名称.this, 要启动的 Activity 名称.class);
```

Step 2 建立一个 Bundle 对象，然后调用 Bundle 对象的 putXXX()方法把数据和数据名称放到 Bundle 对象中，每一种数据类型都有对应的 put 方法可以使用，请参考表 41-1。

我们必须为每一项加入 Bundle 对象的数据取一个名称，对方程式再利用这个名称从 Bundle 对象中取出数据。一个 Bundle 对象可以加入多笔数据，请读者参考下列程序代码：

```
int iVal = 33;
double dVal = 200.5;

Bundle bundle = new Bundle();
bundle.putInt("DATA_INT", iVal);
bundle.putDouble("DATA_DOUBLE", dVal);
```

表 41-1　Bundle 对象的各种数据类型的 put 方法

方法名称	说明
putBoolean(String key, boolean value)	将 boolean 类型的数据放到 Bundle 对象中，第一个自变量是我们帮这个数据所取的名称
putBooleanArray(String key, boolean[] value)	将 boolean 类型的数据数组放到 Bundle 对象中，第一个自变量是我们帮这个数据所取的名称
putByte(String key, byte value)	将 byte 类型的数据放到 Bundle 对象中，第一个自变量是我们帮这个数据所取的名称 putByteArray(String key, byte[] value) 将 byte 类型的数据数组放到 Bundle 对象中，第一个自变量是我们帮这个数据所取的名称
putChar(String key, char value)	将 char 类型的数据放到 Bundle 对象中，第一个自变量是我们帮这个数据所取的名称 putCharArray(String key, char[] value) 将 char 类型的数据数组放到 Bundle 对象中，第一个自变量是我们帮这个数据所取的名称
putDouble(String key, double value)	将 double 类型的数据放到 Bundle 对象中，第一个自变量是我们帮这个数据所取的名称
putDoubleArray(String key, double[] value)	将 double 类型的数据数组放到 Bundle 对象中，第一个自变量是我们帮这个数据所取的名称
putFloat(String key, float value)	将 float 类型的数据放到 Bundle 对象中，第一个自变量是我们帮这个数据所取的名称
putFloatArray(String key, float[] value)	将 float 类型的数据数组放到 Bundle 对象中，第一个自变量是我们帮这个数据所取的名称
putInt(String key, int value)	将 int 类型的数据放到 Bundle 对象中，第一个自变量是我们帮这个数据所取的名称
putIntArray(String key, int[] value)	将 int 类型的数据数组放到 Bundle 对象中，第一个自变量是我们帮这个数据所取的名称
putLong(String key, long value)	将 long 类型的数据放到 Bundle 对象中，第一个自变量是我们帮这个数据所取的名称
putLongArray(String key, long[] value)	将 long 类型的数据数组放到 Bundle 对象中，第一个自变量是我们帮这个数据所取的名称
putShort(String key, short value)	将 short 类型的数据放到 Bundle 对象中，第一个自变量是我们帮这个数据所取的名称
putShortArray(String key, short[] value)	将 short 类型的数据数组放到 Bundle 对象中，第一个自变量是我们帮这个数据所取的名称
putString(String key, String value)	将 String 类型的数据放到 Bundle 对象中，第一个自变量是我们帮这个数据所取的名称
putStringArray(String key, String[] value)	将 String 类型的数据数组放到 Bundle 对象中，第一个自变量是我们帮这个数据所取的名称

Step 3　利用 Intent 对象的 putExtras() 方法把 Bundle 对象传给 Intent 对象，Bundle 对象中的数据就会储存在 Intent 对象中。

```
it.putExtras(bundle);
```

步骤二和步骤三也可以用另一种方式来完成，就是不使用 Bundle 对象，直接利用 Intent 对象的 putExtra()方法把数据逐一存入 Intent 对象，请参考以下范例，这两种方法的效果完全相同。

```
int iVal = 33;
double dVal = 200.5;

it.putExtra("DATA_INT", iVal);
it.putExtra("DATA_DOUBLE", dVal);
```

Step 4 调用 startActivity()方法送出 Intent 对象。

以上就是利用 Intent 对象传送数据的方法，接下来介绍如何从 Intent 对象中取出数据。

41-2 从 Intent 对象中取出数据

Android 系统会根据 Intent 对象中的描述启动适当的 Activity，被启动的 Activity 的 onCreate()方法会先被运行，我们就在 onCreate()方法中利用下列步骤取出附带在 Intent 对象中的数据。

Step 1 调用 getIntent()方法取得传送过来的 Intent 对象。

```
Intent it = getIntent();
```

Step 2 从 Intent 对象中取出数据并包装成 Bundle 对象。

```
Bundle bundle = it.getExtras();
```

Step 3 根据数据名称取出 Bundle 对象中的数据：

```
int iData = bundle.getInt("DATA_INT");
double dData = bundle.getDouble("DATA_DOUBLE");
```

步骤二和步骤三也可以用另一种方式来完成，就是不使用 Bundle 对象，直接利用 Intent 对象的 getXXXExtra()方法把数据逐一取出，其中的 XXX 代表数据类型，例如 Int、Double、Long、……，使用 getXXXExtra()方法时第二个参数必须设置一个内定值，如果指定的数据名称不存在时会使用该内定值，请参考以下范例：

```
int iData = it.getIntExtra("DATA_INT", 0);
double dData = it.getDoubleExtra("DATA_DOUBLE", 0);
```

41-3 范例程序

本单元的范例程序再度请出我们的"计算机猜拳游戏"程序项目，我们希望加上游戏局数的统计资料，看看用户总共玩了几局，其中赢了几局，而计算机赢了几局，还有几局是平手，有关局数统计的数据是交由另一个 Activity 来显示。我们在原来的游戏程序中新增一个"显示局数统计数据"的按钮，按下该按钮时，游戏主程序会将局数统计数据，根据 Intent 对象传送给负责显示的 Activity。用户看完局数统计数据后，再按下"回到游戏"按钮重新回到游戏程序，游戏界面如图 41-1 所示。以下我们逐步列出需要完成的工作。

图 41-1 "计算机猜拳游戏"利用 Intent 对象传递局数统计数据

Step 1 使用 Windows 文件管理器复制 UNIT 19 的程序项目文件夹，复制之后可以变更复制文件夹的名称，然后开启 Eclipse，使用主菜单 File > Import 加载复制的项目文件夹。

Step 2 在原来的游戏接口布局文件中新增一个"显示局数统计数据"的按钮，如以下粗体字的部分：

```
<?xml version="1.0" encoding="utf-8"?>
<RelativeLayout …
    >
…
<Button android:id="@+id/btnShowResult"
    android:layout_width="wrap_content"
    android:layout_height="wrap_content"
    android:text="显示局数统计数据"
    android:layout_below="@id/txtResult"
    android:textSize="20sp"
    android:layout_marginTop="20dp"
    android:layout_centerHorizontal="true"/>
</RelativeLayout>
```

Step 3 在原来的游戏程序中新增局数统计相关变量，并随着游戏的进行持续更新其中的值，有关局数统计的程序代码请参考 UNIT 25 中的说明。

Step 4 建立"显示局数统计数据"按钮的 onClickListener 对象并完成内部的程序代码，然后把它设置给"显示局数统计数据"按钮，请参考以下粗体字的程序代码：

```java
package …

import …

public class Main extends Activity {

private int miCountSet = 0,
          miCountPlayerWin = 0,
          miCountComWin = 0,
          miCountDraw = 0;

    private Button btnShowResult;

    …

    /** Called when the activity is first created. */
    @Override
    public void onCreate(Bundle savedInstanceState) {
        super.onCreate(savedInstanceState);
        setContentView(R.layout.main);

        setupViewComponent();
    }

    private void setupViewComponent() {
        …
        btnShowResult = (Button)findViewById(R.id.btnShowResult);
        btnShowResult.setOnClickListener(btnShowResultLis);
    }

    …

    private Button.OnClickListener btnShowResultLis = new Button.OnClickListener() {
        public void onClick(View v) {
            Intent it = new Intent();
            it.setClass(Main.this, GameResult.class);

            Bundle bundle = new Bundle();
            bundle.putInt("KEY_COUNT_SET", miCountSet);
            bundle.putInt("KEY_COUNT_PLAYER_WIN", miCountPlayerWin);
            bundle.putInt("KEY_COUNT_COM_WIN", miCountComWin);
            bundle.putInt("KEY_COUNT_DRAW", miCountDraw);
            it.putExtras(bundle);
```

```
                startActivity(it);
            }
        };
    }
```

Step 5 在 Android 工程的 res/layout 文件夹下新增一个界面布局文件，这个接口布局文件是用来显示局数统计数据，我们将它取名为 game_result.xml，以下是此接口布局文件的内容：

```xml
<?xml version="1.0" encoding="utf-8"?>
<LinearLayout xmlns:android="http://schemas.android.com/apk/res/android"
    android:orientation="horizontal"
    android:layout_width="match_parent"
    android:layout_height="match_parent">
<LinearLayout
    android:orientation="vertical"
    android:layout_width="0dp"
    android:layout_height="match_parent"
    android:layout_weight="1"/>
<LinearLayout
    android:orientation="vertical"
    android:layout_width="0dp"
    android:layout_height="match_parent"
    android:layout_weight="1">
<TextView
    android:layout_width="match_parent"
    android:layout_height="wrap_content"
    android:text="全部局数："/>
<EditText android:id="@+id/edtCountSet"
    android:layout_width="match_parent"
    android:layout_height="wrap_content"
    android:editable="false"
    android:text=""/>
<TextView
    android:layout_width="match_parent"
    android:layout_height="wrap_content"
    android:text="玩家赢："/>
<EditText android:id="@+id/edtCountPlayerWin"
    android:layout_width="match_parent"
    android:layout_height="wrap_content"
    android:editable="false"
    android:text=""/>
<TextView
    android:layout_width="match_parent"
    android:layout_height="wrap_content"
    android:text="计算机赢："/>
<EditText android:id="@+id/edtCountComWin"
    android:layout_width="match_parent"
    android:layout_height="wrap_content"
```

```xml
        android:editable="false"
        android:text=""/>
<TextView
        android:layout_width="match_parent"
        android:layout_height="wrap_content"
        android:text="平手："/>
<EditText android:id="@+id/edtCountDraw"
        android:layout_width="match_parent"
        android:layout_height="wrap_content"
        android:editable="false"
        android:text=""/>
<Button android:id="@+id/btnBackToGame"
        android:layout_width="match_parent"
        android:layout_height="wrap_content"
        android:text="回到游戏"/>
</LinearLayout>
<LinearLayout
        android:orientation="vertical"
        android:layout_width="0dp"
        android:layout_height="match_parent"
        android:layout_weight="1"/>
</LinearLayout>
```

> **补充说明** 让程序的操作画面同时适用手机和平板电脑的技巧
>
> 本单元的范例程序利用二层 LinearLayout 配合设置 android:layout_weight 属性，让操作界面置于屏幕中央，这种方法适用在大屏幕的平板电脑，但是如果是在手机上运行，就直接使用一个 LinearLayout 即可。为了让程序能够同时适用不同屏幕尺寸的设备，可以参考 UNIT 58 中介绍的资源文件夹命名技术，分别设计应用于不同屏幕大小的接口布局文件。

Step 6 建立一个新的 Activity 负责显示局数统计数据，我们可以将它取名为 GameResult。以下列出这个 Activity 的程序代码。其中要注意的是 showResult()方法，该方法就是负责从 Intent 对象中取出数据并显示出来。另外当用户按下"回到游戏"按钮时，会调用 Activity 对象的 finish()方法结束这个 Activity，请读者注意粗体字的部分：

```java
package ...

import ...

public class GameResult extends Activity {

    private EditText mEdtCountSet,
                mEdtCountPlayerWin,
                mEdtCountComWin,
                mEdtCountDraw;
```

```java
        private Button btnBackToGame;

        @Override
        protected void onCreate(Bundle savedInstanceState) {
            // TODO Auto-generated method stub
            super.onCreate(savedInstanceState);
            setContentView(R.layout.game_result);

            setupViewComponent();
            showResult();
        }

        private void setupViewComponent() {
            mEdtCountSet = (EditText)findViewById(R.id.edtCountSet);
            mEdtCountPlayerWin = (EditText)findViewById(R.id.edtCountPlayerWin);
            mEdtCountComWin = (EditText)findViewById(R.id.edtCountComWin);
            mEdtCountDraw = (EditText)findViewById(R.id.edtCountDraw);
            btnBackToGame = (Button)findViewById(R.id.btnBackToGame);

            btnBackToGame.setOnClickListener(btnBackToGameLis);
        }

        private Button.OnClickListener btnBackToGameLis = new
        Button.OnClickListener() {
            public void onClick(View v) {
                GameResult.this.finish();
            }
        };

        private void showResult() {
            // 从 bundle 中取出数据
            Bundle bundle = getIntent().getExtras();

            int iCountSet = bundle.getInt("KEY_COUNT_SET");
            int iCountPlayerWin = bundle.getInt("KEY_COUNT_PLAYER_WIN");
            int iCountComWin = bundle.getInt("KEY_COUNT_COM_WIN");
            int iCountDraw = bundle.getInt("KEY_COUNT_DRAW");

            mEdtCountSet.setText(Integer.toString(iCountSet));
            mEdtCountPlayerWin.setText(Integer.toString(iCountPlayerWin));
            mEdtCountComWin.setText(Integer.toString(iCountComWin));
            mEdtCountDraw.setText(Integer.toString(iCountDraw));
        }
}
```

Step 7 在 AndroidManifest.xml 文件中新增此 Activity 的信息如下：

```xml
<?xml version="1.0" encoding="utf-8"?>
<manifest xmlns:android="http://schemas.android.com/apk/res/android"
    ...>
```

```xml
<application …>
    <activity …
        …
    </activity>
    <activity android:name=".GameResult"/>
</application>
</manifest>
```

　　以上操作步骤虽然有些复杂,但是概念并不难,而且有许多步骤都是利用前面已经学过的技巧,只是现在我们把它们结合起来一起使用,建议读者可以自己动手实现一次,甚至作些修改和测试,相信一定获益良多。

Android 版本	1.X	2.X	3.X	4.X
适用性	★	★	★	★

UNIT 42
要求被调用的 Activity 返回数据

前一个单元的情况是原来的 Activity 在 Intent 对象中夹带数据传给被调用的 Activity。这个单元我们将学习相反的情况，也就是让被调用的 Activity 返回数据给原来的 Activity。返回数据的方法仍然是利用 Intent 对象，只是运行的机制有些不同。原来的 Activity 传送数据给被调用的 Activity 的过程比较简单，只要把数据放入 Intent 对象中，接着调用 startActivity()方法就可以将数据传给对方。但是要让被调用的 Activity 返回数据并不能使用 startActivity()，因为当调用 startActivity()时，会造成目前的 Activity 处于等待的状态，但是被调用的 Activity 应该在运行完毕后就直接结束而不是处于等待的状态，因此返回数据的机制必然有所不同。

首先是原来的 Activity 必须改用 startActivityForResult()方法替代原来的 startActivity()方法。startActivityForResult()方法中同样是利用一个 Intent 对象来传递数据，可是它还另外多了一个请求代码，这个请求代码是用来确认返回数据的来源。此外在原来的 Activity 中必须新增一个 onActivityResult()方法，这个方法是当被调用的 Activity 结束后，Android 系统会将它返回的 Intent 对象和结果代码，以及前面提到的请求代码一起传入 onActivityResult()方法。至于被调用的 Activity 则必须将运行结果放入一个 Intent 对象，然后调用 setResult()方法传回这个 Intent 对象，另外还必须传回一个结果代码。

以下我们依照惯例使用一个程序项目来示范实现过程。这个程序项目是以 UNIT 37 的范例为基础，原来的程序是在界面上显示一个"运行计算机猜拳游戏"的按钮，用户按下按钮之后就会启动猜拳游戏程序的 Activity。在这个单元中我们将把"计算机猜拳游戏"程序改成具有局数统计的功能（和前一个单元的程序相同），并且加上"完成游戏"和"取消"两个按钮。当使用者玩了几局之后，可以按下"完成游戏"按钮回到主程序界面，同时送回局数统计数据，主程序收到这些返回的数据后会将它们显示出来。

为了方便完成这个程序，读者可以先利用 Windows 文件管理器复制 UNIT 37 的程序项目文件夹，复制后可以变更项目文件夹的名称，再利用 Eclipse 的 File>Import 功能将该复制的程序项目加载 Eclipse，然后依照下列步骤操作：

Step 1 在主程序 Activity 的接口布局文件中新增一个 TextView 接口组件用来显示游戏局数统计数据，如以下粗体字的部分：

```xml
<?xml version="1.0" encoding="utf-8"?>
<LinearLayout xmlns:android="http://schemas.android.com/apk/res/android"
    android:orientation="horizontal"
    android:layout_width="match_parent"
    android:layout_height="match_parent"
    >
<LinearLayout
    android:orientation="vertical"
    android:layout_width="0dp"
    android:layout_height="match_parent"
    android:layout_weight="1"
    />
<LinearLayout
    android:orientation="vertical"
    android:layout_width="0dp"
    android:layout_height="match_parent"
    android:layout_weight="1"
    >
<Button android:id="@+id/btnExecGame"
    android:layout_width="match_parent"
    android:layout_height="wrap_content"
    android:text="运行"计算机猜拳游戏"程序"
    />
<TextView android:id="@+id/txtResult"
    android:layout_width="wrap_content"
    android:layout_height="wrap_content"
    android:text=""
    />
</LinearLayout>
<LinearLayout
    android:orientation="vertical"
    android:layout_width="0dp"
    android:layout_height="match_parent"
    android:layout_weight="1"
    />
</LinearLayout>
```

Step 2 在主程序 Activity 的程序代码中新增一个 onActivityResult()方法，在该方法中依序检查返回的请求代码和结果代码（我们将请求代码定义为私有常数 LAUNCH_GAME），如果返回的请求代码和原来送出的请求代码不同，代表数据的来源有问题，因此程序会放弃运行，否则继续检查结果代码。如果游戏 Activity 传回的结果代码是 RESULT_OK，则从 data 自变量中（data 自变量就是游戏 Activity 传回的 Intent 对象）取出数据然后将它们显示出来。如果游戏 Activity 传回的结果代码是 RESULT_CANCELED，代表使用者是按下"取消"按钮，因此显示"用户取消游戏"的信息，另外记得使用 startActivityForResult()方法替代原来的 startActivity()。以下是实现的程序代码，请留意粗体字的部分：

```java
package ...

import ...

public class Main extends Activity {

    private Button mBtnExecGame;
    private TextView mTxtResult;
    final private int LAUNCH_GAME = 0;

    /** Called when the activity is first created. */
    @Override
    public void onCreate(Bundle savedInstanceState) {
        super.onCreate(savedInstanceState);
        setContentView(R.layout.main);

        setupViewComponent();
    }

    private void setupViewComponent() {
        mTxtResult = (TextView)findViewById(R.id.txtResult);
        mBtnExecGame = (Button)findViewById(R.id.btnExecGame);
        mBtnExecGame.setOnClickListener(btnExecGameOnClkLis);
    }

    private Button.OnClickListener btnExecGameOnClkLis =
                            new Button.OnClickListener() {
        public void onClick(View v) {
            Intent it = new Intent();
            it.setClass(Main.this, Game.class);
            startActivityForResult(it, LAUNCH_GAME);
        }
    };

    protected void onActivityResult(int requestCode, int resultCode,
    Intent   data) {
```

```java
        if (requestCode != LAUNCH_GAME)
            return;

        switch (resultCode) {
        case RESULT_OK:
            Bundle bundle = data.getExtras();

            int iCountSet = bundle.getInt("KEY_COUNT_SET");
            int iCountPlayerWin = bundle.getInt("KEY_COUNT_PLAYER_WIN");
            int iCountComWin = bundle.getInt("KEY_COUNT_COM_WIN");
            int iCountDraw = bundle.getInt("KEY_COUNT_DRAW");

            String s = "游戏结果：你总共玩了" + iCountSet +
                    "局，赢了" + iCountPlayerWin +
                    "局，输了" + iCountComWin +
                    "局，平手" + iCountDraw + "局";
            mTxtResult.setText(s);

            break;
        case RESULT_CANCELED:
            mTxtResult.setText("你选择取消游戏。");
        }
    }
}
```

Step 3 在"计算机猜拳游戏"的 Activity（类名称为 Game）的接口布局文件中新增 2 个按钮，一个是"完成游戏"按钮，另一个则是"取消"按钮，请参考以下范例：

```xml
<?xml version="1.0" encoding="utf-8"?>
<RelativeLayout …
    >
…
<Button android:id="@+id/btnOK"
    android:layout_width="wrap_content"
    android:layout_height="wrap_content"
    android:text="完成游戏"
    android:layout_below="@id/btnNet"
    android:textSize="20sp"
    android:layout_marginTop="20dp"
    android:layout_centerHorizontal="true"
    />
<Button android:id="@+id/btnCancel"
    android:layout_width="wrap_content"
    android:layout_height="wrap_content"
    android:text="取消"
    android:layout_below="@id/btnOK"
    android:textSize="20sp"
    android:layout_marginTop="20dp"
```

```
            android:layout_centerHorizontal="true"
        />
</RelativeLayout>
```

Step 4 在"计算机猜拳游戏"的 Activity 中增加局数统计相关变量,并在游戏进行的过程中累计这些变量的值,读者可以参考 UNIT 25 的游戏程序代码。

Step 5 在"计算机猜拳游戏" Activity 的程序代码中分别完成以上 2 个按钮的 OnClickListener 对象,如以下粗体字的部分:

```java
package ...

import ...

public class Game extends Activity {
    private Button mBtnOK, mBtnCancel;
    …

    public void onCreate(Bundle savedInstanceState) {
        super.onCreate(savedInstanceState);
        setContentView(R.layout.game_activity);

        setupViewComponent();
    }

    private void setupViewComponent() {
        …
        mBtnOK = (Button)findViewById(R.id.btnOK);
        mBtnCancel = (Button)findViewById(R.id.btnCancel);
        mBtnOK.setOnClickListener(btnOKLis);
        mBtnCancel.setOnClickListener(btnCancelLis);
    }

    private Button.OnClickListener btnOKLis = new Button.OnClickListener() {
        public void onClick(View v) {
            Intent it = new Intent();

            Bundle bundle = new Bundle();
            bundle.putInt("KEY_COUNT_SET", miCountSet);
            bundle.putInt("KEY_COUNT_PLAYER_WIN", miCountPlayerWin);
            bundle.putInt("KEY_COUNT_COM_WIN", miCountComWin);
            bundle.putInt("KEY_COUNT_DRAW", miCountDraw);
            it.putExtras(bundle);

            setResult(RESULT_OK, it);
            finish();
        }
```

```
        };

        private Button.OnClickListener btnCancelLis = new Button.
OnClickListener() {
            public void onClick(View v) {
                setResult(RESULT_CANCELED);
                finish();
            }
        };
    }
```

"完成游戏"按钮的 OnClickListener 对象名称是 btnOKLis，它的工作就是建立一个 Intent 对象和一个 Bundle 对象，然后把局数统计相关变量的值放入 Bundle 对象，再把该 Bundle 对象传给 Intent 对象。最后调用 setResult()方法将该 Intent 对象返回给原来的 Activity，同时传送一个结果代码 RESULT_OK 代表运行结果正常，最后调用 finish()方法结束。如果用户按下"取消"按钮，则传回 RESULT_CANCELED 结果代码，然后调用 finish()方法结束。

完成以上步骤之后运行程序，就可以看到如图 42-1 所示的运行界面。以上的范例程序是以平板电脑的屏幕大小来设计，如果是要在手机上运行，只要修改接口布局文件，把第二层的 LinearLayout 删除，只保留第一层的 LinearLayout 并适当的修改排列属性即可。

图 42-1　"计算机猜拳游戏"使用 Intent 对象返回局数统计数据

PART 8

Broadcast Receiver、Service 和 App Widget

UNIT 43　Broadcast Intent 和 Broadcast Receiver
UNIT 44　Service 是幕后英雄
UNIT 45　App Widget 小工具程序
UNIT 46　使用 Alarm Manager 强化 App Widget 程序
UNIT 47　App Widget 程序的其他两种执行模式

Android 版本	1.X	2.X	3.X	4.X
适用性	★	★	★	★

UNIT 43

Broadcast Intent 和 Broadcast Receiver

当 Android 系统发生某种状况，必须通知所有程序进行处理时，例如电池电量不足、收到 email 等，就可以利用 Broadcast Intent 对象的功能来进行信息广播。Broadcast Intent 对象的运行机制包含两个部分，一个是送出 Intent 对象的程序，另一个是监听广播消息的程序（称为 Broadcast Receiver）。Broadcast Receiver 程序本身是一个类，而且必须继承 BroadcastReceiver 类，程序必须向 Android 系统注册为 Broadcast Receiver，并指定要监听的广播消息。当该监听的广播消息被某个程序送出时，Android 系统会启动所有监听该广播消息的 Broadcast Receiver 程序，并运行它们的 onReceive() 方法。

Broadcast Receiver 程序只有在 Android 系统运行它的 onReceive() 方法时才会处于有效状态，一旦 onReceive() 方法运行完毕就有可能被移除，直到下次监听的广播消息出现才会再运行一次。这个特性会影响到 onReceive() 方法中运行的工作，例如在 onReceive() 方法中产生一个 thread 运行某一项比较耗时的工作时，由于 onReceive() 方法运行结束且 Broadcast Receiver 程序被系统移除，此举将导致 Broadcast Receiver 程序所建立的 thread 也有可能被 Android 系统强制清除，类似这种异步的工作并不适合在 onReceive() 方法中运行。

43-1 程序广播 Intent 对象的方法

程序如果要广播 Intent 对象需要完成以下三个步骤：

Step 1 建立一个 Intent 对象，并指定要广播的信息。

```
Intent it = new Intent("tw.android.MY_BROADCAST");
```

广播的信息其实就是一个字符串,每一个程序都可以建立自己的广播消息。为了避免不同的程序误用相同的广播消息名称,一般建议是采用前后颠倒的网址模式来命名,如以上范例。

Step 2 如果需要附带数据,可以将数据放入 Intent 对象中:

```
it.putExtra("sender_name", "Broadcase Receiver 范例程序");
```

或者如同我们在 UNIT 41 中的说明,也可以利用 Bundle 对象将数据存入 Intent 对象中:

```
Bundle bundle = new Bundle();
bundle.putString("sender_name", "Broadcast Receiver 范例程序");
it.putExtras(bundle);
```

Step 3 调用 sendBroadcast()方法广播 Intent 对象。

```
sendBroadcast(it);
```

43-2 建立 Broadcast Receiver 监听广播消息

在本单元一开始的时候我们已经解释过 Broadcast Receiver 的运行机制,以下我们直接来看它的实现步骤:

Step 1 在程序项目中新增一个继承 BroadcastReceiver 类的新类,我们可以把这个新类取名为 MyBroadcastReceiver,在这个类中需要实现 onReceive()方法中的程序代码,这个方法就是当监听的广播消息出现的时候会被 Android 系统启动运行,请参考以下范例,在这个范例中我们只是简单地从接收到的 Intent 对象中取出传送者的数据。注意其中的 Intent 对象是 onReceive()方法的自变量,并不需要调用 getIntent()方法取得。

```
package ...

import ...

public class MyBroadcaseReceiver extends BroadcastReceiver {
    @Override
    public void onReceive(Context context, Intent intent) {
        // TODO Auto-generated method stub
        // 收到监听信息时要运行的程序代码
        String sender = intent.getExtras().getString("sender_name");
    }
}
```

Step 2 在主程序中向 Android 系统注册步骤一所建立的 Broadcast Receiver，以及它要监听的广播消息。注册的方法有两种，第一种是在程序项目的 AndroidManifest.xml 文件中描述这个 Broadcast Receiver 这样就完成注册，请读者参考以下范例的粗体字部分：

```xml
<?xml ...>
<manifest ...>
    <application ...>
        <activity android:name=".Main" ...>
            <intent-filter>
                <action android:name="android.intent.action.MAIN" />
                <category android:name="android.intent.category.LAUNCHER" />
            </intent-filter>
        </activity>
        <receiver android:name=".MyBroadcaseReceiver"
                android:label="@string/app_name">
            <intent-filter>
                <action android:name="tw.android.MY_BROADCAST" />
            </intent-filter>
        </receiver>
    </application>
</manifest>
```

第二种方法是在程序代码中完成注册，这种方法还可以把已经注册过的 Broadcast Receiver 取消，以下先看注册的程序代码：

```
IntentFilter itFilter = new IntentFilter("tw.android.MY_BROADCAST");
MyBroadcaseReceiver  myReceiver = new MyBroadcaseReceiver();
registerReceiver(mMyReceiver, itFilter);
```

如果要取消已经注册的 Broadcast Receiver，则运行以下程序代码：

```
unregisterReceiver(myReceiver);
```

只有使用第二种方式注册的 Broadcast Receiver 才可以取消。读者可以对照以上两种注册的方法，不管使用哪一种方法，在注册时都必须提供两种信息，一是要监听的广播消息，二是 Broadcast Receiver 物件。根据 Android SDK 技术文件的建议，除非某种特殊情况非使用程序代码注册 Broadcast Receiver 不可，否则以使用 AndroidManifest.xml 文件注册为佳。以上就是广播 Intent 和建立 Broadcast Receiver 的方法，接下来用一个程序项目来示范实现过程。

43-3　范例程序

这个范例程序的运行界面如图 43-1 所示，在这个程序中会建立 2 个 Broadcast Receiver 监听不同的广播消息。第一个 Broadcast Receiver（以下称为 BroadcastReceiver1）是利用 AndroidManifest.xml 文件进行注册，第二个 Broadcast Receiver（以下称为 BroadcastReceiver2）是利用程序代码进行注册。程序运行界面的第一个按钮就是注册 BroadcastReceiver2，第二个按钮则是取消注册 BroadcastReceiver2。

图 43-1　Broadcast Intent 和 Broadcast Receiver 范例程序的运行界面

当程序刚开始运行的时候，按下"发送 MY_BROADCAST1"按钮时，屏幕上会出现一个 Toast 信息通知 BroadcastReceiver1 已经收到广播消息，如图 43-2 所示。如果按下"发送 MY_BROADCAST2"按钮则没有任何结果，因为还没有注册 BroadcastReceiver2。请读者按下"注册 BroadcastReceiver2"按钮，然后再按一次"发送 MY_BROADCAST2"按钮，此时就会出现一个 Toast 信息通知 BroadcastReceiver2 已经收到广播消息，如图 43-3 所示。如果读者按下"注销 BroadcastReceiver2"按钮再测试一次，会发现 BroadcastReceiver2 又收不到广播消息。以下是完整的接口布局文件、主程序文件和两个 Broadcast Receiver 类的程序文件，这些程序代码的主要内容都已经在前面讨论过，因此请读者自行参阅。

图 43-2　BroadcastReceiver1 收到广播消息的画面

图 43-3　BroadcastReceiver2 收到广播消息的画面

■　界面布局文件：

```
<?xml version="1.0" encoding="utf-8"?>
<LinearLayout xmlns:android="http://schemas.android.com/apk/res/android"
    android:orientation="horizontal"
```

```xml
        android:layout_width="match_parent"
        android:layout_height="match_parent">
    <LinearLayout
        android:orientation="vertical"
        android:layout_width="0dp"
        android:layout_height="match_parent"
        android:layout_weight="1"/>
    <LinearLayout
        android:orientation="vertical"
        android:layout_width="0dp"
        android:layout_height="match_parent"
        android:layout_weight="1">
    <Button android:id="@+id/btnRegReceiver"
        android:layout_width="match_parent"
        android:layout_height="wrap_content"
        android:text="注册 Broadcase Receiver2"
        android:layout_marginTop="20dp"/>
    <Button android:id="@+id/btnUnregReceiver"
        android:layout_width="match_parent"
        android:layout_height="wrap_content"
        android:text="注销 Broadcase Receiver2"/>
    <Button android:id="@+id/btnSendBroadcase1"
        android:layout_width="match_parent"
        android:layout_height="wrap_content"
        android:text="传送 MY_BROADCAST1"/>
    <Button android:id="@+id/btnSendBroadcase2"
        android:layout_width="match_parent"
        android:layout_height="wrap_content"
        android:text="传送 MY_BROADCAST2"/>
    </LinearLayout>
    <LinearLayout
        android:orientation="vertical"
        android:layout_width="0dp"
        android:layout_height="match_parent"
        android:layout_weight="1"/>
</LinearLayout>
```

■ 主程序文件：

```java
package …

import …

public class Main extends Activity {

    private Button mBtnRegReceiver,
                   mBtnUnregReceiver,
                   mBtnSendBroadcase1,
```

```java
            mBtnSendBroadcase2;

    private MyBroadcaseReceiver2 mMyReceiver2;

    /** Called when the activity is first created. */
    @Override
    public void onCreate(Bundle savedInstanceState) {
        super.onCreate(savedInstanceState);
        setContentView(R.layout.main);

        setupViewComponent();
    }

    private void setupViewComponent() {
        mBtnRegReceiver = (Button)findViewById(R.id.btnRegReceiver);
        mBtnUnregReceiver = (Button)findViewById(R.id.btnUnregReceiver);
        mBtnSendBroadcase1 = (Button)findViewById(R.id.btnSendBroadcase1);
        mBtnSendBroadcase2 = (Button)findViewById(R.id.btnSendBroadcase2);

        mBtnRegReceiver.setOnClickListener(btnRegReceiverOnClickLis);
        mBtnUnregReceiver.setOnClickListener(btnUnregReceiverOnClickLis);
        mBtnSendBroadcase1.setOnClickListener(btnSendBroadcase1OnClickLis);
        mBtnSendBroadcase2.setOnClickListener(btnSendBroadcase2OnClickLis);
    }

    private Button.OnClickListener btnRegReceiverOnClickLis = new Button.OnClickListener() {
        public void onClick(View v) {
            IntentFilter itFilter = new IntentFilter("tw.android.MY_BROADCAST2");
            mMyReceiver2 = new MyBroadcaseReceiver2();
            registerReceiver(mMyReceiver2, itFilter);
        }
    };

    private Button.OnClickListener btnUnregReceiverOnClickLis = new Button.OnClickListener() {
        public void onClick(View v) {
            unregisterReceiver(mMyReceiver2);
        }
    };

    private Button.OnClickListener btnSendBroadcase1OnClickLis = new Button.OnClickListener() {
        public void onClick(View v) {
            Intent it = new Intent("tw.android.MY_BROADCAST1");
            it.putExtra("sender_name", "主程序");
            sendBroadcast(it);
        }
```

```java
        };

        private Button.OnClickListener btnSendBroadcase2OnClickLis = new
        Button.OnClickListener() {
            public void onClick(View v) {
                Intent it = new Intent("tw.android.MY_BROADCAST2");
                it.putExtra("sender_name", "主程序");
                sendBroadcast(it);
            }
        };
    }
```

- BroadcastReceiver1 类：

```java
package …

import …

public class MyBroadcaseReceiver1 extends BroadcastReceiver {
    @Override
    public void onReceive(Context context, Intent intent) {
        // TODO Auto-generated method stub
        String sender = intent.getStringExtra("sender_name");
        Toast.makeText(context, "BroadcastReceiver1 收到" + sender + "
        发送的 Broadcase 信息", Toast.LENGTH_LONG).show();
    }
}
```

- BroadcastReceiver2 类：

```java
package …

import …

public class MyBroadcaseReceiver2 extends BroadcastReceiver {

    @Override
    public void onReceive(Context context, Intent intent) {
        // TODO Auto-generated method stub
        String sender = intent.getStringExtra("sender_name");
        Toast.makeText(context, "BroadcastReceiver2 收到" + sender + "
        发送的 Broadcase 信息", Toast.LENGTH_LONG).show();
    }

}
```

UNIT 44

Android 版本	1.X	2.X	3.X	4.X
适用性	★	★	★	★

Service 是幕后英雄

如果程序需要运行一项比较费时的工作，为了避免让系统处于停滞状态，无法响应使用者的操作，此时可以使用 Service 对象，Service 会和原来启动它的程序一起运行，例如让手机或平板电脑一边播放音乐一边让用户继续操作其他功能，像是上网或是浏览照片。Service 和 Activity 一样也是一个类，它们都是 Android 系统中的运行单元，只是 Activity 会有一个接口布局文件当成它的操作界面，而 Service 没有操作界面，它就是专门负责运行一项指定的工作，直到完成该项工作或是被它的控制程序下令停止。

> **补充说明　Service 和 Thread**
>
> Service 和 Thread 的功能很相似，它们都是可以独立运行的对象，而且都是在幕后运行没有界面，那么二者之间有何分别？简单来说，Android 系统根据它的特性和架构，设计了 Service 这个类，也就是说 Service 类是针对 Android 系统做了最佳的设计，根据 Android SDK 技术文件中的建议，在一般情况下应优先考虑使用 Service。

44-1　Service 的运行方式和生命周期

使用 Service 和使用 Thread 的方式有些不一样。Thread 是 Java 语言内建的类，而 Service 则加入了 Android 系统的特性，所以 Service 自然会用到 Android 系统特有的架构，那就是 Intent。要启动 Service 有两种方法，一种是调用 startService()，一种是调用 bindService()。不管是哪一种方法，主程序都必须利用 Intent 对象告诉 Android 系统所要运行的 Service。

Service 对象从开始到结束的运行过程中会经历不同的状态改变（其实 Activity 也有类似的状态改变过程，我们会在后续的单元再详细讨论），而且不同的启动方式的状态改变过程也有差异，请读者参考图 44-1，左边的状态改变流程图是使用 startService()的结果，右边的状态改变流程图是使用 bindService()的结果。

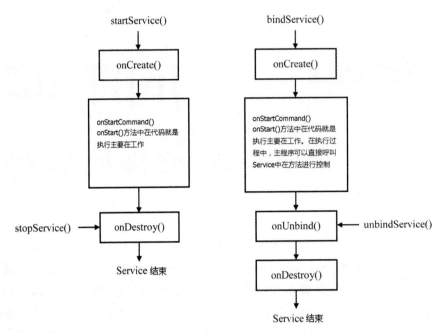

图 44-1　Service 的生命周期

了解 Service 的状态改变过程对于某些类型的应用程序是有必要的。举例来说，当 Service 对象在运行的过程中开启了某一个文件时，必须在它结束的时候（也就是 onDestroy()方法中）关闭该文件，否则会造成该文件被持续占用的情况。编写 Service 的工作之一就是设计在不同状态下应该运行的程序代码，让 Service 对象能够正确地运行。介绍完 Service 的基本概念之后，接下来就让我们学习如何在程序项目中实现 Service。

44-2　在程序项目中建立 Service

建立 Service 的过程就像我们之前在程序项目中新增一个类一样，请读者依照下列步骤操作：

Step 1 在程序项目中新增一个继承 android.app.Service 类的新类，我们可以将这个新类取名为 MyService。刚开始的时候 Android 程序编辑器会自动帮这个 MyService 类加上以下的程序代码：

package …

import …

```
public class MyService extends Service {

    @Override
    public IBinder onBind(Intent arg0) {
        // TODO Auto-generated method stub
        return null;
    }

}
```

这个最原始的程序代码中只有 onBind() 方法,如果读者和图 44-1 的 Service 状态变化流程图对照,会发现还缺少其他状态转变的方法,因此接下来就让我们把这些方法加进来。

Step 2 请在 MyService 类的程序代码编辑窗口中右击,在出现的快捷菜单中选择 Source > Override/Implement Methods…,就会出现图 44-2 的对话框。在对话框的清单中会列出还没有加入 Service 类中的方法,请读者勾选其中的 onCreate()、onStartCommand()、onDestroy() 和 onUnbind() 然后单击 OK 按钮。

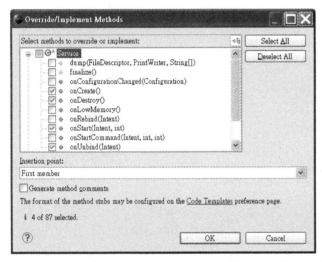

图 44-2 利用 Eclipse 的 Override/Implement Methods 功能加入需要的方法

补充说明 Service 类别的 onStartCommand() 方法和 onStart() 方法

在 Android 2.0 以前,Service 类只有 onStart() 而没有 onStartCommand(),但是在 2.0 版(包含 2.0)以后,为了补充原来 onStart() 的功能,因此增加 onStartCommand(),它是用来取代旧有的 onStart(),因此读者在编写新的程序代码时应该使用 onStartCommand()。

Step 3 请读者编辑 MyService 类的程序代码如下,需要新增和修改的程序代码以粗体字标示。

```java
package ...

import ...

public class MyService extends Service {

    private final String LOG_TAG = "service demo";

    public class LocalBinder extends Binder {
        MyService getService() {
            return    MyService.this;
        }
    }

    private LocalBinder mLocBin = new LocalBinder();

    public void myMethod() {
        Log.d(LOG_TAG, "myMethod()");
    }

    @Override
    public void onCreate() {
        // TODO Auto-generated method stub
        Log.d(LOG_TAG, "onCreate()");
        super.onCreate();
    }

    @Override
    public void onDestroy() {
        // TODO Auto-generated method stub
        Log.d(LOG_TAG, "onDestroy()");
        super.onDestroy();
    }

    @Override
    public int onStartCommand(Intent intent, int flags, int startId) {
        // TODO Auto-generated method stub
        Log.d(LOG_TAG, "onStartCommand()");
        return super.onStartCommand(intent, flags, startId);
    }

    @Override
    public boolean onUnbind(Intent intent) {
        // TODO Auto-generated method stub
        Log.d(LOG_TAG, "onUnbind()");
        return super.onUnbind(intent);
    }
```

```
    @Override
    public IBinder onBind(Intent arg0) {
        // TODO Auto-generated method stub
        Log.d(LOG_TAG, "onBind()");
        return mLocBin;
    }
}
```

需要新增的程序代码不多，主要是利用我们在 UNIT 7 中介绍的程序 log 技巧，在每一个方法中加上产生 log 的程序代码，让我们可以追踪 Service 对象的运行过程。其中要特别注意 onBind() 方法必须传回我们在类中建立的一个 LocalBinder 类的对象。如果没有传回这个对象，就无法完成 bind Service 的动作。LocalBinder 是我们在 MyService 类中定义的内部类（inner class），它是继承 Binder 类，其中有一个 getService() 方法，此方法会传回 MyService 对象。就是根据这个方法我们才能够在调用 bindService() 的时候取得 MyService 对象。另外我们还增加了一个 myMethod() 方法，用来验证 bindService() 的功能。

Step 4 在程序项目的 AndroidManifest.xml 文件中加入描述 Service 类的标签如下：

```xml
<?xml version="1.0" encoding="utf-8"?>
<manifest ...>
    <application ...>
        <activity android:name=".Main"
            …
        </activity>
        <service android:name=".MyService" android:enabled="true" />
    </application>
</manifest>
```

以上就是建立 Service 类的过程，接下来介绍如何在主程序中启动这个 Service。

44-3 启动 Service 的第一种方法

第一种启动 Service 的方法是利用 startService()，它的程序代码很简单，请读者直接参考以下范例：

```
Intent it = new Intent(类名称.this, Service 类名称.class);
startService(it);
```

其中的"类名称"就是启动 Service 的类，"Service 类名称"就是我们要启动的 Service，例如前面范例中的 MyService。当 Service 以这种方式启动之后，就会依照图 44-1 左边的流程运行，如果主程序要停止 Service 的运行，可以运行以下程序代码：

```
Intent it = new Intent(类名称.this, Service 类名称.class);
stopService(it);
```

Service 对象要结束时会先运行 onDestroy()方法，我们把善后的工作在 onDestroy()方法中完成。

44-4　启动 Service 的第二种方法

第二种启动 Service 的方法是利用 bindService()。bindService()方法和 startService()最大的不同是主程序可以取得 Service 对象，因此在 Service 对象运行的过程中，可以直接调用它的方法进行控制，要完成 bindService()需要以下步骤：

Step 1 在主程序中建立一个 ServiceConnection 的对象，请参考以下程序代码：

```
private ServiceConnection mServConn = new ServiceConnection() {
    public void onServiceConnected(ComponentName name,
    IBinder service) {
        // TODO Auto-generated method stub
        mMyServ = ((MyService.LocalBinder)service).
        getService();
    }

    public void onServiceDisconnected(ComponentName name) {
        // TODO Auto-generated method stub
    }
};
```

bindService()方法需要利用这个对象来取得 Service 对象。其中我们必须实现 onServiceConnected()和 onServiceDisconnected()这两个方法。onServiceConnected()方法是在 bind Service 的过程中会由 Android 系统调用运行，其中的程序代码（以粗体字标示）就是取得 Service 对象，并存入定义在主程序中的属性。onServiceDisconnected()方法则是当 Service 对象不正常结束的时候才会运行，一般正常情况下不会运行。

Step 2 建立一个 Intent 对象并指定要启动的 Service 类，然后调用 bindService()如下：

```
Intent it = new Intent(类名称.this, Service 类名称.class);
bindService(it, ServiceConnection 物件, BIND_AUTO_CREATE);
```

其中"ServiceConnection 的对象"就是步骤一中的 mServConn 对象。BIND_AUTO_CREATE 参数是请 Android 系统视情况需要建立一个 Service 对象。

如果要停止 Service 对象则使用下列的程序代码：

```
unbindService(ServiceConnection 物件);
```

其中"ServiceConnection 对象"就是步骤一中的 mServConn 对象。以上就是建立 Service 和使用 Service 的流程，接下来我们用一个完整程序项目来示范。

44-5 范例程序

这个范例程序包含一个主程序类和一个 Service 类，其中的 Service 类就是前面小节所介绍的 MyService，虽然它只是一个 Service 的架构，没有实际的功能，但是我们可以利用它来验证 Service 的运行流程，以便在后续的单元中能够使用 Service 完成具有实际功能的程序代码。我们在主程序的界面加上一些按钮以便测试 Service 的操作，程序的运行界面如图 44-3 所示。这个 Service 范例程序必须配合使用 Eclipse 的 Debug 界面以便观察程序所产生的 log，请读者参考 UNIT 7 中有关 Eclipse 的 Debug 模式的使用说明。读者在启动这个范例程序后，可以利用程序界面中的按钮对 Service 进行控制，并观察所产生的 log 以了解 Service 的运行流程。以下列出主程序类的接口布局文件和程序代码，主程序类中定义的 mMyServ 对象就是用来储存 bind Service 时所得到的 Service 对象，其他有关 Service 控制的程序代码都已经在前面的小节中作过说明，请读者自行参考。

图 44-3 Service 范例程序的运行界面

- 界面布局文件：

```
<?xml version="1.0" encoding="utf-8"?>
<LinearLayout xmlns:android="http://schemas.android.com/apk/res/android"
    android:orientation="horizontal"
    android:layout_width="match_parent"
    android:layout_height="match_parent">
<LinearLayout
    android:orientation="vertical"
    android:layout_width="0dp"
    android:layout_height="match_parent"
    android:layout_weight="1"/>
<LinearLayout
    android:orientation="vertical"
    android:layout_width="0dp"
    android:layout_height="match_parent"
```

```xml
        android:layout_weight="1">
<Button android:id="@+id/btnStartMyService"
    android:layout_width="match_parent"
    android:layout_height="wrap_content"
    android:text="启动 MyService"/>
<Button android:id="@+id/btnStopMyService"
    android:layout_width="match_parent"
    android:layout_height="wrap_content"
    android:text="停止 MyService"/>
<Button android:id="@+id/btnBindMyService"
    android:layout_width="match_parent"
    android:layout_height="wrap_content"
    android:text="连结 MyService"/>
<Button android:id="@+id/btnUnbindMyService"
    android:layout_width="match_parent"
    android:layout_height="wrap_content"
    android:text="断开 MyService"/>
<Button android:id="@+id/btnCallMyServiceMethod"
    android:layout_width="match_parent"
    android:layout_height="wrap_content"
    android:text="调用 MyService 中的 myMethod()"/>
</LinearLayout>
<LinearLayout
    android:orientation="vertical"
    android:layout_width="0dp"
    android:layout_height="match_parent"
    android:layout_weight="1"/>
</LinearLayout>
```

- 主程序文件：

```java
package …

import …

public class Main extends Activity {

    private Button mBtnStartMyService,
                   mBtnStopMyService,
                   mBtnBindMyService,
                   mBtnUnbindMyService,
                   mBtnCallMyServiceMethod;

    private MyService mMyServ = null;

    private final String LOG_TAG = "service demo";

    private ServiceConnection mServConn = new ServiceConnection() {
```

```java
        public void onServiceConnected(ComponentName name, IBinder service) {
            // TODO Auto-generated method stub
            Log.d(LOG_TAG, "onServiceConnected() " + name.getClassName());
            mMyServ = ((MyService.LocalBinder)service).getService();
        }

        public void onServiceDisconnected(ComponentName name) {
            // TODO Auto-generated method stub
            Log.d(LOG_TAG, "onServiceDisconnected()" + name.
            getClassName());
        }
    };

    /** Called when the activity is first created. */
    @Override
    public void onCreate(Bundle savedInstanceState) {
        super.onCreate(savedInstanceState);
        setContentView(R.layout.main);

        mBtnStartMyService = (Button) findViewById(R.id.btnStartMyService);
        mBtnStopMyService = (Button) findViewById(R.id.btnStopMyService);
        mBtnBindMyService = (Button) findViewById(R.id.btnBindMyService);
        mBtnUnbindMyService = (Button) findViewById(R.id.btnUnbindMyService);
        mBtnCallMyServiceMethod = (Button) findViewById(R.id.btnCallMyServiceMethod);

        mBtnStartMyService.setOnClickListener(btnStartMyServiceOnClkLis);
        mBtnStopMyService.setOnClickListener(btnStopMyServiceOnClkLis);
        mBtnBindMyService.setOnClickListener(btnBindMyServiceOnClkLis);
        mBtnUnbindMyService.setOnClickListener(btnUnbindMyServiceOnClkLis);
        mBtnCallMyServiceMethod.setOnClickListener(btnCallMyServiceMethodOnClkLis);
    }

    private OnClickListener btnStartMyServiceOnClkLis = new OnClickListener() {
        public void onClick(View v) {
            mMyServ = null;
            Intent it = new Intent(Main.this, MyService.class);
            startService(it);
        }
    };

    private OnClickListener btnStopMyServiceOnClkLis = new OnClickListener() {
        public void onClick(View v) {
            mMyServ = null;
            Intent it = new Intent(Main.this, MyService.class);
            stopService(it);
        }
    };
```

```java
    private OnClickListener btnBindMyServiceOnClkLis = new OnClickListener() {
        public void onClick(View v) {
            mMyServ = null;
            Intent it = new Intent(Main.this, MyService.class);
            bindService(it, mServConn, BIND_AUTO_CREATE);
        }
    };

    private OnClickListener btnUnbindMyServiceOnClkLis = new OnClickListener() {
        public void onClick(View v) {
            mMyServ = null;
            unbindService(mServConn);
        }
    };

    private OnClickListener btnCallMyServiceMethodOnClkLis = new OnClickListener() {
        public void onClick(View v) {
            if (mMyServ != null)
                mMyServ.myMethod();
        }
    };
}
```

Android 版本	1.X	2.X	3.X	4.X
适用性	★	★	★	★

UNIT 45
App Widget 小工具程序

在开始介绍 App Widget 以前让我们先回想前面已经学过的几种程序类型。首先是 Activity，当它启动之后整个屏幕的程序区域都被它独占直到结束为止。第二是 Service，它和 Activity 完全相反，它是在背景运行而且没有界面。当 Service 启动之后就一直持续运行直到工作完成才结束。第三是 Broadcast Receiver，Broadcast Receiver 运行时也没有界面，当它接收到所监听的信息时便开始运行它的工作，完成之后又再度进入等待状态。本单元的主角 App Widget（简称 Widget）和 Broadcast Receiver 有些类似，它也是靠监听信息的方式运行，并且具有下列两项特色：

1. 它会在手机或平板电脑的操作首页显示程序的运行界面，而且程序的运行界面的大小可以由程序自己决定。
2. 它有下列三种运行方式：
    i. 固定时间间隔运行；
    ii. 在设置的时间点运行；
    iii. 当用户按下程序界面上的按钮时运行。

例如我们经常在手机屏幕上看到的圆形时钟就是一个 App Widget 程序。如果我们按下 Android 4 手机屏幕下方中央的 Apps 按钮，然后在显示的界面单击 Widgets 标签页就会看到如图 45-1 所示的界面，界面中列出所有可以加到 Home screen 的 App Widget 程序，例如可以按住 Bookmarks，屏幕界面就会显示 Home screen，将 Bookmarks 拖曳到想要摆放的位置再放开就会在 Home screen 中加入 Bookmarks，然后切换回到 Home screen 就会看到如图 45-2 所示的界面。如果想要移除首页上的 App Widget，可以按住该 App Widget 就会出现一个"垃圾桶"的图示，如图 45-3 所示，把 App Widget 拖曳到"垃圾桶"上方就完成移除的动作，Android 4 平板电脑的操作方式完全相同。在了解 App Widget 的特性之后，接下来就让我们学习如何建立 App Widget 程序。

图 45-1　选择 App Widget 程序的画面

图 45-2　加入 Bookmarks App Widget 程序后的 Home screen 画面

图 45-3　从 Home screen 移除 App Widget 程序的画面

> **补充说明**　其他 Android 版本模拟器设定 App Widget 的方法
>
> 如果是 Android 3.X 平板电脑，在模拟器的首页界面上按下鼠标左键并持续维持 1 秒钟以上再放开便会切换界面，单击屏幕最左边的 Widgets 标签页就会列出所有的 App Widget 程序，单击想要使用的程序，该程序就会加到模拟器的首页。
>
> 如果是 Android 2.X 手机，同样是在首页界面上按下鼠标左键并持续维持 1 秒钟以上再放开便会显示一个菜单，选择其中的 Widgets 项目就会出现 App Widget 程序列表，从中挑选想要加入首页的 App Widget 即可。

首先介绍的是基本型的 App Widget 程序，它是以固定时间间隔的方式运行，而且间隔时间的长度必须大于 30 分钟（这个限制可以利用程序技术来突破，我们留待下一个单元再介绍），建立 App Widget 程序的步骤如下：

Step 1　请依照之前的方法建立一个新的程序项目，但是在操作过程中不要勾选"Create Activity"，这样就会产生一个空的程序项目，也就是说在 src 文件夹下没有任何程序文件。

Step 2　打开 res/layout/main.xml 界面布局文件和 res/values/strings.xml 字符串资源文件，分别编辑如下，粗体字代表有更动的部分。在接口布局文件中我们改成显示 app_name 字符串，在字符串资源文件中我们把该字符串的内容设置成"App Widget 基本型"。

```
<?xml version="1.0" encoding="utf-8"?>
<LinearLayout xmlns:android="http://schemas.android.com/apk/res/android"
    android:orientation="vertical"
    android:layout_width="match_parent"
    android:layout_height="match_parent"
    >
<TextView
    android:layout_width="match_parent"
```

```
            android:layout_height="wrap_content"
            android:text="@string/app_name"
            android:textColor="#000000"
            />
</LinearLayout>

<?xml version="1.0" encoding="utf-8"?>
<resources>
    <string name="app_name">App Widget 基本型</string>
</resources>
```

Step 3 在 Eclipse 左边的项目检查窗口中右击程序项目的 src/(组件路径名称)文件夹，然后在弹出的菜单中选择"New > Class"就会出现如图 45-4 所示的对话框。在对话框中输入类名称，例如 MyAppWidget，然后按下 Superclass 字段右边的 Browse 按钮，在出现的对话框上方的字段输入 AppWidgetProvider，再单击下方列表中出现的 AppWidgetProvider 项目，最后按下 OK 按钮回到类对话框即可完成 Superclass 的设置，然后按下 Finish 按钮完成类对话框。

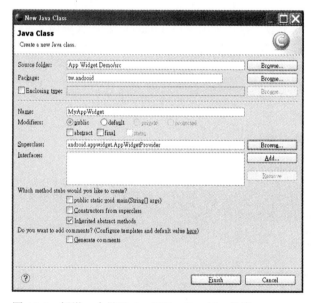

图 45-4　新增一个继承 AppWidgetProvider 的类

Step 4 前一个步骤就是建立 App Widget 程序文件，App Widget 程序的运行流程和上一个单元介绍的 Service 类似，也有不同的状态转换，因此接下来我们继续加入状态转换的相关方法，包括 onDeleted()、onDisabled()、onEnabled()、onReceive() 和 onUpdate()。请开启 MyAppWidget 程序文件，然后在该文件的程序代码编辑窗口中右击，在出现的快选菜单中选择 Source > Override/Implement Methods…，就会出现如图 45-5 所示的对话框。在对话框左边的清单中会列出尚未加入 AppWidgetProvider 类中的方法，请勾选上述 5 个方法然后单击 OK 按钮。

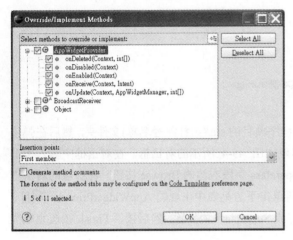

图 45-5　新增 MyAppWidget 类中的状态转换方法

Step 5　仿照上一个单元的作法，在每一个状态转换方法中加上调用 Log.d() 的程序代码如下，这些 log 信息可以用来追踪 App Widget 程序的运行过程。

```
package ...

import ...

public class MyAppWidget extends AppWidgetProvider {

    private final String LOG_TAG = "myAppWidget";

    @Override
    public void onDeleted(Context context, int[] appWidgetIds) {
        // TODO Auto-generated method stub
        super.onDeleted(context, appWidgetIds);
        Log.d(LOG_TAG, "onDeleted()");
    }

    @Override
    public void onDisabled(Context context) {
        // TODO Auto-generated method stub
        super.onDisabled(context);
        Log.d(LOG_TAG, "onDisabled()");
    }

    @Override
    public void onEnabled(Context context) {
```

```java
        // TODO Auto-generated method stub
        super.onEnabled(context);
        Log.d(LOG_TAG, "onEnabled()");
    }

    @Override
    public void onReceive(Context context, Intent intent) {
        // TODO Auto-generated method stub
        super.onReceive(context, intent);
        Log.d(LOG_TAG, "onReceived()");
    }

    @Override
    public void onUpdate(Context context, AppWidgetManager
appWidgetManager,
            int[] appWidgetIds) {
        // TODO Auto-generated method stub
        super.onUpdate(context, appWidgetManager, appWidgetIds);
        Log.d(LOG_TAG, "onUpdate()");
    }
}
```

Step 6 开启程序项目的 AndroidManifest.xml 文件登录这个 App Widget 程序的信息,如以下粗体字的部分:

```xml
<?xml version="1.0" encoding="utf-8"?>
<manifest ...>
    <application ...>
        <receiver android:name="MyAppWidget"
            android:label="App Widget"
            >
            <intent-filter>
                <action android:name="android.appwidget.action.APPWIDGET_UPDATE" />
            </intent-filter>
            <meta-data android:name="android.appwidget.provider"
                android:resource="@xml/appwidget_config"
                />
        </receiver>
    </application>
</manifest>
```

如同本单元开头的说明,App Widget 程序和 Broadcast Receiver 程序一样都是属于监听信息类型的程序,<receiver>标签中的 android:name 属性是设置步骤三所建立的类名称,android:label 属性是设置程序的名称,根据测试目前还不能使用中文。

在<intent-filter>标签中的<action>也是和 Broadcast Receiver 一样是用来指定监听的信息,<meta-data>标签的内容则是描述此 receiver 是属于 AppWidgetProvider,并指定它的组态

配置文件的位置和文件名。

Step 7 App Widget 程序需要一个 xml 格式的组态配置文件来定义它的属性，包括窗口大小、运行间隔、运行界面等，请在程序项目的 res 文件夹右击，选择 New > Folder，在文件夹名称字段输入 xml 然后单击 Finish 按钮。

Step 8 在前一个步骤建立的 xml 文件夹上右击，选择 New > File，在文件名字段输入 appwidget_config.xml（名称可以自定义，但是要使用小写英文，而且要和步骤六中 <meta-data> 标签内设置的 xml 文件同名），单击 Finish 按钮，然后打开这个新文件输入以下程序代码：

```xml
<?xml version="1.0" encoding="utf-8"?>
<appwidget-provider xmlns:android="http://schemas.android.com/apk/res/android"
    android:minWidth="110dp"
    android:minHeight="40dp"
    android:updatePeriodMillis="3000"
    android:initialLayout="@layout/main"
    >
</appwidget-provider>
```

其中的 android:minWidth 和 android:minHeight 属性是定义程序启动后在手机或平板电脑的操作首页上显示的程序界面大小。以手机来说，操作首页一般切割成 4×4 的格子，如果是平板电脑则是 8×7，App Widget 运行时它的程序界面会被缩放到刚好占满整数个格子，而且同时大于 android:minWidth 和 android:minHeight 的设置值，所以在设置这二个属性值的时候，必须知道 dp 长度单位和方格个数的关系。Android SDK 技术文件中提供如表 45-1 所示的计算公式，例如如果想让 App Widget 在首页上的宽度占满 2 个方格，高度占满 1 个方格，那么 android:minWidth 和 android:minHeight 就要分别设置为 110 和 40。

表 45-1　Android SDK 技术文件中关于 App Widget 程序界面大小的计算公式

手机或平板电脑 Home screen 中的方格大小（宽或高）	minWidth 或 minHeight 属性的设置值(dp)
1	40
2	110
3	180
…	…
n	70×n - 30

android:updatePeriodMillis 属性是设置每次运行的间隔时间，以千分之一秒为单位，例如 3000 代表 3 秒。但是请读者注意，Android 系统在运行 App Widget 程序时，最短自动运行间隔是 30 分钟，因此所有小于 30 分钟的设置都视为 30 分钟。如果读者希望运行间隔小于 30 分钟，必须自行

使用 Alarm Manager，下一个单元会示范实现方法，android:initialLayout 属性是设置接口布局文件。

以上就是建立 App Widget 程序的步骤，完成之后请运行程序，读者会发现屏幕界面并没有任何变化，这是正常的情况，因为 App Widget 程序安装之后必须由我们自己启动，请读者参考本单元前面的说明，启动这个"App Widget 基本型"程序就会在模拟器的首页看到它的运行界面，如图 45-6 所示，然后将它移除再把 Eclipse 切换到 Debug 界面（如果读者不熟悉操作方式可以参考 UNIT 7 的说明），然后在 LogCat 窗口新增一个此范例程序的 Log 信息 Filter 以检查程序产生的 log 信息，我们将会看到如图 45-7 所示的结果。读者会发现它并没有依照我们在 appwidget_config.xml 文件中的设置，每 3 秒更新一次（也就是运行 onUpdate()方法）。关于这一点我们已经在前面解释过，这是因为 Android 系统内部对于 App Widget 程序的运行频率有 30 分钟的限制，如果要达到小于 30 分钟的更新频率，必须配合使用 Alarm Manager，我们将在下一个单元继续建立强化版的 App Widget 程序。

图 45-6　启动 App Widget 程序后的界面　　图 45-7　App Widget 程序运行过程中产生的 log 信息

Android 版本	1.X	2.X	3.X	4.X
适用性	★	★	★	★

UNIT 46
使用 Alarm Manager 增强 App Widget 程序

首先提醒读者一个概念，App Widget 程序的运行频率愈高，电池的使用时间就愈短，因此在设计 App Widget 程序时，必须谨慎考虑所需的运行时间间隔。为了提高 App Widget 程序的运行频率，需要使用一个新的系统服务叫作 Alarm Manager，它是 Android 系统提供的一个对象，可以用来定时或是每隔一段时间运行一项工作。除了 Alarm Manager 之外，我们还要加上一个 App Widget 的初始设置程序。这个初始设置程序是一个 Activity，它会在 App Widget 启动时运行一次然后就结束。我们可以在这个 Activity 中设置 Alarm Manager，让它每间隔一段时间就送出一个 App Widget 程序所监听的信息，最后当 App Widget 结束时，再取消 Alarm Manager 的设置。

46-1 建立增强版的 App Widget 程序

在建立强化版的 App Widget 程序的过程中，我们将学会如何使用 Alarm Manager 以及 App Widget 的初始设置程序，请读者依照下列的说明操作：

Step 1 请读者先用 Windows 文件管理器复制上一个单元的 App Widget 程序项目文件夹，复制后可以重新命名，然后开启 Eclipse 程序，再利用主菜单的 File > Import 功能加载复制的项目文件夹。

Step 2 开启 res/values/strings.xml 字符串资源文件，将其中的程序名称字符串改成"App Widget 强化版"。

```xml
<?xml version="1.0" encoding="utf-8"?>
<resources>
    <string name="app_name">App Widget 强化版</string>
</resources>
```

Step 3 在程序项目的"src/(组件路径名称)"文件夹中新增一个继承自 Activity 的类，我们可以将它取名为 AppWidgetConfigAct。

Step 4 利用我们之前学过的操作技巧，在 AppWidgetConfigAct 类中新增一个 onCreate()方法，然后输入以下粗体字的程序代码：

```java
package ...

import ...

public class AppWidgetConfigAct extends Activity {

    int mAppWidgetId;

    @Override
    protected void onCreate(Bundle savedInstanceState) {
        // TODO Auto-generated method stub
        super.onCreate(savedInstanceState);

        Intent itIn = getIntent();
        Bundle extras = itIn.getExtras();
        if (extras != null) {
            mAppWidgetId = extras.getInt(
                    AppWidgetManager.EXTRA_APPWIDGET_ID,
                    AppWidgetManager.INVALID_APPWIDGET_ID);
        }

        if (mAppWidgetId == AppWidgetManager.INVALID_APPWIDGET_ID) {
            finish();
        }

        Intent itOut = new Intent("tw.android.MY_OWN_WIDGET_UPDATE");
        PendingIntent penIt = PendingIntent.getBroadcast(this, 0, itOut, 0);
        AlarmManager alarmMan = (AlarmManager)
            getSystemService(ALARM_SERVICE);
        Calendar calendar = Calendar.getInstance();
        calendar.setTimeInMillis(System.currentTimeMillis());
        calendar.add(Calendar.SECOND, 5);
        alarmMan.setRepeating(AlarmManager.RTC_WAKEUP,
            calendar.getTimeInMillis(), 20*1000, penIt);

        MyAppWidget.SaveAlarmManager(alarmMan, penIt);
```

```
            Intent itAppWidgetConfigResult = new Intent();
            itAppWidgetConfigResult.putExtra
             (AppWidgetManager.EXTRA_APPWI DGET_ID, mAppWidgetId);
            setResult(RESULT_OK, itAppWidgetConfigResult);

            finish();
        }

    }
```

以上的程序代码首先从系统取得 Intent 对象，再从 Intent 对象取得 App Widget 程序的 id，然后存入属性 mAppWidgetId 中。接下来建立一个固定时间间隔运行的 Intent 对象，这个 Intent 对象中就是我们的 App Widget 程序所监听的信息，然后再建立一个广播用的 PendingIntent 对象把它包裹起来。接着再取得系统的 Alarm Manager 对象，然后设置好它的第一次启动时间（范例中是现在算起 5 秒钟之后），后续的间隔运行时间（范例中是每隔 20 秒），以及前面建立好的 PendingIntent 对象，还有一个参数 AlarmManager.RTC_WAKEUP 是要求如果接收的程序是在休眠状态，也要把它唤醒运行。

接下来是把 Alarm Manager 对象和 PendingIntent 对象存入 MyAppWidget 对象中（后续步骤会实现这个 SaveAlarmManager()方法），以便在结束时取消 Alarm Manager 的设置，最后设置这个初始程序的运行结果便结束初始程序。另外请读者注意，当 App Widget 程序使用了初始程序之后，Android 系统就不再运行 App Widget 程序的 onUpdate()方法，如果想要运行该方法必须在初始程序中自行调用。

Step 5 开启 res/xml/appwidget_config.xml 文件，把其中的 android:updatePeriodMillis 属性设置为 "0"，也就是设置 Android 系统不要再自动更新这个 App Widget 程序，换成由我们设置的 Alarm Manager 来处理，另外还要把 android:configure 属性设置为步骤三所建立的 AppWidgetConfigAct 类。

```
<?xml version="1.0" encoding="utf-8"?>
<appwidget-provider xmlns:android="http://schemas.android.com/apk/res/android"
    android:minWidth="110dp"
    android:minHeight="40dp"
    android:updatePeriodMillis="0"
    android:initialLayout="@layout/main"
    android:configure="tw.android.AppWidgetConfigAct"
    >
</appwidget-provider>
```

Step 6 在 AndroidManifest.xml 文件中加入以下粗体字的程序代码，它们用来完成两项工作，一是登录 AppWidgetConfigAct 这个 Activity，并指定要监听 Android 系统的 APPWIDGET_CONFIGURE 信息，二是设置 App Widget 程序要监听 MY_OWN_WIDGET_UPDATE 信息。

```xml
<?xml version="1.0" encoding="utf-8"?>
<manifest ...>
    <application ...>
        <activity android:name="AppWidgetConfigAct">
            <intent-filter>
                <action android:name="android.appwidget.action.APPWIDGET_CONFIGURE" />
            </intent-filter>
        </activity>
        <receiver android:name="MyAppWidget"
            android:label="App Widget">
            <intent-filter>
                <action android:name="android.appwidget.action.APPWIDGET_UPDATE" />
                <action android:name="tw.android.MY_OWN_WIDGET_UPDATE" />
            </intent-filter>
            <meta-data android:name="android.appwidget.provider"
                android:resource="@xml/appwidget_config"/>
        </receiver>
    </application>
</manifest>
```

Step 7 在 MyAppWidget 类中加入以下属性和方法，这个新增加的 SaveAlarmManager()方法在步骤四中使用：

```java
private static AlarmManager mAlarmManager;
private static PendingIntent mPendingIntent;

static void SaveAlarmManager(AlarmManager alarmManager, PendingIntent pendingIntent)
{
    mAlarmManager = alarmManager;
    mPendingIntent = pendingIntent;
}
```

Step 8 在 MyAppWidget 类的 onReceive()方法中加入检查监听信息的程序代码，另外由于我们是使用 Alarm Manager 来启动 App Widget 程序，因此运行的是 onReceive()方法而不是 onUpdate()方法，所以我们要在 onReceive()方法中自己调用 onUpdate()如下：

```java
public void onReceive(Context context, Intent intent) {
    // TODO Auto-generated method stub
    super.onReceive(context, intent);

    if(!intent.getAction().equals("tw.android.MY_OWN_WIDGET_UPDATE"))
        return;

    Log.d(LOG_TAG, "onReceive()");

    AppWidgetManager appWidgetMan = AppWidgetManager.getInstance
```

```
        (context);
        ComponentName thisAppWidget = new ComponentName
        (context.getPackageName(), MyAppWidget.class.getName());
        int[] appWidgetIds = appWidgetMan.getAppWidgetIds(thisAppWidget);

        onUpdate(context, appWidgetMan, appWidgetIds);
    }
```

Step 9 在 MyAppWidget 类的 onDisabled ()方法中加入取消 Alarm Manager 的程序代码：

```
    @Override
    public void onDisabled(Context context) {
        // TODO Auto-generated method stub
        super.onDisabled(context);
        Log.d(LOG_TAG, "onDisabled()");
        mAlarmManager.cancel(mPendingIntent);
    }
```

完成后运行程序，并依照上一个单元的测试方式观察这个新版 App Widget 程序所产生的 log 信息，读者会发现它已经能够在很短的时间间隔内自动重复运行。到目前为止我们建立的两个 App Widget 范例程序都没有用到更新程序界面的功能，因此接下来我们再用一个简单的范例来说明如何更新 App Widget 程序的界面。

46-2　取得并更新 App Widget 程序的界面

App Widget 程序和和它的界面之间的关系和前面学过的 Activity 不同，App Widget 程序的接口布局文件是在 res/xml/appwidget_config.xml 文件中利用 android:initialLayout 属性指定，而不是在程序代码中加载，所以使用接口组件的方式也不一样。另外并非我们学过的所有接口组件和布局都可以用在 App Widget，以前面单元介绍过的接口组件和布局来说，App Widget 可以使用的部分如下：

1. TextView
2. Button
3. ImageButton
4. ImageView
5. ProgressBar
6. LinearLayout
7. RelativeLayout
8. FrameLayout

App Widget 程序的接口组件可以在它的初始设置程序中更新，也可以在 App Widget 程序中更新，首先我们示范如何在初始设置程序中更新 App Widget 程序的接口组件，请读者依照下列步骤操作：

Step 1 开启 res/layout/main.xml，将其中的 TextView 组件改成 ImageView 组件如下：

```
    <ImageView android:id="@+id/imgAppWidget"
        android:layout_width="wrap_content"
        android:layout_height="wrap_content"
```

```
        android:src="@drawable/icon"
    />
```

Step 2 利用 Windows 文件管理器复制一个大小约 150*150 的 png 图像文件到程序项目的 res/drawable-hdpi 文件夹，并将文件名设置成 icon2.png，我们将在程序代码中使用这个新的图像文件取代原来的 icon.png 图像文件成为 App Widget 程序的运行界面。

Step 3 开启 AppWidgetConfigAct 类的程序文件，在 onCreate()方法中加入以下粗体字的程序代码：

```
public class AppWidgetConfigAct extends Activity {
    ...
    @Override
    protected void onCreate(Bundle savedInstanceState) {
        // TODO Auto-generated method stub
        ...
        MyAppWidget.SaveAlarmManager(alarmMan, penIt);

        RemoteViews viewAppWidget = new RemoteViews
          (getPackageName(), R.layout.main);
        viewAppWidget.setImageViewResource(R.id.imgAppWidget,
        R.drawable.icon2);
        AppWidgetManager appWidgetMan = AppWidgetManager.
        getInstance(this);
        appWidgetMan.updateAppWidget(mAppWidgetId, viewAppWidget);

        Intent itAppWidgetConfigResult = new Intent();
        ...
    }
}
```

其中我们使用 RemoteViews 来取得 App Widget 程序的接口组件，然后再用 RemoteViews 对象中的 setImageViewResource()方法来设置 ImageView 中的图像，接着取得 AppWidgetManager 然后利用它更新 App Widget 程序的界面。

在 App Widget 程序中更新界面的方法也和上面的过程类似，我们把更新界面的程序代码写在 onUpdate()方法中，如以下粗体字的程序代码：

```
public void onUpdate(Context context, AppWidgetManager appWidgetManager,
            int[] appWidgetIds) {
    // TODO Auto-generated method stub
    super.onUpdate(context, appWidgetManager, appWidgetIds);
    Log.d(LOG_TAG, "onUpdate()");

    // 更新 App Widget 程序的 view
    RemoteViews viewAppWidget = new RemoteViews
    (context.getPackageName(), R.layout.main);
    viewAppWidget.setImageViewResource(R.id.imgAppWidget, R.drawable.icon2);
```

```
          ComponentName appWidget = new ComponentName
          (context, MyAppWidget.class);
          appWidgetManager.updateAppWidget(appWidget, viewAppWidget);
    }
```

和前面不同的是我们换成使用 ComponentName 对象的方式取得 App Widget，根据上一个小节的讨论，我们会在 onReceive()方法中调用 onUpdate()运行界面更新的工作。测试这个程序的时候，读者可以先取消这些新加入的程序代码，这时候的运行界面如图 46-1 所示，加上更新界面的程序代码之后，运行结果为图 46-2（会依读者使用的 icon2.png 文件而有不同的图示）。本单元所介绍的固定间隔时间运行只是 App Widget 程序的其中一种模式，这种模式在实现上比较复杂。在下一个单元我们将继续介绍另外两种比较容易实现的模式。

图 46-1　未加上更新 view 的程序代码的执行界面

图 46-2　加上更新 view 的程序代码之后的执行界面

Android 版本	1.X	2.X	3.X	4.X
适用性	★	★	★	★

UNIT 47
App Widget 程序的其他两种执行模式

除了上一个单元介绍的固定时间间隔运行的方式之外,还可以让 App Widget 程序在预定的时间点运行,或是等用户按下界面上的按钮后再运行。这两种模式的实现方法比较简单,以下我们依序说明。

47-1 预定运行时间的 App Widget 程序

这种 App Widget 程序不需要使用上一个单元介绍的初始设置 Activity,只要在它的 onEnabled() 方法中设置好 Alarm Manager 的启动时间即可,步骤如下:

Step 1 请读者在 Eclipse 中开启 App Widget 程序项目,如果想要保留原来的程序项目,请先用 Windows 文件管理器复制原来程序项目的文件夹,复制后可以重新命名,再利用 Eclipse 主菜单的 File > Import 功能加载复制的程序项目。

Step 2 开启 res/xml/appwidget_config.xml 文件,把其中的 android:updatePeriodMillis 属性设置为 "0",也就是让 Android 系统不要再主动更新这个 App Widget 程序,换成由我们设置的 Alarm Manager 来处理。

Step 3 开启 MyAppWidget 类的程序文件,新增以下方法:

```
private void setAlarm(Context context, Calendar alarmTime) {
    Intent it = new Intent("tw.android.MY_OWN_WIDGET_UPDATE");
    PendingIntent penIt = PendingIntent.getBroadcast(context, 0, it, 0);
```

```
AlarmManager alarmMan = (AlarmManager)context.getSystemService(context.ALARM_SERVICE);
alarmMan.set(AlarmManager.RTC_WAKEUP, alarmTime.getTimeInMillis(),
   penIt);
}
```

这个方法是用来设置 Alarm Manager 启动 PendingIntent 对象的时间，等时间一到便会送出其中的 Intent 对象，该 Intent 对象就是 MyAppWidget 程序所监听的信息。

Step 4 在 MyAppWidget 类的 onEnabled()方法中加入下列粗体字的程序代码：

```
public void onEnabled(Context context) {
    // TODO Auto-generated method stub
    super.onEnabled(context);
    Log.d(LOG_TAG, "onEnabled()");

    Calendar alarmTime = Calendar.getInstance();
    alarmTime.setTimeInMillis(System.currentTimeMillis());
    alarmTime.add(Calendar.SECOND, 30);
    setAlarm(context, alarmTime);
}
```

这一段程序代码的功能是取得系统现在的时间，然后加上 30 秒，再调用上一个步骤建立的 setAlarm()方法把设置的时间传入，也就是说在 30 秒后 Alarm Manager 会运行我们所设置的动作。

Step 5 在 MyAppWidget 类的 onReceive()方法中加入下列粗体字的程序代码，首先检查是否是我们等待的信息，由于我们是使用 Alarm Manager 来启动 App Widget 程序，因此运行的是 onReceive()方法而不是 onUpdate()，所以我们要在 onReceive()方法中自行调用 onUpdate()。

```
public void onReceive(Context context, Intent intent) {
    // TODO Auto-generated method stub
    super.onReceive(context, intent);

    if(!intent.getAction().equals("tw.android.MY_OWN_WIDGET_UPDATE"))
        return;

    Log.d(LOG_TAG, "onReceived()");

    AppWidgetManager appWidgetMan = AppWidgetManager.getInstance
    (context); ComponentName thisAppWidget = new ComponentName
    (context.getPackageName(), MyAppWidget.class.getName());
    int[] appWidgetIds = appWidgetMan.getAppWidgetIds
    (thisAppWidget);

    onUpdate(context, appWidgetMan, appWidgetIds);
}
```

Step 6 开启 AndroidManifest.xml 文件设置 App Widget 程序要监听 MY_OWN_WIDGET_UPDATE 信息如以下粗体字的程序代码：

```xml
<?xml version="1.0" encoding="utf-8"?>
<manifest xmlns:android="http://schemas.android.com/apk/res/android"
      package="tw.android"
      android:versionCode="1"
      android:versionName="1.0">
    <application android:icon="@drawable/icon" android:label=
"@string/app_name">
        <receiver android:name="MyAppWidget"
            android:label="App Widget"
            >
            <intent-filter>
                <action android:name="android.
                appwidget.action.APPWIDGET_UPDATE" />
                <action android:name="tw.android.MY_
                OWN_WIDGET_UPDATE" />
            </intent-filter>
            <meta-data android:name="android.appwidget.provider"
                android:resource="@xml/appwidget_config"
                />
        </receiver>
    </application>
</manifest>
```

完成以上步骤之后运行程序，然后仿照前面的操作方式观察程序产生的 log 信息，就可以发现 App Widget 确实在预定的时间运行 onReceive() 和 onUpdate() 方法。

47-2 用按钮启动 App Widget 程序

这种方式就是让 App Widget 程序在手机或平板电脑的首页显示一个按钮，当用户按下该按钮时，就会运行一次 App Widget 程序。要实现这种 App Widget 程序的重点在于如何让按下按钮的事件和运行 App Widget 程序的动作链接起来，答案是使用 PengingIntent 对象，请读者依照下列步骤操作：

Step 1 在 Eclipse 中开启 App Widget 程序项目，如果想要保留原来的程序项目，请先用 Windows 文件管理器复制原来程序项目的文件夹，复制后可以重新命名，再利用 Eclipse 主菜单的 File > Import 功能加载复制的程序项目。

Step 2 开启 res/xml/appwidget_config.xml 文件，把其中的 android:updatePeriodMillis 属性设置为 "0"，也就是设置 Android 系统不要再主动更新这个 App Widget 程序，换成由我们的程序自行处理。

Step 3 开启 res/layout/main.xml 文件，把其中的 TextView 组件换成 Button 如下：

```xml
<Button android:id="@+id/btnUpdate"
    android:layout_width="match_parent"
    android:layout_height="wrap_content"
    android:text="更新 App Widget"
/>
```

Step 4 开启 MyAppWidget 类的程序文件，在 onEnabled() 方法中输入以下粗体字的程序代码：

```java
public void onEnabled(Context context) {
    // TODO Auto-generated method stub
    super.onEnabled(context);
    Log.d(LOG_TAG, "onEnabled()");

    Intent it = new Intent();
    it.setAction("tw.android.MY_OWN_WIDGET_UPDATE");
    PendingIntent penIt = PendingIntent.getBroadcast(context, 0, it, 0);
    RemoteViews viewAppWidget = new RemoteViews(context.getPackageName(),R.layout.main);
    viewAppWidget.setOnClickPendingIntent(R.id.btnUpdate, penIt);
    AppWidgetManager appWidgetMan = AppWidgetManager.getInstance(context);
    ComponentName appWidget = new ComponentName(context,
    MyAppWidget.class);
    appWidgetMan.updateAppWidget(appWidget, viewAppWidget);
}
```

首先建立一个包含 App Widget 程序监听信息的 Intent 对象，然后再用一个 PendingIntent 对象把它包裹起来，接下来使用上一个单元介绍过的 RemoteViews 对象取得 App Widget 程序的界面，再用 setOnClickPendingIntent() 方法让按钮按下时运行 PendingIntent 对象，最后取得 AppWidgetManager，把修改后的界面设置给 App Widget 程序。

Step 5 在 MyAppWidget 类的 onReceive() 方法中加入下列粗体字的程序代码，首先检查是否是我们等待的信息，由于我们在步骤二中是设置自己更新 App Widget 程序，因此运行的是 onReceive() 方法而不是 onUpdate()，所以我们要在 onReceive() 方法中自行调用 onUpdate()。

```java
public void onReceive(Context context, Intent intent) {
    // TODO Auto-generated method stub
    super.onReceive(context, intent);

    if(!intent.getAction().equals("tw.android.MY_OWN_WIDGET_UPDATE"))
        return;

    Log.d(LOG_TAG, "onReceived()");

    AppWidgetManager appWidgetMan = AppWidgetManager.getInstance
    (context);
    ComponentName thisAppWidget = new ComponentName
```

```
        (context.getPackageName(), MyAppWidget.class.getName());
        int[] appWidgetIds = appWidgetMan.getAppWidgetIds
        (thisAppWidget);

        onUpdate(context, appWidgetMan, appWidgetIds);
    }
```

Step 6 开启 AndroidManifest.xml 文件设置 App Widget 程序要监听 MY_OWN_WIDGET_UPDATE 信息如以下粗体字的程序代码：

```xml
<?xml version="1.0" encoding="utf-8"?>
<manifest xmlns:android="http://schemas.android.com/apk/res/android"
    package="tw.android"
    android:versionCode="1"
    android:versionName="1.0">
  <application android:icon="@drawable/icon" android:label=
  "@string/app_name">
      <receiver android:name="MyAppWidget"
          android:label="App Widget"
          >
          <intent-filter>
              <action android:name="android.
              appwidget.action.APPWIDGET_UPDATE" />
              <action android:name="tw.android.
              MY_OWN_WIDGET_UPDATE" />
          </intent-filter>
          <meta-data android:name="android.appwidget.provider"
              android:resource="@xml/appwidget_config"
              />
      </receiver>
  </application>
</manifest>
```

完成之后运行程序，并依照上一个单元的操作方式启动 App Widget，在模拟器的首页会出现一个按钮，如图 47-1 所示，请读者先将 Eclipse 切换到 Debug 界面，然后在 LogCat 窗口新增一个此范例程序的 Log 信息 Filter（参考 UNIT 7 的说明），再回到平板电脑模拟器按下首页上的按钮，就可以在 Eclipse 的 LogCat 窗口看到 onReceive()方法所产生的 log 信息。

图 47-1　执行 App Widget 程序后在手机的 Home screen 显示的按钮

PART 9　Activity 的生命周期与进阶功能

UNIT 48　Activity 的生命周期
UNIT 49　帮 Activity 加上菜单
UNIT 50　使用 Context Menu
UNIT 51　在 Action Bar 加上功能选项
UNIT 52　在 Action Bar 上建立 Tab 标签页
UNIT 53　在状态栏显示信息

Android 版本	1.X	2.X	3.X	4.X
适用性	★	★	★	★

UNIT 48
Activity 的生命周期

在单元四十四中我们介绍了 Service 对象的生命周期,也就是它从诞生到结束的过程中所经历的状态转换。本单元让我们换个主角,来看看一直都在使用的 Activity 对象的生命周期。到目前为止,我们所完成的范例程序都把程序代码放在 onCreate()方法中运行,这个 onCreate()方法就是当 Activity 对象被建立的时候所运行的程序代码。除了 onCreate()方法之外,还有许多其他的方法会在 Activity 对象改变状态时被运行。以 UNIT 42 的范例程序来说,当用户按下主程序 Activity 界面上的按钮时,就会启动"计算机猜拳游戏"的 Activity。当使用者想结束游戏时,可以按下"完成游戏"按钮回到主程序 Activity。这两个 Activity 之间的切换会造成它们状态的改变,请读者参考图 48-1。

从图 48-1 我们可以了解在切换 Activity 的过程中所经历的状态改变,以及它们对应的运行方法。了解 Activity 运行过程中的状态变化,可以让我们在设计程序时,知道如何在适当的时机对于使用者的操作做出正确的处理。举例来说,如果用户正在编辑一段文字,突然另一个 Activity 被启动运行(例如用户开启网页浏览程序上网查询数据或是接到电话),这时候如果文字编辑程序没有储存用户已经输入的文字,当用户回到文字编辑程序时,会发现原来已经输入的文字都消失不见,这对于使用者来说会是一个非常糟糕的结果。

为了追踪 Activity 切换时的状态变化过程,我们可以使用和上一个单元相同的程序 log 技巧。这个单元的范例程序是以 UNIT 42 的程序项目为基础,再加上一些状态改变相关的方法以及产生 log 的程序代码,来观察 Activity 的状态变化,请读者依照下列说明进行操作:

Step 1 利用 Windows 文件管理器复制 UNIT 42 的范例程序的项目文件夹,复制后可以将复制的文件夹重新命名。

Step 2 利用 Eclipse 程序主菜单的 File>Import 功能加载步骤一所复制的程序专案。

```
    主程序Activity
    启动程序
        │
        ▼
    ┌─────────┐
    │ onCreate()│
    └─────────┘
        │
        ▼
    ┌─────────┐
    │ onStart() │
    └─────────┘
        │
        ▼
    ┌─────────┐
    │ onResume()│
    └─────────┘
        │
    ┌ ─ ─ ─ ─ ─ ┐
     使用者按下界面上在按
    │钮启动【电脑猜拳游 │
     戏】Activity
    └ ─ ─ ─ ─ ─ ┘
        │                           电脑猜拳游戏Activity
        ▼
    ┌─────────┐                ┌─────────┐
    │ onPause() │ ──────────▶  │ onCreate()│
    └─────────┘                └─────────┘
        │                           │
        ▼                           ▼
    ┌─────────┐                ┌─────────┐
    │ onStop()  │                │ onStart() │
    └─────────┘                └─────────┘
                                    │
                                    ▼
                                ┌─────────┐
                                │ onResume()│
                                └─────────┘
                                    │
                                ┌ ─ ─ ─ ─ ─ ┐
                                 使用者开始运行【电脑猜拳游戏】
                                │直到按下【完成游戏】按钮      │
                                └ ─ ─ ─ ─ ─ ┘
                                    │
                                    ▼
    ┌─────────┐                ┌─────────┐
    │onRestart()│ ◀──────────  │ onPause() │
    └─────────┘                └─────────┘
        │                           │
        ▼                           ▼
    ┌─────────┐                ┌─────────┐
    │ onStart() │                │ onStop()  │
    └─────────┘                └─────────┘
        │                           │
        ▼                           ▼
    ┌─────────┐                ┌─────────┐
    │ onResume()│                │onDestroy()│
    └─────────┘                └─────────┘
        │
        ▼
    •(   回到原来在循环状态   )
```

图 48-1　UNIT 42 范例程序的主程序 Activity 和 "计算机猜拳游戏" Activity 的切换所造成的状态变化流程图

Step 3 在 Eclipse 左边的项目检查窗口中，打开步骤二加载的程序项目，然后开启其中的 src/(组件路径名称)/Main.java 程序文件。

Step 4 在 Main.java 的程序代码中右击，在出现的快捷菜单中选择 Source > Override/Implement Methods…，就会出现图 48-2 的对话框。在对话框左边的清单中会列出 Activity 类中还没有使用的方法，请读者勾选其中的 onDestroy()、onPause()、onRestart()、onResume()、onStart()和 onStop()然后按下 OK 按钮，这样就完成加入我们需要的方法。

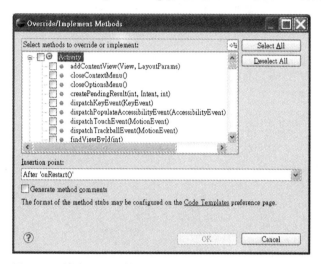

图 48-2　利用 Eclipse 的 Override/Implement Methods 功能加入需要的方法

Step 5 在 onCreate()、onDestroy()、onPause()、onRestart()、onResume()、onStart()和 onStop()等方法的程序代码的第一行插入以下 log 指令：

Log.d(LOG_TAG, "类名称.onXXX()");

其中的 onXXX()代表每一个方法自己的名称，例如在 Main 类的 onDestroy()方法中就是 Main.onDestroy()。另外其中的 LOG_TAG 是我们在这个类的程序代码中定义的一个常数属性如下：

private final String LOG_TAG = "activity lifecycle";

利用这个 Tag 标签可以帮我们过滤出这个程序项目运行时所产生的 log。

Step 6 仿照步骤三到五，在 Game.java 程序文件中加上状态改变相关的方法以及产生 log 的程序代码。

完成程序代码的编辑之后启动程序，并依照上一个单元的运行方式，一边操作范例程序，一边观察 Eclipse 的 Debug 界面右下方所显示的 log（可以设置 log filter 来筛选出我们的程序所产生的 log 以方便观察）。这个范例程序的程序代码并不复杂，而且和前面单元的操作流程很类似，因此我们就不再列出程序代码，如果需要，请读者参考本书光盘中的程序文件。图 48-3 的 log 信息是我们按下主程序界面的“启动计算机猜拳游戏程序”按钮后进行游戏，当游戏完成后又按下“完成游戏”按钮回到主程序 Activity 的结果。

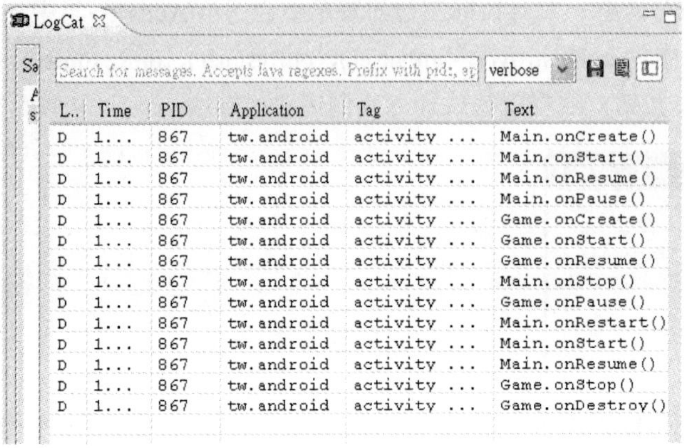
LogCat
Search for messages. Accepts Java regexes. Prefix with pid:, ap
verbose
L.. Time PID Application Tag Text
D 1... 867 tw.android activity ... Main.onCreate()
D 1... 867 tw.android activity ... Main.onStart()
D 1... 867 tw.android activity ... Main.onResume()
D 1... 867 tw.android activity ... Main.onPause()
D 1... 867 tw.android activity ... Game.onCreate()
D 1... 867 tw.android activity ... Game.onStart()
D 1... 867 tw.android activity ... Game.onResume()
D 1... 867 tw.android activity ... Main.onStop()
D 1... 867 tw.android activity ... Game.onPause()
D 1... 867 tw.android activity ... Main.onRestart()
D 1... 867 tw.android activity ... Main.onStart()
D 1... 867 tw.android activity ... Main.onResume()
D 1... 867 tw.android activity ... Game.onStop()
D 1... 867 tw.android activity ... Game.onDestroy()

图 48-3　操作范例程序的过程中所产生的 log 信息

Android 版本	1.X	2.X	3.X	4.X
适用性	★	★	★	★

UNIT 49
帮 Activity 加上菜单

菜单是图形操作接口程序主要的特色之一，如果程序提供许多功能让用户操作，这些功能就会以菜单的形式显示在屏幕上供用户单击。手机程序的菜单是当按下模拟器的 MENU 按钮时出现在屏幕下方，如图 49-1 所示，平板电脑的程序菜单是以动态的方式显示在屏幕的右上方，只有当目前运行的程序有提供菜单功能时才会出现菜单按钮，按下该按钮之后菜单就会出现，如图 49-2 所示。在介绍如何帮程序加上菜单之前，先让我们了解 Android 程序菜单的功能和限制：

图 49-1　手机程序的菜单

图 49-2　平板电脑程序的菜单

1. 菜单最多只能有二层。
2. 菜单只会显示文字而不会显示图标（Android 2.X 以前的手机第一层菜单可以使用文字和图标，第二层菜单只能够使用文字）。

菜单是 Activity 对象的一部分，因此建立菜单的程序代码必须写在 Activity 类中，建立菜单有两种方法，第一种是在程序中逐一加上菜单中的项目，第二种是将菜单写在 res/menu 文件夹下的 xml 格式文件，然后在程序中加载该菜单。不管是使用哪一种方法，建立菜单的过程中都需要用到 onCreateOptionsMenu()和 onOptionsItemSelected()这两个 callback function。

49-1　onCreateOptionsMenu()的工作

Android 系统会自动调用这个函数并传入一个 Menu 类型的对象让程序建立菜单，建立菜单的过程中会使用 Menu 对象的 add()方法，它的功能是在菜单中新增一个选项。调用 add()方法时必须传入选项的群组 id、选项 id、选项的排列次序和选项名称等信息。群组 id 和选项 id 的排列次序都是从 0 开始，但是项目 id 必须从 Menu 类中定义的 FIRST 常数开始，因为有些项目 id 是保留给 Android 系统使用，请读者参考以下程序代码范例：

```java
private static final int MENU_MUSIC = Menu.FIRST,
                        MENU_PLAY_MUSIC = Menu.FIRST + 1;

public boolean onCreateOptionsMenu(Menu menu) {
    // TODO Auto-generated method stub
    menu.add(0, MENU_MUSIC, 0, "背景音乐");
    menu.add(0, MENU_PLAY_MUSIC, 1, "播放背景音乐");
}
```

如果要建立二层式菜单，就必须换成调用 addSubMenu()方法。该方法需要传入的数据和 add()方法完全相同，只是它的传回值是一个 SubMenu 对象，我们必须储存该 SubMenu 对象，然后再调用该 SubMenu 对象的 add()方法在第二层的菜单中加入选项，如以下范例：

```java
public boolean onCreateOptionsMenu(Menu menu) {
    SubMenu subMenu = menu.addSubMenu(0, MENU_MUSIC, 0, "背景音乐");
    subMenu.add(0, MENU_PLAY_MUSIC, 0, "播放背景音乐");
    subMenu.add(0, MENU_STOP_PLAYING_MUSIC, 1, "停止播放背景音乐");
}
```

49-2　onOptionsItemSelected()的工作

当使用者单击了菜单中的某一个选项之后，Android 系统便会运行 onOptionsItemSelected ()，并且传入用户单击的项目，因此这个 callback function 的任务就是根据用户的选择运行对应的程序代码，我们可以使用 switch case 的语法来完成这个工作。

```java
public boolean onOptionsItemSelected(MenuItem item) {
    // TODO Auto-generated method stub
    switch (item.getItemId()) {
    case MENU_PLAY_MUSIC:
        ...
```

```
            break;
    case MENU_STOP_PLAYING_MUSIC:
            …
            break;
    case MENU_ABOUT:
            …
            break;
    case MENU_EXIT:
            …
            break;
    }

    return super.onOptionsItemSelected(item);
}
```

49-3 建立 xml 格式的菜单定义文件

前面的介绍是以程序代码的方式建立菜单，使用程序代码建立菜单的好处是可以根据运行的情况，动态改变菜单中的项目，但缺点是会增加程序的长度。如果程序的菜单是固定的，那么可以根据建立 xml 格式的菜单定义文件来缩短程序代码，程序中只要加载该菜单资源就可以使用，以下是一个菜单定义文件的范例。

```xml
<?xml version="1.0" encoding="utf-8"?>
<menu xmlns:android="http://schemas.android.com/apk/res/android">
    <item android:title="@string/menuItemBackgroundMusic"
          android:icon="@android:drawable/ic_media_ff" >
        <menu>
            <item android:id="@+id/menuItemPlayBackgroundMusic"
                  android:title="@string/menuItemPlayBackgroundMusic" />
            <item android:id="@+id/menuItemStopBackgroundMusic"
                  android:title="@string/menuItemStopBackgroundMusic" />
        </menu>
    </item>
    <item android:id="@+id/menuItemAbout"
          android:title="@string/menuItemAbout"
          android:icon="@android:drawable/ic_dialog_info" />
    <item android:id="@+id/menuItemExit"
          android:title="@string/menuItemExit"
          android:icon="@android:drawable/ic_menu_close_clear_cancel" />
</menu>
```

菜单定义文件的最外层是<menu>标签，里头的每一个<item>标签代表每一个选项，<item>标签中的属性可以指定项目的 id（程序代码会用这个 id 来判断用户所单击的选项）、项目名称以及和项目名称一起显示的小图标，这些属性的设置方式都和前面的接口组件属性用法相同。虽然在上面的范例中我们指定了选项所使用的小图标，可是将这个菜单定义档套用到 Android 3.X/4.X 的平台时，菜单并不会显示小图标。

另外在第一个<item>标签中我们定义了另一个<menu>标签，这就是第二层菜单的建立方式，这时候第一层的<item>标签不需要设置 id，因为它只是第二层菜单的入口，并不是真正的选项，接下来我们列出建立菜单定义档的详细步骤。

Step 1 在 Eclipse 左边的项目检查窗口中展开项目的文件夹，用鼠标右击 res 文件夹，然后运行 New > Folder 就会出现图 49-3 的对话框，在 Folder name 字段输入 menu，再按下 Finish 按钮。

图 49-3 新增文件夹对话框

Step 2 在 Eclipse 左边的项目检查窗口中右击上一个步骤建立的 menu 文件夹，然后运行 New > File 就会出现图 49-4 的对话框，在 File 字段输入菜单定义档的档名（只能使用小写英文字母和底线字符），然后单击 Finish 按钮。

Step 3 这个新增的菜单定义文件会自动开启在程序代码编辑窗口中，我们可以单击程序代码编辑窗口下方最右边的 Tab 标签页，切换到源代码检查模式，然后输入前面菜单定义文件的范例。

Step 4 如果菜单定义文件中使用到字符串资源文件中的字符串，则必须开启程序项目的 res/values/strings.xml 字符串资源文件，然后新增字符串的定义。

Step 5 如果菜单定义文件中使用到图示图像文件，则必须利用 Windows 文件总管，将该图标图像文件复制到程序项目的 res/drawable 文件夹中。

Activity 的生命周期与进阶功能　PART 9

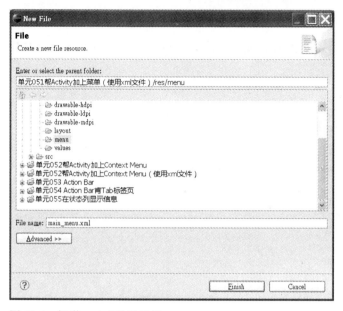

图 49-4　新增 xml 文件对话框

　　以上是建立菜单定义文件的步骤，至于如何在程序代码中加载菜单，我们将在下一小节的实现范例中说明。

49-4　范例程序

　　本单元的范例程序是在一个主程序的 Activity 加上如图 49-1 所示的菜单，其中的"背景音乐"选项是一个二层式的菜单，其中包含"播放背景音乐"和"停止播放背景音乐"两个项目，如图 49-5 所示。如果单击"播放背景音乐"，程序会启动一个 Service 播放储存在 SD 卡中的 song.mp3 文件（运行这个范例程序需要在模拟器的 SD 卡中存有这个文件，有关操作 SD 卡的步骤请参考 UNIT 38 中的介绍），如果单击"停止播放背景音乐"则会结束该 Service 的运行。另外两个选项"关于这个程序…"和"结束"则分别用来显示图 49-6 的程序相关信息和结束程序，以下我们逐一说明完成这个程序项目的步骤。

图 49-5　"背景音乐"的第二层选项

图 49-6　程序相关信息对话框

379

Step 1 请依照之前的方法建立一个新的程序项目。

Step 2 开启程序项目的字符串资源文件 res/values/strings.xml，修改其中用来显示在屏幕上的字符串如下：

```xml
<string name="hello">请开启程序菜单</string>
```

Step 3 开启程序项目的接口布局文件 res/layout/main.xml，将程序代码编辑如下：

```xml
<?xml version="1.0" encoding="utf-8"?>
<LinearLayout xmlns:android="http://schemas.android.com/apk/res/android"
    android:orientation="vertical"
    android:layout_width="match_parent"
    android:layout_height="match_parent"
    >
<TextView
    android:layout_width="match_parent"
    android:layout_height="wrap_content"
    android:text="@string/hello"
    android:textSize="20sp"
    android:gravity="center_horizontal"
    />
</LinearLayout>
```

Step 4 开启程序项目的程序文件 src/(组件路径名称)/Main.java，在程序代码中右击，再利用快捷菜单的 Source > Override/Implement Methods 选项，在程序中加入 onCreateOptionsMenu() 和 onOptionsItemSelected() 两个方法。

Step 5 在以上两个方法的程序代码中加入建立菜单和处理菜单的程序代码如下：

```java
public class Main extends Activity {

    private static final int MENU_MUSIC = Menu.FIRST,
                             MENU_PLAY_MUSIC = Menu.FIRST + 1,
                             MENU_STOP_PLAYING_MUSIC = Menu.FIRST + 2,
                             MENU_ABOUT = Menu.FIRST + 3,
                             MENU_EXIT = Menu.FIRST + 4;

    public boolean onCreateOptionsMenu(Menu menu) {
        // TODO Auto-generated method stub
        SubMenu subMenu = menu.addSubMenu(0, MENU_MUSIC, 0, "背景音乐")
            .setIcon(android.R.drawable.ic_media_ff);
        subMenu.add(0, MENU_PLAY_MUSIC, 0, "播放背景音乐");
        subMenu.add(0, MENU_STOP_PLAYING_MUSIC, 1, "停止播放背景音乐");
        menu.add(0, MENU_ABOUT, 1, "关于这个程序...")
            .setIcon(android.R.drawable.ic_dialog_info);
        menu.add(0, MENU_EXIT, 2, "结束")
            .setIcon(android.R.drawable.ic_menu_close_clear_cancel);
```

```java
        return super.onCreateOptionsMenu(menu);
    }

    @Override
    public boolean onOptionsItemSelected(MenuItem item) {
        // TODO Auto-generated method stub
        switch (item.getItemId()) {
        case MENU_PLAY_MUSIC:
            Intent it = new Intent(Main.this, MediaPlayService.class);
            startService(it);
            break;
        case MENU_STOP_PLAYING_MUSIC:
            it = new Intent(Main.this, MediaPlayService.class);
            stopService(it);
            break;
        case MENU_ABOUT:
            new AlertDialog.Builder(Main.this)
                .setTitle("关于这个程序")
                .setMessage("菜单范例程序")
                .setCancelable(false)
                .setIcon(android.R.drawable.star_big_on)
                .setPositiveButton("确定",
                    new DialogInterface.OnClickListener() {
                        @Override
                        public void onClick
                          (DialogInterface dialog,
                        int which) {
                            // TODO Auto-generated
                            method stub
                        }
                    })
                .show();

            break;
        case MENU_EXIT:
            finish();
            break;
        }

        return super.onOptionsItemSelected(item);
    }
```

Step 6 新增一个继承自 android.app.Service 类的新类，我们可以将它取名为 MediaPlayService，然后利用类似步骤四的操作技巧加入 onStartCommand() 和 onDestroy() 这两个方法。

Step 7 在以上两个方法中加入以下粗体标示的程序代码：

```java
private MediaPlayer player;

public void onDestroy() {
    // TODO Auto-generated method stub
    super.onDestroy();

    player.stop();
}

@Override
public int onStartCommand(Intent intent, int flags, int startId) {
    // TODO Auto-generated method stub

    File file = new File("/sdcard/song.mp3");
    player = MediaPlayer.create(this, Uri.fromFile(file));
    player.start();

    return super.onStartCommand(intent, flags, startId);
}
```

在 onStartCommand()方法中我们建立一个 MediaPlayer 对象并指定播放储存在 SD 卡中的 song.mp3 文件，然后将它启动。在 onDestroy()方法中则停止运行该 MediaPlayer 对象。有关 Service 的用法我们已经在 UNIT 44 作过详细介绍，读者可以参考其中的说明。

以上是利用程序代码的方式建立菜单，如果换成是用菜单定义文件的方式，则只要修改步骤五的程序代码如下，其中粗体字是必须修改的部分，我们换成使用 MenuInflater 对象的 inflate()方法将菜单资源（由菜单定义文件产生）设置给程序中的菜单对象。另外在判断用户单击的选项时，是改成使用菜单定义文件中的<item>标签的 id。

```java
public boolean onCreateOptionsMenu(Menu menu) {
    // TODO Auto-generated method stub
    MenuInflater inflater = getMenuInflater();
    inflater.inflate(R.menu.main_menu, menu);

    return super.onCreateOptionsMenu(menu);
}

@Override
public boolean onOptionsItemSelected(MenuItem item) {
    // TODO Auto-generated method stub
    switch (item.getItemId()) {
    case R.id.menuItemPlayBackgroundMusic:
        Intent it = new Intent(Main.this, MediaPlayService.class);
        startService(it);
        break;
    case R.id.menuItemStopBackgroundMusic:
        it = new Intent(Main.this, MediaPlayService.class);
```

```
            stopService(it);
            break;
        case R.id.menuItemAbout:
            new AlertDialog.Builder(Main.this)
                .setTitle("关于这个程序")
                .setMessage("菜单范例程序")
                .setCancelable(false)
                .setIcon(android.R.drawable.star_big_on)
                .setPositiveButton("确定",
                    new DialogInterface.OnClickListener() {
                        @Override
                        public void onClick(DialogInterface
                        dialog, int which) {
                            // TODO Auto-generated method stub
                        }
                    })
                .show();

            break;
        case R.id.menuItemExit:
            finish();
            break;
        }

        return super.onOptionsItemSelected(item);
    }
```

Android 版本	1.X	2.X	3.X	4.X
适用性	★	★	★	★

UNIT 50

使用 Context Menu

微软窗口程序有一个很方便的功能，就是在不同的位置右击，就会出现一组常用菜单让用户从中选取想要运行的项目。这种快捷菜单是图形接口程序的一种重要操作方式，可是手机和平板电脑上并没有鼠标，所以 Android 程序改用"按久一点"（就是按下触控屏幕界面并维持 1 秒钟）的方式来启动快捷菜单，这种快捷菜单在 Android 程序中叫做 Context Menu，所谓 Context 的意思是说和按下的位置有关，在不同的位置按住屏幕可能出现不同的快捷菜单。

50-1　Context Menu 的用法和限制

使用 Context Menu 并不难，只要在程序中加入 onCreateContextMenu()和 onContextItemSelected() 方法即可。当程序在运行的时候发生"按久一点"的情况时，系统会自动调用程序的 onCreateContextMenu()方法，我们就在这个方法中建立 Context Menu，系统就会将它显示在界面上。当用户从 Context Menu 中选取一个项目时，系统会再调用程序的 onContextItemSelected()方法，我们在这个方法中判断用户选取的项目，然后运行对应的工作。另外我们还要在程序的 onCreate()中调用 registerForContextMenu()方法注册能够接收 Context Menu 事件的 View 组件。

建立 Context Menu 的过程和上一个单元建立菜单的过程非常类似，同样也有两种方式。第一种是利用程序代码建立 Context Menu，第二种是利用 xml 菜单定义档建立 Context Menu，我们将在下一小节中示范实现的方法。另外要提醒读者的是，使用 Context Menu 有下列两项限制：

1. Context Menu 的选项只能使用文字而无法显示图标，就算我们在建立 Context Menu 时利用 setIcon()方法设置好选项的图标，也不会在 Context Menu 中显示。
2. Context Menu 和上一个单元介绍的程序菜单一样最多只能有二层。

接下来我们就以实际范例来示范 Context Menu 的建立过程和运行效果。

50-2 范例程序

我们帮上一个单元的菜单范例程序加上 Context Menu 的功能，请读者依照下列步骤操作：

Step 1 在程序项目的接口布局文件 res/layout/main.xml 中帮<LinearLayout>和<TextView>两个接口组件加上 id 属性如下粗体字的部分，因为在程序中我们要向系统注册这两个接口组件都要使用 Context Menu。

```xml
<?xml version="1.0" encoding="utf-8"?>
<LinearLayout xmlns:android="http://schemas.android.com/apk/res/android"
    android:id="@+id/linLayout"
    android:orientation="vertical"
    android:layout_width="match_parent"
    android:layout_height="match_parent"
    >
<TextView android:id="@+id/txtView"
    android:layout_width="match_parent"
    android:layout_height="wrap_content"
    android:text="@string/hello"
    android:textSize="20sp"
    android:gravity="center_horizontal"
    />
</LinearLayout>
```

Step 2 开启项目的字符串资源文件 res/values/strings.xml，修改用来显示在屏幕上的提示文字。

```xml
<string name="hello">请按住屏幕并维持一秒钟</string>
```

Step 3 开启主程序文件，在 onCreate()方法中加入注册使用 Context Menu 组件的程序代码如下粗体字的部分：

```java
public class Main extends Activity {

    …

    LinearLayout mLinLayout;
    TextView mTxtView;

    public void onCreate(Bundle savedInstanceState) {

        …

        mLinLayout = (LinearLayout) findViewById(R.id.linLayout);
        registerForContextMenu(mLinLayout);
        mTxtView = (TextView)findViewById(R.id.txtView);
        registerForContextMenu(mTxtView);
}
```

...
}

Step 4 在主程序中加入 onCreateContextMenu()和 onContextItemSelected()方法，加入这两个方法的过程就像上一个单元所介绍的操作方式，先在程序代码编辑窗口中右击，然后选择 Source > Override/Implement Methods 就会弹出一个对话框，在对话框中勾选这两个方法再单击 OK 按钮即可，然后在这两个方法中分别输入以下的程序代码：

```java
public void onCreateContextMenu(ContextMenu menu, View v,
    ContextMenuInfo menuInfo) {
    // TODO Auto-generated method stub
    super.onCreateContextMenu(menu, v, menuInfo);

    if (v == mLinLayout) {
        if (menu.size() == 0) {
            SubMenu subMenu = menu.addSubMenu(0, MENU_MUSIC, 0, "背景音乐");
                subMenu.add(0, MENU_PLAY_MUSIC, 0, "播放背景音乐");
                subMenu.add(0, MENU_STOP_PLAYING_MUSIC, 1, "停止播放背景音乐");
            menu.add(0, MENU_ABOUT, 1, "关于这个程序...");
            menu.add(0, MENU_EXIT, 2, "结束");
        }
    }
    else if (v == mTxtView) {
        menu.add(0, MENU_ABOUT, 1, "关于这个程序...");
    }
}

public boolean onContextItemSelected(MenuItem item) {
    // TODO Auto-generated method stub
    onOptionsItemSelected(item);

    return super.onContextItemSelected(item);
}
```

onCreateContextMenu()方法中的工作就是建立想要显示的菜单，由于我们向系统注册 LinearLayout 和 TextView 这两个组件都要使用 Context Menu，如果用户在 TextView 组件以外的区域启动 Context Menu，系统会调用 onCreateContextMenu()然后传入 LinearLayout 对象（自变量 v）。可是如果用户在 TextView 组件上启动 Context Menu，系统会调用 onCreateContextMenu()二次，第一次先传入 TextView 对象，第二次再传入 LinearLayout 对象（因为 TextView 在 LinearLayout 上层），因此我们在 onCreateContextMenu()中根据检查自变量 v（以上程序代码粗体字的部分）来决定究竟使用者是在哪一个区域启动 Context Menu，而且在建立 LinearLayout 的 Context Menu 之前根据调用 size()方法检查是否 menu 中已经含有选项，如果是则代表前面已经运行过 TextView 的 Context Menu 程序代码，因此不再建立 LinearLayout 的 Context Menu 以免重复。

在 onContextItemSelected() 中则是直接调用程序菜单的 onOptionsItemSelected() 方法来处理使用者的选择，因为二者的处理方式完全相同。

以上是利用程序代码的方式建立 Context Menu，如果换成是用菜单定义档的方式，则我们必须先利用上一个单元介绍的方法新增 xml 菜单定义文件如下：

- res/menu/linear_layout_context_menu.xml

```
<?xml version="1.0" encoding="utf-8"?>
<menu xmlns:android="http://schemas.android.com/apk/res/android">
    <item android:title="@string/menuItemBackgroundMusic" >
        <menu>
            <item android:id="@+id/menuItemPlayBackgroundMusic"
                android:title="@string/menuItemPlayBackgroundMusic" />
            <item android:id="@+id/menuItemStopBackgroundMusic"
                android:title="@string/menuItemStopBackgroundMusic" />
        </menu>
    </item>
    <item android:id="@+id/menuItemAbout"
        android:title="@string/menuItemAbout" />
    <item android:id="@+id/menuItemExit"
        android:title="@string/menuItemExit" />
</menu>
```

- res/menu/text_view_context_menu.xml

```
<?xml version="1.0" encoding="utf-8"?>
<menu xmlns:android="http://schemas.android.com/apk/res/android">
    <item android:id="@+id/menuItemAbout"
        android:title="@string/menuItemAbout" />
</menu>
```

以上分别是给 LinearLayout 和 TextView 这两个组件使用的菜单定义文件，然后将步骤四的程序代码修改如下，其中粗体字是必须修改的部分，我们换成使用 **MenuInflater** 对象的 **inflate()** 方法将菜单资源（由菜单定义文件产生）设置给程序中的菜单对象。完成之后运行程序，然后分别在 TextView 组件上和其他区域按住屏幕并维持 1 秒以上，就可以看到不同的 Context Menu，如图 50-1 所示。

```
public void onCreateContextMenu(ContextMenu menu, View v,
    ContextMenuInfo menuInfo) {
    // TODO Auto-generated method stub
    super.onCreateContextMenu(menu, v, menuInfo);

    if (v == mLinLayout) {
        if (menu.size() == 0) {
            MenuInflater inflater = getMenuInflater();
```

```
            inflater.inflate(R.menu.linear_layout_context_menu, menu);
        }
    }
    else if (v == mTxtView) {
        MenuInflater inflater = getMenuInflater();
        inflater.inflate(R.menu.text_view_context_menu, menu);
    }
}

public boolean onContextItemSelected(MenuItem item) {
    // TODO Auto-generated method stub
    onOptionsItemSelected(item);

    return super.onContextItemSelected(item);
}
```

图 50-1　Context Menu 范例程序的运行界面

Android 版本	1.X	2.X	3.X	4.X
适用性			★	★

UNIT 51
在 Action Bar 加上功能选项

在 UNIT 49 中我们学会了如何帮程序加上菜单，Android 程序的菜单正式名称叫做 Options Menu，Options Menu 的作用等同于微软窗口程序的菜单，如果我们回想一下微软窗口程序的操作方式，除了菜单之外，还有一个很常用工具栏。工具栏是最常用的功能项目的集合，它让用户可以直接单击而不需要先拉下菜单。Android 程序同样能够做到这样的操作设计，我们可以将 Options Menu 中常用的选项直接放在屏幕上方的 Action Bar 中，所谓 Action Bar 就是从屏幕左上方的程序小图标延伸到右边的 Options Menu 按钮的那一排区域，如图 51-1 所示。这种直接放在 Action Bar 的功能选项称为 Action Item。除了 Action Item 之外，Action Bar 上还可以建立具有操作接口的 Action View，例如微软 Word 窗口程序工具栏上的字型设置功能，图 51-2 上方的搜寻功能其实就是一个 Action View。在开始学习如何使用 Action Item 和 Action View 以前，先让我们了解如何控制 Action Bar。

图 51-1 Android 程序的 Action Bar

图 51-2 Android 程序的 Action View

51-1 控制 Action Bar

Action Bar 是 Android 3.0 以上版本才具备的功能,如果程序中要使用 Action Bar,必须在新增程序项目的对话框中设置 Min SDK Version 字段为 11 以上(11 是 Android 3.0 的版本编号),或者在 AndroidManifest.xml 程序功能描述文件中设置程序项目的 SDK 版本,如下粗体字的部分:

```xml
<?xml version="1.0" encoding="utf-8"?>
<manifest …>
    <uses-sdk android:minSdkVersion="11" />

    <application android:icon="@drawable/icon" android:label="@string/app_name">
        …
    </application>
</manifest>
```

程序也可以移除或是隐藏 Action Bar 以获取更大的显示空间,如果要移除 Action Bar(移除之后程序就无法再使用),可以在程序功能描述文件中设置 Activity 的 android:theme 属性如下:

```xml
<activity …
    android:theme="@android:style/Theme.Holo.NoActionBar">
```

如果程序只是想暂时隐藏 Action Bar,之后会让它重新显示,则可以在程序代码中取得系统的 ActionBar 对象再对它进行控制。

```java
ActionBar actBar = getActionBar();
actBar.hide();     // 隐藏 Action Bar
…
actBar.show();     // 显示 Action Bar
```

除了隐藏和显示 Action Bar 以外,也可以让程序标题消失(只显示程序的小图标),或是改变程序的小图标,甚至变更 Action Bar 的底图或底色,如以下范例:

```java
actBar.setDisplayShowTitleEnabled(false);     // 隐藏程序标题
actBar.setDisplayUseLogoEnabled(true);        // 改变程序的小图标
actBar.setBackgroundDrawable(new ColorDrawable(0xFF505050));    // 设置 Action Bar 的底色为灰色
```

设置 setDisplayUseLogoEnabled(true) 的时候必须配合修改程序功能描述文件 AndroidManifest.xml,在<application>标签中新增 android:logo 属性,指定程序所使用的图标文件如以下粗体字的程序代码:

```xml
<?xml version="1.0" encoding="utf-8"?>
<manifest …>
    <uses-sdk android:minSdkVersion="11" />
    <application …
        android:logo="@drawable/app_logo">
```

```
            <activity …>
                …
            </activity>
        </application>
</manifest>
```

了解控制 Action Bar 的方法之后,接下来就让我们开始介绍如何使用 Action Item 和 Action View。

51-2　在 Action Bar 加上 Action Item

从前面的说明中我们已经建立一个概念,那就是 Action Item 是来自 Options Menu 的选项,我们将 Options Menu 中常用的功能抽出来放在 Action Bar 上成为 Action Item。要完成这件工作其实很简单,如果我们是使用 xml 菜单定义档的方式建立 Options Menu,只要在<item>标签中增加以下粗体字的属性设置即可:

```
<item …
    android:showAsAction="ifRoom" />
```

这个属性告诉 Android 系统如果 Action Bar 上还有空间,就把这个选项抽出变成 Action Item。Action Item 默认会用图标的方式显示,如果要加上项目名称,可以换成以下的设置值:

```
<item …
    android:showAsAction="ifRoom|withText" />
```

另外我们也可以利用程序代码的方式让选项变成 Action Item,首先在 onCreateOptionsMenu() 中取得选项的 MenuItem 对象,然后调用它的 setShowAsAction()方法如下:

```
MenuItem menuItem = menu.findItem(R.id.MenuItemId);   // MenuItemId 是 xml 菜单定义文件中设置的选项 id
menuItem.setShowAsAction(MenuItem.SHOW_AS_ACTION_IF_ROOM);
```

如果要同时显示选项的图标和名称,可以加上 MenuItem.SHOW_AS_ACTION_WITH_TEXT 参数:

```
MenuItem menuItem = menu.findItem(R.id.MenuItemId);   // MenuItemId 是 xml 菜单定义文件中设置的选项 id
menuItem.setShowAsAction(
    MenuItem.SHOW_AS_ACTION_IF_ROOM|MenuItem.SHOW_AS_ACTION_WITH_TEXT);
```

用户单击 Action Item 后,系统的处理方式就如同单击 Options Menu 中的选项一样,也就是会调用 onOptionsItemSelected()方法,然后传入选项 id,因此程序代码的处理方式和 UNIT 49 中的介绍完全相同。

另外 Android 程序运行时会在 Action Bar 的左边显示一个程序的小图标,其实这个小图标也是一个 Action Item,如果用户单击它,系统同样会调用 onOptionsItemSelected()方法,然后传入

android.R.id.home，如果需要程序也可以对它进行处理。

51-3 在 Action Bar 加上 Action View

前一个小节介绍的 Action Item 只是一个可以单击的按钮，如果想做到类似微软 Word 窗口程序工具栏中的字型设置下拉式菜单，就必须使用 Action View。Action View 可以是各种类型的接口组件所组成的一个操作单元，它的运行方式类似前面介绍过的对话框，也就是说我们可以自己设计操作接口，然后再设置接口组件的事件 listener 让用户可以进行操作。不过由于 Action Bar 的空间有限，因此一个 Action View 通常都只有一、二个接口组件。

建立 Action View 的过程相较于 Action Item 来说需要比较多的步骤，以下是详细的操作过程：

Step 1 在 res/layout 文件夹中建立一个 xml 格式的 Action View 接口布局文件，我们之前学过的各种接口组件语法都可以使用。

Step 2 在 res/menu 文件夹中的菜单定义文件定义一个 <item> 标签，并加上 android:showAsAction 和 android:actionLayout 属性，第一个属性是说这个项目要显示在 Action Bar，第二个属性是指定它所使用的接口布局文件，例如以下范例是假设我们在步骤一中已经建立一个名为 select_region.xml 的界面布局文件。

```
<item android:id="@+id/menuItemRegion"
    android:title="@string/menuItemRegion"
    android:icon="@android:drawable/ic_menu_search"
    android:showAsAction="ifRoom"
    android:actionLayout="@layout/select_region" />
```

Step 3 在程序文件中的 onCreateOptionsMenu() 方法中取得 Action View 中的接口组件并设置好它的事件 listener，另外我们必须在程序文件中自己建立这些事件 listener。

了解 Action Item 和 Action View 的用法之后，接下来我们用一个实际范例来示范实现的过程。

51-4 范例程序

我们将在 UNIT 49 的菜单范例程序中加上 Action Item 和 Action View，并且改变 Action Bar 的外观，程序在平板电脑的运行界面如图 51-3 所示，在手机的运行界面如图 51-4 所示。比较图 51-3 和 51-4 可以发现 Action Item 和 Action View 会自动根据 Action Bar 的可用空间大小自动调整，在手机上有些项目会换成在菜单中显示。在平板电脑的运行界面会显示两个新增的 Action View，最左边的那一个"放大镜"图示是搜寻列，右边的 Action View 是由 TextView 和 Spinner 两个接口组件组成，它的功能是让用户从中选择一个地区，图 51-5 是这两个 Action View 的操作界面。

图 51-3　Action Item 和 Action View 范例程序在平板电脑的运行界面

图 51-4　Action Item 和 Action View 范例程序在手机的运行界面

图 51-5　Action View 的操作界面

　　实现这个范例程序之前可以先用 Windows 文件管理器复制 UNIT 49 的范例程序项目文件夹，复制后可以重新命名，然后开启 Eclipse 并利用主菜单的 File > Import 功能加载复制的程序项目。

Step 1　开启程序项目的 res/menu/main_menu.xml 菜单定义文件，把其中的"关于"和"结束"两个<item>加上 android:showAsAction 属性，让它们成为 Action Item。另外新增两个<item>标签并加上适当的属性让它们成为 Action View，请参考以下粗体字的程序代码，其中我们在最后一个<item>使用 android:actionViewClass 属性而不是上一小节介绍的 android:actionLayout 属性，因为这个 Action View 是利用 Android SDK 所提供的搜寻功能列。

```
<item android:id="@+id/menuItemAbout"
    android:title="@string/menuItemAbout"
    android:icon="@android:drawable/ic_dialog_info"
    android:showAsAction="ifRoom" />
<item android:id="@+id/menuItemExit"
    android:title="@string/menuItemExit"
    android:icon="@android:drawable/ic_menu_close_clear_cancel"
    android:showAsAction="ifRoom|withText" />
```

```xml
<item android:id="@+id/menuItemRegion"
    android:title="@string/menuItemRegion"
    android:icon="@android:drawable/ic_menu_search"
    android:showAsAction="ifRoom"
    android:actionLayout="@layout/select_region" />
<item android:id="@+id/menuItemSearch"
    android:title="@string/menuItemSearch"
    android:showAsAction="ifRoom"
    android:actionViewClass="android.widget.SearchView" />
```

Step 2 在程序项目的 res/layout 文件夹中新增步骤一的 Action View 所使用的接口布局文件 select_region.xml，它的程序代码如下，其中的接口组件和属性都是之前单元学过内容，请读者自行参阅。

```xml
<?xml version="1.0" encoding="utf-8"?>
<LinearLayout xmlns:android="http://schemas.android.com/apk/res/android"
    android:orientation="horizontal"
    android:layout_width="match_parent"
    android:layout_height="match_parent"
    android:layout_gravity="center_horizontal" >
<TextView
    android:layout_width="wrap_content"
    android:layout_height="match_parent"
    android:text="选择地区："
    android:textSize="20sp"
    android:gravity="center_vertical" />
<Spinner android:id="@+id/spnRegion"
    android:layout_width="wrap_content"
    android:layout_height="match_parent"
    android:drawSelectorOnTop="true" />
</LinearLayout>
```

Step 3 开启程序项目的 res/values/strings.xml 字符串资源文件，加入程序项目会用到的所有字符串和 Spinner 接口组件中的选项数组。

```xml
<?xml version="1.0" encoding="utf-8"?>
<resources>
    …
    <string name="menuItemSearch">搜寻...</string>
    <string name="menuItemRegion">选择地区...</string>

    <string-array name="spnRegionList">
        <item>亚洲</item>
        <item>美洲</item>
        <item>欧洲</item>
    </string-array>
</resources>
```

Step 4 开启主程序文件 src/(组件路径名称)/Main.java，根据上一个小节中的说明，针对新增的两个 Action View，我们必须在程序代码中设置好它们的事件 listener，另外在 onOptionsItemSelected()中需要新增"搜索"和"选择地区"这两个 Action View 的处理，因为如果这两个 Action View 被收到菜单中（像是在手机运行的例子），这时候就会变成用选项的方式运行。

```java
public boolean onCreateOptionsMenu(Menu menu) {
    // TODO Auto-generated method stub
    MenuInflater inflater = getMenuInflater();
    inflater.inflate(R.menu.main_menu, menu);

    // 设置 action views
    Spinner spnRegion = (Spinner)
        menu.findItem(R.id.menuItemRegion).getActionView()
        .findViewById(R.id.spnRegion);
    ArrayAdapter<CharSequence> adapRegionList = ArrayAdapter.
    createFromResource(
            this, R.array.spnRegionList, android.R.layout.simple_spinner_item);
    spnRegion.setAdapter(adapRegionList);
    spnRegion.setOnItemSelectedListener(spnRegionItemSelLis);

    SearchView searchView = (SearchView)
        menu.findItem(R.id.menuItemSearch).getActionView();
    searchView.setOnQueryTextListener(searchViewOnQueryTextLis);

    return super.onCreateOptionsMenu(menu);
}

private SearchView.OnQueryTextListener searchViewOnQueryTextLis = new
    SearchView.OnQueryTextListener() {

    @Override
    public boolean onQueryTextChange(String newText) {
        // TODO Auto-generated method stub
        return false;
    }

    @Override
    public boolean onQueryTextSubmit(String query) {
        // TODO Auto-generated method stub
        Toast.makeText(Main.this, query, Toast.LENGTH_LONG).show();

        return true;
    }
};
```

```java
private Spinner.OnItemSelectedListener spnRegionItemSelLis =
        new Spinner.OnItemSelectedListener () {
    public void onItemSelected(AdapterView parent,
                    View v,
                    int position,
                    long id) {
        Toast.makeText(Main.this, parent.getSelectedItem().toString(),
                    Toast.LENGTH_LONG).show();
    }
    public void onNothingSelected(AdapterView parent) {
    }
};

@Override
public boolean onOptionsItemSelected(MenuItem item) {
    // TODO Auto-generated method stub
    switch (item.getItemId()) {
    …(原来项目的程序代码)
    case R.id.menuItemRegion:
        new AlertDialog.Builder(Main.this)
            .setTitle("选择地区")
            .setMessage("这是选择地区对话框")
            .setCancelable(false)
            .setIcon(android.R.drawable.star_big_on)
            .setPositiveButton("确定",
                new DialogInterface.OnClickListener() {
                    @Override
                    public void onClick(DialogInterface
                    dialog, int which) {
                        // TODO Auto-generated method stub
                    }
                })
            .show();

        break;
    case R.id.menuItemSearch:
        new AlertDialog.Builder(Main.this)
            .setTitle("搜索")
            .setMessage("这是搜索对话框")
            .setCancelable(false)
            .setIcon(android.R.drawable.star_big_on)
            .setPositiveButton("确定",
                new DialogInterface.OnClickListener() {
                    @Override
                    public void onClick(DialogInterface
                    dialog, int which) {
                        // TODO Auto-generated method stub
                    }
```

```
            })
            .show();

        break;
    }

    return super.onOptionsItemSelected(item);
}
```

有关于 Spinner 接口组件的设置方法和 UNIT 12 中的说明完全相同，但是请读者注意取得 Action View 的接口组件的过程是先取得选项 id，然后取得其中的 view 对象，最后再取得接口组件。至于 Android SDK 的搜寻功能列，我们只需要设置它的 OnQueryTextListener 即可，当用户在搜寻功能列中输入文字并启动搜寻功能时，系统会自动运行这个事件 listener。这个范例程序会将用户输入的搜寻文字和选择的地区以 Toast 快显信息的方式显示在屏幕上。

Step 5 在主程序文件的 onCreate()方法中修改 Action Bar 的外观：

```
public void onCreate(Bundle savedInstanceState) {
    …
    final ActionBar actBar = getActionBar();
    actBar.setDisplayShowTitleEnabled(false);
    actBar.setDisplayUseLogoEnabled(true);
    actBar.setBackgroundDrawable(new ColorDrawable(0xFF505050));
}
```

然后准备一个高 72 pixel 宽小于 100 pixel 的 png 或 jpg 图像文件，将它复制到程序项目的 res/drawable 文件夹中。最后根据第一小节中的说明修改程序功能描述文件 AndroidManifest.xml，指定该图像文件为程序的图标文件。

完成之后便可以运行程序并加以测试，完整的程序代码请参考光盘中的程序项目源文件。

Android 版本	1.X	2.X	3.X	4.X
适用性			★	★

UNIT 52

在 Action Bar 上建立 Tab 标签页

我们曾经在 UNIT 18 和 UNIT 39 中学过建立 Tab 标签页的方法，那时候使用了 TabHost、TabWidget 和 FrameLayout 三个对象，这种是属于旧式的 Tab 标签页，虽然它可以适用所有的 Android 平台，可是如果在 Android 3.0 以后的手机或平板电脑上运行的话，所建立的 Tab 标签页会在位于 Action Bar 的下方，如图 52-1 所示。如果我们把 UNIT 18 的范例程序改成使用 Action Bar，将会得到如图 52-2 的结果（包括在平板电脑和手机上的运行界面）。和图 52-1 比较可以发现 Tab 标签往上移到 Action Bar 中（手机界面由于 Action Bar 空间不足，因此自动移到下方），因此程序可以使用的界面区域变大了，这就是使用 Action Bar 建立 Tab 标签页的好处。

图 52-1　UNIT 18 Tab 标签页范例程序的运行界面

Activity 的生命周期与进阶功能　PART 9

图 52-2　使用 Action Bar 建立 Tab 标签页的程序运行界面

> **补充说明**　**Android 4.0 的改变**
>
> 旧式的 Tab 标签页类型在 Android 3.X 上可以显示小图标,可是在 Android 4.0 就会将小图标隐藏,只有使用 Action Bar 的 Tab 标签页才会显示小图标。

在 Action Bar 上建立 Tab 标签页需要用到 UNIT 25 中学过的 Fragment 对象,Fragment 对象的用法变化多端,在 UNIT 25 到 UNIT 28 中已经作过详细介绍,不过这里我们只需要用到最基本的功能。接着我们就开始介绍在 Action Bar 上建立 Tab 标签页的过程,它包含以下四项工作:

1. 每一个 Tab 标签页的操作接口和程序代码必须独立建置成一个 Fragment 类。每一个 Fragment 类都有自己的接口布局文件,该接口布局文件就是 Tab 标签页的操作接口,Fragment 类中的程序代码必须各自设置自己的接口组件的事件 listener。
2. 在主程序 Activity 的接口布局文件中建立一个 FrameLayout 组件以供显示 Fragment 对象。
3. 建立一个新类实现 ActionBar.TabListener 接口,当我们在 Action Bar 中加入一个 Tab 标签页时,必须同时设置一个此类的对象,当用户切换 Tab 标签时,系统会自动调用这个对象,我们在它的程序代码中完成 Fragment 切换的动作。
4. 将 Action Bar 设置为 Tab 标签页模式,另外也可以利用上一个单元介绍的 Action Bar 控制技巧来改变 Action Bar 的外观。

接下来我们用一个实际范例来学习实现 Action Bar 的 Tab 标签页程序,我们的目标是建立一个如图 52-2 所示的操作接口,其中包含了 UNIT 16 和 UNIT 17 这两个范例程序的功能。

Step 1　运行 Eclipse 新增一个新的 Android 程序项目,项目属性对话框中的设置依照之前的惯例即可,请读者注意最下面的 Min SDK Version 字段必须设置为 11 或以上,代表这是 Android 3.0 以上的程序项目。

399

Step 2 在 Eclipse 左边的项目检查窗口中展开此程序项目,在"src/(组件路径名称)"文件夹上右击,选择 New > Class 便会出现新增类对话框。在 Name 字段输入"婚姻建议"的 Fragment 类名称,例如 MarriSugFragment,然后按下 Superclass 字段最右边的 Browse 按钮,在对话框上方的字段中输入 fragment,再单击下方清单中的 Fragment 类,最后按下 OK 按钮就完成基础类的设置,按下 Finish 按钮结束新增类对话框。

Step 3 新增的类程序文件会自动显示在程序代码编辑窗口中,先用鼠标单击 Class 内部的程序代码,然后右击,在出现的快捷菜单中选择 Source > Override/Implement Methods…,在对话框左边的清单中会列出 Fragment 类中还没有使用的方法,请勾选其中的 onActivityCreated()和 onCreateView()然后按下 OK 按钮,这样就完成加入我们需要的方法。

Step 4 开启 UNIT 16 的范例程序文件,除了 onCreate()方法的程序代码之外,复制 Class 内所有其他的程序代码,包括变量定义,到上一个步骤的程序文件,读者可以参考光盘中的程序项目。

Step 5 将程序文件中的 onActivityCreated()和 onCreateView()两个方法中的程序代码编辑如下,其中的 inflater.inflate()是设置此 Fragment 类使用的接口布局文件,这里是指定使用 res/layout/marri_sug.xml,我们将在下一个步骤中建立这个文件。另外我们在 onActivityCreated()方法中调用 setupViewComponent()完成所有接口组件的设置。

```java
public View onCreateView(LayoutInflater inflater, ViewGroup container,
        Bundle savedInstanceState) {
    // TODO Auto-generated method stub
    return inflater.inflate(R.layout.marri_sug, container, false);
}

public void onActivityCreated(Bundle savedInstanceState) {
    // TODO Auto-generated method stub
    super.onActivityCreated(savedInstanceState);

    setupViewComponent();
}
```

Step 6 右击程序项目的 res/layout 文件夹,选择 New > Android XML File,在出现的对话框的 File 字段输入 marri_sug 然后按下 Finish 按钮。新增的接口布局文件会显示在程序代码编辑窗口中,单击程序代码编辑窗口下面最右边的 Tab 标签切换成源文件检查模式,然后开启 UNIT 16 范例程序的接口布局文件,将其中的程序代码全部复制过来取代原来的程序代码。

Step 7 开启程序项目的 res/values/strings.xml 字符串资源文件,将 UNIT 16 范例程序的字符串资源文件中程序代码用到的字符串,复制过来如下:

```xml
<?xml version="1.0" encoding="utf-8"?>
<resources>
    <string name="hello">Hello World, Main!</string>
    <string name="app_name">Action Bar 和 Tab 标签页</string>
    .
    .(单元十六范例程序的字符串资源文件中程序代码用到的字符串)
    .
</resources>
```

Step 8 将程序代码编辑窗口切换成 MarriSugFragment 的程序代码，读者会发现在 setupViewComponent() 方法中还有语法错误，这是因为在 Fragment 类中并没有 findViewById()这个方法可以取得接口布局文件中的组件，我们必须先调用 getView()取得 Fragment 的接口，再运行它的 findViewById()取得接口组件如下：

```java
private void setupViewComponent() {
    // 从资源类 R 中取得接口组件
    btnDoSug = (Button)getView().findViewById(R.id.btnDoSug);
    radGSex = (RadioGroup)getView().findViewById(R.id.radGSex);
    radGAge = (RadioGroup)getView().findViewById(R.id.radGAge);
    radBtnAgeRng1 = (RadioButton)getView().findViewById(R.id.radBtnAgeRng1);
    radBtnAgeRng2 = (RadioButton)getView().findViewById(R.id.radBtnAgeRng2);
    radBtnAgeRng3 = (RadioButton)getView().findViewById(R.id.radBtnAgeRng3);
    txtResult = (TextView)getView().findViewById(R.id.txtResult);

    // 设置事件 listener
    btnDoSug.setOnClickListener(btnDoSugOnClick);
    radGSex.setOnCheckedChangeListener(radGSexOnCheChanLis);
}
```

Step 9 仿照步骤二到步骤八建立一个 GameFragment 的新类以及它所使用的接口布局文件，该类的程序代码、接口布局文件和字符串资源都是从 UNIT 17 的范例程序中复制过来。

Step 10 开启程序项目的接口布局文件 res/layout/main.xml，在文件中建立一个 FrameLayout 组件如下，这个 FrameLayout 组件必须设置一个 id 名称，因为程序代码中会使用这个组件。

```xml
<?xml version="1.0" encoding="utf-8"?>
<LinearLayout xmlns:android="http://schemas.android.com/apk/res/android"
    android:orientation="vertical"
    android:layout_width="match_parent"
    android:layout_height="match_parent"
    >
<FrameLayout android:id="@+id/frameLayout"
    android:layout_width="match_parent"
    android:layout_height="match_parent"
    >
</FrameLayout>
</LinearLayout>
```

Step 11 接下来我们还要新增一个类实现 ActionBar.TabListener 接口，在"src/(组件路径名称)"文件夹上右击，选择 New > Class 便会出现新增类对话框，在 Name 字段输入 MyTabListener，然后单击 Finish 按钮结束新增类对话框。

Step 12 新增的类会自动显示在程序代码编辑窗口中，请加上如下粗体字的程序代码：

```
package tw.android;

import android.app.ActionBar;

public class MyTabListener implements ActionBar.TabListener {

}
```

然后类名称下方会出现红色波浪底线标示语法错误，将鼠标光标移到该处便会弹出一个窗口，请单击其中的 Add unimplemented methods，就会在程序代码中加入需要的方法。

Step 13 将程序代码编辑如下：

```
public class MyTabListener implements ActionBar.TabListener {

    private Fragment mFragment;

    public MyTabListener(Fragment fragment) {
        mFragment = fragment;
    }

    @Override
    public void onTabReselected(Tab tab, FragmentTransaction ft) {
        // TODO Auto-generated method stub

    }

    @Override
    public void onTabSelected(Tab tab, FragmentTransaction ft) {
        // TODO Auto-generated method stub
        ft.add(R.id.frameLayout, mFragment, null); // 被单击的 Tab 标签页运行这个方法
    }

    @Override
    public void onTabUnselected(Tab tab, FragmentTransaction ft) {
        // TODO Auto-generated method stub
        ft.remove(mFragment); // 被隐藏的 Tab 标签页运行这个方法
    }

}
```

其中我们新增一个类的建构式将传入的 Fragment 对象储存起来，当用户在 Action Bar 上

切换 Tab 标签页时，系统会调用被单击的 Tab 标签页的 onTabSelected()方法，此时我们将该 Tab 标签页所对应的 Fragment 对象放入接口布局文件中的 FrameLayout 完成显示的工作。至于被隐藏的 Tab 标签页则运行 onTabUnselected()，此时我们将它的 Fragment 对象移出 FrameLayout。

Step 14 最后开启程序项目的主程序文件，加入设置 Action Bar 的程序代码和建立 Tab 标签页的程序代码如下，其中我们利用 ActionBar 对象的 addTab()方法加入 Tab 标签页，每一个加入的 Tab 标签页都可以设置它的标题名称和图标，并且传入一个步骤十一到十三所建立的类的对象，这个对象就是 Tab 标签页的事件 listener。

```java
public class Main extends Activity {
    /** Called when the activity is first created. */
    @Override
    public void onCreate(Bundle savedInstanceState) {
        super.onCreate(savedInstanceState);
        setContentView(R.layout.main);

        setupViewComponent();
    }

    private void setupViewComponent() {
        final ActionBar actBar = getActionBar();

        // 设置 Action Bar 为 Tab 标签页模式
        actBar.setNavigationMode(ActionBar.NAVIGATION_MODE_TABS);

        // 设置第一个 Tab 标签页
        Fragment fragMarriSug = new MarriSugFragment();
        actBar.addTab(actBar.newTab()
                .setText("婚姻建议")
                .setIcon(getResources().getDrawable(android.R.drawable.ic_lock_idle_alarm))
                .setTabListener(new MyTabListener(fragMarriSug)));

        // 设置第二个 Tab 标签页
        Fragment fragGame = new GameFragment();
        actBar.addTab(actBar.newTab()
                .setText("计算机猜拳游戏")
                .setIcon(getResources().getDrawable(android.R.drawable.ic_dialog_alert))
                .setTabListener(new MyTabListener(fragGame)));
    }
}
```

以上就是在 Action Bar 中建立 Tab 标签页的完整步骤，虽然整个过程有些繁复，但其实它的运行原理并不复杂，主要就是完成本单元开头介绍的四项工作，完整的程序代码请参考光盘中的程序项目源文件。

Android 版本	1.X	2.X	3.X	4.X
适用性	★	★	★	★

UNIT 53
在状态栏显示信息

Android 手机和平板电脑屏幕上有一列称为"状态栏"的区域,手机是在屏幕上方,平板电脑是在屏幕右下方。程序可以在"状态栏"中显示信息,如果在平板电脑中单击这个"状态栏",它会开启,如图 53-1 所示,如果是手机必须按住"状态栏"再往下拉。如果状态栏中含有程序显示的信息,则开启后的界面如图 53-2 所示,如果信息中包含启动程序的指令,点击该信息便会启动指定的程序。

图 53-1 开启 System Bar 右边的"状态栏"后的界面

图 53-2 开启的状态栏中含有程序显示的信息

> **补充说明** 平板电脑的 System Bar
>
> 平板电脑的"状态栏"其实是属于 System Bar 的一部份,System Bar 是在平板电脑屏幕下方的那一排操作列,左边有三个按钮,由左至右分别为"回上一页"、"回到 Home Screen"和"列出最近运行的程序",右边才是"状态栏"。如果程序想要隐藏 System Bar 以取得更大的屏幕空间,可以先取得任何一个接口组件,然后调用它的 setSystemUiVisibility() 方法如下:

```
View v = findViewById(R.id.view_id);    // view_id 是定义在接口布局文件中的
                                         // 任何一个接口组件的 id
v.setSystemUiVisibility(View.STATUS_BAR_HIDDEN);
```
设置隐藏的 System Bar 其实不是真的消失，只是每一个按钮都变成一个很不清楚的小点，使用者还是可以单击这些按钮。

程序如果要在"状态栏"上显示信息，必须使用 Android 系统中的 NotificationManager 对象，另外还需要在程序代码中建立 Notification 对象和 PendingIntent 对象。以下我们直接用一个实现范例来说明这些对象的使用方法，这个范例程序是根据 UNIT 41 的"计算机猜拳游戏"程序项目进行修改，请读者依照下列步骤操作：

Step 1 使用 Windows 文件管理器复制 UNIT 41 的程序项目文件夹，复制之后可以变更复制文件夹的名称。

Step 2 开启 Eclipse，使用主菜单 File > Import 加载复制的项目文件夹。

接下来我们把显示状态栏信息的程序代码加在原来显示输赢结果的程序代码之后。

Step 3 在 Eclipse 左边的项目检查窗口中找到项目的主程序文件 Main.java，将它开启后在程序代码中找到第一个显示输赢结果的程序代码，也就是范例程序中的 txtResult.setText()，然后把后续步骤的程序代码加在它的后面。

Step 4 建立一个 Notification 对象，这个对象包含要显示的信息和放在信息前面的小图标，以及指定显示这个信息的时间。

```
Notification noti = new Notification(
                要使用的小图标,
                "要显示的信息",
                System.currentTimeMillis());
```

其中要使用的小图标可以先储存在程序项目中的 res/drawable 文件夹中，再以"R.drawable.文件名"的方式取用，图文件格式必须是 PNG 或是 JPG。另外也可以直接使用 Android SDK 提供的小图标文件，使用的格式为 android.R.drawable.XXX。读者只要在程序代码编辑窗口中输入"android.R.drawable."，就会自动出现候选清单供您选择。System.currentTimeMillis()是指定要立刻显示。

Step 5 建立一个 Intent 对象，这个 Intent 对象后续会和信息链接在一起，当用户单击此信息时，会启动 Intent 对象中指定的 Activity。我们在这个 Intent 对象中设置启动 GameResult 程序，并附带局数统计资料，让用户可以观看局数统计结果。由于这个 Intent 对象是以 PendingIntent 的方式送出运行，因此必须把它的 flag 设置为 Intent.FLAG_ACTIVITY_NEW_TASK。

```
Intent it = new Intent();
it.setClass(this, GameResult.class);
it.setFlags(Intent.FLAG_ACTIVITY_NEW_TASK);
Bundle bundle = new Bundle();
bundle.putInt("KEY_COUNT_SET", miCountSet);
bundle.putInt("KEY_COUNT_PLAYER_WIN", miCountPlayerWin);
bundle.putInt("KEY_COUNT_COM_WIN", miCountComWin);
bundle.putInt("KEY_COUNT_DRAW", miCountDraw);
it.putExtras(bundle);
```

Step 6 建立一个 PendingIntent 对象，指定拥有者和处理方式，并输入上一个步骤建立的 Intent 对象：

```
PendingIntent penIt = PendingIntent.getActivity(
                this, 0, it,
                PendingIntent.FLAG_UPDATE_CURRENT);
```

Step 7 调用 Notification 对象的 setLatestEventInfo()方法，指定展开状态栏后要显示的信息标题和内容。

```
noti.setLatestEventInfo(this, "信息标题", "显示的信息", penIt);
```

Step 8 调用 getSystemService()方法取得系统的 NotificationManager 对象。

```
NotificationManager notiMgr =
                (NotificationManager) getSystemService
                (NOTIFICATION_SERVICE);
```

Step 9 调用 NotificationManager 对象的 notify()方法送出信息，同时指定这个信息的 id 编号。

```
notiMgr.notify(信息 id 编号, noti);
```

Step 10 如果要取消"状态栏"上显示的信息（例如当程序结束时），可以调用 NotificationManager 对象的 cancel()。

　　以上步骤四到九就是在状态栏上显示信息的方法，由于程序中有许多地方都会用到状态栏信息的功能，因此我们把步骤四到九的程序代码写成一个方法如下，请读者注意在送出状态栏信息之前我们先将旧的信息删除，这样新的信息才会完全显示，否则只会进行更新，用户不会看到新的信息。另外 NOTI_ID 是定义在类中的一个属性。

```
private void showNotification(String s) {
    Notification noti = new Notification(
            android.R.drawable.btn_star_big_on,
            s,
            System.currentTimeMillis());

    Intent it = new Intent();
```

```
it.setClass(this, GameResult.class);
it.setFlags(Intent.FLAG_ACTIVITY_NEW_TASK);
Bundle bundle = new Bundle();
bundle.putInt("KEY_COUNT_SET", miCountSet);
bundle.putInt("KEY_COUNT_PLAYER_WIN", miCountPlayerWin);
bundle.putInt("KEY_COUNT_COM_WIN", miCountComWin);
bundle.putInt("KEY_COUNT_DRAW", miCountDraw);
it.putExtras(bundle);

PendingIntent penIt = PendingIntent.getActivity(
        this, 0, it,
        PendingIntent.FLAG_UPDATE_CURRENT);

noti.setLatestEventInfo(this, "游戏结果", s, penIt);

NotificationManager notiMgr =
    (NotificationManager) getSystemService(NOTIFICATION_SERVICE);
    notiMgr.cancel(NOTI_ID);
    notiMgr.notify(NOTI_ID, noti);
}
```

新增这个方法之后，我们可以在需要显示状态栏信息的地方调用这个方法即可，请读者参考以下程序代码。另外我们加入 onDestroy() 方法，当程序结束时删除所显示的状态栏信息，图 53-3 是程序的运行界面。当开启状态栏之后单击其中的信息，就会显示局数统计数据。

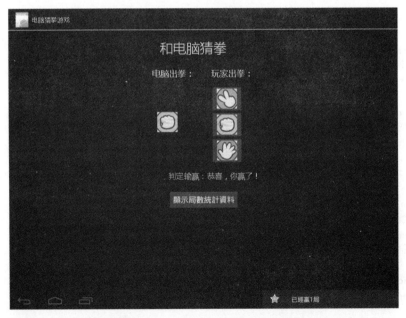

图 53-3 "计算机猜拳游戏"程序加入显示状态栏信息的功能

```java
package …

import …

public class Main extends Activity {

    private static final int NOTI_ID = 100;

    …（同原来程序代码）

    /** Called when the activity is first created. */
    @Override
    public void onCreate(Bundle savedInstanceState) {
        …（同原来程序代码）
    }

    @Override
    protected void onDestroy() {
    // TODO Auto-generated method stub

        ((NotificationManager) getSystemService(NOTIFICATION_SERVICE))
                .cancel(NOTI_ID);

        super.onDestroy();
    }

    …（同原来程序代码）

    private void setupViewComponent() {
        …（同原来程序代码）
    }

    private Button.OnClickListener btnScissorsLin = new Button.OnClickListener() {
        public void onClick(View v) {
            // 决定计算机出拳
            int iComPlay = (int)(Math.random()*3 + 1);

            // 1 - 剪刀，2 - 石头，3 - 布。
            if (iComPlay == 1) {
                mImgComPlay.setImageResource(R.drawable.scissors);
                miCountDraw++;
                txtResult.setText(getString(R.string.result) +
                            getString(R.string.playerDraw));
                showNotification("已经平手" + Integer.
                toString(miCountDraw) + "局");
            }
            else if (iComPlay == 2) {
                mImgComPlay.setImageResource(R.drawable.stone);
```

```
                    miCountComWin++;
                    txtResult.setText(getString(R.string.result) +
                            getString(R.string.playerLose));
                    showNotification("已经输" + Integer.toString
                    (miCountComWin) + "局");
            }
            else {
                    mImgComPlay.setImageResource(R.drawable.net);
                    miCountPlayerWin++;
                    txtResult.setText(getString(R.string.result) +
                            getString(R.string.playerWin));
                    showNotification("已经赢" + Integer.
                    toString(miCountPlayerWin) + "局");
            }
        }
};

private Button.OnClickListener btnStoneLin = new Button.OnClickListener() {
        public void onClick(View v) {
                // 依照以上程序代码修改
                ...
        }
};

private Button.OnClickListener btnNetLin = new Button.OnClickListener() {
        public void onClick(View v) {
                // 依照以上程序代码修改
                ...
        }
};

...

private void showNotification(String s) {
        // 如前面说明
        ...
}
}
```

PART 10 储存程序数据

UNIT 54　使用 SharedPreferences 储存数据
UNIT 55　使用 SQLite 数据库储存数据
UNIT 56　使用 Content Provider 跨程序存取数据
UNIT 57　使用文件储存数据

Android 版本	1.X	2.X	3.X	4.X
适用性	★	★	★	★

UNIT 54
使用 SharedPreferences 储存数据

如果程序需要储存数据，最简单的方法就是使用 SharedPreferences 对象。SharedPreferences 对象可以用来储存基本类型的数据，包括整数、浮点数、字符串和布尔值，每一项数据都必须赋予一个 Key 名称，以下我们依序说明如何使用 SharedPreferences 对象来储存数据、读取数据、删除数据和清空数据。

54-1 储存数据的步骤

Step 1 决定数据文件的名称，只需要主文件名即可，不需要扩展名，例如 GAME_RESULT。

Step 2 调用 getSharedPreferences()方法，传入步骤一的数据文件名称，并指定数据文件的使用范围，有以下三种使用范围设置值：

  MODE_PRIVATE 也就是 0，代表只有这个程序才能使用
  MODE_WORLD_READABLE 所有程序都可以读取
  MODE_WORLD_WRITEABLE 所有程序都可以修改

getSharedPreferences()方法会传回一个 SharedPreferences 对象如下：

  SharedPreferences gameResultData = getSharedPreferences("GAME_RESULT", 0);

Step 3 调用步骤二的 SharedPreferences 对象的 edit()方法取得一个 Editor 对象。

Step 4 调用 Editor 对象的 putXXX()方法将数据写入数据文件中，XXX 代表各种基本数据类型，

例如 Int、Float、String……。每一项数据都必须配合一个 Key 名称，例如 putInt("GAME_SCORE", score)。

Step 5 写入全部的数据之后必须调用 commit()方法才算完成写入数据的动作。

步骤三到五可以利用 Java 程序常见的匿名对象的写法以简化程序代码，例如以下范例：

```
// miCountSet, miCountPlayerWin, miCountComWin, miCountDraw 是 int 类型的变量,
// 其中包含要储存的数据
SharedPreferences gameResultData = getSharedPreferences("GAME_RESULT", 0);

gameResultData.edit()
    .putInt("COUNT_SET", miCountSet)
    .putInt("COUNT_PLAYER_WIN", miCountPlayerWin)
    .putInt("COUNT_COM_WIN", miCountComWin)
    .putInt("COUNT_DRAW", miCountDraw)
    .commit();
```

54-2 读取数据的步骤

Step 1 决定要读取的数据文件名称。

Step 2 调用 getSharedPreferences()方法，传入步骤一的数据文件名称并得到一个 SharedPreferences 对象。

Step 3 调用 SharedPreferences 对象的 getXXX()方法，XXX 代表各种基本数据类型，例如 Int、Float、String……，并指定要读取的数据的 Key 名称，以及当该数据不存在时所要使用的值。

请参考以下范例：

```
// miCountSet, miCountPlayerWin, miCountComWin, miCountDraw 是 int 类型的变量,
// 用来储存取得的数据
SharedPreferences gameResultData = getSharedPreferences("GAME_RESULT", 0);

miCountSet = gameResultData.getInt("COUNT_SET", 0);        // 第二个参数代表如果该项数据不存在，就传回 0
miCountPlayerWin = gameResultData.getInt("COUNT_PLAYER_WIN", 0);   // 第二个参数的功能同上
miCountComWin = gameResultData.getInt("COUNT_COM_WIN", 0);        // 第二个参数的功能同上
miCountDraw = gameResultData.getInt("COUNT_DRAW", 0);             // 第二个参数的功能同上
```

54-3 删除数据的步骤

Step 1 确定要修改的数据文件名称。

Step 2 调用getSharedPreferences()方法,传入步骤一的数据文件名称并得到一个SharedPreferences对象。

Step 3 调用步骤二的 SharedPreferences 对象的 edit()方法取得一个 Editor 对象。

Step 4 调用 Editor 对象的 remove()方法并指定要删除的数据的 Key 名称。

Step 5 调用 commit()方法完成数据的修改。

请参考以下范例:

```
SharedPreferences gameResultData = getSharedPreferences("GAME_RESULT", 0);

gameResultData.edit()
    .remove("COUNT_SET")
    .remove("COUNT_PLAYER_WIN")
    .commit();
```

54-4 清空数据的步骤

Step 1 确定要修改的数据文件名称。

Step 2 调用getSharedPreferences()方法,传入步骤一的数据文件名称并得到一个SharedPreferences对象。

Step 3 调用步骤二的 SharedPreferences 对象的 edit()方法取得一个 Editor 对象。

Step 4 调用 Editor 对象的 clear()方法清除全部数据。

Step 5 调用 commit()方法完成数据的修改。

请参考以下范例:

```
SharedPreferences gameResultData = getSharedPreferences("GAME_RESULT", 0);

gameResultData.edit()
    .clear()
    .commit();
```

54-5 范例程序

接下来我们利用 SharedPreferences 对象帮之前完成的"计算机猜拳游戏"程序加上储存局数统计数据的功能,让用户可以在下一次运行时,读入之前的游戏结果,以下逐步说明修改程序的过程:

Step 1 使用 Windows 文件管理器复制 UNIT 53 的程序项目文件夹,复制之后可以变更复制文件夹的名称。

Step 2 开启 Eclipse，使用主菜单 File > Import 功能加载复制的项目文件夹。

Step 3 开启 res/layout/main.xml 程序接口布局文件，在游戏的操作界面中加入三个按钮分别用来储存、加载、和清除局数统计数据，请参考以下粗体字的程序代码：

```xml
<?xml version="1.0" encoding="utf-8"?>
<RelativeLayout …>
…
<Button android:id="@+id/btnLoadResult"
    android:layout_width="wrap_content"
    android:layout_height="wrap_content"
    android:text="加载局数数据"
    android:layout_below="@id/btnShowResult"
    android:textSize="15sp"
    android:layout_marginTop="5dp"
    android:layout_centerHorizontal="true"
    />
<Button android:id="@+id/btnSaveResult"
    android:layout_width="wrap_content"
    android:layout_height="wrap_content"
    android:text="储存局数数据"
    android:layout_toLeftOf="@id/btnLoadResult"
    android:layout_alignTop="@id/btnLoadResult"
    android:textSize="15sp"
    android:layout_centerHorizontal="true"
    />
<Button android:id="@+id/btnClearResult"
    android:layout_width="wrap_content"
    android:layout_height="wrap_content"
    android:text="清除局数资料"
    android:layout_toRightOf="@id/btnLoadResult"
    android:layout_alignTop="@id/btnLoadResult"
    android:textSize="15sp"
    android:layout_centerHorizontal="true"
    />
</RelativeLayout>
```

Step 4 开启游戏主程序文件 src/(组件路径名称)/Main.java，加入按钮的设置程序代码如下：

```java
package …

import …

public class Main extends Activity {

    private Button mBtnSaveResult,
            mBtnLoadResult,
```

```java
            mBtnClearResult;

    ...

    private void setupViewComponent() {
        ...;
        mBtnSaveResult = (Button)findViewById(R.id.btnSaveResult);
        mBtnLoadResult = (Button)findViewById(R.id.btnLoadResult);
        mBtnClearResult = (Button)findViewById(R.id.btnClearResult);
        ...
        mBtnSaveResult.setOnClickListener(btnSaveResultLis);
        mBtnLoadResult.setOnClickListener(btnLoadResultLis);
        mBtnClearResult.setOnClickListener(btnClearResultLis);
    }

    ...

private Button.OnClickListener btnSaveResultLis = new Button.OnClickListener() {
        public void onClick(View v) {
            SharedPreferences gameResultData = getSharedPreferences("GAME_RESULT", 0);

            gameResultData.edit()
                .putInt("COUNT_SET", miCountSet)
                .putInt("COUNT_PLAYER_WIN", miCountPlayerWin)
                .putInt("COUNT_COM_WIN", miCountComWin)
                .putInt("COUNT_DRAW", miCountDraw)
                .commit();

            Toast.makeText(Main.this, "储存完成", Toast.LENGTH_LONG)
                .show();
        }
    };

    private Button.OnClickListener btnLoadResultLis = new
    Button.OnClickListener() {
        public void onClick(View v) {
            SharedPreferences gameResultData = getSharedPreferences("GAME_RESULT", 0);

            miCountSet = gameResultData.getInt("COUNT_SET", 0);
            miCountPlayerWin = gameResultData.getInt
            ("COUNT_PLAYER_WIN", 0);
            miCountComWin = gameResultData.getInt("COUNT_COM_WIN", 0);
            miCountDraw = gameResultData.getInt("COUNT_DRAW", 0);

            Toast.makeText(Main.this, "载入完成",
            Toast.LENGTH_LONG)
                .show();
        }
```

```
        };

        private Button.OnClickListener btnClearResultLis = new
        Button.OnClickListener() {
            public void onClick(View v) {
                SharedPreferences gameResultData = getSharedPreferences("GAME_RESULT", 0);

                gameResultData.edit()
                    .clear()
                    .commit();

                Toast.makeText(Main.this, "清除完成", Toast.
                LENGTH_LONG)
                .show();
            }
        };
        ...
```

在新增的三个按钮的 onClickListener 对象中,我们分别使用前面介绍的 SharedPreferences 对象来储存、读取和清除游戏局数统计数据,然后利用 Toast 快显信息通知用户,图 54-1 是程序的运行界面。

图 54-1 "计算机猜拳游戏"程序的运行界面

Android 版本	1.X	2.X	3.X	4.X
适用性	★	★	★	★

UNIT 55
使用 SQLite 数据库储存数据

Android 平台内建一个轻量级的数据库系统 SQLite。SQLite 数据库的操作是使用标准的 SQL 语言，而且 Android 应用程序可以很容易地使用 SQLite 数据库来存取数据。不过在介绍如何利用数据库之前，先让我们学习如何使用模拟器的 Linux 系统命令行模式来检查 SQLite 数据库，这种操作技巧可以帮助我们在开发程序的过程中确认程序代码的运行结果。

55-1 进入模拟器的 Linux 命令行模式操作 SQLite 数据库

请读者依照下列步骤操作：

- **Step 1** 运行 Eclipse 程序。
- **Step 2** 从 Eclipse 的工具栏或是主菜单 Window > AVD Manager 中启动手机或平板电脑模拟器。
- **Step 3** 从 Windows 的 "开始>所有程序>附属应用程序" 中运行 "命令提示字符" 程序。
- **Step 4** 将 "命令提示字符" 程序的运行目录切换到 Android SDK 文件夹中的 platform-tools 子文件夹。
- **Step 5** 运行指令 adb -s emulator-5554 shell，其中 emulator-5554 是平板电脑模拟器的名称，注意它不是我们在建立 AVD 时所取的名称。如果想要知道运行中的计算机模拟器的名称，可以切换到 Eclipse 的 DDMS 界面（参考 UNIT 38 中的说明），然后查看左上方的 Devices 窗口。如果目前只有运行一个模拟器，可以将运行指令简化为 adb shell 请读者参考图 55-1。
- **Step 6** 完成之后会显示一个#号提示字符，代表我们已经进入模拟器的 Linux 操作系统。

图 55-1　进入模拟器的 Linux 操作系统

Step 7 运行 cd data/data 进入程序的数据目录，然后利用 ls 指令检查其中的内容，读者将会看到许多组件的名称，其中可以找到我们的程序项目的组件路径名称，例如 tw.android，请运行 cd tw.android 进入组件的目录。

Step 8 运行 ls 指令检查其中的内容，读者会发现只有一个 lib 目录。如果程序中曾经建立数据库文件，将会有一个 databases 目录，如果没有的话，请读者运行 "mkdir databases" 指令建立该目录。

Step 9 运行 cd databases 指令进入该目录。

Step 10 运行 sqlite3 test.db 指令建立一个名为 test.db 的数据库文件，并且进入 SQLite 数据库操作系统，读者将会看到如图 55-2 所示的提示字符。SQLite 数据库的操作指令都是以 "." 开头，以下是几个常用的指令：

图 55-2　SQLite 数据库操作界面

- .databases　　　列出此目录下全部的数据库文件
- .tables　　　　　列出目前所在的数据库文件中全部的数据表格

- .schema   列出数据表格的建构指令
- .help   列出所有指令和说明
- .exit   离开 SQLite 数据库系统

另外我们可以使用 SQL 语法在 SQLite 数据库系统中进行各种数据表格的操作。

Step 11 完成数据库的操作后运行.exit 指令离开数据库系统，然后再运行 exit 离开模拟器的 Linux 操作系统。

55-2 SQLiteOpenHelper 类

要在程序中使用 SQLite 数据库需要利用 SQLiteOpenHelper 类和 SQLiteDatabase 类，我们先介绍 SQLiteOpenHelper 类的使用方法，这个类的主要任务就是帮我们取得操作数据库的对象，我们会指定一个数据库文件，如果该文件不存在，它会自动建立该数据库文件。另外也可以加入建立表格（table）的程序代码，当指定的数据库文件不存在时，就可以同时完成建立数据库文件和表格的工作，以下我们逐步说明如何完成相关的程序代码：

Step 1 在需要使用 SQLite 数据库的程序项目中新增一个继承 SQLiteOpenHelper 的类，例如可以将该类取名为 FriendDbHelper 代表要储存朋友的数据（操作提示：用鼠标右击程序项目的 src/(组件路径名称)文件夹，然后选择 New > Class）。

Step 2 程序编辑窗口中会自动开启该类的程序代码，其中有几个已经自动加入的方法，另外在类名称下方会出现红色波浪底线标示语法错误，请把鼠标光标移到标示错误的地方，便会弹出一个说明窗口，其中有一个建议项目是要新增一个类的构造方法，请单击该项目就会在程序代码中加入一个类的构造方法，然后请参考以下程序代码输入粗体字的部分：

> **补充说明**
>
> 这些由程序代码编辑器自动加入的方法，它们的自变量名称有些是以 arg 命名，这种自变量名称并不利于程序代码的编写，因此建议查询 Android SDK 的说明文件将它们改成有意义的自变量名称。

```
package ...

import ...

public class FriendDbHelper extends SQLiteOpenHelper {

    public String sCreateTableCommand;

    public FriendDbHelper(Context context, String name,
        CursorFactory factory,
```

```
            int version) {
        super(context, name, factory, version);
        // TODO Auto-generated constructor stub
        sCreateTableCommand="";
    }

    @Override
    public void onCreate(SQLiteDatabase db) {
        // TODO Auto-generated method stub
        if (sCreateTableCommand.isEmpty())
            return;

        db.execSQL(sCreateTableCommand);
    }

    @Override
    public void onUpgrade(SQLiteDatabase db, int oldVer, int newVer) {
        // TODO Auto-generated method stub
    }

}
```

我们在类中定义一个 public 的字符串属性 sCreateTableCommand，这个属性可以让外部程序设置建立表格的指令，然后当程序运行 onCreate()方法时，就会依照该指令在数据库中完成建立表格的工作。这个新类会从 SQLiteOpenHelper 类中继承 getWritableDatabase()方法，当外部程序调用该方法时可能出现下列两种情况：

1．如果指定的数据库文件已经存在，则该方法会开启数据库文件并将它包装成一个 SQLiteDatabase 对象传回给调用程序；
2．如果指定的数据库文件不存在，该方法会自动帮我们建立数据库文件，然后运行 onCreate() 方法。我们在 onCreate()方法中先检查 sCreateTableCommand 属性是否包含建立表格的指令，如果没有就直接结束，如果有就运行建立表格的工作，最后将建立好的数据库文件包装成一个 SQLiteDatabase 对象传回给调用程序。

55-3　SQLiteDatabase 类

在前面的说明中提到 SQLiteOpenHelper 类的 getWritableDatabase()方法最后会传回一个 SQLiteDatabase 类型的对象，这个对象提供操作数据库的各种方法，例如 insert()、delete()、query()、replace()、update()、execSQL()、beginTransaction()、endTransaction()等，我们可以利用这个对象对数据库进行各种操作。例如我们可以在主程序类的 onCreate()方法中完成开启数据库的动作，如以下范例：

```java
public class Main extends Activity {

    private static final String DB_FILE = "friends.db",
                                DB_TABLE = "friends";
    private SQLiteDatabase mFriDbRW;

    /** Called when the activity is first created. */
    @Override
    public void onCreate(Bundle savedInstanceState) {
        super.onCreate(savedInstanceState);
        setContentView(R.layout.main);

        // 建立自定义的 FriendDbHelper 对象
        FriendDbHelper friDbHp = new FriendDbHelper(
                    getApplicationContext(), DB_FILE,
                    null, 1);

        // 设置建立 table 的指令
        friDbHp.sCreateTableCommand = "CREATE TABLE " + DB_TABLE + "(" +
                    "_id INTEGER PRIMARY KEY," +
                    "name TEXT NOT NULL," +
                    "sex TEXT," +
                    "address TEXT);";

        // 取得上面指定的文件名数据库，如果该文件名不存在就会自动建立一个数据库文件
        mFriDbRW = friDbHp.getWritableDatabase();
    }
}
```

在 onCreate()方法的最后我们根据调用 getWritableDatabase()方法，取得 SQLiteDatabase 类型的对象 mFriDbRW，后续的程序代码就可以利用这个对象对数据库进行各种操作。当程序即将结束或是不需要再使用数据库时，记得调用 mFriDbRW 对象的 close()方法关闭数据库。

55-4 范例程序

接下来我们根据建立一个程序项目来示范完整的数据库使用流程，这个程序项目是运行类似通讯簿的功能，为了不要让程序代码过于复杂以致模糊了学习主题，我们只简单地使用姓名、性别和住址三个字段，程序的运行界面如图 55-3 所示，使用者可以利用其中的"加入"、"查询"、和"列表"等三个按钮将数据加入通讯簿、查询指定的通讯簿数据、或是列出整个通讯簿的内容。以下列出完成整个程序项目的步骤：

Step 1 运行 Eclipse 新增一个程序项目，项目对话框中的属性设置都依照之前的惯例即可。

Step 2 依照本单元前面的说明，在程序项目中新增一个继承自 SQLiteOpenHelper 类的新类，我

们可以将该类取名为 FriendDbHelper，FriendDbHelper 类中的程序代码如同前面的说明。

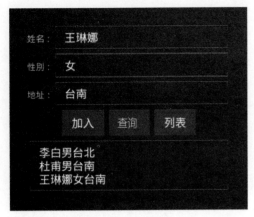

图 55-3　通讯簿程序的运行界面

Step 3 开启程序项目的 res/layout/main.xml 界面布局文件，然后编辑成如下的内容，我们用二层 LinearLayout 将接口组件作适当的排列，并根据设置适当的组件属性让程序操作界面比较整齐美观。

```xml
<?xml version="1.0" encoding="utf-8"?>
<LinearLayout xmlns:android="http://schemas.android.com/apk/res/android"
    android:orientation="vertical"
    android:layout_width="300dp"
    android:layout_height="match_parent"
    android:layout_gravity="center_horizontal"
    >
    <LinearLayout
        android:orientation="horizontal"
        android:layout_width="match_parent"
        android:layout_height="wrap_content"
        >
        <TextView
            android:layout_width="wrap_content"
            android:layout_height="wrap_content"
            android:text="姓名："
            />
        <EditText android:id="@+id/edtName"
            android:layout_width="match_parent"
            android:layout_height="wrap_content"
            />
    </LinearLayout>
    <LinearLayout
        android:orientation="horizontal"
        android:layout_width="match_parent"
```

```xml
        android:layout_height="wrap_content"
        >
    <TextView
        android:layout_width="wrap_content"
        android:layout_height="wrap_content"
        android:text="性别："
        />
    <EditText android:id="@+id/edtSex"
        android:layout_width="match_parent"
        android:layout_height="wrap_content"
        />
</LinearLayout>
<LinearLayout
    android:orientation="horizontal"
    android:layout_width="match_parent"
    android:layout_height="wrap_content"
    >
    <TextView
        android:layout_width="wrap_content"
        android:layout_height="wrap_content"
        android:text="地址："
        />
    <EditText android:id="@+id/edtAddr"
        android:layout_width="match_parent"
        android:layout_height="wrap_content"
        />
</LinearLayout>
<LinearLayout
    android:orientation="horizontal"
    android:layout_width="match_parent"
    android:layout_height="wrap_content"
    android:gravity="center"
    >
    <Button android:id="@+id/btnAdd"
        android:layout_width="wrap_content"
        android:layout_height="wrap_content"
        android:paddingLeft="20dp"
        android:paddingRight="20dp"
        android:text="加入"
        />
    <Button android:id="@+id/btnQuery"
        android:layout_width="wrap_content"
        android:layout_height="wrap_content"
        android:paddingLeft="20dp"
        android:paddingRight="20dp"
        android:text="查询"
        />
    <Button android:id="@+id/btnList"
```

```
                    android:layout_width="wrap_content"
                    android:layout_height="wrap_content"
                    android:paddingLeft="20dp"
                    android:paddingRight="20dp"
                    android:text="列表"
                    />
        </LinearLayout>
        <EditText android:id="@+id/edtList"
            android:layout_width="match_parent"
            android:layout_height="wrap_content"
            />
</LinearLayout>
```

Step 4 开启主程序类的程序文件，在该程序文件中输入前一个小节中的程序代码。另外还必须加上设置接口组件的程序代码，以及设置 button 组件的 onClickListener 的程序代码。我们将这些程序代码一起放在 setupViewComponent()方法中，然后在 onCreate()方法内部调用它。另外在主程序类中加入 onDestroy()方法，在该方法中关闭数据库。加入 onDestroy()方法的操作步骤是先在程序代码编辑窗口中右击，然后选择 Source > Override/Implement Methods 就会弹出一个对话框，在对话框中勾选 onDestroy()再按下 OK 按钮即可，请读者参考以下的程序代码：

```
package ...

import ...

public class Main extends Activity {

    ...

    private EditText mEdtName,
                     mEdtSex,
                     mEdtAddr,
                     mEdtList;

    private Button mBtnAdd,
                   mBtnQuery,
                   mBtnList;

    @Override
    protected void onDestroy() {
        // TODO Auto-generated method stub
        super.onDestroy();

        mFriDbRW.close();
    }

    /** Called when the activity is first created. */
```

```java
@Override
public void onCreate(Bundle savedInstanceState) {
    super.onCreate(savedInstanceState);
    setContentView(R.layout.main);

    setupViewComponent();

    // 建立自定义的 FriendDbHelper 对象
    ...（前一个小节中的程序代码）
}

private void setupViewComponent() {
    mEdtName = (EditText)findViewById(R.id.edtName);
    mEdtSex = (EditText)findViewById(R.id.edtSex);
    mEdtAddr = (EditText)findViewById(R.id.edtAddr);
    mEdtList = (EditText)findViewById(R.id.edtList);

    mBtnAdd = (Button)findViewById(R.id.btnAdd);
    mBtnQuery = (Button)findViewById(R.id.btnQuery);
    mBtnList = (Button)findViewById(R.id.btnList);

    mBtnAdd.setOnClickListener(onClickBtnAdd);
    mBtnQuery.setOnClickListener(onClickBtnQuery);
    mBtnList.setOnClickListener(onClickBtnList);
}

private Button.OnClickListener onClickBtnAdd = new
Button.OnClickListener() {
    @Override
    public void onClick(View v) {
        // TODO Auto-generated method stub

        ContentValues newRow = new ContentValues();
        newRow.put("name", mEdtName.getText().toString());
        newRow.put("sex", mEdtSex.getText().toString());
        newRow.put("address", mEdtAddr.getText().toString());
        mFriDbRW.insert(DB_TABLE, null, newRow);
    }
};

private Button.OnClickListener onClickBtnQuery = new
Button.OnClickListener() {
    @Override
    public void onClick(View v) {
        // TODO Auto-generated method stub

        Cursor c = null;
```

```java
            if (mEdtName.getText().toString().isEmpty()
                    == false) {
                c = mFriDbRW.query(true, DB_TABLE, new
                        String[]{"name", "sex", "address"},
                        "name=" + "\"" + mEdtName.getText().
                        toString() + "\"", null, null, null,
                        null, null);
            }
            else if (mEdtSex.getText().toString().
                    isEmpty() == false) {
                c = mFriDbRW.query(true, DB_TABLE, new
                        String[]{"name", "sex",
                        "address"}, "sex=" + "\"" + mEdtSex.getText().toString()
                        + "\"", null, null, null, null);
            }
            else if (mEdtAddr.getText().toString().
                    isEmpty() == false) {
                c = mFriDbRW.query(true, DB_TABLE, new
                        String[]{"name", "sex", "address"},
                        "address=" + "\"" + mEdtAddr.
                        getText().toString() + "\"", null,
                        null, null, null, null);
            }

            if (c == null)
                return;

            if (c.getCount() == 0) {
                mEdtList.setText("");
                Toast.makeText(Main.this, "没有这笔资料",
                        Toast.LENGTH_LONG)
                        .show();
            }
            else {
                c.moveToFirst();
                mEdtList.setText(c.getString(0) +
                        c.getString(1)   + c.getString(2));

                while (c.moveToNext())
                    mEdtList.append("\n" +
                            c.getString(0) + c.getString(1)   +
                            c.getString(2));
            }
        }
    };

    private Button.OnClickListener onClickBtnList = new
            Button.OnClickListener() {
```

```java
            @Override
            public void onClick(View v) {
                // TODO Auto-generated method stub
                Cursor c = mFriDbRW.query(true, DB_TABLE, new String[]{"name", "sex",
                        "address"},       null, null, null, null, null, null);

                if (c == null)
                    return;

                if (c.getCount() == 0) {
                    mEdtList.setText("");
                    Toast.makeText(Main.this, "没有资料",
                            Toast.LENGTH_LONG)
                            .show();
                }
                else {
                    c.moveToFirst();
                    mEdtList.setText(c.getString(0) +
                            c.getString(1)   + c.getString(2));

                    while (c.moveToNext())
                        mEdtList.append("\n" +
                                c.getString(0) + c.getString(1)   +
                                c.getString(2));
                }
            }
        };
    }
```

在"加入"按钮的 OnClickListener 中，我们先把域名和对应的数据放入 ContentValues 类型的对象中，然后调用数据库对象的 insert() 方法把新的数据写入数据库。

在"查询"按钮的 OnClickListener 中，我们根据字段的内容决定查询的依据，然后调用数据库对象的 query() 方法进行查询，再把查询结果存入一个 Cursor 类型的对象，最后检查 Cursor 对象的结果并进行适当的处理。

在"列表"按钮的 OnClickListener 中，我们利用类似"查询"按钮的 OnClickListener 的程序代码查询数据表中的全部内容，然后根据得到的 Cursor 对象进行适当的处理。

完成以上步骤之后，就可以运行程序并测试每一个按钮的功能。

Android 版本	1.X	2.X	3.X	4.X
适用性	★	★	★	★

UNIT 56
使用 Content Provider 跨程序存取数据

　　在开始介绍本单元的主角之前，先让我们回顾一下已经学过的几种 Android 程序类型。首先是 Activity，它是最常使用的一种程序类型，可以独立运行并且有操作界面。其次是 Broadcast Receiver，当它启动之后会在后台监听特定的广播消息，等到该信息出现的时候才会开始运行。App Widget 程序基本上也是属于一种 Broadcast Receiver。第三种程序类型是 Service，它是在背景运行，没有运行界面，而且一旦启动之后就会持续运行直到工作结束为止。

　　除了以上三种程序类型之外，还有一种就是 Content Provider，也就是这个单元的主角。就字面上的意义来说，它是所谓的"内容提供者"。"内容"其实就是代表数据，因此换句话说 Content Provider 就是一种 Data Server，它负责储存和提供数据。读者心中或许会有一个疑问，前一个单元介绍的 SQLite 数据库的功能也是储存和提供数据，那么二者又有何不同？这二者的功能确实一样，但是运行的条件并不相同。SQLite 数据库中的数据只能由原来建立数据文件的程序存取，其他程序不能使用。但是 Content Provider 没有这项限制，所有程序都可以向 Content Provider 要求存取数据。另外如果读者对于神出鬼没的 Intent 对象已经感觉有些厌烦的话，一个好消息是这个单元我们不会再看到 Intent 对象的踪影。不过 Content Provider 必须用到另一种对象，那就是 Uri，Uri 可以说是 Content Provider 的令牌或是身份证，必须通过它才可以找到 Content Provider 并完成资料的存取。

56-1　Activity 和 Content Provider 之间的运行机制

每一个 Content Provider 都有一个独一无二的 Uri 名称，当 Activity 程序需要使用 Content Provider 时，必须准备一个 Uri 对象，其中包含 Content Provider 的 Uri 名称。Uri 对象只是一个信息的载体，还要加上一个传送的动作才能够完成和 Content Provider 之间的互动，这个传送的任务就需要借助 ContentResolver 对象，这个 ContentResolver 对象可以利用 getContentResolver()方法从 Android 系统取得，图 56-1 展示 Activity 和 Content Provider 之间的运行流程。

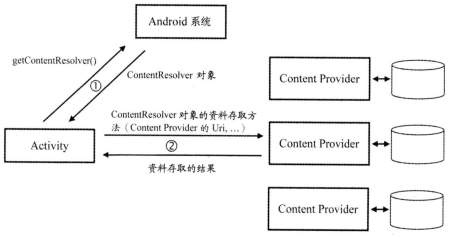

图 56-1　Activity 和 Content Provider 之间的运行流程图

以上的讨论是从 Activity 端来看 Content Provider 的使用流程，基本上只需要两个步骤就可以完成资料的存取。但是如果从 Content Provider 端来考虑程序的实现，需要完成的工作就比较复杂，以下我们逐项说明。

1. Content Provider 必须设置自己的 Uri 名称，这个 Uri 名称的格式如下：

   content://(Content Provider 的组件路径名称).(Content Provider 的类名称)/(数据表名称)

   Content Provider 的组件路径名称一般都是以 providers 结尾，例如：tw.android.providers。依照 Android SDK 技术文件中的规定,这个 Uri 名称必须以 public 属性的方式定义在 Content Provider 类中，而且属性的名称必须叫做 CONTENT_URI 以供外部程序使用。

2. Uri 名称中的"(Content Provider 的组件路径名称).(Content Provider 的类名称)"称为 authority，这个 authority 在 AndroidManifest.xml 文件中登录 Content Provider 时也会用到。另外为了方便编写程序代码，我们也会将数据表名称定义成一个常数，因此综合第 1 点和第 2 点的说明，假设我们想建立一个用来存取朋友数据的 Content Provider，我们会在该类中定义如下的常数，注意其中的 CONTENT_URI 常数是 public：

```
private static final String AUTHORITY =
            "tw.android.providers.FriendsContentProvider";
private static final String DB_TABLE = "friends";
public static final Uri CONTENT_URI = Uri.parse("content://"
            + AUTHORITY + "/" + DB_TABLE);
```

3. 当 Activity 需要 Content Provider 来操作数据时，必须指定第 1 点说明中的 Uri 名称。当该 Uri 名称被送到 Content Provider 时，Content Provider 必须先解析该 Uri 名称再决定要运行的工作，这个步骤需要借助一个 UriMatcher 类型的对象来完成，而 Content Provider 必须在启动的时候就设置好这个 UriMatcher 对象，请参考以下范例：

```
private static final int URI_ROOT = 0,
                         DB_TABLE_FRIENDS = 1;
private static final UriMatcher sUriMatcher = new UriMatcher(URI_ROOT);
static {
    sUriMatcher.addURI(AUTHORITY, DB_TABLE, DB_TABLE_FRIENDS);
}
```

建立 UriMatcher 对象的时候必须指定它的 root 所对应的传回值，一般都设置为 0，接下来的 addURI() 方法就是加入每一个 table（数据表）所对应的传回值，一般 table 所对应的传回值都是从 1 开始递增编号。

4. Content Provider 必须实现以下方法以提供 client 端程序对于数据操作的需求：

- onCreate()　　　　当 Content Provider 被建立时运行
- getType()　　　　 传回 Content Provider 中的数据类型(MIME type)
- delete()　　　　　删除 Content Provider 中的数据
- insert()　　　　　在 Content Provider 中新增资料
- query()　　　　　查询 Content Provider 中的数据
- update()　　　　 更新 Content Provider 中的数据

当 Content Provider 收到存取数据的要求时，必须先调用 UriMatcher 对象的 match() 方法解析附带的 Uri 名称。举例来说，根据第 3 点说明中的程序代码，如果解析后的 Uri 是 DB_TABLE 定义的值，match() 方法就会传回 DB_TABLE_FRIENDS 定义的值，因此我们可以根据 match() 方法的传回值来决定如何进行数据的操作，读者可以参考后面范例程序的实现程序代码。

5. 以上的讨论都是有关 Activity 和 Content Provider 之间的互动机制，接下来要考虑的是 Content Provider 如何实现数据存取的核心功能。Content Provider 内部要如何完成数据的储存和管理完全由程序设计者决定，前提是必须满足第 4 点说明中所列出的数据操作方法。如果读者熟悉 SQL 数据库语言，或是对照前一个单元的 SQLite 数据库的使用方式，将会发现这些数据操作方法和 SQL 数据库的语法非常类似，因此在 Content Provider 中使用 SQLite 数据库来储存和管理数据是最方便的做法。

在介绍完 Content Provider 的基本概念和运行细节之后，接下来就让我们用一个实际的程序范例来学习如何实现 Content Provider。

56-2 范例程序

在前一个单元中我们利用一个通讯录程序来示范 SQLite 数据库的用法。该程序的功能虽然有点普通，却清楚且完整地示范了 SQLite 数据库的使用流程，同样地这个程序也很适合用来学习建立 Content Provider，接下来就让我们将原来的通讯簿程序改成使用 Content Provider 来存取数据。

Step 1 使用 Windows 文件管理器复制上一个单元的程序项目文件夹，复制后可以变更项目文件夹名称，再利用 Eclipse 的 File>Import 功能将该复制的程序项目加载 Eclipse。

Step 2 在 Eclipse 左边的项目检查窗口中，用鼠标右击程序项目的 src 文件夹，在弹出的菜单中选择 New>Package，然后在出现的组件对话框中输入新组件的名称。我们可以先输入原来的组件名（例如 tw.android），后面再加上 .providers，完成后单击 OK 按钮。

Step 3 在项目检查窗口中找到新建立的组件右击，在弹出的菜单中选择 New>Class。

Step 4 在类对话框中输入类名称，例如 FriendsContentProvider，然后单击 Superclass 右边的 Browse 按钮，在出现的对话框上方的字段中输入 ContentProvider，再单击下方列表中出现的 ContentProvider 项目，最后单击 OK 按钮回到类对话框即可完成 Superclass 的设置，然后按下 Finish 按钮完成新增类的操作。

Step 5 编辑上一个步骤所建立的类程序文件，输入下列粗体字的程序代码，在原来内建的程序代码中有些自变量名称是以 arg 命名，读者可以参考以下的程序范例将它们改成有意义的名称。

```
package …;

import …;

public class FriendsContentProvider extends ContentProvider {

    private static final String AUTHORITY =
            "tw.android.providers.FriendsContentProvider";
    private static final String DB_FILE = "friends.db",
                                DB_TABLE = "friends";
    private static final int URI_ROOT = 0,
                             DB_TABLE_FRIENDS = 1;
    public static final Uri CONTENT_URI = Uri.parse("content://"
```

```java
            + AUTHORITY + "/" + DB_TABLE);
private static final UriMatcher sUriMatcher = new
UriMatcher(URI_ROOT);
static {
    sUriMatcher.addURI(AUTHORITY, DB_TABLE, DB_TABLE_FRIENDS);
}
private SQLiteDatabase mFriDb;

@Override
public int delete(Uri   uri, String   selection, String[]
selectionArgs) {
    // TODO Auto-generated method stub
    return 0;
}

@Override
public String getType(Uri uri) {
    // TODO Auto-generated method stub
    return null;
}

@Override
public Uri insert(Uri uri, ContentValues values) {
    // TODO Auto-generated method stub
    if (sUriMatcher.match(uri) != DB_TABLE_FRIENDS) {
        throw new IllegalArgumentException("Unknown
        URI " + uri);
    }

    long rowId = mFriDb.insert(DB_TABLE, null, values);
    Uri insertedRowUri = ContentUris.withAppendedId
    (CONTENT_URI, rowId);

    getContext().getContentResolver().notifyChange
    (insertedRowUri, null);
    return insertedRowUri;
}

@Override
public boolean onCreate() {
    // TODO Auto-generated method stub
    FriendDbHelper friDbHp = new FriendDbHelper(
            getContext(), DB_FILE,
            null, 1);

    friDbHp.sCreateTableCommand = "CREATE TABLE " + DB_TABLE +
            "(" + "_id INTEGER PRIMARY KEY," +
            "name TEXT NOT NULL," +
```

```java
                    "sex TEXT," +
                    "address TEXT);";

            mFriDb = friDbHp.getWritableDatabase();

            return true;
        }

        @Override
        public Cursor query(Uri uri, String[] projection, String selection,
                String[] selectionArgs, String sortOrder) {
            // TODO Auto-generated method stub
            if (sUriMatcher.match(uri) != DB_TABLE_FRIENDS) {
                throw new IllegalArgumentException("Unknown
                    URI " + uri);
            }

            Cursor c = mFriDb.query(true, DB_TABLE, projection,
            selection, null, null, null, null);

            c.setNotificationUri(getContext().getContentResolver(), uri);

            return c;
        }

        @Override
        public int update(Uri uri, ContentValues values, String selection,
                String[] selectionArgs) {
            // TODO Auto-generated method stub
            return 0;
        }
    }
```

最前面的部分是定义我们前面解释过的 Uri 相关属性，以及使用的 SQLite 数据库文件和其中的数据表名称。接下来在 insert() 方法中先使用 UriMatcher 对象的 match() 方法解析收到的 uri 名称，如果 uri 名称没问题就调用 SQLiteDatabase 对象的 insert() 方法加入新的数据项，然后建立一个新的 Uri 对象并加上新加入的数据 id，通知其他程序已经有新的数据加入，并传回新建立的 Uri 对象。

在 onCreate() 方法中则是仿照原来使用 SQLite 数据库的程序代码，建立一个 FriendDbHelper 类型的对象，然后建立新的数据表。在 query() 方法中同样先使用 UriMatcher 对象的 match() 方法解析收到的 uri 名称，如果 uri 名称没问题就调用 SQLiteDatabase 对象的 query() 方法进行数据查询，然后在得到的 Cursor 对象中设置好 Uri 对象，最后传回该 Cursor 对象。

Step 6 重复步骤三和四，在同样的组件中新增一个继承自 SQLiteOpenHelper 类的新类，该类可以取名为 FriendDbHelper（和上一个单元的程序项目相同），然后复制上一个单元的程序项目的同名程序文件中的内容，但是注意不要改变程序代码第一行的 package 名称，因为它们位于不同的组件中。

Step 7 删除程序项目中的 src/tw.android/ FriendDbHelper.java 程序文件（操作提示：在 Eclipse 左边的项目检查窗口中，用鼠标右击该程序文件然后选择 Delete）。

Step 8 开启 src/tw.android/Main.java 程序文件，将它编辑成如下的内容，我们将原来使用 SQLite 数据库的程序代码换成使用 ContentResolver 对象来进行数据的存取。有关程序接口组件的设置都和原来的程序项目相同，因此我们将其省略，请读者留意标示粗体字的程序代码。

```java
package …;

import tw.android.providers.FriendsContentProvider;
import …;

public class Main extends Activity {

    private static ContentResolver mContRes;

    private EditText …

    private Button …

    @Override
    protected void onCreate(Bundle savedInstanceState) {
        // TODO Auto-generated method stub
        super.onCreate(savedInstanceState);
        setContentView(R.layout.main);
        setupViewComponent();
        mContRes = getContentResolver();
    }

    private void setupViewComponent() {
        …
    }

    private Button.OnClickListener onClickBtnAdd = new Button.OnClickListener() {
        @Override
        public void onClick(View v) {
            // TODO Auto-generated method stub

            ContentValues newRow = new ContentValues();
            newRow.put("name", mEdtName.getText().toString());
```

```java
            newRow.put("sex", mEdtSex.getText().toString());
            newRow.put("address", mEdtAddr.getText().toString());
            mContRes.insert(FriendsContentProvider.CONTENT_URI,
                newRow);
        }
    };

    private Button.OnClickListener onClickBtnQuery = new
    Button.OnClickListener() {
        @Override
        public void onClick(View v) {
            // TODO Auto-generated method stub

            Cursor c = null;

            String[] projection = new String[]{"name", "sex", "address"};

            if (mEdtName.getText().toString().isEmpty() == false) {
                c = mContRes.query(FriendsContentProvider.
                    CONTENT_URI, projection, "name=" + "\"" +
                    mEdtName.getText().toString() + "\"",
                    null, null);
            }
            else if (mEdtSex.getText().toString().isEmpty() ==
            false) {
                c = mContRes.query(FriendsContentProvider.
                    CONTENT_URI, projection, "sex=" + "\"" +
                    mEdtSex.getText().toString() + "\"", null,
                    null);
            }
            else if (mEdtAddr.getText().toString().isEmpty() ==
            false) {
                c = mContRes.query(FriendsContentProvider.
                    CONTENT_URI, projection, "address=" + "\"
                    " + mEdtAddr.getText().toString() + "\"",
                    null, null);
            }

            if (c == null)
                return;

            if (c.getCount() == 0) {
                    … (同前一单元的程序代码)
            }
            else {
                …; (同前一单元的程序代码)
            }
        }
```

```
        };

        private Button.OnClickListener onClickBtnList = new
        Button.OnClickListener() {
            @Override
            public void onClick(View v) {
                // TODO Auto-generated method stub

                String[] projection = new String[]{"name",
                "sex", "address"};

                Cursor c = mContRes.query(FriendsContentProvider.
                CONTENT_URI, projection, null, null, null);

                if (c == null)
                    return;

                if (c.getCount() == 0) {
                    … (同前一单元的程序代码)
                }
                else {
                    … (同前一单元的程序代码)
                }
            }
        };
}
```

Step 9 开启程序项目的 AndroidManifest.xml 文件，加入以下粗体字的部分，这一段程序代码是向 Android 系统注册表我们的 Content Provider 程序：

```xml
<?xml version="1.0" encoding="utf-8"?>
<manifest …>
    <application android:icon="@drawable/icon" android:label="@string/app_name">
        <activity …>
            …
        </activity>
        <provider android:name=".providers.FriendsContentProvider"
            android:authorities="tw.android.providers.
            FriendsContentProvider">
        </provider>
    </application>
</manifest>
```

Step 10 开启程序项目的字符串资源文件 res/values/strings.xml，修改其中的程序标题字符串 app_name 如下：

```xml
<?xml version="1.0" encoding="utf-8"?>
<resources>
    <string name="app_name">使用 Content Provider</string>
</resources>
```

完成之后运行程序并测试每一个按钮的功能，读者可以发现和上一个单元的范例程序完全相同，二者的操作接口也没有任何改变，但是核心的数据存取方式却不相同，一个是直接使用 SQLite 数据库，一个则是借助 Content Provider 程序来完成数据的存取。

Android 版本	1.X	2.X	3.X	4.X
适用性	★	★	★	★

UNIT 57 使用文件储存数据

除了前面介绍的三种方法之外，Android 程序也可以像个人计算机的应用程序一样使用文件储存数据，只是程序自己建立的文件必须放在自己的文件夹中，不可以随意放在其他路径。Android 程序读写文件的方式和个人计算机上的 Java 程序相同，都是使用 FileInputStream 和 FileOutputStream 类，另外为了提升读写大型文件的效率，可以配合使用 BufferedInputStream 和 BufferedOutputStream 这两个类，以下我们分别介绍将数据写入文件和从文件读取数据的方法。

57-1 将数据写入文件的方法

将数据写入文件的过程分成以下几个步骤：

Step 1 调用 openFileOutput()方法从 Android 系统取得一个 FileOutputStream 类型的对象。

FileOutputStream fileOut = openFileOutput("文件名(不能指定路径)"，写入模式);

在这个步骤中必须决定是要使用覆盖的模式或是附加的模式将数据写入文件，覆盖模式是传入 MODE_PRIVATE 或是 0，它会将原来文件中的数据清除后再写入新的数据，附加模式是传入 MODE_APPEND，它会将新的数据加在文件的最后。

Step 2 建立一个 BufferedOutputStream 类型的对象，把步骤一得到的 FileOutputStream 对象包裹起来。

BufferedOutputStream bufFileOut = new BufferedOutputStream(fileOut);

这个步骤可以提升大型文件的读写效率，但不是必要的步骤。如果没有使用

BufferedOutputStream 对象也可以利用 FileOutputStream 对象来写入数据，这两个对象写入数据的方法很类似。

Step 3 调用 BufferedOutputStream 对象的 write()方法将数据写入文件。

```
String sData = "写入测试"
bufFileOut.write(sData.getBytes());
```

要写入文件的数据必须储存在 byte 类型的数组中，以上的范例是利用 String 类的 getBytes() 方法取得字符串的 byte 数组。

Step 4 将数据全部写入文件之后再调用 BufferedOutputStream 对象的 close ()方法关闭文件。

```
bufFileOut.close();
```

Step 5 以上存取文件的程序代码必须加上例外处理，也就是说要用 try…catch…语法将它们包起来如下：

```
try {
    // 存取文件的程序代码
    …
} catch (Exception e) {
    // TODO Auto-generated catch block
    // 处理错误的程序代码
    …
}
```

57-2 从文件读取数据的方法

从文件读取数据的过程分成以下几个步骤：

Step 1 调用 openFileInput()方法从 Android 系统取得一个 FileInputStream 类型的对象。

```
FileInputStream fileIn = openFileInput("文件名(不能指定路径)");
```

Step 2 建立一个 BufferedInputStream 类型的对象把步骤一得到的 FileInputStream 对象包裹起来。

```
BufferedInputStream bufFileIn = new BufferedInputStream(fileIn);
```

这个步骤可以提升大型文件的读写效率，但不是必要的步骤。如果没有使用 BufferedInputStream 对象也可以利用 FileInputStream 对象来读取数据，这两个对象读取数据的方法很类似。

Step 3 调用 BufferedInputStream 对象的 read()方法读取数据。

```
byte[] bufBytes = new byte[10];
int c = bufFileIn.read(bufBytes);
```

我们必须准备一个 byte 类型的数组来存放读取的数据，read ()方法会传回读取的 byte 数，如果传回-1 代表数据已经读取完毕。我们可以使用一个循环连续读取文件中的数据，直到 read ()方法传回-1 为止，读者可以参考后面的实现范例。

Step 4 文件使用完毕后调用 BufferedInputStream 对象的 close ()方法关闭文件。

```
bufFileIn.close();
```

Step 5 以上存取文件的程序代码必须加上例外处理，也就是说要用 try…catch…语法将它们包起来如下：

```
try {
        // 存取文件的程序代码
        …
} catch (Exception e) {
        // TODO Auto-generated catch block
        // 处理错误的程序代码
        …
}
```

57-3　范例程序

接下来我们用一个实际的程序项目来示范读写文件的功能，这个程序项目的运行界面如图 57-1 所示，用户在程序界面上方的 EditText 组件中输入任何文字，然后按下"加入文件"按钮就会把输入的文字写入文件中，按下"列出文件内容"按钮就会将文件的内容显示在下方的文本块，按下"清除文件内容"按钮则会删除文件中的所有数据，完成此程序项目的步骤如下：

图 57-1　读写文件程序的运行界面

Step 1 运行 Eclipse 新增一个 Android 程序项目，项目的属性设置都依照之前的惯例即可。

Step 2 开启程序项目的 res/layout/main.xml 接口布局文件，编辑成如下的内容：

```
<?xml version="1.0" encoding="utf-8"?>
<LinearLayout xmlns:android="http://schemas.android.com/apk/res/android"
```

```xml
    android:orientation="vertical"
    android:layout_width="match_parent"
    android:layout_height="match_parent"
    android:gravity="center_horizontal"
    >
<EditText android:id="@+id/edtIn"
    android:layout_width="300dp"
    android:layout_height="wrap_content"
    android:text=""
    />
<LinearLayout
    android:orientation="horizontal"
    android:layout_width="300dp"
    android:layout_height="wrap_content"
    android:gravity="center"
    >
    <Button android:id="@+id/btnAdd"
        android:layout_width="100dp"
        android:layout_height="wrap_content"
        android:paddingLeft="20dp"
        android:paddingRight="20dp"
        android:text="加入文件"
        />
    <Button android:id="@+id/btnRead"
        android:layout_width="100dp"
        android:layout_height="wrap_content"
        android:paddingLeft="20dp"
        android:paddingRight="20dp"
        android:text="列出文件内容"
        />
    <Button android:id="@+id/btnClear"
        android:layout_width="100dp"
        android:layout_height="wrap_content"
        android:paddingLeft="20dp"
        android:paddingRight="20dp"
        android:text="清除文件内容"
        />
</LinearLayout>
<TextView
    android:layout_width="300dp"
    android:layout_height="wrap_content"
    android:layout_marginTop="10dp"
    android:text="文件内容"
    />
<EditText android:id="@+id/edtFileContent"
    android:layout_width="400dp"
    android:layout_height="wrap_content"
    android:editable="false"
```

```
        />
    </LinearLayout>
```

Step 3 开启主程序文件,将程序代码编辑如下:

```java
package ...

import ...

public class Main extends Activity {

    private static final String FILE_NAME = "file io.txt";

    private EditText mEdtIn,
                mEdtFileContent;

    private Button mBtnAdd,
                mBtnRead,
                mBtnClear;

    /** Called when the activity is first created. */
    @Override
    public void onCreate(Bundle savedInstanceState) {
        super.onCreate(savedInstanceState);
        setContentView(R.layout.main);

        setupViewComponent();
    }

    private void setupViewComponent() {
        mEdtIn = (EditText)findViewById(R.id.edtIn);
        mEdtFileContent = (EditText)findViewById
          (R.id.edtFileContent);

        mBtnAdd = (Button)findViewById(R.id.btnAdd);
        mBtnRead = (Button)findViewById(R.id.btnRead);
        mBtnClear = (Button)findViewById(R.id.btnClear);

        mBtnAdd.setOnClickListener(onClickBtnAdd);
        mBtnRead.setOnClickListener(onClickBtnRead);
        mBtnClear.setOnClickListener(onClickBtnClear);
    }

    private Button.OnClickListener onClickBtnAdd = new
    Button.OnClickListener() {
        @Override
        public void onClick(View v) {
            // TODO Auto-generated method stub
```

```java
            FileOutputStream fileOut = null;
            BufferedOutputStream bufFileOut = null;

            try {
                fileOut = openFileOutput(FILE_NAME, MODE_APPEND);
                bufFileOut = new BufferedOutputStream(fileOut);
                bufFileOut.write(mEdtIn.getText().toString().getBytes());
                bufFileOut.close();
            } catch (Exception e) {
                // TODO Auto-generated catch block
                e.printStackTrace();
            }
        }
    };

    private Button.OnClickListener onClickBtnRead = new
    Button.OnClickListener() {
        @Override
        public void onClick(View v) {
            // TODO Auto-generated method stub

            FileInputStream fileIn = null;
            BufferedInputStream bufFileIn = null;

            try {
                fileIn = openFileInput("file io.txt");
                bufFileIn = new BufferedInputStream(fileIn);

                byte[] bufBytes = new byte[10];

                mEdtFileContent.setText("");

                do {
                    int c = bufFileIn.read(bufBytes);

                    if (c == -1)
                        break;
                    else
                        mEdtFileContent.append(new String(bufBytes), 0, c);
                } while (true);

                bufFileIn.close();
            } catch (Exception e) {
                // TODO Auto-generated catch block
                e.printStackTrace();
            }
        }
    };
```

```java
        private Button.OnClickListener onClickBtnClear = new
        Button.OnClickListener() {
            @Override
            public void onClick(View v) {
                // TODO Auto-generated method stub

                FileOutputStream fileOut = null;

                try {
                    fileOut = openFileOutput(FILE_NAME, MODE_PRIVATE);
                    fileOut.close();
                } catch (Exception e) {
                    // TODO Auto-generated catch block
                    e.printStackTrace();
                }
            }
        };
    }
```

其中我们利用自行建立的 setupViewComponent() 方法完成所有 View 组件的设置。在"加入文件"按钮的 onClickListener 中，我们利用前面介绍的方法将程序界面上方的 EditText 中的文字写入文件。在"列出文件内容"按钮的 onClickListener 中，则是利用前面介绍的方法从文件中读出数据并显示在程序界面下方的 EditText 中。在"清除文件内容"的 onClickListener 中，则是先以覆写模式开启文件，然后将它关闭就可以清空文件中的数据。

PART 11　程序项目的整备工作和发布

UNIT 58　支持多语系和屏幕模式
UNIT 59　开发不同 Android 版本程序的考虑
UNIT 60　取得屏幕的宽度、高度和分辨率
UNIT 61　将程序安装到实体设备或在网络上发布

Android 版本	1.X	2.X	3.X	4.X
适用性	★	★	★	★

UNIT 58
支持多语系和屏幕模式

如果我们开发的程序只要安装在自己的平板电脑上使用，就可以完全依照自己的需要和屏幕的分辨率来设计程序接口。可是如果程序是要在网络上公开，让其他人也能够下载使用，那么程序操作接口的排列方式和所使用的语言，就必须能够满足不同语系的用户和屏幕尺寸及方向的变化。针对不同语系和屏幕的变化，Android 系统在开发时就已经考虑到相关的问题，因此特别设计了一套程序资源的配置法则。在程序项目的 res 文件夹下的各种资源文件夹，例如 layout、values 和 drawable，都可以在它们的名称后面加上特定的代表字，以提供特定的设备环境使用。这些特定代表字的种类包括移动通讯的国家代码和网络商代码、语系、屏幕大小、屏幕外观和方向、底座链接状态、夜间模式、屏幕分辨率、屏幕触控模式、键盘模式、输入方式、导引相关设置和系统版本等，以下我们针对比较常用的语系、屏幕大小、屏幕方向和屏幕像素密度等四类列表说明，请读者参考表 58-1。

表 58-1 资源文件夹名称后面可以加上的关键词

资源文件夹关键词的分类	资源文件夹名称的关键词	说明
语系	zh-rTW	语系的关键词可以是一层或是二层，例如 zh 代表中文语系，zh-rTW 代表中文语系中的繁体字，zh-rCN 则代表中文语系中的简体字。又例如 en 代表英文语系，en-rUS 则代表英文语系中的美国语，en-rUK 代表英文语系中的英国语，ja 是日本语等
	zh-rCN	
	en	
	en-rUS	
	en-rUK	
	ja	
	...	
屏幕大小	small	small：小尺寸的屏幕
	normal	normal：正常尺寸的屏幕

续表

资源文件夹关键词的分类	资源文件夹名称的关键词	说明
屏幕大小	Large	large：大尺寸的屏幕
	xlarge	xlarge：超大尺寸的屏幕
		屏幕尺寸大小的决定方式请参考表 58-2
屏幕方向	port	port：直式屏幕（高大于宽）
	land	Land：横式屏幕（宽大于高）
屏幕像素密度(dpi)ldpi	mdpi	ldpi：低屏幕像素密度，约 120dpi
	hdpi	mdpi：中屏幕像素密度，约 160dpi
	xhdpi	hdpi：高屏幕像素密度，约 240dpi
	nodpi	xhdpi：超高屏幕像素密度，约 320dpi

其中的屏幕大小取决于屏幕分辨率和屏幕像素密度的配合情况，例如以同样的屏幕分辨率来说，当屏幕像素密度低的时候屏幕尺寸就会变大，当屏幕像素密度高的时候屏幕尺寸就会变小，表 58-2 列出 Android SDK 技术文件中对于屏幕尺寸的分类。接下来我们介绍如何使用这些资源文件夹名称的关键词，让程序支持多语系和不同的屏幕模式。

表 58-2 根据屏幕的宽度、高度和分辨率决定屏幕尺寸的分类

屏幕尺寸的分类	Low density (120), *ldpi*	Medium density (160), *mdpi*	High density (240), *hdpi*	Extra high density (320), *xhdpi*
Small	QVGA (240×320)		480×640	
Normal	WQVGA400 (240×400) WQVGA432 (240×432)	HVGA (320×480)	WVGA800 (480×800) WVGA854 (480×854) 600×1024	640×960
Large	WVGA800 (480×800) WVGA854 (480×854)	WVGA800 (480×800) WVGA854 (480×854) 600×1024		
Extra Large	1024×600	WXGA (1280×800) 1024×768 1280×768	1536×1152 1920×1152 1920×1200	2048×1536 2560×1536 2560×1600

58-1 让程序支持多语系的方法

在 UNIT 9 中我们曾经讨论过何谓良好的程序架构，其中一项要求是把程序中使用的字符串定义在 res/values/strings.xml 文件中。如果我们在开发程序的时候确实遵守这项要求，那么在建立程序的多语系功能时就很容易，因为 Android 系统可以根据目前平板电脑或手机的语系设置（操作提示：按下模拟器屏幕上的 Apps 按钮，选择 Settings > Language & input > Language 来改变目前使用

的语系），自动从 res 文件夹中挑选适合的 values 文件夹下的字符串资源文件使用。也就是说，我们可以建立多个对应到不同语系的 values 文件夹，让 Android 系统从中择一使用。例如在图 58-1 的程序项目文件夹中，我们新增了三个对应到特定语系的 values 文件夹。当平板电脑或手机设置为中文繁体语系时，Android 系统就会使用 values-zh-rTW 文件夹中的文件，如果设置为美式英文语系，Android 系统就会使用 values-en-rUS 文件夹中的文件，如果设置为其他英文语系，Android 系统就会使用 values-en 文件夹中的文件，如果设置为日本语系，Android 系统就会使用 values 文件夹中的文件，因为当找不到指定语系所对应的 values 文件夹时，则一律使用内定的 values 文件夹。

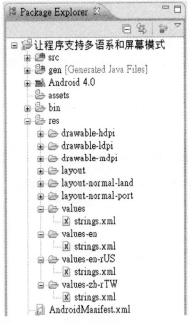

图 58-1　在程序项目的 res 文件夹中建立多个对应到不同语系的 values 文件夹

58-2　让程序支持多种屏幕模式

让程序支持多种屏幕模式的方法和上述支持多语系的方法非常类似，Android 系统的接口布局文件是放在 res/layout/main.xml 文件中，我们同样可以建立多个对应到不同屏幕模式的 layout 文件夹，让 Android 系统从中择一使用。和屏幕相关的属性包括屏幕大小、方向和像素密度，这些属性关键词的排列必须依照表 58-1 的顺序。例如图 58-2 的程序项目中建立了两个对应到不同屏幕模式的 layout 文件夹，当平板电脑或手机屏幕是正常大小且为直式时会使用 layout-normal-port 文件夹中的文件，当平板电脑或手机屏幕是正常大小且为横式时会使用 layout-normal-land 文件夹中的文件，其他的屏幕模式则会使用内定的 layout 文件夹中的文件。

图 58-2　在程序项目的 res 文件夹下建立多个对应到不同屏幕模式的 layout 文件夹

58-3　范例程序

依照惯例我们用一个实际的程序项目来示范如何让程序支持多语系和屏幕模式，建立这个范例程序的步骤如下：

Step 1 请读者依照之前的方法在 Eclipse 中新增一个 Android 程序项目。

Step 2 在程序项目的 res 文件夹中新增如图 58-1 所示的三个不同语系的 values 文件夹，新增文件夹的步骤是先用鼠标右击 res 文件夹，然后选择 New > Folder 再输入文件夹的名称。

Step 3 分别在新增的文件夹中建立一个名为 strings.xml 的文件，新增文件的步骤类似新增文件夹，只是换成选择 New > File 选项，或者读者可以利用按下鼠标右键后的菜单中的 Copy 和 Paste 功能，从原来的 values 文件夹中复制 strings.xml 文件到新的文件夹。

Step 4 将不同的 values 文件夹下的 strings.xml 文件编辑成如下的内容：

- values/strings.xml

```
<?xml version="1.0" encoding="utf-8"?>
<resources>
    <string name="hello">使用字符串资源文件：values/strings.xml</string>
    <string name="app_name">多语系和屏幕模式</string>
</resources>
```

- values-en/strings.xml

```
<?xml version="1.0" encoding="utf-8"?>
<resources>
    <string name="hello">使用字符串资源文件：values-en/strings.xml</string>
```

```xml
    <string name="app_name">多语系和屏幕模式</string>
</resources>
```

■ values-en-rUS/strings.xml

```xml
<?xml version="1.0" encoding="utf-8"?>
<resources>
    <string name="hello">使用字符串资源文件：values-en-rUS/strings.xml</string>
    <string name="app_name">多语系和屏幕模式</string>
</resources>
```

■ values-zh-rTW/strings.xml

```xml
<?xml version="1.0" encoding="utf-8"?>
<resources>
    <string name="hello">使用字符串资源文件：values-zh-rTW/strings.xml</string>
    <string name="app_name">多语系和屏幕模式</string>
</resources>
```

在这些字符串资源文件中，我们让 hello 字符串显示出它的文件路径，这样就可以知道 Android 系统目前究竟使用哪一个字符串资源文件。

Step 5 仿照步骤三和四的方法新增两个对应到不同屏幕模式的 layout 文件夹（参考图 58-2），并在其中建立 main.xml 文件，这些文件的内容如下。在这些接口布局文件中我们显示它们所在的路径以及 hello 字符串的内容，这样就可以知道程序目前使用哪一个接口布局文件和字符串资源文件。

■ layout/main.xml

```xml
<?xml version="1.0" encoding="utf-8"?>
<LinearLayout xmlns:android="http://schemas.android.com/apk/res/android"
    android:orientation="vertical"
    android:layout_width="300dp"
    android:layout_height="match_parent"
    android:layout_gravity="center_horizontal">
<TextView
    android:layout_width="match_parent"
    android:layout_height="wrap_content"
    android:text="使用接口布局文件：layout/main.xml"/>
<TextView
    android:layout_width="match_parent"
    android:layout_height="wrap_content"
    android:text="@string/hello"/>
</LinearLayout>
```

■ layout-normal-land/main.xml

```xml
<?xml version="1.0" encoding="utf-8"?>
<LinearLayout xmlns:android="http://schemas.android.com/apk/res/android"
    android:orientation="vertical"
```

```
        android:layout_width="300dp"
        android:layout_height="match_parent"
        android:layout_gravity="center_horizontal">
<TextView
        android:layout_width="match_parent"
        android:layout_height="wrap_content"
        android:text="使用接口布局文件：layout-normal-land/main.xml"/>
<TextView
        android:layout_width="match_parent"
        android:layout_height="wrap_content"
        android:text="@string/hello"/>
</LinearLayout>
```

- layout-normal-port/main.xml

```
<?xml version="1.0" encoding="utf-8"?>
<LinearLayout xmlns:android="http://schemas.android.com/apk/res/android"
        android:orientation="vertical"
        android:layout_width="300dp"
        android:layout_height="match_parent"
        android:layout_gravity="center_horizontal">
<TextView
        android:layout_width="match_parent"
        android:layout_height="wrap_content"
        android:text="使用接口布局文件：layout-normal-port/main.xml"/>
<TextView
        android:layout_width="match_parent"
        android:layout_height="wrap_content"
        android:text="@string/hello"/>
</LinearLayout>
```

完成以上步骤之后运行这个程序项目，手机模拟器界面会显示正在使用 layout-normal-port/main.xml 界面布局文件和 values-en-rUS/strings.xml 字符串资源文件（如图 58-3 所示），因为目前模拟器是直式界面并且使用美式英文。如果我们将模拟器的语言改成繁体中文，则会显示使用的字符串资源文件为 values-zh-rTW/strings.xml，如图 58-4 所示。如果要切换模拟器屏幕的直式和横式状态，可以在运行模拟器的时候，同时按下计算机键盘左边的 **Ctrl** 和 **F12** 键，此时模拟器的界面会变成横式类型如图 58-5 所示，程序界面也会切换成使用 layout-normal-land/main.xml 界面布局文件。利用本单元介绍的多语系和动态调整屏幕模式的功能，读者可以让辛苦开发的程序站上国际舞台，让全世界的 Android 用户都能使用您的作品。

图 58-3　程序运行时显示所使用的接口布局文件和字符串资源文件

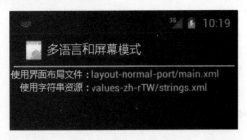

图 58-4　将模拟器的语言改成繁体中文后程序显示正在使用字符串资源文件
　　　　　values-zh-rTW/strings.xml

图 58-5　利用 Ctrl+F12 将手机模拟器屏幕切换成横式状态并观察程序接口的改变

Android 版本	1.X	2.X	3.X	4.X
适用性			★	★

UNIT 59
开发不同 Android 版本程序的考虑

　　Google 刚开始发布 Android 平台的时候是以智能手机为目标，后来当平板电脑出现之后，便顺应趋势加入支持平板电脑的相关技术。但是以现阶段而言，Android 系统已经不再限制只能用在手机和平板电脑，在 2011 年 5 月举办的 Google IO Conference 中，Google 对外公开代表，未来 Android 的目标是成为所有智能型设备的核心，例如现在正在推广中的网络电视，以及智能家电像是洗衣机、电冰箱、空调系统等。当然以目前来说，这些智能型家电或许还有些遥远，Google TV 也尚未普及，当前最热门的 IT 产品还是手机和平板电脑，而且这二者的特性和用途也最接近。当程序设计人员在开发 Android 应用程序的时候，最常见的困扰是如何让程序能够同时适用于手机和平板电脑。这个问题牵涉不同 Android 版本的功能支持，还有就是如何让程序的操作接口能够同时适用不同屏幕尺寸的设备。

　　平板电脑屏幕的大小通常是手机屏幕的二倍以上，因此有些时候我们必须针对平板电脑程序采用不同的操作接口排列方式。前一个单元介绍的 res 文件夹命名技术可以用来区分手机和平板电脑的接口描述文件，对于一些程序而言这个方法已经可以满足它们的需要，但是有些情况可能还要更复杂一些。举例来说，如果手机程序因为屏幕太小必须将操作接口分成二页来显示，像是我们前面的 "计算机猜拳游戏" 范例，必须将游戏界面和局数统计数据分开显示。如果换成在平板电脑上运行，就可以让游戏界面和局数统计数据同时显示，这种情况就不是单纯使用 res 文件夹命名技术就能够解决，因为它还牵涉程序代码运行流程的改变，在手机上必须切换界面，但是在平板电脑运行时就不用。

　　根据以上的讨论，读者应该能够体会要开发同时适用多种 Android 平台版本和不同屏幕尺寸的程序并不是一个单纯的问题。对于操作接口比较简单的程序来说也许很容易,可是如果程序的操作接口比较复杂,那么就需要使用一些程序设计的技巧。为了让读者有基本的法则可以遵循,笔者特别整理出一个树状的决策流程如图 59-1 所示。首先第一个要考虑的问题就是程序有没有需要在不同的 Android 平台上运行,如果只是一个学校的程序作业,或是单纯自己用来研究实验的程序,就不需要考虑不同 Android 平台的适用问题,只要依照自己的需求选定一个 Android 平台,然后在上面开发程序即可。但是如果要将开发的程序公布在网络上让世界各地的人下载,就必须谨慎考虑如何处理多种不同 Android 平台和屏幕尺寸的问题。我们可以利用前一个单元介绍的资源文件夹命名技术,或是本单元随后介绍的 Fragment 技巧,让程序的操作接口能够适应不同设备的屏幕大小。另外也可以根据限制程序可以安装的 Android 平台版本和设备屏幕尺寸,这一个部分留待下一个单元再作介绍。

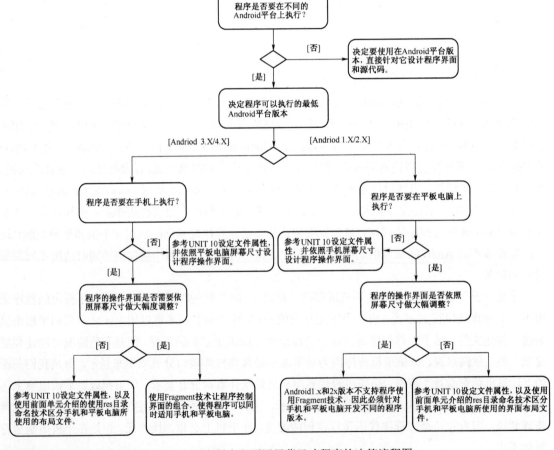

图 59-1　开发同时适用多种 Android 平台版本和不同屏幕尺寸程序的决策流程图

59-1 利用 Fragment 控制分页或单页显示

有关 Fragment 的用法我们已经在 UNIT 25～UNIT 28 中介绍过，包括最基本的静态 Fragment 和利用程序代码动态显示和隐藏 Fragment。如果程序的操作界面包含比较多的组件，那么有可能在手机屏幕上必须将操作界面分成二页显示，但是如果换成在平板电脑上运行就可以用单页显示，这个问题可以利用动态 Fragment 的技巧解决，也就是 UNIT 28 的范例程序中的 Fragment 控制技术以及 callback 函数，但是程序的架构不需要像该范例那么复杂，或者可以说是 UNIT 28 范例程序的简化版。

读者可以回头参考 UNIT 28 的范例程序，它在运行的过程中，使用者可以随意控制"局数统计界面"的显示和隐藏，而且为了示范完整的动态 Fragment 控制技巧，我们还特别实现了两种"局数统计界面"，并且加上 Back Stack 的功能。如果只是要利用 Fragment 让程序能够依照屏幕尺寸，控制操作界面的单页或分页显示，并不需要用到 UNIT 28 范例程序这么复杂的技巧，只要在程序启动时检查屏幕大小，然后决定要显示的 Fragment 数目即可，接着在程序运行的过程中，再依照使用者的操作适当地变换 Fragment。以下我们以 UNIT 28 的"计算机猜拳游戏"为例，将它修改成能够依照屏幕尺寸自动调整 Fragment 的显示个数，请读者依照以下步骤操作：

Step 1 运行文件管理器，复制 UNIT 28 的"计算机猜拳游戏"范例程序的项目文件夹，复制后可以重新命名。

Step 2 利用 Eclipse 主菜单的 File > Import 功能加载上一个步骤复制的程序专案。

Step 3 在 Eclipse 左边的项目检查窗口中展开 res/layout 文件夹，开启其中的接口布局文件 main.xml。原来的程序接口是直接建立一个 fragment 组件用来显示游戏界面，再利用一个 FrameLayout 组件用来动态加载"局数统计界面"的 fragment。现在我们要将其中的 fragment 组件也改成使用 FrameLayout，也就是说程序界面中有两个 FrameLayout 组件，这样我们就可以利用程序代码控制每一个 FrameLayout 中显示的 Fragment。当程序在平板电脑上运行时，会同时显示这两个 FrameLayout 组件，并将游戏程序的 Fragment 放在第一个 FrameLayout，"局数统计界面"的 fragment 则显示在第二个 FrameLayout。如果程序是在手机上运行，同样会将游戏程序的 Fragment 放在第一个 FrameLayout，"局数统计界面"的 fragment 显示在第二个 FrameLayout，但刚开始只会显示第一个 FrameLayout，并将第二个 FrameLayout 隐藏。等到用户按下"显示结果"按钮时，再将第一个 FrameLayout 隐藏，换成显示第二个 FrameLayout，以下是修改后的接口布局文件：

```
<?xml version="1.0" encoding="utf-8"?>
<LinearLayout xmlns:android="http://schemas.android.com/apk/res/android"
    android:orientation="horizontal"
    android:layout_width="match_parent"
```

```xml
            android:layout_height="match_parent"
            android:paddingLeft="30dp"
            android:paddingRight="30dp"
            android:paddingTop="10dp"
            >
    <FrameLayout android:id="@+id/frame1"
        android:layout_weight="1"
        android:layout_width="0px"
        android:layout_height="wrap_content"
        />
    <FrameLayout android:id="@+id/frame2"
        android:layout_weight="1"
        android:layout_width="0px"
        android:layout_height="wrap_content"
        android:background="?android:attr/detailsElementBackground"
        />
</LinearLayout>
```

Step 4 在 Eclipse 左边的项目检查窗口中展开 src/(组件路径名称)文件夹，开启其中的主程序文件 Main.java。原来的程序是利用 GameFragment 类中定义的 CallbackInterface 接口让游戏程序（也就是 GameFragment）能够通知更新后的局数统计资料，这里我们还是继续采用这种作法。当主程序开始运行时，我们调用 getResources()方法取得 Resource 对象，再调用 Resource 对象的 getConfiguration()方法取得 Configuration 对象，然后利用位 mask 的运算检查 screenLayout 属性，以取得屏幕大小的分类。如果屏幕是属于 xlarge 类（也就是平板电脑的屏幕），就将 UITypeFlag 设置为 TWO_FRAMES 让后续的程序同时显示两个 fragment，如果屏幕是属于 small、normal 或 large 类（也就是手机屏幕），则将 UITypeFlag 设置为 ONE_FRAME 让后续的程序只显示一个 fragment。

在 onResume()方法中（当程序界面即将显示时运行）我们先将游戏程序的 fragment 和 "局数统计界面" 的 fragment 分别放到两个 FrameLayout 中，然后判断程序界面是否已经设置好，如果不是就依照 UITypeFlag 中的值决定是要同时显示两个 FrameLayout 或是只显示第一个 FrameLayout，另外我们也不再需要 enableGameResult()这个方法，修改后的程序代码如下：

```java
package ...

import ...

public class Main extends Activity implements GameFragment.
CallbackInterface {

    private GameFragment fragGame;
    private GameResultFragment fragGameResult;
```

```java
private boolean bUISettedOK = false;

enum UIType {
    ONE_FRAME, TWO_FRAMES;
}

public UIType UITypeFlag;

/** Called when the activity is first created. */
@Override
public void onCreate(Bundle savedInstanceState) {
    super.onCreate(savedInstanceState);
    setContentView(R.layout.main);

    // 取得设备屏幕的大小的分类
    switch (getResources().getConfiguration().screenLayout &
            Configuration.SCREENLAYOUT_SIZE_MASK) {
    case Configuration.SCREENLAYOUT_SIZE_SMALL:
        UITypeFlag = UIType.ONE_FRAME;
        break;
    case Configuration.SCREENLAYOUT_SIZE_NORMAL:
        UITypeFlag = UIType.ONE_FRAME;
        break;
    case Configuration.SCREENLAYOUT_SIZE_LARGE:
        UITypeFlag = UIType.ONE_FRAME;
        break;
    case Configuration.SCREENLAYOUT_SIZE_XLARGE:
        UITypeFlag = UIType.TWO_FRAMES;
        break;
    }

    fragGame = new GameFragment();
    fragGameResult = new GameResultFragment();
}

@Override
protected void onResume() {
    // TODO Auto-generated method stub

    FragmentTransaction fragTran = getFragmentManager().
    beginTransaction();
    fragTran.replace(R.id.frame1, fragGame, "Game");
    fragTran.replace(R.id.frame2, fragGameResult, "Game Result");
    fragTran.commit();

    if (bUISettedOK == false) {
        bUISettedOK = true;
```

```
                switch (UITypeFlag) {
                case ONE_FRAME:
                        findViewById(R.id.frame1).setVisibility
                                (View.VISIBLE);
                        findViewById(R.id.frame2).setVisibility
                                (View.GONE);
                        break;
                case TWO_FRAMES:
                        findViewById(R.id.frame1).setVisibility
                                (View.VISIBLE);
                        findViewById(R.id.frame2).setVisibility
                                (View.VISIBLE);
                        break;
                }
        }

        super.onResume();
    }

    @Override
    public void updateGameResult(int iCountSet, int iCountPlayerWin,
            int iCountComWin, int iCountDraw) {
        // TODO Auto-generated method stub

        if (findViewById(R.id.frame2).isShown()) {
            fragGameResult.updateGameResult(iCountSet,
                iCountPlayerWin, iCountComWin, iCountDraw);
        }

    }
}
```

Step 5 开启游戏程序文件 GameFragment.java 并依照下列项目修改：

1. 删除接口 CallbackInterface 中的 enableGameResult()方法，因为我们不再让使用者直接控制局数统计数据的显示和隐藏。另外也只提供一种局数统计资料界面，因此删除 GameResultType 的定义。

2. 程序中使用的 Button 数目也会随着操作的简化而减少（参考下一个步骤的界面布局文件的内容）。

3. 接口布局文件中的"显示结果"按钮必须随着 FrameLayout 显示的数目而改变，当两个 FrameLayout 都显示时，"显示结果"按钮就要隐藏，因为局数统计界面已经显示在屏幕上（在平板电脑上运行时），如果只有显示一个 FrameLayout，"显示结果"按钮就要启动（在手机上运行时）。

4. 当在手机上运行时，如果用户按下"显示结果"按钮，程序必须切换两个 FrameLayout

的显示状态，也就是隐藏游戏程序换成显示局数统计界面。
以下列出相关的程序代码：

```
package ...

import ...

public class GameFragment extends Fragment {

    // 所属的 Activity 必须实现以下接口中的 Callback 方法
    public interface CallbackInterface {
        public void updateGameResult(int iCountSet,
                                     int iCountPlayerWin,
                                     int iCountComWin,
                                     int iCountDraw);
    };

    private CallbackInterface mCallback;
    private Button mBtnScissors,
            mBtnStone,
            mBtnNet,
            mBtnShowResult;

    private TextView mTxtComPlay,
            mTxtResult;

    private int miCountSet = 0,
            miCountPlayerWin = 0,
            miCountComWin = 0,
            miCountDraw = 0;

    @Override
    public View onCreateView(LayoutInflater inflater, ViewGroup container,
        Bundle savedInstanceState) {
      // TODO Auto-generated method stub
      return inflater.inflate(R.layout.game, container, false);
    }

    @Override
    public void onActivityCreated(Bundle savedInstanceState) {
      // TODO Auto-generated method stub
      super.onActivityCreated(savedInstanceState);

      setupViewComponent();
    }

    @Override
```

```java
public void onAttach(Activity activity) {
    // TODO Auto-generated method stub
    super.onAttach(activity);

    try {
        mCallback = (CallbackInterface) activity;
    } catch (ClassCastException e) {
        throw new ClassCastException(activity.toString() +
                "must implement GameFragment.CallbackInterface.");
    }
}

private void setupViewComponent() {
    mTxtComPlay = (TextView)getView().findViewById(R.id.txtComPlay);
    mTxtResult = (TextView)getView().findViewById(R.id.txtResult);
    mBtnScissors = (Button)getView().findViewById(R.id.btnScissors);
    mBtnStone = (Button)getView().findViewById(R.id.btnStone);
    mBtnNet = (Button)getView().findViewById(R.id.btnNet);
    mBtnShowResult = (Button)getView().findViewById(R.id.btnShowResult);

    mBtnScissors.setOnClickListener(btnScissorsLin);
    mBtnStone.setOnClickListener(btnStoneLin);
    mBtnNet.setOnClickListener(btnNetLin);

    mBtnShowResult.setOnClickListener(btnShowResultLin);

    if (((Main)getActivity()).UITypeFlag == Main.UIType.
    TWO_FRAMES) {
        mBtnShowResult.setVisibility(View.GONE);
    } else {
        mBtnShowResult.setVisibility(View.VISIBLE);
    }
}

private Button.OnClickListener btnShowResultLin = new Button.OnClickListener() {
    public void onClick(View v) {
        getActivity().findViewById(R.id.frame1).setVisibility
        (View.GONE);
        getActivity().findViewById(R.id.frame2).setVisibility
        (View.VISIBLE);

        mCallback.updateGameResult(miCountSet, miCountPlayerWin,
                miCountComWin, miCountDraw);
    }
};

private Button.OnClickListener btnScissorsLin = new Button.
```

```
            OnClickListener() {
                public void onClick(View v) {
                    …（和原来项目的程序代码相同）
                }
            };

        private Button.OnClickListener btnStoneLin = new Button.
            OnClickListener() {
                public void onClick(View v) {
                    …（和原来项目的程序代码相同）
                }
            };

        private Button.OnClickListener btnNetLin = new Button.
            OnClickListener() {
                public void onClick(View v) {
                    …（和原来项目的程序代码相同）
                }
            };
}
```

Step 6 开启游戏程序的接口布局文件 game.xml 并修改按钮如下：

```
<?xml version="1.0" encoding="utf-8"?>
<RelativeLayout xmlns:android="http://schemas.android.com/apk/res/android"
    android:layout_width="400dp"
    android:layout_height="match_parent"
    android:layout_gravity="center_horizontal"
    >
…（和原来项目的程序代码相同）
<TextView android:id="@+id/txtResult"
    android:layout_width="wrap_content"
    android:layout_height="wrap_content"
    android:text="@string/result"
    android:layout_below="@id/btnNet"
    android:layout_alignLeft="@id/txtCom"
    android:textSize="20sp"
    android:textColor="#0FFFFF"
    android:layout_marginTop="20dp"
    />
<Button android:id="@+id/btnShowResult"
    android:layout_width="wrap_content"
    android:layout_height="wrap_content"
    android:text="@string/btnShowResult"
    android:layout_below="@id/txtResult"
    android:layout_centerHorizontal="true"
    android:textSize="20sp"
    android:layout_marginTop="10dp"
```

```xml
        />
    </RelativeLayout>
```

Step 7 开启"局数统计界面"的接口布局文件 game_result.xml,在最后新增一个"回到游戏"的按钮。当程序只显示单一 FrameLayout 时用户可以按下这个按钮回到游戏界面。

```xml
<?xml version="1.0" encoding="utf-8"?>
<LinearLayout xmlns:android="http://schemas.android.com/apk/res/android"
    android:orientation="vertical"
    android:layout_width="match_parent"
    android:layout_height="match_parent"
    >
…(和原来项目的程序代码相同)
<Button android:id="@+id/btnBackToGame"
    android:layout_width="wrap_content"
    android:layout_height="wrap_content"
    android:layout_gravity="center_horizontal"
    android:text="@string/btnBackToGame"
    android:textSize="20sp"
    />
</LinearLayout>
```

Step 8 开启"局数统计界面"的程序文件 GameResultFragment.java,在程序代码中加入对"回到游戏"按钮的控制如下,这个按钮的控制方式和前面步骤讨论过的"显示结果"按钮类似:

```java
package ...

import ...

public class GameResultFragment extends Fragment {

    private TextView mEdtCountSet,
                     mEdtCountPlayerWin,
                     mEdtCountComWin,
                     mEdtCountDraw;
    private Button mBtnBackToGame;

    @Override
    public View onCreateView(LayoutInflater inflater, ViewGroup container,
        Bundle savedInstanceState) {
        // TODO Auto-generated method stub
        return inflater.inflate(R.layout.game_result, container,
            false);
    }

    @Override
```

```java
public void onResume() {
    // TODO Auto-generated method stub
    super.onResume();

    mEdtCountSet = (EditText)getActivity().findViewById
    (R.id.edtCountSet);
    mEdtCountPlayerWin = (EditText)getActivity().findViewById
    (R.id.edtCountPlayerWin);
    mEdtCountComWin = (EditText)getActivity().findViewById
    (R.id.edtCountComWin);
    mEdtCountDraw = (EditText)getActivity().findViewById
    (R.id.edtCountDraw);
    mBtnBackToGame = (Button)getActivity().findViewById
    (R.id.btnBackToGame);

    mBtnBackToGame.setOnClickListener(btnBackToGameLin);

    if ((((Main)getActivity()).UITypeFlag == Main.UIType.
    TWO_FRAMES) {
        mBtnBackToGame.setVisibility(View.GONE);
    } else {
        mBtnBackToGame.setVisibility(View.VISIBLE);
    }
}

private Button.OnClickListener btnBackToGameLin = new Button.
OnClickListener() {
    public void onClick(View v) {
        getActivity().findViewById(R.id.frame1).
        setVisibility(View.VISIBLE);
        getActivity().findViewById(R.id.frame2).
        setVisibility(View.GONE);
    }
};

public void updateGameResult(int iCountSet,
                             int iCountPlayerWin,
                             int iCountComWin,
                             int iCountDraw) {
    mEdtCountSet.setText(new Integer(iCountSet).toString());
    mEdtCountDraw.setText(new Integer(iCountDraw).toString());
    mEdtCountComWin.setText(new Integer(iCountComWin).toString());
    mEdtCountPlayerWin.setText(new Integer(iCountPlayerWin).
    toString());
}
}
```

完成以上修改之后可以分别在平板电脑模拟器和手机模拟器上运行程序，当在平板电脑运行时，程序会同时显示游戏界面和局数统计界面，如图 59-2 所示，当在手机上运行时程序刚开始只会显示游戏界面，用户必须按下"显示结果"按钮才会切换到局数统计界面，如图 59-3 所示。

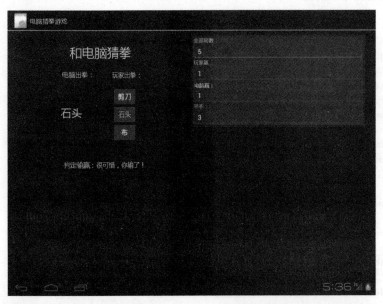

图 59-2　程序在平板电脑运行时会同时显示游戏界面和局数统计界面

图 59-3　程序在手机上运行时只会显示游戏界面，用户必须利用界面下方的按钮进行切换

　　在这个范例中我们用到侦测屏幕大小分类的程序代码，在开发能够同时适用不同 Android 设备的程序时，屏幕的尺寸和分辨率是很重要的考虑，因为它会影响程序接口的排列，下一个单元我们将继续介绍如何在程序中侦测屏幕的实际宽度、高度和分辨率。

Android 版本	1.X	2.X	3.X	4.X
适用性	★[①]	★[②]	★	★

UNIT 60
取得屏幕的宽度、高度和分辨率

在设计程序的操作接口时,屏幕的尺寸和分辨率是很重要的参考依据。就像前一个单元的范例程序,我们必须根据屏幕的大小来决定是否显示全部的接口组件,或是采用分页显示的方式。该范例是利用 Android 系统的屏幕尺寸分类作为依据,但是有时候我们可能需要实际的屏幕宽度和高度,甚至是分辨率,或者是程序界面的宽和高(也就是扣掉屏幕上方的 Action Bar 和下方的 System Bar 的区域),这个单元就让我们来学习如何在程序中取得这些屏幕相关的信息。

60-1 取得屏幕的宽高和分辨率

如果程序需要取得屏幕的尺寸规格,通常会在 onCreate()方法中进行,以便在后续的程序中决定操作接口的显示方式。如果想要知道屏幕的实际宽度和高度,可以运行下列的程序代码:

```
Display defDisp = getWindowManager().getDefaultDisplay();
int dispWidth = defDisp.getWidth();        // 屏幕的宽度,以 pixel 为单位
int dispHeight = defDisp.getHeight();      // 屏幕的高度,以 pixel 为单位
```

如果想要知道屏幕的分辨率,必须先取得 DisplayMetrics 对象,再从它内部的 xdpi 和 ydpi 字段得到水平和垂直分辨率如下:

[①] 有些常数和属性只适用 3.X 以上的版本,请参考文中的说明。

```
DisplayMetrics dm = new DisplayMetrics();
getWindowManager().getDefaultDisplay().getMetrics(dm);
float horiDpi = dm.xdpi;        // 屏幕的水平分辨率
float vertDpi = dm.ydpi;        // 屏幕的垂直分辨率
```

另外我们也可以获得 Android 系统对于目前的屏幕分辨率的分类：

```
int screenSizeClass = dm.densityDpi;
switch(screenSizeClass){
case DisplayMetrics.DENSITY_LOW:
    // 屏幕属于低分辨率，运行对应的程序代码
    …
    break;
case DisplayMetrics.DENSITY_MEDIUM:
    // 屏幕属于中分辨率，运行对应的程序代码
    …
    break;
case DisplayMetrics.DENSITY_HIGH:
    // 屏幕属于高分辨率，运行对应的程序代码
    …
    break;
case DisplayMetrics.DENSITY_XHIGH:
    // 屏幕属于超高分辨率，运行对应的程序代码
    …
    break;
}
```

在 Android SDK 技术文件中有一份表格说明如何根据屏幕的宽度、高度以及分辨率，来决定屏幕尺寸的分类，该表格就是表 58-2。例如如果屏幕的宽和高是 480×800，在中分辨率（medium density）的情况下该屏幕是属于 large 尺寸，可是在高分辨率（high density）的情况下该屏幕是属于 normal 尺寸。利用以上的屏幕相关信息就可以适当地决定程序界面的显示和排列方式，以提供使用者更方便美观的操作接口。

60-2 取得程序界面的宽和高

一般程序运行的时候只能使用介于 Action Bar 和 System Bar 之间的区域，如果程序需要知道这个可用区域的大小，必须根据 View 对象来取得，可是当程序在运行 onCreate() 的时候，程序界面的 View 还没有建立完成，因此如果要在 onCreate() 中取得程序界面区域的宽度和高度，就运行的顺序来说是不可能的，可是我们可以通过程序设计的技巧来达成。简单地说，就是在主程序开始运行之前，先启动一个暂时的 Activity，它的任务就是利用一个自定义的 LinearLayout 类建立一个空的程序界面，以取得可用区域的宽度和高度。这个暂时性的 Activity 一启动之后就会立刻运行主程序，然后便结束。后续运行的主程序再根据该自定义的 LinearLayout 类取得程序可用区域的宽和

高，以下利用一个实现的程序项目来示范：

Step 1 新增一个程序项目，主程序类名称取名为 MeasureAppWindowSize，项目的其他属性依照惯例设置即可，这个主程序类就是用来当成暂时的 Activity。

Step 2 在 Eclipse 左边的项目检查窗口中展开 res 文件夹，再用鼠标右键单击其中的 layout 文件夹并选择 New > File，然后在 File name 字段输入文件名 measure_app_window_size.xml。这个文件是用来当成前一个步骤的主类的接口布局文件，开启这个接口布局文件然后编辑如下，其中只有一个 view 组件，该组件指定使用我们项目中的 MyLayout 类：

```xml
<?xml version="1.0" encoding="utf-8"?>
<LinearLayout xmlns:android="http://schemas.android.com/apk/res/android"
    android:orientation="vertical"
    android:layout_width="match_parent"
    android:layout_height="match_parent"
    >
<view
    class="tw.android.MyLayout"
    android:layout_width="match_parent"
    android:layout_height="match_parent"
    />
</LinearLayout>
```

Step 3 在 Eclipse 左边的项目检查窗口中展开 src 文件夹，用鼠标右击其中的组件路径文件夹，然后选择 New > Class。在 Name 字段输入类名称 MyLayout，然后单击 Superclass 字段最右边的 Browse 按钮，在出现的对话框上方的文字字段输入 LinearLayout，然后在下方的清单中单击该类再单击 OK 按钮，最后单击 Finish 按钮完成新增类的操作。

Step 4 以上建立的新类会自动开启在程序代码编辑窗口中，并出现一个红色波浪底线标示语法错误。将鼠标光标移到红色波浪底线上方，会弹出一个信息窗口提供修正建议，请单击加入一个含有两个参数的构造方法（根据笔者的测试，其他格式的构造方法会造成程序运行时的例外错误），然后在程序代码编辑窗口中右击，单击快捷菜单中的 Source > Override/Implement Methods…叫出方法清单对话框，勾选其中的 onMeasure() 方法再按下 OK 按钮，最后将程序代码编辑如下，在程序代码中我们定义两个 static 变量以储存程序界面区域的宽度和高度，这样当第一个暂时性的 Activity 结束的时候，变量的值还会继续存在以供后续运行的 Activity 读取。

```
package …

import …

public class MyLayout extends LinearLayout {
```

```java
        static public int appWindowWidth;
        static public int appWindowHeight;

        // 一定要使用这个 constructor，否则会发生例外错误
        public MyLayout(Context context, AttributeSet attrs) {
            super(context, attrs);
            // TODO Auto-generated constructor stub
        }

        @Override
        protected void onMeasure(int widthMeasureSpec, int heightMeasureSpec) {
            // TODO Auto-generated method stub
            appWindowWidth = getMeasuredWidth();
            appWindowHeight = getMeasuredHeight();

            super.onMeasure(widthMeasureSpec, heightMeasureSpec);
        }
    }
```

Step 5 从 Eclipse 左边的项目检查窗口中开启 src/(组件路径名称)/ MeasureAppWindowSize.java 程序文件，将它的内容编辑如下，其中我们建立一个 Handler 对象以便利用 post Runnable 对象的方式启动后续的程序，之后便结束目前的 Activity，请读者参考程序中的批注说明。

```java
package …

import …

public class MeasureAppWindowSize extends Activity {

    final Handler mHandler = new Handler();

    @Override
    protected void onCreate(Bundle savedInstanceState) {
        // TODO Auto-generated method stub
        super.onCreate(savedInstanceState);
        setContentView(R.layout.measure_app_window_size);

        // 必须用 post Runnable 对象的方式启动主程序，这个
        // 测试程序窗口大小的 Activity 才会继续完成建立 view 的工作
        mHandler.post(new Runnable() {
            @Override
            public void run() {
                // TODO Auto-generated method stub
                Intent it = new Intent(
                        MeasureAppWindowSize.this, Main.class);
```

```
            startActivity(it);
            finish();
         }
      });
   }
}
```

Step 6 接下来请读者仿照步骤三的方法新增一个 Main 类，这个类是继承 Activity，然后再仿照步骤四的操作加入 onConfigurationChanged()方法，最后将程序代码编辑如下。在 onCreate()中只是利用 Toast 快显信息，显示 MyLayout 类中记录的程序界面的宽和高。至于 onConfigurationChanged()是当手机或平板电脑被转动的时候，Android 系统会自动运行这个函式，由于转动屏幕会造成宽度和高度对换，因此我们再利用 Intent 启动 MeasureAppWindowSize 程序，并结束目前的程序，这样侦测程序可用区域的动作就会重新运行一遍。

```
package ...

import ...

public class Main extends Activity {
    /** Called when the activity is first created. */
    @Override
    public void onCreate(Bundle savedInstanceState) {
        super.onCreate(savedInstanceState);
        setContentView(R.layout.main);

        // 显示程序可用区域的宽和高
        Toast.makeText(Main.this,
                       "appw = " + MyLayout.appWindowWidth +
                       " apph = " + MyLayout.appWindowHeight,
                       Toast.LENGTH_LONG)
             .show();
    }

    @Override
    public void onConfigurationChanged(Configuration newConfig) {
        // TODO Auto-generated method stub
        super.onConfigurationChanged(newConfig);

        Intent it = new Intent(this, MeasureAppWindowSize.class);
        startActivity(it);
        finish();
    }
}
```

Step 7 我们必须在程序功能描述文件 AndroidManifest.xml 中加上自行新增的 Main Activity 的信息，另外也要设置 Activity 的 android:configChanges 属性，这样当手机或平板电脑被转动的时候，Android 系统才会运行 onConfigurationChanged()。

```xml
<?xml version="1.0" encoding="utf-8"?>
<manifest xmlns:android="http://schemas.android.com/apk/res/android"
    … />

    <application android:icon="@drawable/icon" android:label=
"@string/app_name">
        <activity android:name=".MeasureAppWindowSize"
                android:label="@string/app_name">
            <intent-filter>
                <action android:name="android.intent.action.MAIN" />
                <category android:name="android.intent.category.
LAUNCHER" />
            </intent-filter>
        </activity>
        <activity android:name=".Main"
                android:label="@string/app_name"
                android:configChanges="orientation"
                >
        </activity>

    </application>
</manifest>
```

完成这个程序项目之后运行程序就可以看到屏幕上显示程序可用区域的宽度和高度，如图 60-1 所示，读者可以同时按下键盘左边的 Ctrl 和 F12 按键，模拟器的屏幕会旋转 90 度，这时候程序会重新运行显示新的宽度和高度，如图 60-2 所示。

图 60-1　程序显示可用区域的宽度和高度

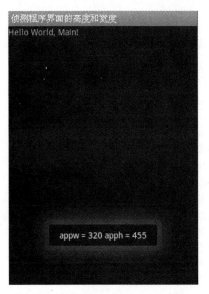

图 60-2　将模拟器旋转 90 度后程序重新显示可用区域的宽度和高度

60-3　利用 AndroidManifest.xml 文件设置程序运行的屏幕条件

如果我们要设置程序运行的屏幕条件，可以在程序功能描述文件 AndroidManifest.xml 中进行，在<manifest>标签中可以加入<supports-screens>标签，它提供许多属性让我们控制程序如何在不同的屏幕尺寸上运行，表 60-1 是这些相关属性的使用说明。

表 60-1　用于<supports-screens>标签中的屏幕相关属性设置

属性名称	设置值	功能说明	
android:smallScreens	true/false	默认值是 true，代表程序可以在屏幕尺寸分类为 small 的设备上运行。	Android 2.X/3.X/4.X
android:normalScreens	true/false	默认值是 true，代表程序可以在屏幕尺寸分类为 normal 的设备上运行。	Android 2.X/3.X/4.X
android:largeScreens	true/false	不同的 Android 版本会有不同的默认值，true 代表程序可以在屏幕尺寸分类为 large 的设备上运行，如果设置为 false，Android 会用放大程序界面的方式运行这个程序。	Android 2.X/3.X/4.X
android:xlargeScreens	true/false	不同的 Android 版本会有不同的默认值，true 代表程序可以在屏幕尺寸分类为 extra large 的设备上运行，如果设置为 false，Android 会用放大程序界面的方式运行这个程序。	Android 2.3.X/3.X/4.X

续表

属性名称	设置值	功能说明	
android:anyDensity	true/false	true/false 默认值是 true,代表程序可以在任何屏幕分辨率的设备上运行。	Android 2.X/3.X/4.X
android:requiresSmallestWidthDp	整数值,例如 320(4 寸屏幕)、600(7 寸屏幕)、720(10 寸屏幕)	设置程序能够运行的最小屏幕,虽然它的名称是 Width,但是不管宽高都必须大于这个设置值(因为手持设备翻转后宽和高会互换)。	Android 3.2/4.X
android:compatibleWidthLimitDp	整数值	设置程序能够运行的最大屏幕,虽然它的名称是 Width,但是不管宽高都必须小于这个设置值(因为手持设备翻转后宽和高会互换)。如果屏幕的宽和高中比较小的那一个值大于这个设置值,Android 会用放大程序界面的方式运行这个程序,并且在 System Bar 中会显示一个选项可以取消界面放大功能。	Android 3.2/4.X
android:largestWidthLimitDp	整数值	如果屏幕的宽和高中比较小的那一个值大于这个设置值,Android 会用放大程序界面的方式运行这个程序,而且用户不能取消界面放大功能。	Android 3.2/4.X

以上的属性可以和 UNIT 58 中介绍的支持多屏幕模式技巧一起配合使用,例如程序的接口布局文件如果只针对 small、normal 和 large 类型的屏幕设计,那么我们可以在程序功能描述文件 AndroidManifest.xml 中加上下列的<supports-screens>标签,让属于 extra large 屏幕的 Android 设备(例如平板电脑)采用放大界面的方式运行这个程序:

```
<?xml version="1.0" encoding="utf-8"?>
<manifest xmlns:android="http://schemas.android.com/apk/res/android"
        package="tw.android"
        android:versionCode="1"
        android:versionName="1.0">
    …
    <supports-screens android:smallScreens="true"
                android:normalScreens="true"
                android:largeScreens="true"
                android:xlargeScreens="false" />

    <application …>
        …
    </application>
</manifest>
```

以 UNIT 10 的手机程序为例,我们先将程序项目使用的 Android 版本改为 2.3.X 以上,然后开启程序功能描述文件 AndroidManifest.xml，将上述范例中的<supports-screens>标签的内容复制过去，另外记得取消 android:targetSdkVersion="11"的设置，因为我们是要用手机程序模式运行这个项目。修改后在平板电脑模拟器上就可以看到图 60-3 的运行界面，在屏幕右下方会出现一个四个方向箭头的按钮，它是用来切换程序的运行模式，按下它就会显示图 60-4 的选项，如果切换成 Zoom to fill screen 就会变成图 10-1 的模式。

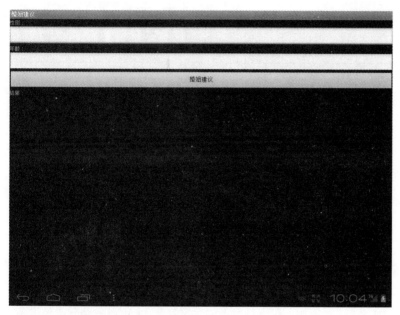

图 60-3　手机程序加上<supports-screens>设置后在平板电脑上的运行界面

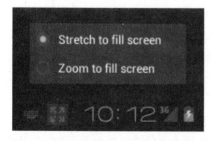

图 60-4　平板电脑运行手机程序时提供的运行模式选项

UNIT 61 将程序安装到设备或在网络上发布

Android 版本	1.X	2.X	3.X	4.X
适用性	★	★	★	★

　　程序要在模拟器或是在实体手机或平板电脑上安装时必须有一个数字签名，在模拟器上运行程序时，Android SDK 开发工具会先自动帮程序加上 debug 模式的数字签名，所以程序才可以安装在模拟器中运行。但是程序开发完成之后，必须加上程序开发者的数字签名，才可以安装到实体设备。为程序加上数字签名必须注意下列事项：

1. 同一个开发者的程序应该加上同一个数字签名，因为当程序要进行更新时，Android 系统会核对新版程序和旧版程序的签名是否相同。其次是 Android 系统可以让具有相同数字签名的程序使用同一个 process，也就是说你可以将自己开发的程序分成不同的模块发行，但是它们要具有相同的数字签名，而且将来也可以个别更新。另外具有相同数字签名的程序可以互相使用对方的功能和数据。

2. 在数字签名中有一栏是有效日期（Validity），这个字段会影响程序将来是否可以更新，如果超过这个有效期限程序将不能更新，因此 Android SDK 技术文件建议至少大于 25 年。如果程序要上传到 Android Market 网站，该网站要求程序的数字签名的有效日期必须在 2033 年 10 月 22 日之后。

　　除了帮开发好的程序加上数字签名之外，还需要对它进行所谓"数据对齐"（zipalign）的动作，这是为了增加程序的运行效率，最后整个程序项目会被包装成一个 apk 文件，这个 apk 文件就可以上传到实体设备进行安装。以上的操作步骤可以借助 Eclipse ADT plugin 中的 Export Wizard 来完成。

61-1　利用 Export Wizard 帮程序加上数字签名和完成 zipalign

Step 1 右击 Eclipse 左边的项目检查窗口中要发布的程序项目，然后选择 Android Tools > Export Signed Application Package 就会出现图 61-1 的对话框。要注意的是这里的项目名称不可以有中文，如果读者选择的项目名称中包含中文，可以先利用 Windows 文件管理器复制该程序项目的文件夹，再将复制的项目文件夹改成英文，然后利用 Eclipse 的 File > Import 功能加载英文名称的程序项目。

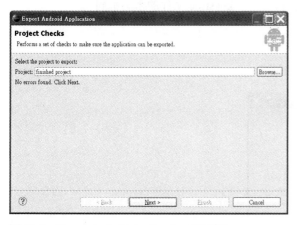

图 61-1　Export Android Application 对话框

Step 2 确定程序项目的名称后单击 Next 按钮就会出现图 61-2 的对话框，如果是第一次使用请选择 Create new keystore，然后输入文件的储存路径和文件名，扩展名是 keystore，然后设置密码，密码长度必须大于 6 个字符。

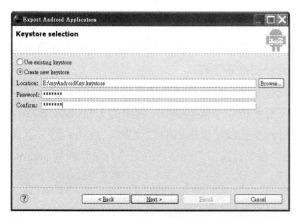

图 61-2　设置 Export Android Application 对话框

> **补充说明**
>
> 如果以前已经建立过 keystore 文件，可以单击 Use existing keystore，然后在 Location 字段输入文件路径和名称，或是利用右边的 Browse 按钮挑选，接着在 Password 字段输入密码，按下 Next 按钮后再输入另一个验证密码。

Step 3 输入完成后单击 Next 按钮就会出现图 61-3 的对话框，接着输入 key 的信息。

- Alias：输入 key 的名称
- Password：设置 key 的密码，可以和上一个 keystore 文件的密码不同
- Confirm：再输入一次密码
- Validity(years)：建议至少 25
- First and Last Name：程序开发者的姓名

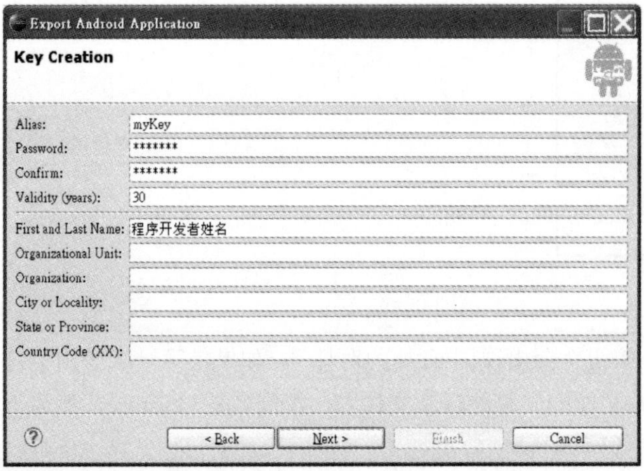

图 61-3　输入 key 的信息

Step 4 输入完成后单击 Next 按钮就会出现图 61-4 的对话框，请输入最后产生的 apk 文件路径和名称，然后单击 Finish 按钮就完成发布程序的准备工作。

最后得到的 apk 文件就可以上传到实体设备进行安装。如果读者手边有实体手机或平板电脑，可以先自行测试一下，首先在开发程序的 PC 上安装手机或平板电脑原厂提供的 USB 驱动程序，安装好之后把手机或平板电脑连接到 PC，然后进入手机或平板电脑的"设置 > 应用程序"界面启用"安装未知来源的程序"。接着运行 Windows"命令行窗口"程序，将工作目录切换到 android sdk 文件夹中的 platform-tools 子文件夹，然后运行：

　　adb -d install (apk 文件路径和完整文件名)

指定的 apk 文件就会通过 USB 上传到实体手机或平板电脑并完成安装。

程序项目的整备工作和发布　PART 11

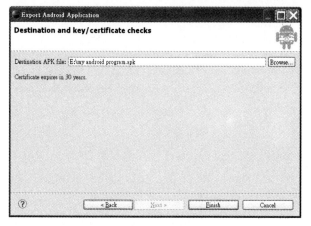

图 61-4　输入最后产生的 apk 文件路径和名称

> **补充说明　移除安装过的程序**
>
> 如果之前已经安装过该程序会导致安装失败，这时候必须先将原来安装的程序移除。请进入手机或平板电脑的应用程序列表界面，单击"设置 > 应用程序 > 管理应用程序"，从列表中单击要移除的程序，然后按下"卸载"按钮。

61-2　将程序上传到 Google 的 Android Market 网站

程序的 apk 文件也可以在网络上公开让其他人下载使用，为了便于 Android 程序的流通和交易，Google 建立了一个官方的 Android 程序网站，名称叫做 Android Market。如果要将程序上传到该网站，必须完成以下工作：

Step 1　注册一个 Google 账号，这个账号和 Gmail 共享，如果读者已经拥有 Gmail 的账号，就不需要再注册。

Step 2　必须拥有一张信用卡，因为要在 Android Market 网站贩卖程序必须先缴交美金 25 元的费用，这笔费用只在第一次使用时收取。

Step 3　上传的 apk 文件必须符合前面介绍的条件。

Step 4　输入程序的介绍和说明。

Step 5　在 Android Market 网站上的程序可以是付费下载或是免费使用。如果必须付费购买，贩卖程序所得的 30% 必须支付给 Google。

Step 6　Android Market 网站可以让程序用户进行评价和提供使用意见供程序作者参考。

PART 12　2D 和 3D 绘图

UNIT 62　使用 Drawable 对象
UNIT 63　使用 Canvas 绘图
UNIT 64　使用 View 在 Canvas 上绘制动画
UNIT 65　使用 SurfaceView 进行高速绘图
UNIT 66　3D 绘图

Android 版本	1.X	2.X	3.X	4.X
适用性	★	★	★	★

UNIT 62
使用 Drawable 对象

Android 程序的绘图功能是来自 SGL 和 OpenGL ES 链接库，然后经由上层的 Application Framework 加以包装成各种绘图类供应用程序使用。在 Android 程序中有两种绘制图形的方法：
1. 使用 Drawable 对象
2. 直接在 Canvas 上绘图

使用 Drawable 对象绘图的方式和我们之前的程序架构很类似，基本上就是在接口布局文件中建立 ImageView 组件，然后在程序中建立 Drawable 对象，再把 Drawable 对象设置给 ImageView 组件完成绘图的工作。Android 程序提供三种建立 Drawable 对象的方法：
1. 使用 res/drawable 文件夹中的图像文件
2. 在 res/drawable 文件夹中建立 xml 文件格式的 Drawable 对象定义文件
3. 在程序中建立 Drawable 相关类的对象

以下我们依序介绍这三种方法。

62-1 从 res/drawable 文件夹的图像文件建立 Drawable 对象

这种方式最简单，只需要 3 个步骤：

Step 1 取得程序项目的资源对象：

```
Resources res = getResources();
```

Step 2 调用资源对象的 getDrawable() 方法取得程序项目中的图像资源：

```
Drawable drawImg = res.getDrawable(R.drawable.img);
```

Step 3 将 Drawable 对象设置为 ImageView 对象的 background：

```
ImageView imgView = (ImageView)findViewById(R.id.imgView);
imgView.setBackgroundDrawable(drawImg);
```

> **补充说明**
>
> 有些 Drawable 对象设置为 ImageView 的 foreground 时（利用 setImageDrawable()方法）会无法正常显示，因此建议使用 setBackgroundDrawable()。

62-2 在 res/drawable 文件夹中建立 xml 文件格式的 Drawable 对象定义文件

我们在 UNIT 22 已经学过如何在程序中建立 Frame animation 的 xml 动画资源文件，这种动画资源文件放在程序项目的 res/drawable 文件夹中，而且它就是一种 Drawable 对象。除了 Frame animation 之外，我们还可以建立其他类型的 Drawable 对象定义文件，例如颜色、形状、图像切换效果等，表 62-1 列出一些常用的 Drawable 对象类型，包括所使用的 xml 标签和对应的类名称，其中的<animation-list>就是我们在 UNIT 22 中介绍过的 Frame animation。以下我们再以形状和图像切换效果的 Drawable 对象为例，示范如何在程序项目中使用 xml 文件建立不同类型的 Drawable 对象。

表 62-1　Drawable 对象类型所使用的 xml 标签和对应的类名称

xml 标签名称	类名称	说明
<animation-list>	AnimationDrawable	建立由多个图像文件组成的动画，可以用来当成 View 对象的背景
<bitmap>	BitmapDrawable	可以让图像重复排列并适当地缩放和对齐
<clip>	ClipDrawable	用来剪裁 Drawable 对象
<color>	ColorDrawable	用来填上颜色
<rotate>	RotateDrawable	让 Drawable 对象旋转一个角度
<scale>	ScaleDrawable	缩放 Drawable 物件
<shape>	GradientDrawable	绘制特定形状，注意<shape>的对应类不是 ShapeDrawable 类
<transition>	TransitionDrawable	以淡入淡出的方式轮流显示一组图像

Step 1 在 res/drawable 文件夹中新增 xml 文件格式的 Drawable 对象定义文件（操作提示：还不熟悉新增文件操作步骤的读者请参考前面相关单元的说明），例如以下是用 xml 文件建立 Transition 和椭圆形 Drawable 对象的范例：

- res/drawable/tran_drawable.xml

```xml
<?xml version="1.0" encoding="utf-8"?>
<transition xmlns:android="http://schemas.android.com/apk/res/android">
    <item android:drawable="@drawable/img01"/>
    <item android:drawable="@drawable/img02"/>
</transition>
```

其中 img01 和 img02 是置于 res/drawable 文件夹的两个图像文件。

- res/drawable/shap_drawable.xml

```xml
<?xml version="1.0" encoding="utf-8"?>
<shape xmlns:android="http://schemas.android.com/apk/res/android"
    android:shape="oval"
/>
```

Step 2 在程序中先调用 getResources()方法取得资源对象,再调用资源对象的 getDrawable()方法取得 xml 文件格式的 Drawable 对象,然后把取得的 Drawable 对象转换成适当的类型,我们也可以进一步设置 Drawable 对象的属性,请参考以下程序代码：

```java
Resources res = getResources();

TransitionDrawable drawTran =
        (TransitionDrawable)res.getDrawable(R.drawable.tran_drawable);

GradientDrawable gradShape =
        (GradientDrawable)res.getDrawable(R.drawable.shap_drawable);
gradShape.setColor(0xffffff00);
```

Step 3 把得到的 Drawable 对象设置给 ImageView 完成显示。有些 Drawable 对象使用 ImageView 的 setImageDrawable()方法会无法正常显示,此时可以换成使用 setBackgroundDrawable()。
另外请读者注意步骤二和三的程序代码要避免放在 onXXX()方法中运行（包括 onCreate()、onResume()和 onStart()),因为当程序还未完成起始动作之前,有些 Drawable 对象会无法正常作用。

62-3　在程序中建立 Drawable 类型的对象

除了以上两种方法之外,我们也可以在程序中自己建立 Drawable 对象,例如以下程序代码会建立和前一个小节的 xml 文件相同的椭圆形 Drawable 对象,有关程序代码中使用的 Drawable 相关类和 xml 文件中的标签名称的对应关系请参考表 62-1。

```java
GradientDrawable gradShape = new GradientDrawable();
gradShape.setShape(GradientDrawable.OVAL);
```

建立 Drawable 对象所调用的方法,和前面 xml 文件中用到的属性有清楚的对应关系,读者自行对照就可以了解它们的用法,另外也可以利用单元五所介绍的方法查询 Android SDK 说明文件。读者也可以尝试在程序中建立和前一个小节的 xml 文件相同的 TransitionDrawable 对象,建立 Drawable 对象后再设置给 ImageView 对象完成显示。

62-4 范例程序

我们用一个完整的程序项目来示范 Drawable 对象的绘图效果,请读者依照下列步骤操作:

Step 1 依照前面惯用的操作方法新增一个 Android 程序项目。

Step 2 在项目检查窗口中右击 res/drawable-hdpi 文件夹,然后选择 New > File,新增一个名为 tran_drawable.xml 的文件,并将该文件编辑如下:

```xml
<?xml version="1.0" encoding="utf-8"?>
<transition xmlns:android="http://schemas.android.com/apk/res/android">
    <item android:drawable="@drawable/img01"/>
    <item android:drawable="@drawable/img02"/>
</transition>
```

其中会用到两个名为 img01 和 img02 的图像文件,读者可以自行选择两个宽和高都不超过 300 pixel 的 png 或 jpg 图像文件,利用 Windows 文件总管把它们复制到程序项目的 res/drawable-hdpi 文件夹中。

Step 3 开启 res/layout/main.xml 接口布局文件并编辑成如下内容,其中我们建立一个 Button 组件和三个 ImageView 组件,并设置 LinearLayout 的 android:gravity 属性,让操作界面显示在屏幕的中央。

```xml
<?xml version="1.0" encoding="utf-8"?>
<LinearLayout xmlns:android="http://schemas.android.com/apk/res/android"
    android:orientation="vertical"
    android:layout_width="match_parent"
    android:layout_height="match_parent"
    android:gravity="center_horizontal"
    >
<Button android:id="@+id/btnStart"
    android:layout_width="wrap_content"
    android:layout_height="wrap_content"
    android:text="开始绘图"
    />
<ImageView android:id="@+id/imgView1"
    android:layout_width="wrap_content"
    android:layout_height="wrap_content"
    android:layout_marginTop="10dp"
    />
<ImageView android:id="@+id/imgView2"
```

```
            android:layout_width="wrap_content"
            android:layout_height="wrap_content"
            android:layout_marginTop="10dp"
            />
        <ImageView android:id="@+id/imgView3"
            android:layout_width="200dp"
            android:layout_height="100dp"
            android:layout_marginTop="10dp"
            />
</LinearLayout>
```

Step 4 开启"src/(组件路径名称)"文件夹中的主程序文件并编辑成如下内容：

```
package …

import …

public class Main extends Activity {

    private ImageView mImgView1,
                      mImgView2,
                      mImgView3;
    private Button mBtnStart;

    /** Called when the activity is first created. */
    @Override
    public void onCreate(Bundle savedInstanceState) {
        super.onCreate(savedInstanceState);
        setContentView(R.layout.main);

        setupViewComponent();
    }

    private void setupViewComponent() {
        mImgView1 = (ImageView)findViewById(R.id.imgView1);
        mImgView2 = (ImageView)findViewById(R.id.imgView2);
        mImgView3 = (ImageView)findViewById(R.id.imgView3);

        mBtnStart = (Button)findViewById(R.id.btnStart);
        mBtnStart.setOnClickListener(btnStartOnClick);
    }

    private Button.OnClickListener btnStartOnClick = new Button.OnClickListener() {
        @Override
        public void onClick(View v) {
            // TODO Auto-generated method stub
            Resources res = getResources();
```

```
            Drawable drawImg = res.getDrawable(R.drawable.img01);
            mImgView1.setBackgroundDrawable(drawImg);

            TransitionDrawable drawTran =
                (TransitionDrawable)res.getDrawable(R.drawable.
                tran_drawable);
            mImgView2.setImageDrawable(drawTran);
            drawTran.startTransition(5000);

            GradientDrawable gradShape = new GradientDrawable();
            gradShape.setShape(GradientDrawable.OVAL);
            gradShape.setColor(0xffffff00);
            mImgView3.setBackgroundDrawable(gradShape);
        }
    };
}
```

在按钮的 onClickListener 中，我们利用前面介绍的方法产生一个图像的 Drawable 对象、一个 TransitionDrawable 对象和一个椭圆形 Drawable 对象，并且将它们设置给不同的 ImageView。请读者注意在设置 TransitionDrawable 对象后必须调用它的 startTransition() 方法才会开始播放图像转场效果。

完成程序代码的编辑之后运行程序，然后按下"开始绘图"按钮就会看到如图 62-1 所示的运行界面。

图 62-1　Drawable 对象绘图程序的执行画面

Android 版本	1.X	2.X	3.X	4.X
适用性	★	★	★	★

UNIT 63
使用 Canvas 绘图

除了使用 Drawable 对象绘图之外，也可以利用 Canvas 对象来进行绘图，而且 Canvas 的绘图功能比 Drawable 对象更强大，那么什么是 Canvas？其实程序的运行界面就是一个 Canvas 对象，之前我们是在 Canvas 上建立接口组件像是 ImageView，然后把图像或图形设置给 ImageView 显示，现在我们直接把图形画在 Canvas 上，不需要再借助接口组件。

Canvas 对象中其实还有一个 Bitmap 对象。我们可以把 Canvas 对象想象成是一个绘图桌，上面有许多绘图工具，还有一个 Bitmap 让我们在上面进行绘图。要直接在 Canvas 上绘图其实不难，只要建立一个继承 View 的新类，然后把它设置为程序的界面，当程序运行时就会调用这个新类的 onDraw() 方法，同时传入程序界面的 Canvas 对象让我们在上面绘图，以下我们利用一个程序项目来示范。

Step 1 依照前面惯用的操作方法新增一个 Android 程序项目。

Step 2 在程序项目中新增一个继承 View 的新类，我们可以将它取名为 ShapeView。开启这个新类的程序文件，其中会标示一个红色波浪底线，将鼠标光标移到红色波浪底线上方就会弹出一个信息要求加入一个构造方法，请读者选择加入第一个构造方法。接下来我们要新增一个 onDraw() 方法，请在程序代码编辑窗口中右击，在出现的快选菜单中选择 Source > Override/Implement Methods…，就会出现一个方法清单的对话框，勾选其中的 onDraw() 方法然后按下 OK 按钮，接着将程序编辑如下：

```
package …

import …
```

```java
public class ShapeView extends View {

    private ShapeDrawable mShapeDraw;
    private Paint mPaint;

    public ShapeView(Context context) {
        super(context);
        // TODO Auto-generated constructor stub

        mShapeDraw = new ShapeDrawable(new OvalShape());
        mShapeDraw.getPaint().setColor(0xffffff00);

        mPaint = new Paint();
        mPaint.setAntiAlias(true);
        mPaint.setColor(Color.CYAN);
    }

    @Override
    protected void onDraw(Canvas canvas) {
        // TODO Auto-generated method stub
        super.onDraw(canvas);

        mShapeDraw.setBounds(10, 10,
            canvas.getWidth()/2 - 10, canvas.getHeight()/2 - 20);
        mShapeDraw.draw(canvas);

        canvas.drawOval(new RectF(canvas.getWidth()/2 + 10, 10,
            canvas.getWidth() - 10, canvas.getHeight()/2 - 20), mPaint);
        canvas.drawText(this.getContext().getString(R.string.hello),
            10, canvas.getHeight()/2, mPaint);
        canvas.drawLine(canvas.getWidth()/2 + 10, canvas.
        getHeight()/2 -10 ,
            canvas.getWidth() - 10, canvas.getHeight()/2, mPaint);

        Resources res = getResources();
        Drawable drawImg = res.getDrawable(R.drawable.img01);
        drawImg.setBounds(10, canvas.getHeight()/2 + 10,
            canvas.getWidth()/2 - 10, canvas.getHeight() - 20);
        drawImg.draw(canvas);

        RotateDrawable drawRota =
            (RotateDrawable)res.getDrawable(R.drawable.rota_drawable);
        drawRota.setLevel(1000);
        drawRota.setBounds(canvas.getWidth()/2 + 30, canvas.getHeight()/2,
            canvas.getWidth() + 10, canvas.getHeight() - 30);
        drawRota.draw(canvas);
    }
}
```

我们在类的构造方法中建立一个椭圆的 Drawable 对象和一支画笔并指定它们的颜色。在 onDraw()方法中先绘制该椭圆 Drawable 对象,要在 Canvas 中绘制 Drawable 对象时,必须先调用 setBounds()方法设置绘制的位置,然后再调用 draw()方法并传入 Canvas 对象完成绘制。

接下来我们调用 Canvas 对象的 drawOval()方法绘制另一个椭圆,再调用 drawText()显示文字,然后调用 drawLine()画出一条直线,在这些方法的自变量中必须指定绘制的位置。

接下来从资源类 R 中取出一张图像并显示,最后再从程序资源中取得一个由 rota_drawable.xml 文件定义的 RotateDrawable 对象并显示在 Canvas 中,这个 rota_drawable.xml 文件的内容如下:

```xml
<?xml version="1.0" encoding="utf-8"?>
<rotate xmlns:android="http://schemas.android.com/apk/res/android"
        android:fromDegrees="0"
        android:toDegrees="90"
        android:pivotX="0%"
        android:pivotY="0%"
        android:drawable="@drawable/img02"
/>
```

读者可以利用上一个单元介绍的方法,将它建立在程序项目的 res/drawable-hdpi 文件夹中。另外在这个文件中需要用到储存在 res/drawable-hdpi 文件夹中的图像文件 img02,读者可以利用文件总管将上一个单元的程序项目中的图像文件复制到这个程序项目中。

Step 3 开启 "src/(组件路径名称)" 文件夹中的主程序文件 Main.java,在 onCreate()方法中建立一个步骤二类的对象,并将它设置成为程序的界面:

```java
package tw.android;

import android.app.Activity;
import android.os.Bundle;

public class Main extends Activity {
    /** Called when the activity is first created. */
    @Override
    public void onCreate(Bundle savedInstanceState) {
        super.onCreate(savedInstanceState);

        ShapeView shapeView = new ShapeView(this);
        setContentView(shapeView);
    }
}
```

完成之后运行程序就可以看到图 63-1 的运行界面。

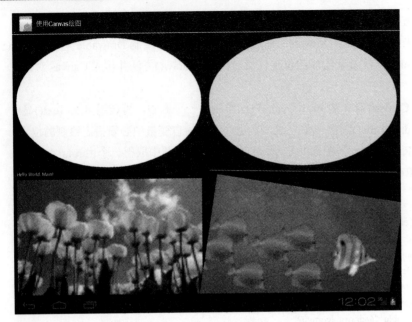

图 63-1　Canvas 绘图程序在平板电脑运行的界面

Android 版本	1.X	2.X	3.X	4.X
适用性	★	★	★	★

UNIT 64
使用 View 在 Canvas 上绘制动画

在上一个单元我们已经学会如何在程序的 Canvas 中绘图，主要步骤就是自己建立一个继承 View 的类，然后将绘图的程序代码写在该类的 onDraw()方法中，最后在主程序建立一个该类的对象，再把该对象设置为程序的运行界面。在这一个单元我们将进一步学习如何让绘制的图形动起来，也就是所谓的动画效果。

64-1 程序绘制动画的原理

让程序产生动画的原理就是在每一次绘制图形的时候，让图形的状态随着时间改变，图形的状态包括大小、颜色、位置、形状等。根据上一个单元的程序架构，只要我们能够连续运行 onDraw()方法，然后在 onDraw()方法中让每一次绘制的图形状态都有些改变，这样就可以产生动画的效果。

要改变绘制图形的状态，只需要把绘图程序代码使用的图形参数作一些变化即可，另一个比较值得考虑的问题是如何连续运行 onDraw()方法。这个 onDraw()方法并不是由我们的程序直接调用，而是由 Android 系统调用，因为它需要一个 Canvas 对象，而这个 Canvas 对象是由 Android 系统设置。当程序需要更新界面时，可以调用 invalidate()要求 Android 系统运行 onDraw()。但是当 Android 系统收到重绘界面的要求时，却不一定会立刻运行 onDraw()，因为绘制界面是比较耗时的工作，所以当系统忙碌时会被忽略。举例来说，如果一部车子每秒前进 10 米，如果程序在一秒钟之内调用 10 次 invalidate()方法要求更新 10 次车子的位置，但是由于系统正处于繁忙的运算工作，它可以只绘制其中3次车子的界面而忽略其他7次,这时候我们看到的车子是以不连续地跳动方式前进。

因为invalidate()方法具有以上的特性,所以我们不能直接在程序中以循环的方式连续调用它,这样Android系统会直接把这一连串的invalidate()调用运行完毕,最后再运行一次onDraw()。为了让系统在收到invalidate()的请求后有时间运行onDraw(),我们必须使用Multi-thread程序架构,也就是说把连续调用invalidate()的程序代码独立写成一个thread,这样当系统在thread之间切换的时候,Android才会运行onDraw()。了解绘制动画的原理之后,接下来让我们用一个程序项目来示范如何实现。

64-2 范例程序

这个范例程序会在界面上显示一个由大逐渐变小再由小逐渐变大的椭圆,如图64-1所示,请读者依照下列步骤操作。

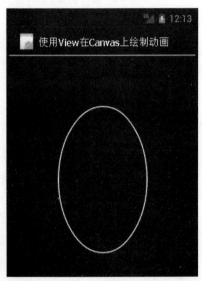

图64-1 绘制椭圆动画程序在手机运行的界面

Step 1 依照前面惯用的操作方法新增一个Android程序项目。

Step 2 在程序项目中新增一个继承View的新类,我们可以将它取名为ShapeView。开启这个新类的程序文件,其中会标示一个红色波浪底线,将鼠标光标移到红色波浪底线上方就会弹出一个信息要求加入一个构造方法,请读者选择加入第一个构造方法。接下来我们要新增一个onDraw()方法,请在程序代码编辑窗口中右击,在出现的快捷菜单中选择Source > Override/Implement Methods…就会出现一个方法清单的对话框,请勾选其中的onDraw()和onSizeChanged()然后按下OK按钮,再将程序编辑如下:

```java
package ...

import ...

public class ShapeView extends View implements Runnable {

    private Paint mPaintForeColor, mPaintBackColor;
    private static final int INT_STROCK_THICK = 2;
    private int mIntXMaxLen, mIntYMaxLen,
            mIntXCent, mIntYCent,
            mIntXCurLen, mIntYCurLen,
            mIntSign;

    private Handler mHandler = new Handler();

    public ShapeView(Context context) {
        super(context);
        // TODO Auto-generated constructor stub

        setFocusable(true);

        mPaintForeColor = new Paint();
        mPaintForeColor.setAntiAlias(true);
        mPaintForeColor.setColor(Color.CYAN);

        mPaintBackColor = new Paint();
        mPaintBackColor.setAntiAlias(true);
        mPaintBackColor.setColor(Color.BLACK);

        new Thread(this).start();
    }

    @Override
    protected void onDraw(Canvas canvas) {
        // TODO Auto-generated method stub
        super.onDraw(canvas);

        Log.d("TEST_VIEW", "onDraw()");

        canvas.drawOval(new RectF(mIntXCent - mIntXCurLen,
                        mIntYCent - mIntYCurLen,
                        mIntXCent + mIntXCurLen,
                        mIntYCent + mIntYCurLen),
                mPaintForeColor);
        canvas.drawOval(new RectF(mIntXCent - mIntXCurLen +
                        INT_STROCK_THICK,
```

```
                                    mIntYCent - mIntYCurLen +
                                    INT_STROCK_THICK,
                                    mIntXCent + mIntXCurLen –
                                    INT_STROCK_THICK,
                                    mIntYCent + mIntYCurLen –
                                    INT_STROCK_THICK),
                    mPaintBackColor);

            if (mIntXCurLen + mIntSign * INT_STROCK_THICK < 0 ||
                mIntYCurLen + mIntSign * INT_STROCK_THICK < 0) {
                mIntSign = 1;
            }
            else if (mIntXCurLen + mIntSign * INT_STROCK_THICK >
            mIntXMaxLen ||
                mIntYCurLen + mIntSign * INT_STROCK_THICK >
                mIntYMaxLen) {
                mIntSign = -1;
            }

            mIntXCurLen += mIntSign * INT_STROCK_THICK;
            mIntYCurLen += mIntSign * INT_STROCK_THICK;
        }

        @Override
        protected void onSizeChanged(int w, int h, int oldw, int oldh) {
            // TODO Auto-generated method stub
            super.onSizeChanged(w, h, oldw, oldh);

            mIntXMaxLen = w / 2 - 10;
            mIntYMaxLen = h / 2 - 10;
            mIntXCent = w / 2;
            mIntYCent = h / 2;

            mIntXCurLen = mIntXMaxLen;
            mIntYCurLen = mIntYMaxLen;
            mIntSign = -1;
        }

        public void run() {
            for (int i = 0; i < 10000; i++) {
                Log.d("TEST_VIEW", "run() " + i);
                mHandler.post(new Runnable() {
                    public void run() {
                        Log.d("TEST_VIEW", "invalidate()");
                        invalidate();
                    }
                });
            }
```

          }
      }

首先我们让这个 ShapeView 类实现 Runnable 接口以建立 Multi-thread 程序，然后在其中的 run()方法根据 mHandler 对象（关于 Handler 对象的功能读者可以参考 UNIT 33 中的说明）连续以 post Runnable 对象的方式调用 invalidate()，因此 invalidate()将会以 background thread 的方式运行，这样一来主程序（main thread）就有机会更新程序界面。在 ShapeView 类的构造方法中我们建立二个 Paint 对象，一个用来绘制对象的颜色，一个是当成背景颜色，然后根据这个 ShapeView 类建立一个 Thread 对象并将它启动开始运行。

当程序界面的 View 对象产生的时候会运行 onSizeChanged()方法，在这个方法中我们计算椭圆的中心点和最大长度，以及设置第一次要绘制的椭圆长度。由于椭圆有变大和变小两种模式，我们用一个 Sign 变量来控制。最后在 onDraw()方法中先绘制一个实心椭圆，再绘制一个较小的背景颜色的椭圆，这样就可以做到空心椭圆形的效果，然后检查是否需要切换变大和变小模式，最后设置下一次绘制的椭圆大小。

Step 3 开启主程序文件 Main.java，在 onCreate()方法中建立一个步骤二类的对象，并将它设置为程序的界面：

```
package tw.android;

import android.app.Activity;
import android.os.Bundle;

public class Main extends Activity {
    /** Called when the activity is first created. */
    @Override
    public void onCreate(Bundle savedInstanceState) {
        super.onCreate(savedInstanceState);

        ShapeView shapeView = new ShapeView(this);
        setContentView(shapeView);
    }
}
```

完成之后就可以运行程序观察动画的效果，读者会发现椭圆变化的过程不是很平顺，因为程序绘图的时间点是由系统控制，我们可以根据上面程序代码中所加入的 log 信息来观察调用 invalidate()方法和运行 onDraw()方法之间的对应关系。请读者依照 UNIT 7 介绍的方式，一边运行程序，一边观察 Eclipse 的 Debug 界面右下方所显示的 log（操作提示：可以设置 log filter 来筛选出我们的程序所产生的 log 以方便观察），读者可以发现大部分的重绘界面要求都被系统忽略，因此造成动画效果不佳。为了能够达到更好的动画效果，我们可以使用下一个单元介绍的 SurfaceView 类。

UNIT 65
使用 SurfaceView 进行高速绘图

Android 版本	1.X	2.X	3.X	4.X
适用性	★	★	★	★

从上一个单元的范例程序可以发现，使用 View 的绘图速度不太能够令人满意，主要的原因在于 Android 系统忽略了大部分的 invalidate() 重绘程序界面的要求。这并不是 Android 系统独有的问题，所有图形接口的操作系统都有相同的特性，包括 MS Windows 也是如此。因为重绘界面的工作在操作系统中的优先权比较低，必要时系统会忽略重绘界面的要求以节省处理时间。这样的逻辑对于一般应用程序来说是对的，因为一般应用程序的核心功能并不是界面的显示，而是内部的运算过程。但是对于以绘图为主要任务的程序来说（例如计算机游戏、动画制作软件），这样的绘图效率就无法满足要求，为了这一类需要高速绘图的应用程序，Android 系统提供另一种绘图的方法，那就是使用 SurfaceView。

65-1 使用 SurfaceView 的步骤

SurfaceView 的概念其实很简单，相较于上一个单元的 View 类是由 Android 系统代为调用 onDraw() 方法进行绘图，SurfaceView 的作法是把程序界面的 Canvas 对象直接开放给程序自由存取，因此程序可以自己决定什么时候要绘图，当程序完成绘图后再把 Canvas 对象交给 Android 系统显示，要使用 SurfaceView 必须完成以下步骤：

Step 1 在程序项目中新增一个继承 SurfaceView 的新类。
Step 2 这个新类必须实现 SurfaceHolder.Callback 接口和 Runnable 接口如下：

```java
public class ShapeSurfaceView extends SurfaceView
                implements SurfaceHolder.Callback, Runnable {

    @Override
    public void surfaceChanged(SurfaceHolder holder, int format, int width, int height) {
        // TODO Auto-generated method stub
        // 当程序界面的大小改变时运行
    }

    @Override
    public void surfaceCreated(SurfaceHolder holder) {
        // TODO Auto-generated method stub
        // 当程序界面的 Canvas 被建立时运行
    }

    @Override
    public void surfaceDestroyed(SurfaceHolder holder) {
        // TODO Auto-generated method stub
        // 当程序界面的 Canvas 被销毁时运行
    }

    public void run() {

    }
}
```

实现 SurfaceHolder.Callback 接口的原因是我们需要使用 SurfaceHolder 对象来存取程序的 Canvas，而建立 SurfaceHolder 对象的时候需要把 surfaceChanged()、surfaceCreated()和 surfaceDestroyed()这三个方法当成是 SurfaceHolder 对象的 callback 函数（也就是 SurfaceHolder 对象会主动回来调用这三个函数），这三个 callback 函数的运行时机请参考以上程序代码范例中的批注。至于实现 Runnable 接口是因为要建立 Multi-thread 程序，我们把调用绘图函数的程序代码写在 run()方法中让它连续运行以产生动画效果。

Step 3 在这个新类中建立一个 SurfaceHolder 对象，这个 SurfaceHolder 对象将用来取得程序的 Canvas。另外我们必须将步骤二的 3 个 callback 函式设置给该 SurfaceHolder 对象，请读者参考后面的程序项目。

Step 4 在主程序中建立一个步骤一类的对象，并将它设置成程序的 View。

接下来我们用一个完整的程序项目来示范 SurfaceView 的绘图效果。

65-2 范例程序

为了比较 View 和 SurfaceView 的绘图效率，我们将建立和上一个单元相同的椭圆动画程序，请读者依照下列步骤操作。

Step 1 依照前面惯用的操作方法新增一个 Android 程序项目。

Step 2 在程序项目中新增一个继承 SurfaceView 的新类，我们可以将它取名为 ShapeSurfaceView。开启这个新类的程序文件，其中会标示一个红色波浪底线，将鼠标光标移到上面会弹出一个信息要求加入一个构造方法，请读者选择加入第一个构造方法。接下来我们要新增一个 draw() 方法，请在程序代码编辑窗口中右击，在出现的快捷菜单中选择 Source > Override/Implement Methods…就会出现一个方法清单的对话框，勾选其中的 draw() 然后按下 OK 按钮。

接下来依照上一个小节步骤二的说明，让 ShapeSurfaceView 类实现 SurfaceHolder.Callback 接口和 Runnable 接口，再依照程序代码编辑窗口的错误修正建议，加入需要实现的方法，完成后我们将此类的程序代码编辑如下：

```
package …

import …

public class ShapeSurfaceView extends SurfaceView
implements SurfaceHolder.Callback, Runnable {

    private Paint mPaintForeColor,
            mPaintBackColor;
    private static final int INT_STROCK_THICK = 2;
    private int mIntXMaxLen, mIntYMaxLen,
            mIntXCent, mIntYCent,
            mIntXCurLen, mIntYCurLen,
            mIntSign;

    private SurfaceHolder mSurfHold;

    public ShapeSurfaceView(Context context) {
        super(context);
        // TODO Auto-generated constructor stub

        mSurfHold = getHolder();
        mSurfHold.addCallback(this);

        setFocusable(true);

        mPaintForeColor = new Paint();
        mPaintForeColor.setAntiAlias(true);
        mPaintForeColor.setColor(Color.CYAN);

        mPaintBackColor = new Paint();
```

```java
        mPaintBackColor.setAntiAlias(true);
        mPaintBackColor.setColor(Color.BLACK);
    }

    public void draw(Canvas canvas) {
        // TODO Auto-generated method stub
        super.draw(canvas);

        Log.d("TEST_SURFACEVIEW", "draw()");

        canvas.drawOval(new RectF(mIntXCent - mIntXCurLen,
                                  mIntYCent - mIntYCurLen,
                                  mIntXCent + mIntXCurLen,
                                  mIntYCent + mIntYCurLen),
                mPaintForeColor);
        canvas.drawOval(new RectF(mIntXCent - mIntXCurLen +
                                  INT_STROCK_THICK,
                                  mIntYCent - mIntYCurLen +
                                  INT_STROCK_THICK,
                                  mIntXCent + mIntXCurLen -
                                  INT_STROCK_THICK,
                                  mIntYCent + mIntYCurLen -
                                  INT_STROCK_THICK),
                mPaintBackColor);

        if (mIntXCurLen + mIntSign * INT_STROCK_THICK < 0 ||
            mIntYCurLen + mIntSign * INT_STROCK_THICK < 0) {
            mIntSign = 1;
        }
        else if (mIntXCurLen + mIntSign * INT_STROCK_THICK >
                mIntXMaxLen ||
                mIntYCurLen + mIntSign * INT_STROCK_THICK >
                mIntYMaxLen) {
            mIntSign = -1;
        }

        mIntXCurLen += mIntSign * INT_STROCK_THICK;
        mIntYCurLen += mIntSign * INT_STROCK_THICK;
    }

    public void run() {
        for (int i = 0; i < 500; i++) {
            Log.d("TEST_SURFACEVIEW", "run() " + i);

            Canvas c = null;
            try {
                c = mSurfHold.lockCanvas();
                synchronized(mSurfHold) {
```

```
                    draw(c);
                }
            }
            finally {
                if (c != null)
                    mSurfHold.unlockCanvasAndPost(c);
            }
        }
    }

    @Override
    public void surfaceChanged(SurfaceHolder holder, int format,
    int width, int height) {
        // TODO Auto-generated method stub

        mIntXMaxLen = width / 2 - 10;
        mIntYMaxLen = height / 2 - 10;
        mIntXCent = width / 2;
        mIntYCent = height / 2;

        mIntXCurLen = mIntXMaxLen;
        mIntYCurLen = mIntYMaxLen;
        mIntSign = -1;
    }

    @Override
    public void surfaceCreated(SurfaceHolder holder) {
        // TODO Auto-generated method stub
        new Thread(this).start();
    }

    @Override
    public void surfaceDestroyed(SurfaceHolder holder) {
        // TODO Auto-generated method stub

    }
}
```

首先如上一小节步骤三的说明，我们定义一个 SurfaceHolder 对象，然后在类的构造方法中调用 getHolder() 方法从系统取得该对象，接着调用 addCallback() 方法设置 callback 函式。另外我们把绘制椭圆的程序代码写在 draw() 方法中，这个方法将由我们的程序直接调用运行。绘制椭圆的程序代码都和上一个单元相同，因此不再赘述请读者自行参考。

补充说明

使用 SurfaceView 类时 Android 系统就不再调用 onDraw() 方法，所有绘图都由程序自己负责。

在run()方法中我们用一个循环连续调用draw()方法进行绘图。要取得程序的Canvas只要调用SurfaceHolder对象的lockCanvas()方法即可。另外还有两点要注意，在Canvas绘图的过程中不可以有其他存取Canvas的行为，因此我们加上synchronized指令。另外我们用try-catch-finally语法结构把Canvas绘图程序代码包裹起来进行例外处理。

在surfaceChanged()方法中设置椭圆形的绘制参数，在surfaceCreated()方法中启动负责绘图的Thread（也就是开始运行run()方法）。

Step 3　开启"src/(组件路径名称)"文件夹中的主程序文件Main.java，在onCreate()方法中建立一个步骤二类的对象，并将它设置为程序的界面：

```
package tw.android;

import android.app.Activity;
import android.os.Bundle;

public class Main extends Activity {
    /** Called when the activity is first created. */
    @Override
    public void onCreate(Bundle savedInstanceState) {
        super.onCreate(savedInstanceState);

        ShapeSurfaceView shapeSurfView = new ShapeSurfaceView(this);
        setContentView(shapeSurfView);
    }
}
```

完成程序代码编辑之后运行程序将会看到图65-1的运行界面，和上一个单元的程序比较绘图速度，读者将会发现二者的绘图效率有很大的差别。如果将Eclipse切换到Debug界面观察程序产生的log信息，会发现这个使用SurfaceView的程序运行draw()方法的频率远比上一个单元使用View的程序要高出许多，这就是让程序自己控制Canvas的好处。

图65-1　绘图程序在手机运行的界面

Android 版本	1.X	2.X	3.X	4.X
适用性	★	★	★	★

UNIT 66

3D 绘图

由于计算机处理器和绘图芯片的效能日益精进，3D 绘图程序的应用越来越普遍，除了 3D 游戏的数量大幅增加之外，甚至连一些操作系统和应用程序的操作接口也开始使用 3D 技术，这个风潮也蔓延到手机和平板电脑平台，像 Android 这种新兴的操作系统自然也内建了 3D 绘图功能，因此这个单元就让我们来探索一下 3D 绘图的基本概念和程序架构。Android 系统的 3D 绘图是由 OpenGL ES 链接库负责，OpenGL 是一套非常有名的 3D 绘图组件，它最早的版本是在 1992 年由 SGI 公司发布，SGI 公司是以研发高性能的绘图工作站闻名，后来该公司将 OpenGL 公开成为开放源代码的软件，经过开放源代码社群的努力，现在 OpenGL 已经成为一套可以在多种平台使用的 3D 绘图组件。目前 OpenGL 的最新版本是 4.2，而 OpenGL ES 是专供 Embeded System（嵌入式系统）使用的版本，它是由 OpenGL 标准版精简而来，Android 4 目前已经支持 OpenGL ES 2.0。在简单介绍了 OpenGL 的发展史之后，接下来就让我们开始进入 3D 绘图的领域，不过在学习 3D 绘图程序之前，我们必须先了解 3D 绘图的基本概念。

66-1　3D 绘图的基本概念

如果读者以前没有写过 3D 绘图程序，对于究竟如何建立 3D 场景和对象一定觉得很好奇，我们用图 66-1 来解释程序如何使用 3D 场景。

1. 世界空间（world space）

    图 66-1 中的 xyz 坐标轴代表世界空间，世界空间就是程序中使用 OpenGL 时所建立的 3D 场景，在初始状态下，程序所看见的投影界面是在 xyz 坐标轴的原点，投影界面的右手边是 x 轴的正向，上方是 y 轴的正向，后方是 z 轴的正向。

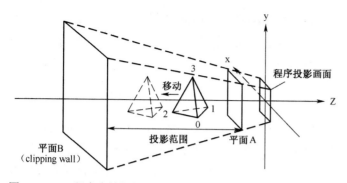

图 66-1　3D 程序中的场景

2. 投影范围（可视范围）

在世界空间中只有位于投影范围内的物体才会出现在程序的投影界面中，世界空间中的投影范围是一个立体的梯形，请读者参考图 66-1，最前面是平面 A（靠近程序的投影界面），最后面是平面 B（离程序的投影界面最远），把平面 A 和平面 B 以直线将对应的顶点连接起来就是投影范围，平面 A 和平面 B 离原点的距离可由程序设置。

3. 3D 对象

世界空间中的 3D 对象是用构成它的表面的顶点（vertex）来建立。所有 3D 对象的表面都是用 polygon（多边形）一片一片连结而成，polygon 愈细小所建立的 3D 对象表面就愈精细。而 triangle（三角形）是 polygon 中最简单的一种，OpenGL ES 限制只能使用 triangle。每一个 triangle 都有三个顶点，图 66-1 中的三角锥的表面是由 4 个 triangle 所组成，为了要代表这 4 个 triangle，我们必须记录 12 个顶点的信息（4 个 triangle 乘上每一个 triangle 需要 3 个顶点），但是实际的顶点数目只有 4 个，也就是说会有顶点数据重复的情况。为了解决这个问题，3D 对象实际上是利用一个顶点索引数组来建立，通过这个顶点索引数组来取用实际的顶点数组，如此就不必重复记录顶点数据。每一个顶点的数据除了坐标之外，还可以包括颜色、法向量等，以表现物体的颜色、光线反射效果等特性。

4. triangle 的正面和反面

在世界空间中的 triangle 还有正面（front face）和反面（backface）的区别，只有正面朝向投影界面的 triangle 才会显示在投影界面中。决定 triangle 正反面的方式是看它的顶点排列顺序，当我们面对一个 triangle 时，如果它的顶点排列顺序是逆时针则代表正面朝向我们，如果是顺时针则代表现在是反面朝向我们，例如图 66-1 中的三角锥的 triangle (1, 3, 0)对于投影界面来说是正面，因此它会出现在投影界面中，如果在程序中错误地将它写成(1, 0, 3)就会看不见它，因而在程序的投影界面中显示不完整的三角锥。

5. 物体的移动和转动

如果读者对于 3D 空间的数学运算还有印象（如果没有也无大碍不必担心），就能了解 3D 物体的移动、转动、缩放和投影都是二维矩阵运算的结果，例如在图 66-1 中将立体三角锥往远处移动（减少 z 坐标的值）就可以利用 translation matrix 来完成，如果要旋转则可以利用 rotation matrix 来完成，另外如果要把物体投影到程序的界面，则是利用 projection matrix，这些转换矩阵在 OpenGL 中都有对应的函数可以使用，我们只要在调用这些函数的时候传入适当的参数即可。

6. z-buffer（或称为 depth-buffer）

   要计算多个 3D 物体在投影界面上所形成的图像时是一个一个物体依序计算，但是物体在 3D 空间中会有前面物体遮蔽后面物体的情况，所以在投影的过程中必须记录每一个投影界面的像素它的图像来源的远近，这样当有下一个物体投影到该像素时才能判断是否要覆盖原来的值，这个记录投影距离的内存空间称为 z-buffer，z-buffer 的精确度（bit 数）会影响计算物体遮蔽效果的正确性。

根据以上说明相信读者对于程序如何建立 3D 场景和进行相关运算已经具备基本的概念，接下来就让我们开始学习 3D 绘图程序的实现方法。

66-2　3D 绘图程序

3D 绘图程序的架构和前一个单元的 SurfaceView 绘图程序有些类似，甚至简单一些。在 SurfaceView 绘图程序中必须另外建立一个 thread 来连续调用我们自己的 draw()方法，但是 3D 绘图程序是由系统自动连续运行我们建立的 onDrawFrame()方法，因此不必再自己建立 thread 来运行绘图的工作。另外和 SurfaceView 绘图程序不同的是，3D 绘图程序必须换成使用 GLSurfaceView 类并且实现 GLSurfaceView.Renderer 接口，以下是建立一个 3D 绘图程序项目的完整过程。

Step 1　依照前面惯用的操作方法新增一个 Android 程序项目。

Step 2　在程序项目中新增一个继承 GLSurfaceView 的新类，我们可以将它取名为 MyGLSurfaceView。开启这个新类的程序文件，其中会标示一个红色波浪底线，将鼠标光标移到上面会弹出一个信息要求加入一个构造方法，请读者选择加入第一个构造方法。接下来让 MyGLSurfaceView 类实现 GLSurfaceView.Renderer 接口，当读者在类名称那一行输入 "implements GLSurfaceView.Renderer" 之后会出现红色波浪底线提示语法错误，请将鼠标光标移到错误的地方就会弹出一个信息窗口，然后单击其中的 "Add unimplemented methods"，就会在程序代码中自动加入需要实现的方法。不过这些自动加入的方法所用的自变量名称有些是以 arg 命名，建议读者参考 SDK 在线文件中的函式原型，把这些自变量改成有意义的名称，然后将程序代码编辑成如下的内容，这个程序文件中包含所有 3D 绘图的程序代码，因此我们要好好检查一番：

```
package …

import …
```

```java
public class MyGLSurfView extends GLSurfaceView
                implements GLSurfaceView.Renderer {

    // 储存 OpenGL 对象用到的顶点坐标和颜色
    private FloatBuffer mVertBuf,
                mVertColorBuf;

    // 储存 OpenGL 对象的顶点在 mVertBuf 中的索引
    private ShortBuffer mIndexBuf;

    // OpenGL 对象用到的顶点坐标，在程序中会将它们设置给 mVertBuf
    // 每一列是一个顶点的 xyz 坐标
    private float[] m3DObjVert = {
        -0.5f, -0.5f, 0.5f,
        0.5f, -0.5f, 0.5f,
        0f, -0.5f, -0.5f,
        0f, 0.5f, 0f
    };

    // OpenGL 对象用到的顶点的颜色，在程序中会将它们设置给 mVertColorBuf
    // 每一列是一个顶点的颜色，颜色值的顺序为 rgba，a 是 alpha 值
    private float[] m3DObjVertColor = {
        1.0f, 1.0f, 0.0f, 1.0f,
        1.0f, 0.0f, 1.0f, 1.0f,
        0.0f, 1.0f, 1.0f, 1.0f,
        1.0f, 1.0f, 0.5f, 1.0f
    };

    // OpenGL 对象的顶点在 m3DObjVert 中的索引，每一列代表一个 triangle
    // 程序中会将它们设置给 mIndexBuf
    private short[] m3DObjVertIndex = {
        0, 2, 1,
        3, 2, 0,
        3, 1, 2,
        1, 3, 0
    };

    // 最远方的底色
    private float backColorR = 0.3f,
                backColorG = 0.3f,
                backColorB = 0.3f,
                backColorA = 1.0f;

    private float mfRotaAng = 0f;

    private void setup() {
        // 建立 OpenGL 专用的 vertex buffer
        ByteBuffer vertBuf = ByteBuffer.allocateDirect(
```

```
            4 * m3DObjVert.length);
        vertBuf.order(ByteOrder.nativeOrder());
        mVertBuf = vertBuf.asFloatBuffer();

        // 建立 OpenGL 专用的 vertex color buffer
        ByteBuffer vertColorBuf = ByteBuffer.allocateDirect(
            4 * m3DObjVertColor.length);
        vertColorBuf.order(ByteOrder.nativeOrder());
        mVertColorBuf = vertColorBuf.asFloatBuffer();

        // 建立 OpenGL 专用的 index buffer
        ByteBuffer indexBuf = ByteBuffer.allocateDirect(
            2 * m3DObjVertIndex.length);
        indexBuf.order(ByteOrder.nativeOrder());
        mIndexBuf = indexBuf.asShortBuffer();

        mVertBuf.put(m3DObjVert);
        mVertColorBuf.put(m3DObjVertColor);
        mIndexBuf.put(m3DObjVertIndex);

        mVertBuf.position(0);
        mVertColorBuf.position(0);
        mIndexBuf.position(0);
    }

    public MyGLSurfView(Context context) {
        super(context);
        // TODO Auto-generated constructor stub

        setRenderer(this);
    }

    @Override
    public void onDrawFrame(GL10 gl) {
        // TODO Auto-generated method stub

        // 清除场景填上背景颜色,并且清除 z-buffer
        gl.glClear(GL10.GL_COLOR_BUFFER_BIT |
            GL10.GL_DEPTH_BUFFER_BIT);

        // 设置对象用到的顶点坐标
        gl.glVertexPointer(3, GL10.GL_FLOAT, 0, mVertBuf);

        // 设置对象用到的顶点的颜色
        gl.glColorPointer(4, GL10.GL_FLOAT, 0, mVertColorBuf);

        gl.glLoadIdentity();
```

```java
        // 将对象沿指定的轴移动
        gl.glTranslatef(0f, 0f, -4f);

        // 将对象沿指定的轴转动
        gl.glRotatef(mfRotaAng, 0f, 1f, 0f);
        mfRotaAng += 1f;

        // 设置对象顶点的索引并绘制对象
        gl.glDrawElements(GL10.GL_TRIANGLES, m3DObjVertIndex.length,
                    GL10.GL_UNSIGNED_SHORT, mIndexBuf);
    }

    @Override
    public void onSurfaceChanged(GL10 gl, int width, int height) {
        // TODO Auto-generated method stub

        // 设置透视投影参数
        // 设置最近和最远的可视范围和左右视角
        // 上下可视范围由左右视角和屏幕的宽高比来计算
        final float fNEAREST = .01f,
                    fFAREST = 100f,
                    fVIEW_ANGLE = 45f;
        gl.glMatrixMode(GL10.GL_PROJECTION);  // 切换到投影矩阵模式
        float fViewWidth = fNEAREST * (float) Math.tan(Math.
        toRadians(fVIEW_ANGLE) / 2);
        float aspectRatio = (float)width / (float)height;
        gl.glFrustumf(-fViewWidth, fViewWidth,
                    -fViewWidth / aspectRatio, fViewWidth /
                    aspectRatio, fNEAREST, fFAREST);
        gl.glMatrixMode(GL10.GL_MODELVIEW);    // 切换到原来模式

        gl.glViewport(0, 0, width, height);
    }

    @Override
    public void onSurfaceCreated(GL10 gl, EGLConfig config) {
        // TODO Auto-generated method stub

        // 设置 3D 场景的背景颜色，也就是 clipping wall 的颜色
        gl.glClearColor(backColorR, backColorG, backColorB,
        backColorA);

        // 设置 OpenGL 的功能
        gl.glEnable(GL10.GL_DEPTH_TEST); // 物体远近的遮蔽效果
        gl.glEnable(GL10.GL_CULL_FACE);  // 区分 Triangle 的正反面
        gl.glFrontFace(GL10.GL_CCW);     // 逆时针顶点顺序为正面
        gl.glCullFace(GL10.GL_BACK);     // 反面的 Triangle 不显示
        gl.glEnableClientState(GL10.GL_VERTEX_ARRAY);
```

```
    // 使用顶点数组
    gl.glEnableClientState(GL10.GL_COLOR_ARRAY);
    // 使用颜色数组

    setup();
    }
}
```

为了让读者比较容易了解每一段程序代码的功能，我们特别加入适当的程序批注，其中比较重要的部分以粗体字标示，接下来我们依照程序代码的先后顺序来加以说明。首先在类的开头定义需要用到的顶点坐标和颜色的内存以及顶点索引内存，接下来是列出所有对象用到的顶点坐标，每一列是一个顶点的 xyz 坐标（提示：在 OpenGL 中所有的浮点数都用 float 的格式储存以节省内存空间）。接下来是指定每一个顶点的颜色，然后就是 3D 对象表面的每一个 triangle 的顶点索引，这里再一次提醒读者，必须用逆时针的方式列出顶点顺序。接着是指定投影范围最远方的底色，也就是图 66-1 中的平面 B 的颜色，然后是程序中用来控制物体旋转角度的变量。

setup() 方法是用来设置 OpenGL 使用的顶点数组、顶点颜色数组和顶点索引数组，这些数组必须依照 OpenGL 的规定用 byte 的原始类型代表，程序代码中的常数 4 代表 float 类型的数据是占 4 个 byte，2 代表 short 类型的数据是占 2 个 byte。接下来在类的构造方法中是将目前这个类设置成负责产生投影界面的对象（称为 renderer），因为我们已经让这个类实现 GLSurfaceView.Renderer 界面。

onDrawFrame() 方法会由系统自动连续运行，它的工作就是在世界空间中建立 3D 对象。首先我们清除界面和 z-buffer，然后设置使用的顶点坐标数组和顶点颜色数组，接下来调用 glLoadIdentity() 方法清除之前设置的转换矩阵，然后再设置新的转换矩阵，如果没有运行 glLoadIdentity() 则会累积之前的转换矩阵。请读者注意转换矩阵设置的先后顺序会造成不同的运行结果（这是数学矩阵运算的特性），OpenGL 会先运行后面设置的转换矩阵。以上面的程序代码为例，会先对物体进行旋转再移动。如果将转换矩阵的次序对调，运行结果将大不相同，读者可自行测试。最后利用 glDrawElements() 方法传入 3D 物体的顶点索引数组把该 3D 物体绘制出来。

接下来我们先说明后面的 onSurfaceCreated() 方法，因为当建立 3D 场景时它会先运行。首先是指定 3D 场景在投影界面的背景颜色，然后设置 OpenGL 的功能，最后调用前面的 setup() 方法完成 3D 场景的建立。系统运行完 onSurfaceCreated() 之后会再调用 onSurfaceChanged() 方法传入屏幕的宽和高，我们根据屏幕的宽和高设置好投影参数，最后设置观看窗口的位置和大小。

Step 3 开启 "src/(组件路径名称)" 文件夹中的主程序文件 Main.java，在 onCreate() 方法中建立一个步骤二类的对象，并将它设置为程序的界面：

```
package tw.android;

import android.app.Activity;
import android.os.Bundle;

public class Main extends Activity {
    /** Called when the activity is first created. */
    @Override
    public void onCreate(Bundle savedInstanceState) {
        super.onCreate(savedInstanceState);

        MyGLSurfaceView glSurfView = new MyGLSurfaceView(this);
        setContentView(glSurfView);
    }
}
```

完成程序代码的编辑之后运行程序项目就会看到模拟器的屏幕上出现一个彩色的立体三角锥，如图 66-2 所示，该三角锥会以 y 轴为中心自转。读者看到自己辛苦完成的 3D 程序界面，心中是不是感受到一股成就感！在 3D 绘图的领域里，这个程序只不过是一个开端，还有许多有趣的主题值得进一步学习，像是建立光源、设置物体表面的反射、镜射效果、材质贴图、碰撞侦测等，由于本书的篇幅有限只能够点到为止，如果读者有兴趣可以进一步阅读相关的数据。

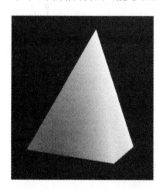

图 66-2　旋转立体三角锥的 3D 绘图程序

PART 13　Google 地图程序

UNIT 67　使用 Google 地图
UNIT 68　Google 地图的进阶用法
UNIT 69　帮地图加上标记
UNIT 70　加上定位让地图活起来

Android 版本	1.X	2.X	3.X	4.X
适用性	★	★	★	★

UNIT 67

使用 Google 地图

Google 的地图服务功能非常强大，包括传统的街道图和高科技的卫星图像地图，而且可以根据不同的交通方式做路径规划，也可以让使用者自行加上标示，甚至还有 3D 地图的版本。如果将地图的应用和手机以及平板电脑的便利性结合，将可以创造出许多生活化的应用，包括食衣住行各方面，一切就看程序设计者的巧思。

67-1 开发 Google 地图应用程序的准备工作

在开始建立 Google 地图应用程序之前，我们必须先了解如何启用 Google 地图服务以及查询地理位置坐标的方法。

1. Google 地图应用程序必须使用 Google APIs

   一般的 Android 程序只需要 Android 平台就可以运行，但是如果要使用 Google 地图功能，就必须再加上 Google APIs，所以我们必须在 Eclipse 中安装 Google APIs 项目才能够开发 Google 地图程序。请读者从 Eclipse 的主菜单中选取 Window > Android SDK Manager 就会出现图 67-1 的对话框，在对话框中每一个 Android 版本都有 SDK Platform 和 Google APIs by Google Inc.两个项目，如果 Google APIs by Google Inc.项目的右边显示 Installed，代表已经安装好 Google APIs 功能，否则请读者勾选所需版本的 Google APIs by Google Inc.项目，然后按下右下角的 Install package 按钮进行安装。

2. 建立运行 Google APIs 应用程序的模拟器

   Google 地图程序必须在包含 Google APIs 的模拟器上才能运行，请读者从 Eclipse 主菜单的 Window > AVD Manager 开启 AVD Manager 对话框，按下对话框右上角的 New 按钮新增一个模拟器，请将此模拟器的 Target 设置为想要的 Google APIs 版本，其他项目依照以前的惯例设置即可。

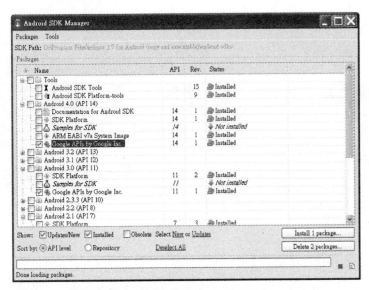

图 67-1　检查 Eclipse 中已经安装的 Android 组件

3. **KML 和 KMZ**

   KML（Keyhole Markup Language）是 Google 针对地图的应用所制订的 xml 信息交换格式，主要的功能是记录地理位置坐标、位置名称等信息供 Google 地图使用。KMZ 则是将 KML 文件和它所用到的图像文件一起压缩成一个文件，再交给 Google 地图使用。

4. **取得使用 Google 地图的 API Key**

   如果程序要使用 Google 地图服务，必须先申请一个 API Key。申请 API Key 需要两个数据：一是程序开发者的数字签名，二是程序开法者的 Google 账号。在程序开发的过程中我们可以暂时使用 Debug 模式的数字签名，也就是 debug.keystore 文件。首先我们必须找出该文件的位置，然后取得它的 MD5 编码，最后到 Google 网站申请 API Key，完整的操作步骤如下：

   a. 请选择 Eclipse 主菜单的 Window > Preferences，在对话框的左边展开 Android 项目，再单击其中的 Build 项目就会在右边的 Default debug keystore 字段中显示 debug.keystore 文件文件的位置。

   b. 请运行 Windows 文件管理器，把 debug.keystore 文件复制到 JRE 安装目录下的 bin 子目录中（默认路径为 c:\Program Files\Java\jre6\bin）。

   c. 从"开始"菜单的"所有程序 > 附属应用程序 > 命令提示字符"开启"命令行窗口"。

   d. 在"命令行窗口"中输入下列指令，把所在路径切换到 JRE 目录下的 bin 子目录：

   ```
   cd c:\Program Files\Java\jre6\bin
   ```

接着再运行以下指令：

> keytool -list -keystore debug.keystore

并输入默认密码 android 就会看到如图 67-2 所示的结果，请读者将"认证指纹（MD5）"该行的整串编码记录下来（操作提示：可以单击"命令行窗口"左上角的小图标再从菜单中选择"编辑 > 标记"，然后用鼠标按住拖曳的方式将该串编码反白，最后按下 Enter键完成复制）。

图 67-2　取得 debug.keystore 数字签名文件 MD5 编码的界面

补充说明

如果是 JRE 7 的版本，运行 keytool 指令时必须加上 -v 参数，也就是：
> keytool -v -list -keystore debug.keystore。

e. 开启网页浏览器，在网址列输入：

> http://code.google.com/android/add-ons/google-apis/maps-api-signup.html

就会出现图 67-3 的网页，请勾选下方的同意项目，然后在 My certificate's MD5 fingerprint 字段中输入上一个步骤所得到的 MD5 编码（可以用贴上功能），最后按下 Generate API Key 按钮。

f. 在出现的网页中输入 Google 账号和密码便会显示图 67-4 的界面，其中的 Your key is 下方的那一串编码就是程序需要的 API Key，请将它复制下来并储存备用。

5. 查询全球各地经纬度地理坐标的方法

最基本的 Google 地图使用方式就是设置一个经纬度地理坐标，然后 Google 地图上就会显示该坐标的平面图。那么要如何得到特定地点的地理坐标呢？以下网址可以提供地理坐标的查询功能：

http://www.mygeoposition.com/

请在网页浏览器的网址列中输入以上网址就会看到如图 67-5 所示的界面，在上方的文字编辑字段中输入要查询的地点名称，然后按下 Calculate geodata 按钮，就会在下方的地图中显示该地点的经纬度资料。其中 Latitude（简称 Lat）是纬度坐标、Longitude（简称 Lon）是经度坐标。由于经纬度数字有一长串的小数，如果要用复制的功能可以把地图上方的标签页切换到 KML，然后复制其中的<coordinates>标签的内容。

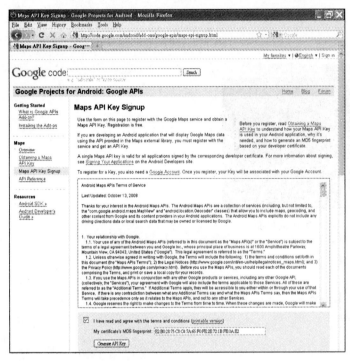

图 67-3　申请 Google API Key 的网页

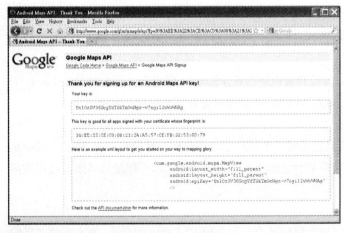

图 67-4　最后获得 Google API Key 的网页

Google 地图程序 PART 13

图 67-5　查询全球各地经纬度地理坐标的网站

以上的步骤有些冗长，但是只需要做一次，以后就可以重复使用您的 API Key。接下来我们开始介绍如何建立 Google 地图应用程序。

67-2　建立 Google 地图应用程序的步骤

建立 Google 地图应用程序的过程比前面的准备工作简单一些，请读者依照下列步骤操作：

Step 1 新增一个 Android 程序项目，项目属性对话框中的 Build Target 字段必须选择一个 Google APIs 版本而不是之前我们惯用的 Android 版本。

Step 2 在 res/layout/main.xml 界面布局文件中新增一个 <com.google.android.maps.MapView> 组件，并设置好它的属性如下：

```
<com.google.android.maps.MapView android:id="@+id/（自定义的组件 id 名称）"
    android:layout_width="match_parent"
    android:layout_height="match_parent"
    android:apiKey="（前面得到的 API Key 编码）"
/>
```

Step 3 开启"src/(组件路径名称)"文件夹中的主程序文件，将它的继承类由 Activity 改成 MapActivity，然后依照编辑器的语法错误修正提示新增一个 isRouteDisplayed() 方法如下：

```
public class Main extends MapActivity {
    public void onCreate(Bundle savedInstanceState) {
        …
```

513

```
    }

    protected boolean isRouteDisplayed() {
        // TODO Auto-generated method stub
        return false;
    }
}
```

Step 4 在 onCreate()方法中取得接口布局文件中的 MapView 组件，然后再取得 MapView 组件的 Controller 并设置地图的缩放比例，接着将要显示的地理位置坐标建立成一个 GeoPoint 对象，地理位置坐标必须乘上一百万倍，也就是 1,000,000 或是 1e6 并转成整数类型，再调用 animateTo()方法传给 Map Controller，如以下程序代码范例：

```
public void onCreate(Bundle savedInstanceState) {
...
    MapView map = (MapView) findViewById(R.id.map);

    MapController mapCtrl = map.getController();
    mapCtrl.setZoom(18);

    double dLat = 25.019943,
            dLon = 121.542353;
    GeoPoint gp = new GeoPoint((int)(dLat * 1e6), (int)(dLon * 1e6));
    mapCtrl.animateTo(gp);
}
```

Step 5 开启程序项目的功能描述文件 AndroidManifest.xml，新增以下粗体标示的程序代码，这二行程序代码的功能是允许程序使用网络以及设置所使用的链接库。

```
<?xml version="1.0" encoding="utf-8"?>
<manifest ...>
    <uses-permission android:name="android.permission.INTERNET" />
    <application ...>
        <uses-library android:name="com.google.android.maps" />
        <activity ...>
            ...
        </activity>

    </application>
</manifest>
```

完成以上步骤之后就可以让程序显示 Google 地图，接下来我们用一个实际的程序项目来示范 Google 地图的运行效果。

67-3 范例程序

这个 Google 地图程序会显示一个 spinner 下拉式菜单，其中有 4 所大学的选项如图 67-6 所示，当使用者选择其中一所学校之后，程序界面会显示该学校的卫星图像地图，建立这个程序项目的步骤如下：

图 67-6　Google 地图程序的地点菜单

Step 1　依照前一小节介绍的方法新增一个 Android 程序项目，在项目对话框中的 Build Target 项目选择一个 Google APIs 版本，其他设置依照惯例即可。

Step 2　开启接口布局文件并建立一个 Spinner 和一个 com.google.android.maps.MapView 组件，然后设置好属性如下：

```
<?xml version="1.0" encoding="utf-8"?>
<LinearLayout …
    android:orientation="vertical"
    android:layout_width="match_parent"
    android:layout_height="match_parent"
    android:gravity="center_horizontal"
    >
<Spinner android:id="@+id/spnLocation"
    android:layout_width="wrap_content"
    android:layout_height="wrap_content"
    android:drawSelectorOnTop="true"
android:prompt="@string/prompt_select_location"
    />
<com.google.android.maps.MapView android:id="@+id/map"
    android:layout_width="match_parent"
    android:layout_height="match_parent"
    android:apiKey="（您的 API Key 编码）"
    />
</LinearLayout>
```

在 Spinner 的 prompt 属性中我们使用一个定义在 res/values/strings.xml 字符串资源文件中名为 prompt_select_location 的字符串：

```xml
<?xml version="1.0" encoding="utf-8"?>
<resources>
    …
    <string name="prompt_select_location" >请选择地点</string>
</resources>
```

> **补充说明**
>
> 在测试程序的时候可以使用 debug.keystore 的 API Key，如果是要发布已经开发好的程序，就必须换成程序设计者自己的数字签名的 API Key，要得到自己的 API Key 只要在运行 keytool 指令的时候使用在 UNIT 61 产生的 keystore 数字签名文件即可。

Step 3 开启"src/(组件路径名称)"文件夹中的主程序文件，把它的继承类改成 MapActivity，程序编辑器会标示一个红色波浪底线的语法错误，将鼠标光标移到该错误的地方，然后在弹出的信息窗口中选择 Add unimplemented methods，就会自动在程序代码中插入一个 isRouteDisplayed()方法，然后将程序代码编辑如下：

```java
package …

import …

public class Main extends MapActivity {

    private static String[][] locations = {
            {"台湾大学", "25.019943,121.542353"},
            {"清华大学", "24.795621,120.998153"},
            {"交通大学", "24.791704,121.003341"},
            {"成功大学", "23.000875,120.218017"}};

    private Spinner mSpnLocation;
    private MapView mMapView;
    private MapController mMapCtrl;

    /** Called when the activity is first created. */
    @Override
    public void onCreate(Bundle savedInstanceState) {
        super.onCreate(savedInstanceState);
        setContentView(R.layout.main);

        setupViewComponent();
```

```java
        setMapLocation();
    }

    private void setupViewComponent() {
        mSpnLocation = (Spinner) this.findViewById(R.id.spnLocation);
        mMapView = (MapView) findViewById(R.id.map);

        mMapCtrl = mMapView.getController();
        mMapCtrl.setZoom(18);

        ArrayAdapter<String> adapter = new ArrayAdapter<String>(
                this, android.R.layout.simple_spinner_item);

        for (int i = 0; i < locations.length; i++)
            adapter.add(locations[i][0]);

        adapter.setDropDownViewResource(android.R.layout.simple_spinner_dropdown_item);
        mSpnLocation.setAdapter(adapter);
        mSpnLocation.setOnItemSelectedListener(mSpnLocationOnItemSelLis);
    }

    private OnItemSelectedListener mSpnLocationOnItemSelLis = new OnItemSelectedListener() {
        @Override
        public void onItemSelected(AdapterView parent, View v,
                int position, long id) {
            // TODO Auto-generated method stub

            setMapLocation();
        }

        @Override
        public void onNothingSelected(AdapterView<?> arg0) {
            // TODO Auto-generated method stub
        }
    };

    public void setMapLocation() {
        int iSelect = mSpnLocation.getSelectedItemPosition();
        String[] sLocation = locations[iSelect][1].split(",");
        double dLat = Double.parseDouble(sLocation[0]);    // 南北纬
        double dLon = Double.parseDouble(sLocation[1]);    // 东西经
        GeoPoint gp = new GeoPoint((int)(dLat * 1e6), (int)(dLon * 1e6));
        mMapCtrl.animateTo(gp);
    }

    @Override
    protected boolean isRouteDisplayed() {
        // TODO Auto-generated method stub
```

```
            return false;
        }
    }
```

首先我们建立一个二维字符串数组储存地点名称和它们的地理坐标,接下来是定义程序中会用到的接口对象。我们把设置接口对象的程序代码写在 setupViewComponent()方法中,然后在 onCreate()内调用该方法。另外我们把设置地图的地理位置坐标的程序代码写在 setMapLocation()方法中,该方法会从 Spinner 组件中取得目前用户选定的地点,然后根据它的地理坐标设置 Google 地图并完成显示。

Step 4 依照上一小节步骤五的说明修改项目的功能描述文件 AndroidManifest.xml。

完成后可以在连上 Internet 的计算机运行程序,就可以看到如图 67-7 所示的 Google 地图界面,如果和 Google 官方的地图网站比较,这个范例程序的功能显得有些阳春,它无法切换地图模式和缩小放大,下一个单元我们将继续为这个 Google 地图程序增加一些功能。

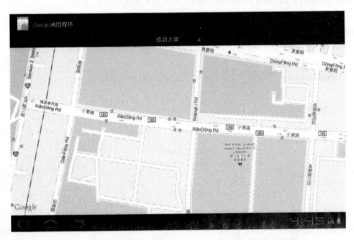

图 67-7　在平板电脑模拟器运行 Google 地图程序的界面

Android 版本	1.X	2.X	3.X	4.X
适用性	★	★	★	★

UNIT 68

Google 地图的进阶用法

上一个单元的 Google 地图程序只有最基本的显示功能，无法放大和缩小，也不能切换成卫星图像模式或是移动地图察看邻近的区域，这一个单元我们将继续加强这个地图程序的功能，也借此机会学习更多操作 Google 地图的方法。

68-1 地图的缩放和拖曳功能

当读者对于 Android 程序设计累积了足够的经验之后，就能逐渐掌握 Android 程序架构的精神，也就是本书一开始就提示的面向对象和事件驱动。当我们着手开发一项新功能时，第一个念头就是寻找相关对象的信息，再查阅对象所提供的方法，从中找到需要的功能，然后在适当的事件处理程序中写入程序代码。回顾上一个单元的地图程序，其中使用了和地图相关的三个对象：MapView、MapController 和 GeoPoint。MapView 用来显示地图，MapController 是负责控制地图的状态和功能，GeoPoint 则是用来指定地图上的位置坐标。当需要加上地图的缩放功能时，从对象所负责的工作来思考，应该和负责显示的 MapView 对象有关，如果读者查阅 Android SDK 的联机帮助文件，或是在 Eclipse 的程序代码编辑器中先输入 MapView 对象的名称再输入"点字符"(.)，就会出现如图 68-1 所示的可用方法和属性列表，我们可以在列表中仔细检查每一个方法的名称，当单击其中一个方法时还会出现第二层的方法说明窗口，利用这种方式我们可以学习更多对象的方法。

利用以上的操作技巧可以发现 MapView 对象中有一个叫做 setBuiltInZoomControls()的方法，可以用来运行地图缩放的功能，只要在程序中调用这个方法并传入 true 参数，就可以启用"地图缩放控制组件"。这个"地图缩放控制组件"是当用户在地图上按下按键，或是在触控屏幕的界面上按下地图后才会出现，所以我们还要启用地图可以接受"点按"的功能，这项功能需要在接口布局文件中进行设置，请参考以下程序代码范例。另外启用地图的"点按"功能之后，用户也可以利用按住拖曳的方式移动地图。

图 68-1　Eclipse 程序代码编辑器的对象方法提示列表和说明

■　界面布局文件：

```
<?xml version="1.0" encoding="utf-8"?>
<LinearLayout …
    >
<Spinner …
    />
<com.google.android.maps.MapView android:id="@+id/map"
    …
    android:clickable="true"
    />
</LinearLayout>
```

■　程序代码：

```
package …

import …

public class Main extends MapActivity {

    …

    /** Called when the activity is first created. */
    @Override
    public void onCreate(Bundle savedInstanceState) {
        …
    }
```

520

```
private void setupViewComponent() {
    …
    mMapCtrl = mMapView.getController() ;
    mMapView.setBuiltInZoomControls(true) ;
    mMapCtrl.setZoom(18) ;
    …
}
    …
}
```

68-2 加上键盘控制功能

这里我们也趁机学习一下 Android 程序的键盘控制技巧，虽然在 Android 设备上使用键盘的机会不多，但是必要时还是可以使用。假设我们希望用户按下键盘上的 I 键时可以 zoom in（放大）地图，按下 O 键时可以 zoom out（缩小）地图。为了处理按键按下的动作，我们可以在主程序中加上 onKeyDown() 事件处理函式。如果程序在运行时有按键被按下，Android 系统便会运行这个事件处理函式。

在程序代码中加入事件处理函数的第一步就是在程序编辑窗口中右击，然后在出现的菜单中选择 Source > Override/Implement Methods…就会出现图 68-2 的对话框，在对话框左边的清单中会列出相关类和可以加入的事件处理函式。但是请读者注意，虽然我们现在的类是继承 MapActivity，可是 onKeyDown() 是属于 MapActivity 的基础类 Activity，因此必须在对话框左边选择 Activity 类，才可以找到 onKeyDown()，请读者勾选它然后单击 OK 按钮，然后在 onKeyDown() 中请输入以下粗体标示的程序代码。

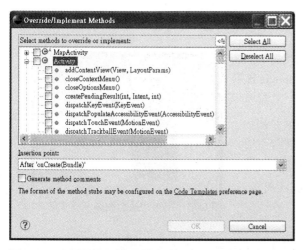

图 68-2 选择要加入的事件处理函数

```
public boolean onKeyDown(int keyCode, KeyEvent event) {
    // TODO Auto-generated method stub

    int nextZoom, zoom;
    switch (keyCode) {
    case KeyEvent.KEYCODE_I:
        nextZoom = mMapView.getZoomLevel() + 1;
        zoom = nextZoom > mMapView.getMaxZoomLevel() ?
                    mMapView.getMaxZoomLevel() : nextZoom;
        mMapCtrl.setZoom(zoom);
        break;
    case KeyEvent.KEYCODE_O:
        nextZoom = mMapView.getZoomLevel() - 1;
        zoom = nextZoom < 1 ?
                    1 : nextZoom;
        mMapCtrl.setZoom(zoom);
        break;
    }

    return super.onKeyDown(keyCode, event);
}
```

这一段程序代码的功能是判断用户按下的按键，判断的方式是将 keyCode 参数和 KeyEvent 中所定义的常数比较，然后运行对应的程序代码。Google 地图的最小缩放比例是 1（缩到最小的情况，也就是显示全世界地图），最大缩放比例则不固定，依地图模式和区域可能有所不同，我们可以调用 MapView 对象的 getMaxZoomLevel()方法取得目前可以设置的最大缩放值。为了让设置的地图缩放值在合法的范围，我们在上述的程序代码中加入适当的判断和处理。

68-3 切换地图显示模式

依照前面介绍的 Android 程序架构逻辑，地图显示模式也是由 MapView 对象控制。在预设情况下，Google 地图是传统的街道模式，如果要换成卫星图像地图，只需要调用 MapView 对象的 setSatellite()方法并传入 true 即可。为了让用户可以在街道模式和卫星图像地图之间自由切换，我们在程序的操作接口中增加一个 Spinner 下拉式组件，其中包含街道图和卫星图两个选项，如图 68-3 所示，修改后的程序接口布局文件如下：

图 68-3　用来切换地图显示模式的选项

```
<?xml version="1.0" encoding="utf-8"?>
<LinearLayout xmlns:android="http://schemas.android.com/apk/res/android"
    ...
```

```xml
>
<LinearLayout
    android:orientation="horizontal"
    android:layout_width="match_parent"
    android:layout_height="wrap_content"
    android:gravity="center_horizontal"
    >
<Spinner android:id="@+id/spnMapType"
    android:layout_width="wrap_content"
    android:layout_height="wrap_content"
    android:drawSelectorOnTop="true"
    android:prompt="@string/prompt_select_map_type"
    />
<Spinner android:id="@+id/spnLocation"
    android:layout_width="wrap_content"
    android:layout_height="wrap_content"
    android:drawSelectorOnTop="true"
    android:prompt="@string/prompt_select_location"
    />
</LinearLayout>
<com.google.android.maps.MapView android:id="@+id/map"
    …
    />
</LinearLayout>
```

我们新增了一个<LinearLayout>标签，把新增的地图模式 Spinner 组件和原来的地单击项 Spinner 组件包裹起来让它们形成水平排列，Spinner 组件的标题同样使用定义在字符串资源文件中的字符串。当用户开启地图模式的 Spinner 组件并单击其中一个项目后，程序便根据用户所选择的项目，调用 MapView 对象的 setSatellite()方法进行地图模式的设置，以下是修改上一个单元的程序项目后的程序代码，有更动的部分以粗体字代表，程序在手机模拟器的运行界面如图 68-4 所示。

```java
package …

import …

public class Main extends MapActivity {

    …（同原来的程序代码）

    private static String[] mapType = {
                "街道图",
                "卫星图"};

    /** Called when the activity is first created. */
    @Override
    public void onCreate(Bundle savedInstanceState) {
        …（同原来的程序代码）
```

图 68-4 具有地图模式切换和缩放控制功能的地图程序

```
    }

    @Override
    public boolean onKeyDown(int keyCode, KeyEvent event) {
        // TODO Auto-generated method stub

        // 请参考本单元前面的说明
        …
    }

    private void setupViewComponent() {
        mSpnLocation = (Spinner) this.findViewById(R.id.spnLocation);
        mSpnMapType = (Spinner) this.findViewById(R.id.spnMapType);
        mMapView = (MapView) findViewById(R.id.map);

        mMapCtrl = mMapView.getController();
        mMapView.setBuiltInZoomControls(true);
        mMapCtrl.setZoom(18);

        ArrayAdapter<String> adapter = new ArrayAdapter<String>(
                this, android.R.layout.simple_spinner_item);

        for (int i = 0; i < locations.length; i++)
            adapter.add(locations[i][0]);
```

```java
        adapter.setDropDownViewResource(android.R.layout.simple_spinner_
        dropdown_item);
        mSpnLocation.setAdapter(adapter);
        mSpnLocation.setOnItemSelectedListener(mSpnLocationOnItemSelLis);

        adapter = new ArrayAdapter<String>(
                this, android.R.layout.simple_spinner_item);

        for (int i = 0; i < mapType.length; i++)
            adapter.add(mapType[i]);

        adapter.setDropDownViewResource(android.R.layout.simple_spinner_
        dropdown_item);
        mSpnMapType.setAdapter(adapter);
        mSpnMapType.setOnItemSelectedListener(mSpnMapTypeOnItemSelLis);
    }

    private OnItemSelectedListener mSpnLocationOnItemSelLis = new
    OnItemSelectedListener() {
        …（同原来的程序代码）
    };

    private OnItemSelectedListener mSpnMapTypeOnItemSelLis = new
    OnItemSelectedListener() {
        @Override
        public void onItemSelected(AdapterView parent, View v, int
        position, long id) {
            // TODO Auto-generated method stub

            switch (position) {
            case 0:
                mMapView.setSatellite(false);
                break;
            case 1:
                mMapView.setSatellite(true);
                break;
            }
        }

        @Override
        public void onNothingSelected(AdapterView<?> arg0) {
            // TODO Auto-generated method stub
        }
    };

    …（同原来的程序代码）
}
```

Android 版本	1.X	2.X	3.X	4.X
适用性	★	★	★	★

UNIT 69
帮地图加上标记

在 Google 地图网站上查看地图时，会发现地图上有许多地点的标记（mark），像是建筑物名称、机关名称、商店名称等。这些地标并不是直接画在地图上，因为这样做的话会破坏原来的地图数据，而且这些地标也可能随着时间变动，因此最好的作法是用附加的方式显示在地图上，也就是说类似绘图软件使用的图层概念，地图是在底层，地标则放在地图上面，这样就可以随时改变地标显示的数据，这种方式在 Google 地图程序中叫做 Overlay。

69-1 地图程序使用 Overlay 的步骤

我们将地图程序使用 Overlay 的步骤整理说明如下：

Step 1 在地图程序项目中新增一个继承 ItemizedOverlay 的类，ItemizedOverlay 是一个泛型类（generic class），它可以处理各种类型的对象。这里我们指定使用 OverlayItem 类型的对象，因此定义时使用 ItemizedOverlay<OverlayItem>。修改好类的定义后程序代码编辑器会以红色波浪底线标记语法错误，把鼠标光标移到错误的位置就会出现一个信息窗口，请依照其中的建议加入类的构造方法和需要的方法，然后请将程序代码编辑如下：

```
public class MapItemizedOverlay extends ItemizedOverlay<OverlayItem> {

//    private ArrayList<OverlayItem> mOverlayItems =
//        new ArrayList<OverlayItem>();
    private OverlayItem mOverlayItem;

    Context mContext;

    public MapItemizedOverlay(Drawable defaultMarker, Context context) {
```

```
            super(defaultMarker);
            // TODO Auto-generated constructor stub
            mContext = context;
        }

        @Override
        protected OverlayItem createItem(int i) {
            // TODO Auto-generated method stub
//          return mOverlayItems.get(i);
            return mOverlayItem;
        }

        @Override
        public int size() {
            // TODO Auto-generated method stub
//          return mOverlayItems.size();
            return 1 ;
        }
    }
```

读者看了以上的程序代码可能觉得有些奇怪,有些部分被批注掉了,而这些被批注掉的程序代码又和它们下面的程序代码很类似。其实这些被批注掉的程序代码是 Google 官方 SDK 的范例程序,可是在笔者深入追踪这一份程序代码后,发现原来的作法有画蛇添足的情况,因为我们建立的这个类是用来储存地目标数据,而每一个地标都是一个独立的对象,那么在这个类中定义一个 ArrayList 对象似乎多此一举,取而代之只需要定义一个 OverlayItem 对象即可,因此笔者把相关的程序代码加以修改,修改后的程序代码和原来的功能完全相同。

在类的构造方法中我们新增一个参数用来传入程序的 context 并加以储存。createItem()方法是当 Android 系统需要一个地标数据的时候会自动调用,我们直接把 OverlayItem 对象传给系统即可。size()方法也是由 Android 系统调用以取得地标数据个数,我们直接传回 1 。

Step 2 在步骤一的类中需要再增加以下几个以粗体字标记的方法,其中的 draw()和 onTap()是 ItemizedOverlay 类内部的方法,因此我们必须叫出程序代码编辑窗口的快捷菜单,从中选择 Source > Override/Implement Methods……,再从方法清单中选取并完成加入,至于 addOverlay()则是我们自行加入的方法:

```
package ...

import ...

public class MapItemizedOverlay extends ItemizedOverlay<OverlayItem> {

    …(同步骤一的程序代码)
```

```java
    public MapItemizedOverlay(Drawable d, Context context) {
        …（同步骤一的程序代码）
    }

    @Override
    protected OverlayItem createItem(int i) {
        …（同步骤一的程序代码）
    }

    @Override
    public int size() {
        …（同步骤一的程序代码）
    }

    @Override
    public void draw(Canvas canvas, MapView mapView, boolean shadow) {
        // TODO Auto-generated method stub
        super.draw(canvas, mapView, false);
    }

    @Override
    protected boolean onTap(int index) {
        // TODO Auto-generated method stub
//      OverlayItem item = mOverlayItems.get(index);
        AlertDialog.Builder dialog = new AlertDialog.Builder(mContext);
//      dialog.setTitle(item.getTitle());
//      dialog.setMessage(item.getSnippet());
        dialog.setTitle(mOverlayItem.getTitle());
        dialog.setMessage(mOverlayItem.getSnippet());
        dialog.show();
//      return super.onTap(index);
        return super.onTap(0);
    }

    public void addOverlay(OverlayItem overlayItem) {
//      mOverlayItems.add(overlayItem);
        mOverlayItem = overlayItem;
        populate();
    }
}
```

第一个 draw()方法是用来避免地标数据的自动阴影效果（把 shadow 参数设为 false）。onTap()方法是当用户按下地标时系统会调用的方法，我们取得地目标标题和说明文字，然后以对话框的形式把它显示出来，有关 AlertDialog 对话框的使用方式请读者参考 UNIT 34。最后 addOverlay()是唯一由我们程序调用的方法，它的功能是把建立好的地标对象存入这个类的对象中，设置好地标对象后记得要调用 populate()方法通知系统。

Step 3 在主程序中先从 MapView 对象取得地图的 Overlay，地图的 Overlay 是一个 List 容器对象，我们会把建立好的地标对象，也就是步骤一类的对象一个一个加入地图的 Overlay 中。要建立地标对象首先从项目资源中取得用来当作地目标图像，另外我们特别设计一个 setDrawableBounds()方法用来设置地标图像在地图上的对齐方式，总共分成左上、中上、右上、左中、中央、右中、左下、中下、右下 9 种对齐模式。如果要让地标有透明的效果，则可以调用 setAlpha()方法（请参考以下范例）。然后我们建立一个步骤一类的对象并传入地标图像，接着将地标在地图上的位置（GeoPoint）、地标标题、说明文字建立成一个 OverlayItem 对象，再把该对象加入地标对象中，最后把地标对象加入地图的 Overlay。重复上述动作在地图的 Overlay 中加入所有要在地图上显示的地标。

```
package ...

import ...

public class Main extends MapActivity {

    …

    public enum BasePoint {
        TOP_LEFT, TOP_CENTER, TOP_RIGHT,
        MIDDLE_LEFT, MIDDLE_CENTER, MIDDLE_RIGHT,
        BOTTOM_LEFT, BOTTOM_CENTER, BOTTOM_RIGHT
    }

    /** Called when the activity is first created. */
    @Override
    public void onCreate(Bundle savedInstanceState) {
        …
    }

    private void setupViewComponent() {
        mSpnLocation = (Spinner) this.findViewById(R.id.spnLocation);
        mSpnMapType = (Spinner) this.findViewById(R.id.spnMapType);
        mMapView = (MapView) findViewById(R.id.map);

        mMapCtrl = mMapView.getController();

        mMapView.setBuiltInZoomControls(true);
        mMapCtrl.setZoom(18);

        List<Overlay> mapOverlays = mMapView.getOverlays();

        MapItemizedOverlay mapItemizedOverlay;
```

```java
OverlayItem overlayitem;
Drawable drawable;
GeoPoint gp;
double dLat, dLon;
String[] sLocation;

// 第一个地标
drawable = getResources().getDrawable(R.drawable.地标图像文件名称);
setDrawableBounds(drawable, BasePoint.MIDDLE_CENTER);
drawable.setAlpha(100);              // 0(全透明)~255
mapItemizedOverlay = new MapItemizedOverlay(drawable, this);
gp = new GeoPoint((int)(地标位置的纬度 * 1e6), (int)(地标位置的经度 * 1e6));
overlayitem = new OverlayItem(gp, "地标标题", "地标说明文字");
mapItemizedOverlay.addOverlay(overlayitem);
mapOverlays.add(mapItemizedOverlay);

// 继续加上其他地标
…

ArrayAdapter<String> adapter = new ArrayAdapter<String>(
        this, android.R.layout.simple_spinner_item);


…
}

private void setDrawableBounds(Drawable d, BasePoint bp) {
    switch (bp) {
    case TOP_LEFT:
        d.setBounds(0,
                0,
                d.getMinimumWidth(),
                d.getMinimumHeight());
        break;
    case TOP_CENTER:
        d.setBounds(-d.getMinimumWidth()/2,
                0,
                d.getMinimumWidth()/2,
                d.getMinimumHeight());
        break;
    case TOP_RIGHT:
        d.setBounds(-d.getMinimumWidth(),
                0,
                0,
                d.getMinimumHeight());
        break;
    case MIDDLE_LEFT:
        d.setBounds(0,
                -d.getMinimumHeight()/2,
```

```
                            d.getMinimumWidth()/2,
                            d.getMinimumHeight()/2);
                break;
            case MIDDLE_CENTER:
                d.setBounds(-d.getMinimumWidth()/2,
                            -d.getMinimumHeight()/2,
                            d.getMinimumWidth()/2,
                            d.getMinimumHeight()/2);
                break;
            case MIDDLE_RIGHT:
                d.setBounds(-d.getMinimumWidth(),
                            -d.getMinimumHeight()/2,
                            0,
                            d.getMinimumHeight()/2);
                break;
            case BOTTOM_LEFT:
                d.setBounds(0,
                            -d.getMinimumHeight(),
                            d.getMinimumWidth(),
                            0);
                break;
            case BOTTOM_CENTER:
                d.setBounds(-d.getMinimumWidth()/2,
                            -d.getMinimumHeight(),
                            d.getMinimumWidth()/2,
                            0);
                break;
            case BOTTOM_RIGHT:
                d.setBounds(-d.getMinimumWidth(),
                            -d.getMinimumHeight(),
                            0,
                            0);
                break;
        }
    }
}
```

69-2 范例程序

我们将以上学到的方法应用在前一个单元完成的地图程序，在地图中加上四所大学的地标。读者可以自行准备 4 个 png 或 jpg 图像文件，分辨率大约 50*50 即可，然后利用 Windows 文件管理器，把图像文件复制到项目文件夹中的 res/drawable-hdpi 目录下。程序代码依照上一个小节介绍的步骤依序完成即可，其中把建立地目标程序代码（在 setupViewComponent()方法内），修改如下：

```
drawable = getResources().getDrawable(R.drawable.图像文件名称);
```

```
setDrawableBounds(drawable, BasePoint.MIDDLE_CENTER);
drawable.setAlpha(100);              // 0(全透明)~255
mapItemizedOverlay = new MapItemizedOverlay(drawable, this);
sLocation = locations[0][1].split(",");
dLat = Double.parseDouble(sLocation[0]);       // 南北纬
dLon = Double.parseDouble(sLocation[1]);       // 东西经
gp = new GeoPoint((int)(dLat * 1e6), (int)(dLon * 1e6));
overlayitem = new OverlayItem(gp, locations[0][0], locations[0][0]);
mapItemizedOverlay.addOverlay(overlayitem);
mapOverlays.add(mapItemizedOverlay);
```

主要的改变是使用读者自己建立的图像文件名称，另外有关地标所使用的经纬度坐标和标题及说明则是从原来程序的地点数组中取得。完成程序代码的编辑之后运行程序项目就可以看到如图69-1所示的界面，读者可以参考本书光盘中的程序项目原始文件。

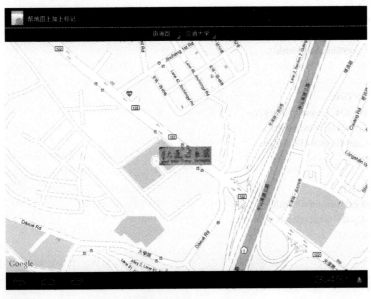

图 69-1　加上地标后的地图程序界面

Android 版本	1.X	2.X	3.X	4.X
适用性	★	★	★	★

UNIT 70
加上定位让地图活起来

地图本身是一项非常有用的数据，它可以让我们预先浏览想要去的地方，也能够让我们事先规划旅游的行程和路径，如果我们处在一个陌生的环境，像是在旅行途中或是刚到一个新的地点，地图还能够提供当地的地理信息。可是这些都是被动的查询功能，也就是说必须由使用者自己先找出目前所在的位置，才能够开始查阅地图。如果要让地图自己主动提供信息，就要加上"自动定位"的功能。"自动定位"功能就像是地图的灵魂，让地图拥有生命可以知道目前所在的位置。加上定位功能的地图不但可以适时提供"行"的信息，还能够提供"食、衣、住、育、乐"等各种生活情报，因此现在已经成为智能手机和平板电脑最热门的卖点之一。这种和"自动定位"功能相关的生活化应用称为"适地性服务"（Location-Based Service, LBS），它可以让传统被动的商业模式（静静地等顾客上门），进化成为主动出击的促销服务，图 70-1 是 LBS 的基本架构示意图和一些应用的范例。

图 70-1　从"自动定位"功能衍生出来的"适地性服务"

70-1　移动设备的定位技术

手机或平板电脑有两种定位的方法,第一种是使用目前已经非常普遍的 GPS(Global Positioning System)全球定位系统,如图 70-2 所示,它是使用地球上空的卫星所发射的坐标信号以及信号上所记载的发射时间,根据计算接收到不同卫星的 GPS 信号的时间差,以三角几何公式算出接收器目前的位置,在一般情况下的定位误差距离约在 5 至 15 公尺之间。GPS 定位方式有两个缺点,第一是当所在的位置接收不到 GPS 信号时(例如室内)就无法定位,第二是有些设备的 GPS 定位功能需要比较长的处理时间才能得到结果。第二种定位方法是利用手机基地台的地理位置来定位,这种方法的好处是速度快,而且只要能够接收到手机基地台信号的地方就能够完成定位,但缺点是误差范围比较大,从数十公尺到上百公尺都有可能。

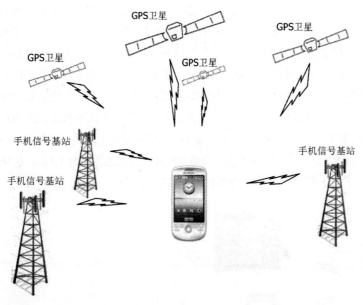

图 70-2　手机或平板电脑的两种定位方法

> **补充说明**　**GPS 相关知识**
> GPS 是由美国国防部建立的全球卫星定位系统,它是由二十四颗 GPS 卫星组成,每天绕行地球二圈,不论在任何时间和地点,至少会有四颗 GPS 卫星出现在我们的上空。从技术面来说,三颗 GPS 卫星可以做 2D 定位(经度和纬度),四颗 GPS 卫星则可以进一步达成 3D 定位(经纬度加上高度)。如果 GPS 信号接收器的周围有高耸庞大的障碍物,例如非常高大的建筑物或是山丘,就有可能阻碍定位信号的接收而影响定位的准确性。

对于程序开发者来说，并不需要深入了解硬件设备内部的定位设备和运行方法，比较重要的是学习如何取得系统提供的定位信息。Android 系统总共有两种让程序取得定位信息的方法，一种是使用 MyLocationOverlay 对象，这种方式比较简单，另一种则是使用 LocationManager，这种方式可以让我们设置比较多的定位参数，但是程序代码相对就比较复杂。当我们着手开发定位相关程序时，记得要先启动模拟器的定位功能，首先进入手机或平板电脑系统设置的 Settings 界面，然后选择"Location & security"项目，再勾选其中的"Use wireless networks"和"Use GPS satellites"二项，接下来就让我们依序介绍这二种定位程序的实现方式。

70-2　第一种定位方法——使用 MyLocationOverlay

MyLocationOverlay 类内部封装了定位流程的细节，它让我们可以很容易地取得系统最新的定位数据，使用 MyLocationOverlay 对象只需要下列四个步骤：

Step 1 在程序中定义并建立一个 MyLocationOverlay 对象，然后建立一个 Runnable 对象，把每一次定位信息更新时要做的工作写在这个 Runnable 对象中，再调用 MyLocationOverlay 对象的 runOnFirstFix()方法把它传给 Android 系统，于是当每一次 Android 系统的定位更新时便会运行一次这个 Runnable 对象，实现的程序代码请读者参考后面的范例程序。

Step 2 在主程序加入 onResume() 方法，在该方法中调用 MyLocationOverlay 对象的 enableMyLocation()方法启动定位功能。

Step 3 在主程序加入 onStop()方法，在该方法中调用 MyLocationOverlay 对象的 disableMyLocation()方法停止定位功能。

Step 4 在程序项目的功能描述文件 AndroidManifest.xml 中新增一行程序代码允许程序使用定位功能：

```
<?xml version="1.0" encoding="utf-8"?>
<manifest …>
    <uses-permission android:name="android.permission.INTERNET" />
    <uses-permission android:name="android.permission.ACCESS_
    FINE_LOCATION" />
    <application …>
        …
    </application>
</manifest>
```

如上一小节的说明，程序有两种定位的方式：使用 GPS 或是手机基地台。GPS 在 Android 系统中称之为 ACCESS_FINE_LOCATION，另一种利用手机基地台的定位方式则称为 ACCESS_COARSE_LOCATION。如果程序指定了 ACCESS_FINE_LOCATION 就自动包含 ACCESS_COARSE_LOCATION，但是如果只有指定 ACCESS_COARSE_LOCATION

则不包含 ACCESS_FINE_LOCATION。

我们把以上介绍的定位方法结合 UNIT 67 的地图程序，程序代码和功能描述文件只要依照上述步骤进行小幅修改即可，功能描述文件的修改部分请参考上述步骤四的说明，程序文件的修改部分以粗体标记如下。在 Runnable 对象中我们只做一件事，就是把地图移到新的定位坐标，调用 runOnFirstFix()方法的下一行程序代码就是把 MyLocationOverlay 对象设置给地图成为其中的一个 Overlay，也就是上一个单元所说的地标。

■ 程序代码：

```
package …

import …

public class Main extends MapActivity {

    private MyLocationOverlay mMyLocation;

    /** Called when the activity is first created. */
    @Override
    public void onCreate(Bundle savedInstanceState) {
        …
    }

    private void setupViewComponent() {
        mSpnLocation = (Spinner) this.findViewById(R.id.spnLocation);
        mMapView = (MapView) findViewById(R.id.map);

        mMapCtrl = mMapView.getController();
        mMapCtrl.setZoom(18);

        List<Overlay> mapOverlays = mMapView.getOverlays();
        mMyLocation = new MyLocationOverlay(this, mMapView);
        mMyLocation.runOnFirstFix(new Runnable() {
            @Override
            public void run() {
                // TODO Auto-generated method stub
                mMapCtrl.animateTo(mMyLocation.getMyLocation());
            }
        });
        mapOverlays.add(mMyLocation);

        ArrayAdapter<String> adapter = new ArrayAdapter<String>(
            this, android.R.layout.simple_spinner_item);

        for (int i = 0; i < locations.length; i++)
```

```
        adapter.add(locations[i][0]);

        adapter.setDropDownViewResource(android.R.layout.simple_spinner_
        dropdown_item);
        mSpnLocation.setAdapter(adapter);
        mSpnLocation.setOnItemSelectedListener(mSpnLocationOnItemSelLis);
    }

    @Override
    protected void onResume() {
        // TODO Auto-generated method stub
        super.onResume();
        mMyLocation.enableMyLocation();
    }

    @Override
    protected void onStop() {
        // TODO Auto-generated method stub
        mMyLocation.disableMyLocation();
        super.onStop();
    }

    …
}
```

编辑好程序代码之后就可以运行程序，记得要让计算机连上 Internet 而且必须建立具备 Google APIs 功能的模拟器。当程序在模拟器上运行的时候，请将 Eclipse 切换到 DDMS 界面（参考"补充说明"）。在 DDMS 界面左边中间有一个叫做 Emulator Control 的标签页可以用来控制模拟器（参考图 70-3），不过我们必须先在 Emulator Control 标签页上方的 Devices 标签页中选择要控制的模拟器，然后回到 Emulator Control 中把滚动条往下拉直到出现 Location Controls，然后在它下方的 Manual 中有 Longitude（经度）和 Latitude（纬度）两个字段，读者可以在这两个字段中输入想要传给模拟器的定位坐标，输入完毕后按下下方的 Send 按钮就会看到如图 70-4 所示的程序界面，在这个范例中我们是输入纽约自由女神像的位置，中央的圆点就是我们指定的定位坐标。接下来我们介绍第二种定位程序的作法。

补充说明　将 Eclipse 切换到 DDMS 画面

在 Eclipse 主菜单下方的工具栏最右边有一个名称叫做 Open Perspective 的按钮，将鼠标光标停在 Eclipse 工具栏上的按钮时就会弹出一个按钮名称的信息，读者可以利用这个方法找到指定的按钮，单击 Open Perspective 按钮然后选择其中的 DDMS 选项。

图 70-3　Eclipse 的 DDMS 界面中的 Emulator Control

图 70-4　使用 Emulator Control 仿真定位的程序执行界面

70-3　第二种定位方法——使用 LocationManager

使用 LocationManager 的好处是可以让我们指定使用 GPS 定位或是手机基地台定位, 甚至可以设置选择条件像是精确度或是耗电率, 让 Android 系统自动帮我们选择一个定位方式。以下是使用 LocationManager 的步骤：

Step 1　让主程序类实现 LocationListener 如以下程序代码, 此时程序代码编辑器会在主程序类名称下方标记红色波浪底线代表语法错误, 请将鼠标光标移到该错误上就会弹出一个信息窗口, 其中有一项建议是 Add unimplemented methods, 请单击它就会在类中自动加入 onLocationChanged()、onProviderDisabled()、onProviderEnabled()和 onStatusChanged()四个方法。

```
public class 主程序类 extends 父类 implements LocationListener {
    …
}
```

Step 2　在主程序类中定义一个 LocationManager 和一个 String 对象, 然后在 onCreate()方法中使用 LocationManager 的程序代码如下：

```
public class 主程序类 extends 父类 implements LocationListener {
    private LocationManager mLocationMgr;
    private String mBestLocationProv;
    public void onCreate(Bundle savedInstanceState) {
```

```
        ...
        mLocationMgr = (LocationManager)getSystemService(LOCATION_SERVICE);
        Criteria c = new Criteria();
        mBestLocationProv = mLocationMgr.getBestProvider(c, true);
    }
}
```

我们从系统取得 LocationManager 对象，然后建立一个 Criteria 对象，这个 Criteria 对象可以利用它内部的方法像是 setAccuracy()或是 setPowerRequirement()等，设置挑选定位功能的条件，然后调用 LocationManager 的 getBestProvider()并传入 Criteria 对象，就可以得到适合的定位方式的名称，以上的范例是没有设置任何条件让系统自行决定。

Step 3 在主程序的 onLocationChanged()、onProviderDisabled()、onProviderEnabled()和 onStatusChanged()四个方法中加入以下的程序代码：

```java
@Override
public void onLocationChanged(Location location) {
    // TODO Auto-generated method stub
    mMapCtrl.animateTo(new GeoPoint(
            (int)(location.getLatitude() * 1e6),
            (int)(location.getLongitude() * 1e6)));
}

@Override
public void onProviderDisabled(String provider) {
    // TODO Auto-generated method stub
    mLocationMgr.removeUpdates(this);
}

@Override
public void onProviderEnabled(String provider) {
    // TODO Auto-generated method stub
    mLocationMgr.requestLocationUpdates(mBestLocationProv, 60000,
        1, this);
}

@Override
public void onStatusChanged(String provider, int status, Bundle  extras) {
    // TODO Auto-generated method stub
    Criteria c = new Criteria();
    mBestLocationProv = mLocationMgr.getBestProvider(c, true);
}
```

onLocationChanged()是当定位更新时会自动运行的方法，我们所做的工作就是移动地图的位置到新的定位坐标（从传入的 location 参数中取得）。onProviderDisabled()是当使用的定位功能被取消时会运行的方法，此时我们就停止使用定位。onProviderEnabled()是当定位功能启动时运行的方法，此时我们启动程序的定位功能。最后 onStatusChanged()是定位状态改变时运行的方法，此时我们重新选取一个适合的定位方式。

Step 4 最后在主程序中加入 onStop()和 onResume()方法，操作方式请参考上一个小节的操作提示，然后输入以下程序代码。onStop()是当程序即将结束时运行，此时我们停止定位功能。onResume()是当程序开始作用时运行，此时我们启动定位功能。

```
@Override
protected void onStop() {
    // TODO Auto-generated method stub
    mLocationMgr.removeUpdates(this);
    super.onStop();
}

@Override
protected void onResume() {
    // TODO Auto-generated method stub
    super.onResume();
    mLocationMgr.requestLocationUpdates(mBestLocationProv, 60000, 1, this);
}
```

Step 5 在程序项目的功能描述文件 AndroidManifest.xml 中新增一行程序代码允许程序使用定位功能如下，有关 ACCESS_FINE_LOCATION 和 ACCESS_COARSE_LOCATION 的使用方式如同前一小节的说明。

```
<?xml version="1.0" encoding="utf-8"?>
<manifest …>
    <uses-permission android:name="android.permission.INTERNET" />
    <uses-permission android:name="android.permission.ACCESS_FINE_LOCATION" />
    <application …>
        …
    </application>
</manifest>
```

我们把以上介绍的定位方法结合 UNIT 67 的地图程序，功能描述文件的修改如同上述步骤五，程序文件的修改部分我们以粗体标记如下。当读者编辑好程序代码之后就可以运行程序，然后利用 Emulator Control 传送定位坐标给模拟器，图 70-5 是传送定位坐标后的程序界面。读者可以和图 70-4 作比较，二者的差别在于图 70-5 中少了一个圆点，因为使用 MyLocationOverlay 的方法就如同前一个单元介绍的地标，它会在地图上画记号。如果需要我们也可以把 LocationManager 的定位方法加上前一个单元的地标就可以得到和图 70-4 一样的结果，读者如果有兴趣不妨试看看。

图 70-5　使用 Emulator Control 传送定位坐标后的程序界面

```
package …

import …

public class Main extends MapActivity implements LocationListener {

    private LocationManager mLocationMgr;
    private String mBestLocationProv;

    …

    /** Called when the activity is first created. */
    @Override
    public void onCreate(Bundle savedInstanceState) {
        …

        mLocationMgr = (LocationManager)getSystemService(LOCATION_SERVICE);
        Criteria c = new Criteria();
        mBestLocationProv = mLocationMgr.getBestProvider(c, true);

        setMapLocation();
    }

    …
```

```java
@Override
public void onLocationChanged(Location location) {
    // TODO Auto-generated method stub
    mMapCtrl.animateTo(new GeoPoint(
            (int)(location.getLatitude() * 1e6),
            (int)(location.getLongitude() * 1e6)));
}

@Override
public void onProviderDisabled(String provider) {
    // TODO Auto-generated method stub
    mLocationMgr.removeUpdates(this);
}

@Override
public void onProviderEnabled(String provider) {
    // TODO Auto-generated method stub
    mLocationMgr.requestLocationUpdates(mBestLocationProv, 60000, 1, this);
}

@Override
public void onStatusChanged(String provider, int status, Bundle   extras) {
    // TODO Auto-generated method stub
    Criteria c = new Criteria();
    mBestLocationProv = mLocationMgr.getBestProvider(c, true);
}

@Override
protected void onStop() {
    // TODO Auto-generated method stub
    mLocationMgr.removeUpdates(this);
    super.onStop();
}

@Override
protected void onResume() {
    // TODO Auto-generated method stub
    super.onResume();
    mLocationMgr.requestLocationUpdates(mBestLocationProv, 60000,
        1, this);
}
}
```

PART 14　拍照、录音、录像与多媒体播放

UNIT 71　使用 MediaPlayer 建立音乐播放器
UNIT 72　播放背景音乐和 Audio Focus
UNIT 73　录音程序
UNIT 74　播放视频
UNIT 75　拍照程序
UNIT 76　录像程序

Android 版本	1.X	2.X	3.X	4.X
适用性	★	★	★	★

UNIT 71
使用 MediaPlayer 建立音乐播放器

播放音乐和影片是智能手机和平板电脑最吸引人的应用之一，当我们使用系统内建的多媒体播放程序时，心中或许会有一些好奇，想知道究竟究竟程序是如何播放影片和音乐，这个单元就让我们来学习建立一个音乐播放器。

71-1 音乐播放程序的架构

音乐播放程序和一般应用程序最大的不同就是会发出声音，而且是悦耳的旋律。这项工作牵涉从数据至声音信号的转换和发声设备的控制，其中包含许多专业技术。如果我们将一个音乐播放程序拆开来看，它可以分成图 71-1 所示的四个部分。

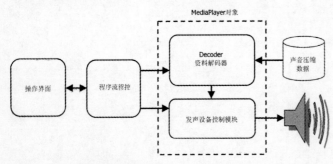

图 71-1　音乐播放程序的架构

1. 资料译码器（Decoder）

   数据译码器负责将数字的压缩数据还原成可以播放的声音信号，声音信号的译码是一门专业的学问，一般人很难自己写程序来完成这项工作，因此 Android 内建一个 MediaPlayer 类帮助程序开发者完成声音数据的解码。

> **补充说明** **Audio Codec**
>
> Codec 是由 coder 和 decoder 这二个单字结合而成，它的意思是编码器和译码器。当我们要将声音信号储存成文件的时候，必须先将声音信号经过取样处理（sampling）再将它压缩（更正确的说法是破坏性压缩，代表压缩后的数据无法再还原成和原来一模一样）以节省后储存空间。如何在提高压缩率的同时保有近似原来声音的质量，还有考虑运行时的速度，这就是研发 codec 算法的重点。

2. 发声设备控制模块

   程序必须控制手机或平板电脑的发声设备，然后将译码后的声音数据传送给发声设备进行播放。控制发声设备需要了解硬件设备驱动器的运行，因此也是一件复杂的工作，幸好 MediaPlayer 类也将这一部分纳入，我们只需要利用它提供的方法就可以同时完成声音文件的译码和播放，也能够设置声音的大小。

3. 程序流程控制

   在播放音乐的过程中我们可以暂停或是完全停止，也可以跳至指定的位置。这些操作可以任意排列组合，例如播放→暂停→播放→停止→…，或是播放→停止→跳至指定位置→播放→…。读者或许认为这很正常，但是当核心 decoder（也就是 MediaPlayer）在运行时，必须遵守一定的规则，否则会出现运行时期错误（或称为例外错误），因此程序必须根据用户的操作和目前的运行状态，适当的控制 MediaPlayer 以维持正常运行。

4. 程序的操作接口

   也就是程序运行时用户看到的界面，这个部分就看程序设计者的巧思，重点是让界面清楚、美观，以及考虑操作的便利性。

从以上的介绍我们可以了解 MediaPlayer 类在建立音乐播放器的过程中扮演非常重要的角色，因此在着手开发程序之前，我们必须先学会使用 MediaPlayer 类。

71-2　MediaPlayer 类的用法

从图 71-1 中可以了解 MediaPlayer 类的功能包括数据的译码和发声设备的控制，也就是说它帮我们完成音乐播放程序中最困难和麻烦的工作，而且支持的声音文件格式也很完整，读者可以参考表 71-1。另外如前一个小节的介绍，MediaPlayer 的运行有既定的法则，如果没有遵守它的规定，程序马上会回报一个例外错误。图 71-2 是 Android SDK 技术文件提供的 MediaPlayer 对象运行状态流程图，它是操作 MediaPlayer 对象时很重要的参考数据，举例来说，当我们开始播放声音文件时，必须先运行 prepare()或是 prepareAsync()完成播放前的准备工作，然后 MediaPlayer 便会进入 Prepared 状态，之后才能开始播放音乐。又例如 MediaPlayer 进入 Stopped 状态之后，也必须重新运行 prepare()或是 prepareAsync()才能再次播放。

表 71-1　MediaPlayer 类支持的声音文件格式

Codec 技术名称	文件格式	编码器（Coder）	译码器（Decoder）
AAC LC/LTP AAC+ (HE-AACv1) enhanced AAC+ (HE-AACv2)	.aac (Android 3.1+) .mp4 .m4a .3gp .ts (Android 3.0+)	AAC LC/LTP (Android 4.0+)	全部支持
AMR-NB AMR-WB	.3gp	全部支持	全部支持
FLAC	.flac	不支持	支持(Android 3.1+)
MP3	.mp3	不支持	支持
MIDI	.mid, .xmf, .mxmf .rtttl, .rtx, .ota, .imy	不支持	支持
Vorbis	.ogg .mkv (Android 4.0+)	不支持	支持
PCM/WAVE	.wav	不支持	支持

　　读者心中也许还有一个疑问，究竟 prepare()和 prepareAsync()有何不同？这是因为如果遇到大型文件或是需要从网络下载数据时，MediaPlayer 需要一些时间来完成播放前的准备工作，如果这个准备工作是在主程序的 main thread 中运行，程序将会停止响应用户的操作，这对使用者来说会是一个负面的感受，甚至会认为系统出现问题。如果换成使用 prepareAsync()方法，它会另外建立一个 background thread 来运行准备工作，这样主程序就可以继续响应用户的操作。

　　由于 prepareAsync()方法是以 multi-thread 异步方式运行，它必须使用 callback method 的方式来通知主程序最后的运行结果，包括遇到错误的时候。另外一个类似的情况就是当 MediaPlayer 对象播放完文件时，也会以 callback method 通知主程序。根据以上的讨论，音乐播放主程序必须完成下列三项工作：

1. 让主程序类实现相关接口。
2. 完成接口中所订定的方法的程序代码。
3. 将主程序类设置给 MediaPlayer 对象，于是它就会在适当的时机调用这些方法。

拍照、录音、录像与多媒体播放 **PART 14**

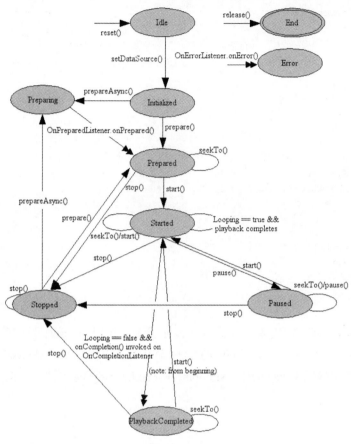

图 71-2 MediaPlayer 对象的运行状态流程图

以下是主程序的范例：

```
// 让主程序类实现相关接口
public class Main extends Activity
    implements MediaPlayer.OnPreparedListener,
               MediaPlayer.OnErrorListener,
               MediaPlayer.OnCompletionListener {

    @Override
    protected void onResume() {
        // TODO Auto-generated method stub
        super.onResume();

        mMediaPlayer = new MediaPlayer();

        // 将主程序类设置给 MediaPlayer 对象
        mMediaPlayer.setOnPreparedListener(this);
        mMediaPlayer.setOnErrorListener(this);
```

```
            mMediaPlayer.setOnCompletionListener(this);
}

@Override
public void onPrepared(MediaPlayer mp) {
    // TODO Auto-generated method stub
    (prepareAsync()方法运行完成后会调用这个方法)
}

@Override
public boolean onError(MediaPlayer mp, int what, int extra) {
    // TODO Auto-generated method stub
    (prepareAsync()方法运行过程中出现错误时会调用这个方法)
}

@Override
public void onCompletion(MediaPlayer arg0) {
    // TODO Auto-generated method stub
    (MediaPlayer 对象播放完文件后会调用这个方法)
}
}
```

当程序不再需要 MediaPlayer 对象时（例如程序将要结束运行），必须马上调用 release()方法释放所占用的系统资源。因此我们必须在主程序类中加入 onStop()方法，然后在其中调用 MediaPlayer 对象的 release()。另外 MediaPlayer 类也提供 isPlaying()、isLooping()和 setLooping()等方法，让程序可以侦测播放状态和设置播放模式。

71-3　范例程序

在了解 MediaPlayer 的功能和用法之后，我们就可以开始着手建立音乐播放程序。这个音乐播放程序的重点在于示范如何正确地控制 MediaPlayer，让使用者可以随心所欲地播放、暂停、停止或是跳至指定的时间位置，而不会出现例外错误。图 71-3 是程序的运行界面，最上面的"播放"按钮会随着播放状态的改变显示"暂停"或是"播放"，完成这个范例程序的步骤如下：

图 71-3　音乐播放程序的运行界面，最上面的"播放"按钮会自动切换"暂停"或"播放"功能

Step 1 新增一个 Android 程序项目，项目属性的设置依照前面的惯例即可。

Step 2 在 Eclipse 左边的项目检查窗口中展开此程序项目的 res/layout 文件夹，开启其中的接口布局文件 main.xml 编辑如下，其中有一个新的 ToggleButton 组件，它可以用来切换 on 和 off 两种状态，读者参考后续的程序代码就可以了解它的用法。

```xml
<?xml version="1.0" encoding="utf-8"?>
<LinearLayout xmlns:android="http://schemas.android.com/apk/res/android"
    android:orientation="vertical"
    android:layout_width="match_parent"
    android:layout_height="match_parent"
    android:gravity="center_horizontal"
    >
<ImageButton android:id="@+id/btnMediaPlayPause"
    android:layout_width="wrap_content"
    android:layout_height="wrap_content"
    android:src="@android:drawable/ic_media_play"
    />
<ImageButton android:id="@+id/btnMediaStop"
    android:layout_width="wrap_content"
    android:layout_height="wrap_content"
    android:src="@android:drawable/ic_menu_close_clear_cancel"
    />
<ToggleButton android:id="@+id/btnMediaRepeat"
    android:layout_width="wrap_content"
    android:layout_height="wrap_content"
    android:background="@android:drawable/ic_menu_revert"
    />
<EditText android:id="@+id/edtMediaGoto"
    android:layout_width="60sp"
    android:layout_height="wrap_content"
    />
<ImageButton android:id="@+id/btnMediaGoto"
    android:layout_width="wrap_content"
    android:layout_height="wrap_content"
    android:src="@android:drawable/ic_menu_directions"
    />
<Button android:id="@+id/btnInsertMediaToMediaStore"
    android:layout_width="wrap_content"
    android:layout_height="wrap_content"
    android:text="加入 mp3 档"
    />
</LinearLayout>
```

Step 3 在 Eclipse 左边的项目检查窗口右击程序项目的 res 文件夹，然后选择 New > Folder，在对话框的 Folder name 字段输入 raw，按下 Finish 按钮。这个步骤是建立一个用来储存音乐文件的文件夹，我们将把音乐文件存放在程序项目的 res/raw 文件夹中，Android 程序编译程序会保留这个文件夹中的文件不会进行任何处理。

Step 4 利用 Windows 文件管理器将某个 mp3 音乐文件复制到前一个步骤建立的文件夹中并更名为 song.mp3。

Step 5 在 Eclipse 左边的项目检查窗口中展开程序项目的 src/(组件路径名称)文件夹，开启其中的主程序文件 Main.java，然后让它实现 MediaPlayer.OnPreparedListener、MediaPlayer.OnErrorListener 和 MediaPlayer.OnCompletionListener 三个接口，并根据程序编辑窗口的错误修正提示加入所需的方法。

> **补充说明　Eclipse 操作技巧**
>
> 先在类定义的第一行（也就是含有类名称的那一行）输入 implements MediaPlayer.OnPreparedListener, MediaPlayer.OnErrorListener, MediaPlayer.OnCompletionListener，类名称下方会出现红色波浪底线，再把鼠标光标移到红色波浪底线上方就会弹出信息窗口，选择其中的 Add unimplemented methods 就会自动加入所需的方法。

```
package …

import …

public class Main extends Activity
            implements MediaPlayer.OnPreparedListener,
                        MediaPlayer.OnErrorListener,
                        MediaPlayer.OnCompletionListener {
    /** Called when the activity is first created. */
    @Override
    public void onCreate(Bundle savedInstanceState) {
        super.onCreate(savedInstanceState);
        setContentView(R.layout.main);
    }

    @Override
    public void onCompletion(MediaPlayer mp) {
        // TODO Auto-generated method stub

    }

    @Override
    public boolean onError(MediaPlayer mp, int what, int extra) {
        // TODO Auto-generated method stub
        return false;
    }
```

```
    @Override
    public void onPrepared(MediaPlayer mp) {
        // TODO Auto-generated method stub

    }
}
```

Step 6 在主程序的编辑窗口中右击，然后选择 Source > Override/Implement Methods…，从弹出的方法清单对话框中勾选 onResume() 和 onStop() 两个方法，完成后按下 OK 按钮。

Step 7 最后将主程序的程序代码编辑如下：

```
public class Main extends Activity
    implements MediaPlayer.OnPreparedListener,
               MediaPlayer.OnErrorListener,
               MediaPlayer.OnCompletionListener {

    …（接口组件的定义请参考光盘中的程序项目）

    // 程序使用的 MediaPlayer 对象
    private MediaPlayer mMediaPlayer = null;

    // 用来记录是否 MediaPlayer 对象需要运行 prepareAsync()
    private boolean mBoolIsInitial = true;

    /** Called when the activity is first created. */
    @Override
    public void onCreate(Bundle savedInstanceState) {
        super.onCreate(savedInstanceState);
        setContentView(R.layout.main);

        setupViewComponent();
    }

    @Override
    protected void onResume() {
        // TODO Auto-generated method stub
        super.onResume();

        mMediaPlayer = new MediaPlayer();

        Uri uri = Uri.parse("android.resource://" +
            getPackageName() + "/" + R.raw.song);

        try {
            mMediaPlayer.setDataSource(this, uri);
        } catch (Exception e) {
            // TODO Auto-generated catch block
```

```java
            Toast.makeText(Main.this, "指定的播放文件错误！",
            Toast.LENGTH_LONG)
                .show();
        }

        mMediaPlayer.setOnPreparedListener(this);
        mMediaPlayer.setOnErrorListener(this);
        mMediaPlayer.setOnCompletionListener(this);
    }

    @Override
    protected void onStop() {
        // TODO Auto-generated method stub
        super.onStop();

        mMediaPlayer.release();
        mMediaPlayer = null;
    }

    private void setupViewComponent() {
        …（设置接口组件的程序代码，请参考光盘中的程序项目）
    }

    private Button.OnClickListener btnMediaPlayPauseLis = new
                                                Button.OnClickListener() {
        public void onClick(View v) {
            if (mMediaPlayer.isPlaying()) {
mBtnMediaPlayPause.setImageResource(android.R.drawable.ic_media_play);
                mMediaPlayer.pause();
            }
            else {
mBtnMediaPlayPause.setImageResource(android.R.drawable.ic_media_pause);
                if (mBoolIsInitial) {
                    mMediaPlayer.prepareAsync();
                    mBoolIsInitial = false;
                }
                else
                    mMediaPlayer.start();
            }
        }
    };

    private Button.OnClickListener btnMediaStopLis = new Button.OnClickListener() {
        public void onClick(View v) {
            mMediaPlayer.stop();

            // 停止播放后必须再运行 prepareAsync()
```

```java
            // 或 prepare() 才能重新播放。
            mBoolIsInitial = true;
            mBtnMediaPlayPause.setImageResource
            (android.R.drawable.ic_media_play);
        }
    };

    private Button.OnClickListener btnMediaRepeatLis = new
    Button.OnClickListener() {
        public void onClick(View v) {
            if (((ToggleButton)v).isChecked())
                mMediaPlayer.setLooping(true);
            else
                mMediaPlayer.setLooping(false);
        }
    };

    private Button.OnClickListener btnMediaGotoLis = new
    Button.OnClickListener() {
        public void onClick(View v) {
            if(mEdtMediaGoto.getText().toString().equals("")) {
                Toast.makeText(Main.this,
                    "请先输入要播放的位置（以秒为单位）",
                    Toast.LENGTH_LONG)
                    .show();
                return;
            }

            int seconds = Integer.parseInt(mEdtMediaGoto.
            getText().toString());
            mMediaPlayer.seekTo(seconds * 1000); // 以毫秒
            （千分之一秒）为单位
        }
    };

    @Override
    public void onPrepared(MediaPlayer mp) {
        // TODO Auto-generated method stub
        mp.seekTo(0);
        mp.start();

        Toast.makeText(Main.this, "开始播放", Toast.LENGTH_LONG)
            .show();
    }

    @Override
    public boolean onError(MediaPlayer mp, int what, int extra) {
        // TODO Auto-generated method stub
```

```
            mp.release();
            mp = null;

            Toast.makeText(Main.this, "发生错误，停止播放", Toast.LENGTH_LONG)
                .show();

            return true;
        }

        @Override
        public void onCompletion(MediaPlayer arg0) {
            // TODO Auto-generated method stub
            mBtnMediaPlayPause.setImageResource(android.R.
            drawable.ic_media_play);
        }
    }
```

我们定义一个 mBoolIsInitial 变量用来记录是否 MediaPlayer 对象需要运行 prepareAsync()。当程序启动的时候 Android 系统会先运行 onResume()方法，我们在该方法中设置好 MediaPlayer 对象，并利用 setDataSource()指定播放的文件。在 onStop()中我们运行 release()清除 MediaPlayer 对象。接下来是按下不同的按钮后要运行的程序代码，编写这些程序代码时要注意使用者可能出现的各种操作流程，务必让 MediaPlayer 对象的状态符合图 71-2 的运行规则。最后的 onPrepared()、onError()和 onCompletion()就是前面解释过的 callback method，我们根据 MediaPlayer 对象的状态进行适当的处理。

完成程序代码的编辑之后就可以运行程序测试播放音乐的功能，读者可以试看看不同的操作流程。这个音乐播放程序有一个美中不足之处，就是 MediaPlayer 对象是在主程序中运行，因此一离开程序音乐就会停止，无法做到背景播放音乐的效果。下一个单元我们将继续改良这个程序，让它实现播放背景音乐的功能。

Android 版本	1.X	2.X	3.X	4.X
适用性	★	★	★	★

UNIT 72

播放背景音乐和 Audio Focus

延续上一个单元的程序,我们希望将它改成使用后台运行的方式,这样就可以一边运行其他工作,像是收发电子邮件或是浏览网页,一边听音乐。要能够做到背景播放音乐的功能必须完成下列四项工作:

1. MediaPlayer 对象必须在背景运行。其实在 UNIT 49 的范例程序中我们已经实现播放背景音乐的功能,当时是采用 Service 对象的方式来运行 MediaPlayer,这个单元我们将使用同样的方法。
2. 当 MediaPlayer 在后台运行时必须能够随时加以控制,像是停止播放,这项功能需要用到状态栏信息。我们在 UNIT 53 曾经学过状态栏信息的用法,这个单元我们将采用类似的方式来控制背景运行的 MediaPlayer。
3. 正在播放背景音乐的时候如果刚好有电话打进来,音乐必须自动暂停,否则可能会听不到铃声,这项功能叫做 Audio Focus。Android 系统会在必要的时候主动通知运行中的程序,告诉它们目前有重要的声音要播放,当播放音乐的程序收到这个信息时,必须采取适当的动作。
4. 使用背景运行的方式播放音乐时,如果用户没有继续操作其他功能,经过一段时间之后,手机或平板电脑就会自动进入休眠状态以节省电力,此时 CPU 也会停止运行造成音乐播放终止。为了避免这种情况,程序必须告知系统继续维持 CPU 的运行。

接下来我们依序介绍完成这四项工作的方法。

72-1 用 Service 的方式运行 MediaPlayer

如果要把 MediaPlayer 放到 Service 对象中运行,只要在程序项目中新增一个继承 Service 的类,例如可以将它取名为 MediaPlayerService，然后在该类中建立 MediaPlayer 对象即可，而且同样要让这个类实现 MediaPlayer.OnPreparedListener、MediaPlayer.OnErrorListener 和 MediaPlayer.OnCompletionListener 等三个接口和相关方法。然后我们在 onCreate()中设置好 MediaPlayer 对象，在 onDestroy()中清除 MediaPlayer 对象，在 onStartCommand()方法中根据主程序传送过来的 Intent 对象中的指令控制 MediaPlayer 的运行，请读者参考以下程序代码：

```java
public class MediaPlayerService extends Service
        implements MediaPlayer.OnPreparedListener,
                   MediaPlayer.OnErrorListener,
                   MediaPlayer.OnCompletionListener {

    // 程序使用的 MediaPlayer 对象
    private MediaPlayer mMediaPlayer = null;

    // 用来记录是否 MediaPlayer 对象需要运行 prepareAsync()
    private boolean mBoolIsInitial = true;

    @Override
    public IBinder onBind(Intent arg0) {
        // TODO Auto-generated method stub
        return null;
    }

    @Override
    public void onCreate() {
        // TODO Auto-generated method stub
        super.onCreate();

        …(设置 MediaPlayer 对象的功能，如同前一个单元范例程序中的 onResume())
    }

    @Override
    public void onDestroy() {
        // TODO Auto-generated method stub
        super.onDestroy();

        …(清除 MediaPlayer 对象，如同前一个单元范例程序中的 onStop())
    }

    @Override
    public int onStartCommand(Intent intent, int flags, int startId) {
        // TODO Auto-generated method stub

        …(根据主程序传送过来的 Intent 对象中的指令控制 MediaPlayer 的运行)

        return super.onStartCommand(intent, flags, startId);
```

```
        }

        @Override
        public void onPrepared(MediaPlayer mp) {
            // TODO Auto-generated method stub

            …(启动 MediaPlayer 对象开始播放音乐,和前一个单元的范例程序相同)
        }

        @Override
        public boolean onError(MediaPlayer mp, int what, int extra) {
            // TODO Auto-generated method stub

            …(显示 MediaPlayer 对象运行时发生错误,和前一个单元的范例程序相同)
        }

        @Override
        public void onCompletion(MediaPlayer arg0) {
            // TODO Auto-generated method stub
            mMediaPlayer.release();
            mMediaPlayer = null;

            stopForeground(true);

            mBoolIsInitial = true;
        }
}
```

72-2 使用状态栏信息控制 Foreground Service

一般的 Service 对象是启动之后就在后台运行直到工作完成就自动结束,如果我们希望在 Service 运行的过程中可以随时通过 Activity 重新设置 Service,那么可以借助状态栏信息(提示:有关状态栏信息的用法请参考 UNIT 53 的介绍),而这种具有控制端 Activity 的 Service 称为 Foreground Service,让 Service 成为 Foreground Service 的步骤如下:

Step 1 在程序项目中建立一个用来控制 Service 的 Activity 类。

Step 2 在 Service 类中完成以下程序代码:

1. 建立一个 Intent 对象,并指定前一个步骤所建立的主控 Activity 类。
2. 建立一个 PendingIntent 对象,设置它的运行环境,并传入前一个步骤建立的 Intent 对象,然后指定处理方式为 FLAG_UPDATE_CURRENT。
3. 建立一个 Notification 对象,这个对象包含要显示的信息和放在信息前面的小图标,以及指定显示这个信息的时间。

```
Notification noti = new Notification(
        要使用的小图标,
        "要显示的信息",
        System.currentTimeMillis());
```

4. 调用 Notification 对象的 setLatestEventInfo()方法，指定展开状态栏后要显示的信息标题和内容。另外也可以根据设置 Notification 对象的 flags 属性指定它的类型。

```
noti.setLatestEventInfo(this, "信息标题", "显示的信息", penIt);
noti.flags |= Notification.FLAG_ONGOING_EVENT;
```

5. 运行 startForeground()传入此状态栏信息的 id 和前面建立的 Notification 对象。

```
startForeground(1, noti)
```

Step 3 在 Service 类的 onDestroy()方法中运行 stopForeground(true)取消显示中的状态栏信息。我们可以把步骤二的程序代码写在 Service 类的 onPrepared()方法中，如以下范例。当开始播放音乐的时候就会显示状态栏信息，如果是在平板电脑运行可以单击屏幕右下方的信息小图标就可以展开完整的信息，如图 72-1 所示。如果是在手机上运行则可以按住屏幕上方的状态栏往下拉就会显示信息，然后点一下该信息就可以启动控制用的 Activity。

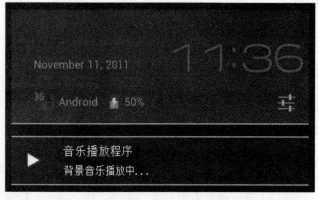

图 72-1 Foreground Service 运行时显示的状态栏信息

```java
public class MediaPlayerService extends Service
    implements MediaPlayer.OnPreparedListener,
            MediaPlayer.OnErrorListener,
            MediaPlayer.OnCompletionListener {

    …（其他程序代码）

    @Override
    public void onPrepared(MediaPlayer mp) {
        // TODO Auto-generated method stub
```

```java
        …（其他程序代码）

        Intent it = new Intent(getApplicationContext(),
        Main.class);
        PendingIntent pendIt = PendingIntent.getActivity(
            getApplicationContext(), 0, it,
            PendingIntent.FLAG_UPDATE_CURRENT);
        Notification noti = new Notification(
            android.R.drawable.ic_media_play,
            "背景音乐播放中...",
            System.currentTimeMillis());
        noti.setLatestEventInfo(getApplicationContext(),
            "音乐播放程序", "背景音乐播放中...", pendIt);
        noti.flags |= Notification.FLAG_ONGOING_EVENT;
        startForeground(1, noti);
    }

    …（其他程序代码）
}
```

72-3 使用 Audio Focus 和 Wake Lock

Audio Focus 的目的是要接受 Android 系统的通知，适时改变 MediaPlayer 的状态，例如当接到来电时必须停止播放音乐，当通话结束时再自动启动播放。使用 Audio Focus 的方式是让类实现 AudioManager.OnAudioFocusChangeListener 接口，然后编写 onAudioFocusChange() 方法中的程序代码，Android 系统会用 focusChange 自变量通知程序 Audio Focus 改变的情况，程序再依此做适当的处理，请读者参考以下范例和程序代码批注：

```java
public class MediaPlayerService extends Service
    implements AudioManager.OnAudioFocusChangeListener {

    …（其他程序代码）

    @Override
    public void onAudioFocusChange(int focusChange) {
        // TODO Auto-generated method stub

        if (mMediaPlayer == null)
            return;

        switch (focusChange) {
        case AudioManager.AUDIOFOCUS_GAIN:
            // 程序取得声音播放权
            mMediaPlayer.setVolume(0.8f, 0.8f);
```

```
                    mMediaPlayer.start();
                    break;
            case AudioManager.AUDIOFOCUS_LOSS:
                    // 程序尚失声音播放权，而且时间可能很久
                    stopSelf();         // 结束这个 Service
                    break;
            case AudioManager.AUDIOFOCUS_LOSS_TRANSIENT:
                    // 程序丧失声音播放权，但预期很快就会再取得
                    if (mMediaPlayer.isPlaying())
                    mMediaPlayer.pause();
                    break;
            case AudioManager.AUDIOFOCUS_LOSS_TRANSIENT_CAN_DUCK:
                    // 程序丧失声音播放权，但是可以用很小的音量继续播放
                    if (mMediaPlayer.isPlaying())
                    mMediaPlayer.setVolume(0.1f, 0.1f);
                    break;
            }
        }
    …（其他程序代码）
}
```

另外如果要让 CPU 持续运行不要进入休眠状态而导致播放中断，可以调用 MediaPlayer 对象的 setWakeMode()方法并指定适当的参数，但是这个方法只能在实际的手机或平板电脑上运行，如果在模拟器运行会发生错误。

72-4　播放不同来源的文件

为了简化程序，上一个单元的音乐文件是直接储存在程序项目中，其实 MediaPlayer 对象可以播放下列四种文件来源：

1. 在程序项目的 res/raw 文件夹中的音乐文件，如同上一个单元的范例程序。
2. 在手机或平板电脑的 sd card 中的音乐文件，我们可以用以下程序代码指定播放档的来源：

   ```
   File file = new File("/sdcard/song.mp3");
   Uri uri = Uri.fromFile(file);
   ```

3. 从 "http://.../(完整文件名)" 网址取得播放文件，此时可以用以下程序代码指定播放文件的来源：

   ```
   Uri uri = Uri.parse("http://.../song.mp3");
   mMediaPlayer.setAudioStreamType(AudioManager.STREAM_MUSIC);
   ```

4. 从 Android 系统内建的数据库取得播放文件，这种方式需要使用 ContentResolver 对象（提示：有关 ContentResolver 和 ContentProvider 的用法可以参考 UNIT 56 的说明）。以下我们直接列出实现的程序代码，这一段程序代码的功能是取得数据库中的第一个音乐文件：

```
ContentResolver contRes = getContentResolver();
Cursor c = contRes.query(
            android.provider.MediaStore.Audio.Media.EXTERNAL_
            CONTENT_URI, null, null, null, null);

Uri uri = null;
if (c == null) {
    Toast.makeText(MediaPlayerService.this, "Content Resolver 错误！",
        Toast.LENGTH_LONG) .show();
}
else if (!c.moveToFirst()) {
    Toast.makeText(MediaPlayerService.this, "数据库中没有数据！",
        Toast.LENGTH_LONG) .show();
}
else {
    int idColumn = c.getColumnIndex(android.provider.MediaStore.
        Audio.Media._ID);
    long id = c.getLong(idColumn);
    uri = ContentUris.withAppendedId(
        android.provider.MediaStore.Audio.Media.EXTERNAL_CONTENT_URI, id);
}

mMediaPlayer.setAudioStreamType(AudioManager.STREAM_MUSIC);
```

> **补充说明** **Android 系统的多媒体数据库**
>
> Android 系统中的多媒体数据库可以用来管理图片、音乐和影片文件，所有应用程序都可以利用 ContentResolver 对象来取用多媒体数据库中的文件，也可以将多媒体文件登录到数据库。

72-5　范例程序

这个单元的范例程序是把上一个单元的音乐播放器改成背景运行模式，并且换成从 Android 系统的数据库中取得播放文件。我们在程序中增加一个"加入 mp3 文件"按钮，用来把 sd card 中的 song.mp3 音乐文件加入数据库，然后就可以按下"开始播放"按钮播放音乐（提示：有关模拟器的 sd card 操作方式请参考 UNIT 38 的说明）。在主程序中我们用另一种方式来设置 Button 的 OnClickListener。首先让主程序类实现 OnClickListener 接口，然后依照程序代码编辑器的要求新增一个 onClick() 方法，在该方法中我们利用检查 View 对象的 id 的方式决定目前被按下的按钮，然后根据 Intent 对象来控制 Service，最后将这个类设置给所有 Button 当成是 OnClickListener。另外在程序项目中建立 Service 类后，也要记得在 AndroidManifest.xml 文件中加入 Service 的记录才能正常运行。图 72-2 是完成后的程序界面，以下是完整的接口布局文件、主程序文件和

Service 程序文件。

图 72-2　背景音乐播放程序的运行界面

■　界面布局文件：

```xml
<?xml version="1.0" encoding="utf-8"?>
<LinearLayout xmlns:android="http://schemas.android.com/apk/res/android"
    android:orientation="vertical"
    android:layout_width="match_parent"
    android:layout_height="match_parent"
    android:gravity="center_horizontal"
    >
    <Button android:id="@+id/btnStart"
        android:layout_width="wrap_content"
        android:layout_height="wrap_content"
        android:text="开始播放"
        />
    <Button android:id="@+id/btnPause"
        android:layout_width="wrap_content"
        android:layout_height="wrap_content"
        android:text="暂停播放"
        />
    <Button android:id="@+id/btnStop"
        android:layout_width="wrap_content"
        android:layout_height="wrap_content"
        android:text="停止播放"
        />
    <Button android:id="@+id/btnSetRepeat"
        android:layout_width="wrap_content"
        android:layout_height="wrap_content"
```

```xml
        android:text="重复播放"
        />
    <Button android:id="@+id/btnCancelRepeat"
        android:layout_width="wrap_content"
        android:layout_height="wrap_content"
        android:text="取消重复播放"
        />
    <EditText android:id="@+id/edtGoto"
        android:layout_width="60sp"
        android:layout_height="wrap_content"
        />
    <Button android:id="@+id/btnGoto"
        android:layout_width="wrap_content"
        android:layout_height="wrap_content"
        android:text="跳至指定位置（秒）"
        />
    <Button android:id="@+id/btnAddToMediaStore"
        android:layout_width="wrap_content"
        android:layout_height="wrap_content"
        android:text="加入 mp3 檔"
        />
</LinearLayout>
```

- 主程序文件：

```java
package ...

import ...

public class Main extends Activity
    implements OnClickListener{

    private Button mBtnAddToMediaStore,
            mBtnStart, mBtnPause,
            mBtnStop, mBtnSetRepeat,
            mBtnCancelRepeat, mBtnGoto;

    private EditText mEdtGoto;

    /** Called when the activity is first created. */
    @Override
    public void onCreate(Bundle savedInstanceState) {
        super.onCreate(savedInstanceState);
        setContentView(R.layout.main);

        setupViewComponent();
    }
```

```java
private void setupViewComponent() {
    mBtnStart = (Button)findViewById(R.id.btnStart);
    mBtnPause = (Button)findViewById(R.id.btnPause);
    mBtnStop = (Button)findViewById(R.id.btnStop);
    mBtnSetRepeat = (Button)findViewById(R.id.btnSetRepeat);
    mBtnCancelRepeat = (Button)findViewById(R.id.btnCancelRepeat);
    mBtnGoto = (Button)findViewById(R.id.btnGoto);
    mBtnAddToMediaStore = (Button)findViewById(R.id.
    btnAddToMediaStore);
    mEdtGoto = (EditText)findViewById(R.id.edtGoto);

    mBtnStart.setOnClickListener(this);
    mBtnPause.setOnClickListener(this);
    mBtnStop.setOnClickListener(this);
    mBtnSetRepeat.setOnClickListener(this);
    mBtnCancelRepeat.setOnClickListener(this);
    mBtnGoto.setOnClickListener(this);
    mBtnAddToMediaStore.setOnClickListener(btnAddToMediaStoreLis);
}

@Override
public void onClick(View v) {
    // TODO Auto-generated method stub

    Intent it;

    switch(v.getId()) {
    case R.id.btnStart:
        it = new Intent(Main.this, MediaPlayerService.class);
        it.setAction(MediaPlayerService.ACTION_PLAY);
        startService(it);
        break;
    case R.id.btnPause:
        it = new Intent(Main.this, MediaPlayerService.class);
        it.setAction(MediaPlayerService.ACTION_PAUSE);
        startService(it);
        break;
    case R.id.btnStop:
        it = new Intent(Main.this, MediaPlayerService.class);
        stopService(it);
        break;
    case R.id.btnSetRepeat:
        it = new Intent(Main.this, MediaPlayerService.class);
        it.setAction(MediaPlayerService.ACTION_SET_REPEAT);
        startService(it);
        break;
    case R.id.btnCancelRepeat:
        it = new Intent(Main.this, MediaPlayerService.class);
```

```java
            it.setAction(MediaPlayerService.ACTION_CANCEL_REPEAT);
            startService(it);
            break;
        case R.id.btnGoto:
            if (mEdtGoto.getText().toString().equals("")) {
                Toast.makeText(Main.this,
                    "请先输入要播放的位置（以秒为单位）",
                    Toast.LENGTH_LONG)
                    .show();
                break;
            }

            int seconds = Integer.parseInt(mEdtGoto.getText().toString());

            it = new Intent(Main.this, MediaPlayerService.class);
            it.setAction(MediaPlayerService.ACTION_GOTO);
            it.putExtra("GOTO_POSITION_SECONDS", seconds);
            startService(it);
            break;
        }
    }

    private Button.OnClickListener btnAddToMediaStoreLis = new Button.OnClickListener() {
        public void onClick(View v) {
            ContentValues val = new ContentValues();
            val.put(MediaColumns.TITLE, "my mp3");
            val.put(MediaColumns.MIME_TYPE, "audio/mp3");
            val.put(MediaColumns.DATA, "/sdcard/song.mp3");
            ContentResolver contRes = getContentResolver();
            Uri newUri = contRes.insert(
                android.provider.MediaStore.Audio.Media.EXTERNAL_CONTENT_URI,
                val);
            sendBroadcast(new Intent(Intent.ACTION_MEDIA_SCANNER_SCAN_FILE,
                newUri));
        }
    };
}
```

■ Service 程序文件：

```java
package ...

import ...

public class MediaPlayerService extends Service
    implements MediaPlayer.OnPreparedListener,
        MediaPlayer.OnErrorListener,
```

```java
            MediaPlayer.OnCompletionListener,
            AudioManager.OnAudioFocusChangeListener {

    public static final String
        ACTION_PLAY = "tw.android.mediaplayer.action.PLAY",
        ACTION_PAUSE = "tw.android.mediaplayer.action.PAUSE",
        ACTION_SET_REPEAT = "tw.android.mediaplayer.action.SET_REPEAT",
        ACTION_CANCEL_REPEAT = "tw.android.mediaplayer.action.CANCEL_REPEAT",
        ACTION_GOTO = "tw.android.mediaplayer.action.GOTO";

    // 程序使用的 MediaPlayer 对象
    private MediaPlayer mMediaPlayer = null;

    // 用来记录是否 MediaPlayer 对象需要运行 prepareAsync()
    private boolean      mBoolIsInitial = true,
                         mBoolAudioFileFound = false;

    @Override
    public IBinder onBind(Intent arg0) {
        // TODO Auto-generated method stub
        return null;
    }

    @Override
    public void onCreate() {
        // TODO Auto-generated method stub
        super.onCreate();

        // MediaStore 是用来指定图像、音频或影片类型的数据
        // 这种方式要设置 stream type,
        // 这个播放文件必须先主本程序的"加入 mp3 文件"按钮加入
        ContentResolver contRes = getContentResolver();
        String[] columns = {
                MediaColumns.TITLE,
                MediaColumns._ID};
        Cursor c = contRes.query(
                android.provider.MediaStore.Audio.Media.EXTERNAL_
                CONTENT_URI, columns, null, null, null);

        Uri uri = null;
        if (c == null) {
            Toast.makeText(MediaPlayerService.this, "Content
                Resolver 错误！", Toast.LENGTH_LONG)
                    .show();
            return;
        }
        else if (!c.moveToFirst()) {
            Toast.makeText(MediaPlayerService.this, "数据库中没有资
```

```
                料！", Toast.LENGTH_LONG)
                        .show();
                return;
        }
        else {
                do {
                        String title = c.getString(c.getColumnIndex
                        (MediaColumns.TITLE));
                        if (title.equals("my mp3")) {
                                mBoolAudioFileFound = true;
                                break;
                        }
                } while(c.moveToNext());

                if (! mBoolAudioFileFound) {
                        Toast.makeText(MediaPlayerService.this, "找不
                        到指定的 mp3 文件！", Toast.LENGTH_LONG)
                            .show();
                        return;
                }

                int idColumn = c.getColumnIndex(android.provider.
                MediaStore.Audio.Media._ID);
                long id = c.getLong(idColumn);
                uri = ContentUris.withAppendedId(
                        android.provider.MediaStore.Audio.
                        Media.EXTERNAL_CONTENT_URI, id);
        }

        mMediaPlayer = new MediaPlayer();
        mMediaPlayer.setAudioStreamType(AudioManager.STREAM_MUSIC);

        try {
                mMediaPlayer.setDataSource(this, uri);
        } catch (Exception e) {
                // TODO Auto-generated catch block
                Toast.makeText(MediaPlayerService.this, "指定的播放文件错误！", Toast.LENGTH_LONG).show();
        }

        mMediaPlayer.setOnPreparedListener(this);
        mMediaPlayer.setOnErrorListener(this);
        mMediaPlayer.setOnCompletionListener(this);

        // 设置 Media Player 在背景运行时，让 CPU 维持运转
        // 如果播放的是来自网络的 streaming audio，
        // 还要设置网络维持运行
        // 只能在实体设备上使用，模拟器运行时会产生错误
        //   mMediaPlayer.setWakeMode(getApplicationContext(),
```

```java
    //     PowerManager.PARTIAL_WAKE_LOCK);
}

@Override
public void onDestroy() {
    // TODO Auto-generated method stub
    super.onDestroy();

    if (mBoolAudioFileFound) {
        mMediaPlayer.release();
        mMediaPlayer = null;
    }

    stopForeground(true);
}

@Override
public int onStartCommand(Intent intent, int flags, int startId) {
    // TODO Auto-generated method stub

    if (! mBoolAudioFileFound) {
        stopSelf();
        return super.onStartCommand(intent, flags, startId);
    }

    if (intent.getAction().equals(ACTION_PLAY))
        if (mBoolIsInitial) {
            mMediaPlayer.prepareAsync();
            mBoolIsInitial = false;
        }
        else
            mMediaPlayer.start();
    else if (intent.getAction().equals(ACTION_PAUSE))
        mMediaPlayer.pause();
    else if (intent.getAction().equals(ACTION_SET_REPEAT))
        mMediaPlayer.setLooping(true);
    else if (intent.getAction().equals(ACTION_CANCEL_REPEAT))
        mMediaPlayer.setLooping(false);
    else if (intent.getAction().equals(ACTION_GOTO)) {
        int seconds = intent.getIntExtra
          ("GOTO_POSITION_SECONDS", 0);
        mMediaPlayer.seekTo(seconds * 1000); // 以毫秒（千分之一秒）为单位
    }

    return super.onStartCommand(intent, flags, startId);
}

@Override
```

```java
public void onPrepared(MediaPlayer mp) {
    // TODO Auto-generated method stub

    // 是否取得 audio focus
    AudioManager audioMgr =
        (AudioManager)getSystemService(Context.AUDIO_SERVICE);
    int r = audioMgr.requestAudioFocus(this, AudioManager.
    STREAM_MUSIC, AudioManager.AUDIOFOCUS_GAIN);
    if (r != AudioManager.AUDIOFOCUS_REQUEST_GRANTED)
        mp.setVolume(0.1f, 0.1f);      // 降低音量

    mp.start();

    Intent it = new Intent(getApplicationContext(), Main.class);
    PendingIntent pendIt = PendingIntent.getActivity(
            getApplicationContext(), 0, it,
            PendingIntent.FLAG_UPDATE_CURRENT);
    Notification noti = new Notification(
            android.R.drawable.ic_media_play,
            "背景音乐播放中...",
            System.currentTimeMillis());
    noti.flags |= Notification.FLAG_ONGOING_EVENT;
    noti.setLatestEventInfo(getApplicationContext(),
            "音乐播放程序", "背景音乐播放中...", pendIt);
    startForeground(1, noti);

    Toast.makeText(MediaPlayerService.this, "开始播放",
    Toast.LENGTH_LONG)
    .show();
}

@Override
public boolean onError(MediaPlayer mp, int what, int extra) {
    // TODO Auto-generated method stub
    mp.release();
    mp = null;

    Toast.makeText(MediaPlayerService.this, "发生错误,停止播放",
    Toast.LENGTH_LONG)
        .show();

    return true;
}

@Override
public void onAudioFocusChange(int focusChange) {
    // TODO Auto-generated method stub
```

```
        if (mMediaPlayer == null)
            return;

    switch (focusChange) {
    case AudioManager.AUDIOFOCUS_GAIN:
        // 程序取得声音播放权
        mMediaPlayer.setVolume(0.8f, 0.8f);
        mMediaPlayer.start();
        break;
    case AudioManager.AUDIOFOCUS_LOSS:
        // 程序丧失声音播放权，而且时间可能很久
        stopSelf();           // 结束这个 Service
        break;
    case AudioManager.AUDIOFOCUS_LOSS_TRANSIENT:
        // 程序丧失声音播放权，但预期很快就会再取得
        if (mMediaPlayer.isPlaying())
            mMcdiaPlaycr.pause();
        break;
    case AudioManager.AUDIOFOCUS_LOSS_TRANSIENT_CAN_DUCK:
        // 程序丧失声音播放权，但是可以用很小的音量继续播放
        if (mMediaPlayer.isPlaying())
            mMediaPlayer.setVolume(0.1f, 0.1f);
        break;
    }
}

@Override
public void onCompletion(MediaPlayer arg0) {
    // TODO Auto-generated method stub
    mMediaPlayer.release();
    mMediaPlayer = null;

    stopForeground(true);

    mBoolIsInitial = true;
}
}
```

Android 版本	1.X	2.X	3.X	4.X
适用性	★	★	★	★

UNIT 73

录音程序

录音程序的架构和音乐播放程序的架构很类似，请读者参考图73-1，并和UNIT 71中的图71-1比较，二者的差别只在于换成从麦克风取得声音信号，再将声音数据通过编码器压缩成特定的格式，然后存入文件中。录音程序中最复杂的工作是控制麦克风收录声音信号和运行声音数据的编码，还好Android系统提供MediaRecorder类帮我们完成这两件最麻烦的工作。

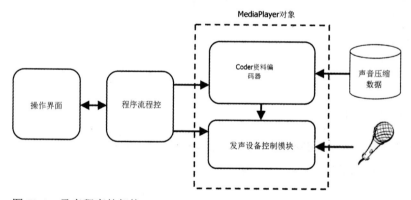

图 73-1　录音程序的架构

73-1　MediaRecorder 类的用法

MediaRecorder 在运行的过程中和 MediaPlayer 一样也会有不同的状态转换，图 73-2 是 MediaRecorder 对象的操作流程图，在使用 MediaRecorder 对象时务必要遵守其中的规则，否则会出现异常错误。另外读者会发现这个 MediaRecorder 类也有录制 video 的功能，不过这里先将焦点集中在录音的部分即可,有关影片的录制和播放留到后续单元再作介绍。一般使用 MediaRecorder 进行录音的步骤如下:

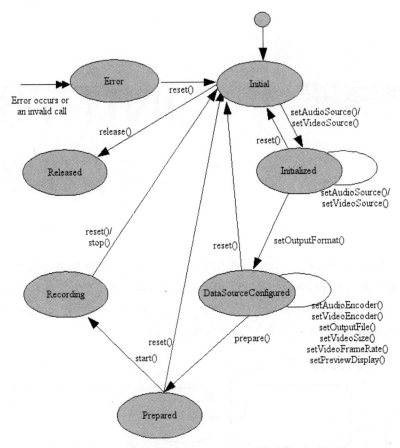

图 73-2　MediaRecorder 对象的操作流程图

Step 1　在程序项目的功能描述文件 AndroidManifest.xml 中加入开启录音功能的描述:

<uses-permission android:name="android.permission.RECORD_AUDIO" />

Step 2　建立一个 MediaRecorder 类的对象。

Step 3　调用 MediaRecorder 对象的 setAudioSource() 方法设置收音设备,例如指定 MediaRecorder.AudioSource.MIC。

Step 4　调用 setOutputFormat() 方法设置声音信号的编码(压缩)格式,可以使用的格式是定义成 MediaRecorder.OutputFormat 类中的常数如下:

- AMR_NB:AMR NB 格式
- AMR_WB:AMR WB 格式

- DEFAULT：预设格式
- MPEG_4：MPEG4 格式
- RAW_AMR：AMR NB 格式
- HREE_GPP：3GPP 格式

Step 5 调用 setOutputFile()方法设置储存的文件名，我们可以将录音文件储存在 sd card 中。

Step 6 调用 setAudioEncoder()方法设置声音信号的编码器，可以使用的格式是定义成 MediaRecorder.AudioEncoder 类中的常数如下：

- AAC：AAC 编码器
- AMR_NB：AMR Narrowband 编码器
- AMR_WB：AMR Wideband 编码器
- DEFAULT：默认编码器

Step 7 调用 prepare()方法完成录音前的准备工作，程序中必须将这个方法放在 try…catch…中运行以进行例外处理。

Step 8 调用 start()方法开始录音，同时进行数据压缩和储存的工作。

Step 9 调用 stop()方法停止录音。

Step 10 当不再需要录音时，必须调用 release()方法释放系统资源。

以上步骤二到步骤九的程序代码范例如下，其中我们调用 Environment.getExternalStorageDirectory().getAbsolutePath()取得 sd card 的路径：

```
mRecorder = new MediaRecorder();
mRecorder.setAudioSource(MediaRecorder.AudioSource.MIC);

mRecorder.setOutputFormat(MediaRecorder.OutputFormat.THREE_GPP);
mRecorder.setOutputFile(
        Environment.getExternalStorageDirectory().getAbsolutePath() +
        "/" + mFileName);
mRecorder.setAudioEncoder(MediaRecorder.AudioEncoder.AMR_NB);

try {
    mRecorder.prepare();
} catch (Exception e) {
    Toast.makeText(Main.this, "MediaRecorder 错误!", Toast.LENGTH_LONG)
        .show();
}

mRecorder.start();
```

73-2 范例程序

这个程序同时使用 MediaRecorder 和 MediaPlayer 让使用者可以先录音，然后再播放录音文件。程序的操作界面只有两个按钮，如图 73-3 所示，按下"开始录音"按钮后，该按钮会变成"停止录音"功能，用户便可以对着麦克风说话，程序会完成录音和存盘的工作。完成录音后按下"开始播放"按钮就可以播放录制的声音文件。完成这个范例程序的步骤如下：

图 73-3 录音和播放程序的运行界面

Step 1 新增一个 Android 程序项目，项目属性的设置依照前面的惯例即可。

Step 2 开启程序功能描述文件 AndroidManifest.xml，加入录音功能和 sd card 写入功能：

```
<?xml version="1.0" encoding="utf-8"?>
<manifest ...>
    <uses-sdk android:minSdkVersion="11" />
    <uses-permission android:name="android.permission.WRITE_
    EXTERNAL_STORAGE" />
    <uses-permission android:name="android.permission.RECORD_AUDIO" />

    <application ...>
        <activity ...>
            ...
        </activity>

    </application>
</manifest>
```

Step 3 展开此程序项目的 res/layout 文件夹，开启其中的接口布局文件 main.xml 编辑如下：

```
<?xml version="1.0" encoding="utf-8"?>
<LinearLayout xmlns:android="http://schemas.android.com/apk/res/android"
    android:orientation="vertical"
    android:layout_width="match_parent"
    android:layout_height="match_parent"
```

```xml
            android:gravity="center_horizontal"
            >
    <Button android:id="@+id/btnAudioRecoOnOff"
            android:layout_width="wrap_content"
            android:layout_height="wrap_content"
            android:text="开始录音"
            />
    <Button android:id="@+id/btnPlayAudioOnOff"
            android:layout_width="wrap_content"
            android:layout_height="wrap_content"
            android:text="拨放录音"
            />
</LinearLayout>
```

Step 4 展开程序项目的 src/(组件路径名称)文件夹，开启其中的主程序文件，然后让它实现 MediaPlayer.OnPreparedListener、MediaPlayer.OnErrorListener 和 MediaPlayer.OnCompletionListener 三个接口，然后依照程序编辑窗口的语法修正建议加入所需的方法，最后将程序代码编辑如下：

```java
package …

import …

public class Main extends Activity
        implements MediaPlayer.OnPreparedListener,
                MediaPlayer.OnErrorListener,
                MediaPlayer.OnCompletionListener {

    private final String mFileName = "my_recorded_audio.3gp";

    private Button mBtnAudioRecoOnOff,
                mBtnPlayAudioOnOff;

    private boolean mBoolRecording = false,
                mBoolPlaying = false;

    private MediaRecorder mRecorder = null;

    private MediaPlayer mPlayer = null;

    /** Called when the activity is first created. */
    @Override
    public void onCreate(Bundle savedInstanceState) {
        super.onCreate(savedInstanceState);
        setContentView(R.layout.main);

        setupViewComponent();
```

```java
    }

    private void setupViewComponent() {
        mBtnAudioRecoOnOff = (Button)findViewById(R.id.btnAudioRecoOnOff);
        mBtnPlayAudioOnOff = (Button)findViewById(R.id.btnPlayAudioOnOff);

        mBtnAudioRecoOnOff.setOnClickListener(btnClickLisAudioRecoOnOff);
        mBtnPlayAudioOnOff.setOnClickListener(btnClickLisPlayAudioOnOff);
    }

    private Button.OnClickListener btnClickLisAudioRecoOnOff = new
                                                    Button.OnClickListener() {
        public void onClick(View v) {
            if (mBoolRecording) {
                mRecorder.stop();
                mRecorder.release();
                mRecorder = null;

                mBoolRecording = false;
                mBtnAudioRecoOnOff.setText("开始录音");
            } else {
                mRecorder = new MediaRecorder();
                mRecorder.setAudioSource(MediaRecorder.
                AudioSource.MIC);

                mRecorder.setOutputFormat(MediaRecorder.
                OutputFormat.THREE_GPP);
                mRecorder.setOutputFile(
                Environment.getExternalStorageDirectory().
                getAbsolutePath() + "/" + mFileName);
                mRecorder.setAudioEncoder(MediaRecorder.
                AudioEncoder.AMR_NB);

                try {
                    mRecorder.prepare();
                } catch (Exception e) {
                    Toast.makeText(Main.this,
                    "MediaRecorder 错误!",
                    Toast.LENGTH_LONG)
                        .show();
                }

                mRecorder.start();

                mBoolRecording = true;
                mBtnAudioRecoOnOff.setText("停止录音");
            }
        }
```

```java
    };

    private Button.OnClickListener btnClickLisPlayAudioOnOff = new
                                    Button.OnClickListener() {
        public void onClick(View v) {
            if (mBoolRecording) {
                mRecorder.stop();
                mRecorder.release();
                mRecorder = null;

                mBoolRecording = false;
                mBtnAudioRecoOnOff.setText("开始录音");
            }

            if (mBoolPlaying) {
                mPlayer.stop();
                mPlayer.release();
                mPlayer = null;

                mBoolPlaying = false;
                mBtnPlayAudioOnOff.setText("开始拨放");
            } else {
                mPlayer = new MediaPlayer();

                try {
                    mPlayer.setDataSource(
                        Environment.getExternalStorageDirectory().
                        getAbsolutePath() + "/" + mFileName);
                } catch (Exception e) {
                    Toast.makeText(Main.this,
                        "MediaPlayer 错误!",
                        Toast.LENGTH_LONG)
                            .show();
                }

                mPlayer.setOnPreparedListener(Main.this);
                mPlayer.setOnErrorListener(Main.this);
                mPlayer.setOnCompletionListener(Main.this);

                mPlayer.prepareAsync();

                mBoolPlaying = true;
                mBtnPlayAudioOnOff.setText("停止拨放");
            }
        }
    };

    @Override
```

```java
        public void onCompletion(MediaPlayer mp) {
            // TODO Auto-generated method stub

            mPlayer.release();
            mPlayer = null;

            mBoolPlaying = false;
            mBtnPlayAudioOnOff.setText("开始拨放");

        }

        @Override
        public boolean onError(MediaPlayer mp, int what, int extra) {
            // TODO Auto-generated method stub
            mPlayer.release();
            mPlayer = null;

            Toast.makeText(Main.this, "MediaPlayer 错误!",
            Toast.LENGTH_LONG)
                .show();

            return true;
        }

        @Override
        public void onPrepared(MediaPlayer mp) {
            // TODO Auto-generated method stub

            mPlayer.setVolume(1.0f, 1.0f);
            mPlayer.start();

            Toast.makeText(Main.this, "开始拨放...", Toast.LENGTH_LONG)
                .show();
        }
    }
```

在程序开头先定义文件名，然后定义两个操作接口的按钮，以及用来记录程序目前是否正在录音状态或是播放状态的变量，另外还有一个 MediaRecorder 对象和一个 MediaPlayer 对象。在 onCreate()方法中我们调用自己的 setupViewComponent()完成程序接口的设置，然后程序便能根据用户按下的按钮和目前的状态运行适当的操作。有关录音的程序代码如同前一个小节的范例，有关播放声音文件的程序代码如同 UNIT 71 的范例程序。

Android 版本	1.X	2.X	3.X	4.X
适用性	★	★	★	★

UNIT 74 播放影片

从这个单元开始我们将介绍 Android 系统的图像相关功能，包括如何播放影片、如何使用内建摄影机拍照和录像。但是在学习这些功能之前，先让我们了解一下 Android 系统支持的图像和影片文件格式。

74-1　Android 支持的图像和影片文件格式

目前计算机可以使用的图像和影片文件格式很多，例如以图像文件而言就包含 bmp、jpeg、gif、png、tiff 等超过十种格式，以影片档而言也包括 mpeg1/2/4、avi、wmv、3gp、dvix、flv 等十几种格式。在这些琳琅满目的文件格式中，可以在 Android 系统中使用的如表 74-1 所示。虽然 Android 系统支持的图像格式比例并不算多，不过这些支持的格式都是很常见的类型，因此实用性还是很高。

表 74-1　Android 支持的图像和影片文件格式

数据类型	Codec 技术名称	文件格式	编码器（Coder）	译码器（Decoder）
图像文件	JPEG	.jpg	支持	支持
	PNG	.png	支持	支持
	GIF	.gif	不支持	支持
	BMP	.bmp	不支持	支持
	WEBP	.webp	支持（Android 4.0+）	支持（Android 4.0+）
影片档	H.263	.3gp .mp4	支持	支持
	H.264 AVC	.3gp .mp4 .ts（只支援 audio, Android 3.0+）	支持（Android 3.0+）	支持
	MPEG-4 SP	.3gp	不支持	支持
	VP8	.webm .mkv (Android 4.0+)	不支持	支持（Android2.3.3+）

> **补充说明**　影像和影片文件的 codec
>
> 计算机处理图像和影片数据的方式和处理声音资料的方式很类似，都必须将原来的数据进行压缩（编码）以减少储存空间。不同的图像压缩格式代表不同的压缩技术，当要显示图像或影片数据时，必须将它们还原（译码）成原来的数据格式，各种 codec 技术追求的目标是要达到更高的压缩率和更好的图像质量。

74-2　使用 VideoView 和 MediaController

要在程序中编写播放影片的程序代码时，利用 VideoView 接口组件是最方便的作法。读者或许会想到前面单元介绍过的 MediaPlayer 类，虽然它也可以用来播放影片，但是在播放影片的时候，通常还需要用到暂停、往前快转、往后快转和显示进度列等控制钮，MediaPlayer 类并没有提供这项功能，因此还需要自行设计播放控制钮。如果换成使用 VideoView 接口组件，只要建立一个 MediaController 对象就可以提供播放控制钮的功能。图 74-1 是 VideoView 的功能架构图，从图中可以看出其实它的内部就是利用 MediaPlayer 对象来进行影片的译码和播放。VideoView 也支持多种影片文件来源，包括从项目资源文件、从 sd card、从指定网址、或是根据 ContentResolver 从系统内建的数据库中取得播放文件，读者可以参考 UNIT 72 中的相关说明。接下来就让我们用一个实际范例来学习如何建立影片播放程序，请读者依照下列步骤操作：

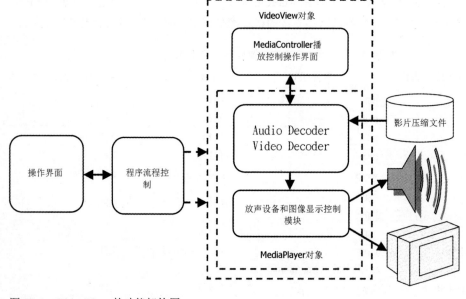

图 74-1　VideoView 的功能架构图

> **补充说明**
>
> 利用单元三十八介绍的 Intent 也可以请求外部程序帮忙播放影片,这样就不需要自己编写播放影片的程序代码,甚至在单元七十一中介绍的音乐播放功能也可以通过 Intent 请求外部程序代劳,读者可以视情况决定究竟要使用哪一种方法。

Step 1 建立一个新的 Android 程序项目,项目的属性设置依照之前的惯例即可。

Step 2 在 Eclipse 左边的项目检查窗口中展开此项目的 res/layout 文件夹,开启其中的接口布局文件 main.xml 并编辑如下,我们根据设置 android:layout_gravity 属性让 VideoView 组件置于屏幕的中央。

```xml
<?xml version="1.0" encoding="utf-8"?>
<LinearLayout xmlns:android="http://schemas.android.com/apk/res/android"
    android:orientation="vertical"
    android:layout_width="match_parent"
    android:layout_height="match_parent"
    >
<VideoView android:id="@+id/videoView"
    android:layout_width="match_parent"
    android:layout_height="match_parent"
    android:layout_gravity="center" />
</LinearLayout>
```

Step 3 在 Eclipse 左边的项目检查窗口中展开此项目的 src/(组件路径名称)文件夹,开启其中的程序文件,让主程序类实现 MediaPlayer.OnErrorListener 和 MediaPlayer.OnCompletionListener 接口,然后依照程序编辑窗口的错误提示和修正建议,在程序中加入需要的方法,完成后的程序代码如下:

```java
package tw.android;

import android.app.Activity;
import android.media.MediaPlayer;

public class Test extends Activity
    implements MediaPlayer.OnErrorListener,
               MediaPlayer.OnCompletionListener {

    /** Called when the activity is first created. */
    @Override
    public void onCreate(Bundle savedInstanceState) {
        super.onCreate(savedInstanceState);
        setContentView(R.layout.main);
    }
```

```java
        @Override
        public void onCompletion(MediaPlayer mp) {
            // TODO Auto-generated method stub

        }

        @Override
        public boolean onError(MediaPlayer mp, int what, int extra) {
            // TODO Auto-generated method stub
            return false;
        }

    }
```

Step 4 在程序编辑窗口中右击，选择 Source > Override/Implement Methods…，从弹出的方法清单对话框中勾选 onResume()和 onPause()两个方法，完成后按下 OK 按钮，然后将程序编辑如下：

```java
package ...;

import ...;

public class Main extends Activity
        implements MediaPlayer.OnErrorListener,
                MediaPlayer.OnCompletionListener {

    private VideoView mVideoView;

    /** Called when the activity is first created. */
    @Override
    public void onCreate(Bundle savedInstanceState) {
        super.onCreate(savedInstanceState);
        setContentView(R.layout.main);

        setupViewComponent();
    }

    @Override
    protected void onResume() {
        // TODO Auto-generated method stub

        mVideoView.start();
        super.onResume();
    }

    @Override
    protected void onPause() {
```

```java
        // TODO Auto-generated method stub

        mVideoView.stopPlayback();
        super.onPause();
    }

    @Override
    public void onCompletion(MediaPlayer mp) {
        // TODO Auto-generated method stub

        Toast.makeText(Main.this, "播放完毕！ ", Toast.LENGTH_LONG)
            .show();
    }

    @Override
    public boolean onError(MediaPlayer mp, int what, int extra) {
        // TODO Auto-generated method stub
        Toast.makeText(Main.this, "运行错误！ ", Toast.LENGTH_LONG)
            .show();

        return true;
    }

    private void setupViewComponent() {
        mVideoView = (VideoView)findViewById(R.id.videoView);
        MediaController mediaController = new MediaController(this);
        mVideoView.setMediaController(mediaController);
        mVideoView.setOnCompletionListener(this);
        mVideoView.setOnErrorListener(this);

        Uri uri = Uri.parse("android.resource://" +
                    getPackageName() + "/" + R.raw.video);
        mVideoView.setVideoURI(uri);
    }
}
```

我们在 onCreate()方法中调用自定义的 setupViewComponent()方法完成接口组件的设置，setupViewComponent()方法中的工作就是取得接口布局文件中的 VideoView 组件，然后加入一个 MediaController 对象，并设置好播放完毕和发生错误时的 callback 函式（也就是在步骤三中实现那两个接口的目的），最后调用 VideoView 的 setVideoURI()设置要播放的影片文件。

在 onResume()方法中我们调用 VideoView 对象的 start()开始播放影片。在 onPause()方法中则调用 stopPlayback()停止播放（当用户切换到其他程序时会运行 onPause()）。onCompletion()和 onError()就是播放完毕和发生错误时的 callback 函式，当中的程序代码只是单纯利用 Toast 对象显示信息。

Step 5 在 Eclipse 左边的项目检查窗口中右击 res 文件夹,选择 New > Folder 建立一个名为 raw 的文件夹,然后开启 Windows 文件总管,将一个 mpeg4 格式的影片文件复制到这个 raw 文件夹中并将它取名为 video.mp4。

完成以上步骤之后在平板电脑模拟器上运行程序就可以看到如图 74-2 所示的界面,如果要叫出播放控制列,只要用鼠标在屏幕上点一下即可。

图 74-2　影片播放程序在平板电脑运行的界面

Android 版本	1.X	2.X	3.X	4.X
适用性	★	★	★	★

UNIT 75 拍照程序

现在的智能手机和平板电脑都内建摄影机,而且通常都有两个。一个在屏幕的上方,可以用来自拍和打视频电话,一个则是在屏幕的背面,专门用来拍摄前方的景物。这些内建的摄影机兼具拍照和录像两项功能,这个单元先让我们学习如何编写拍摄照片的程序代码,录制影片的部分留待下一个单元再作介绍。要设计拍照程序需要用到 Camera 对象和前面学过的 SurfaceView,我们必须先了解它们的用法和之间的关系。

> **补充说明**
>
> 利用 Intent 也可以请求外部程序帮忙拍照并传回图像,这样就不需要自己编写拍照程序,这种方式就像是 UNIT 42 介绍的要求 Intent 返回数据的作法,读者可以视情况决定适合使用哪一种方法。

75-1 Camera 和 SurfaceView

程序在开始使用手机或平板电脑的内建摄影机之前,必须先取得摄影机的控制权。在程序中摄影机是用 Camera 对象代表,取得 Camera 对象就代表取得摄影机的使用权。当程序不需要再使用摄影机时,必须释放它的控制权让其他程序可以使用。以下是使用 Camera 对象的流程:

Step 1 调用 Camera 类的 open() 方法取得 Camera 对象,open() 方法有两个版本,一个需要传入参数,另一个则不需要,请参考下列程序代码范例和批注:

```
Camera cam1, cam2;
cam1 = Camera.open();        // 取得屏幕背面的摄影机,这个摄影机的 id 编号是 0
cam2 = Camera.open(1);       // 取得和屏幕同方向的摄影机,这个摄影机的 id 编号是 1
```

如果要知道有几个内建摄影机，可以调用 getNumberOfCameras()。

Step 2 如果需要知道摄影机提供的功能，可以调用 getParameters()，它会传回一个 Camera.Parameters 物件。根据该对象提供的方法，可以检查摄影机所具备的功能，也可以更改摄影机的设置，然后再调用 setParameters()并传入变更后的 Parameters 对象。

```
Camera cam = Camera.open();
Camera.Parameters camParas = cam. getParameters();
if (camParas.getFlashMode().equals(FLASH_MODE_OFF)
    camParas.setFlashMode(FLASH_MODE_AUTO);

cam.setParameters(camParas);
```

Step 3 由于手机和平板电脑可以自由翻转，摄影机取得的图像可能和用户观看的方向不一致，此时可以调用 setDisplayOrientation()设置图像的旋转角度。

Step 4 要让摄影机拍摄的图像显示在程序界面，必须在程序中建立一个 SurfaceView 对象，并设置好它的 callback 函数，再将 SurfaceView 物件的 SurfaceHolder 传给 Camera 对象。

Step 5 完成以上设置工作之后，就可以运行 Camera 对象的 startPreview()将摄影机拍摄到的实时图像显示在程序界面。

Step 6 摄影机在 preview 状态下可以随时调用 takePicture()进行拍照。拍摄到的图像会利用 callback 函数传回原始图像数据以及 jpeg 压缩后的图像数据，另外也可以设置按下快门时的 callback 函数。运行 takePicture()之后摄影机会停止 preview 状态，程序必须重新运行 startPreview()。

Step 7 当程序不需要再使用摄影机时必须调用 stopPreview()，然后运行 release()释放摄影机的控制权。

在步骤四中必须使用 SurfaceView 对象，关于 SurfaceView 的用法可以参考 UNIT 65 中的说明，不过针对 Camera 对象的需要，这里建立的 SurfaceView 类有下列两点不同：

1. 我们不必实现 Runnable 接口，因为这个 SurfaceView 对象会直接传给 Camera 使用，而不是由我们的控制。
2. 必须调用 setType()将 Surface 的类型设置为 SurfaceHolder.SURFACE_TYPE_PUSH_BUFFERS。

另外我们还要在程序功能描述文件 AndroidManifest.xml 中设置启用摄影机：

```
<uses-permission android:name="android.permission.CAMERA" />
<uses-feature android:name="android.hardware.camera" />
<uses-feature android:name="android.hardware.camera.autofocus" />
```

以上就是使用 Camera 和 SurfaceView 对象的方法，看起来有些复杂，但是如果再配合以下的实现范例，相信可以进一步理清程序的运行流程。

> **补充说明**
>
> 如果要在程序中侦测手机或平板电脑是否具有摄影机,可以运行下列指令:
>
> getPackageManager().hasSystemFeature(PackageManager.FEATURE_CAMERA)
>
> 如果传回 true 代表有摄影机。

75-2 范例程序

我们利用一个程序项目来示范拍照程序的建立过程,请读者依下列步骤操作:

Step 1 建立一个新的 Android 程序项目,项目的属性设置依照之前的惯例即可。

Step 2 从 Eclipse 左边的项目检查窗口中开启程序功能描述文件 AndroidManifest.xml,加入摄影机的使用设置:

```
<?xml version="1.0" encoding="utf-8"?>
<manifest …>
    …
    <uses-permission android:name="android.permission.CAMERA" />
    <uses-feature android:name="android.hardware.camera" />
    <uses-feature android:name="android.hardware.camera.autofocus" />

    <application …>
        …
    </application>
</manifest>
```

Step 3 在 Eclipse 左边的项目检查窗口中展开此项目的 "src/(组件路径名称)" 文件夹,用鼠标右击 "(组件路径名称)" 文件夹,然后选择 New > Class 建立一个新的类。我们可以按下对话框中 Superclass 字段右边的 Browse 按钮,设置此类继承 SurfaceView,这个新类可以取名为 CameraPreview,最后按下 Finish 按钮完成新增类的操作。

Step 4 编辑 CameraPreview 类的程序代码,让它实现 SurfaceHolder.Callback,然后依照程序编辑窗口的语法修正提示,新增一个类构造方法,以及 surfaceChanged()、surfaceCreated()和 surfaceDestroyed()等三个方法,最后将程序代码编辑如下:

```
package …

import …

public class CameraPreview extends SurfaceView
    implements SurfaceHolder.Callback {
```

```java
private Camera mCamera;
private SurfaceHolder mSurfHolder;
private Activity mActivity;

public CameraPreview(Context context) {
    super(context);
    // TODO Auto-generated constructor stub

    mSurfHolder = getHolder();
    mSurfHolder.addCallback(this);
    mSurfHolder.setType(SurfaceHolder.SURFACE_TYPE_PUSH_BUFFERS);
}

public void set(Activity activity, Camera camera) {
    mActivity = activity;
    mCamera = camera;
}

@Override
public void surfaceChanged(SurfaceHolder holder, int format,
                        int width, int height) {
    // TODO Auto-generated method stub
}

@Override
public void surfaceCreated(SurfaceHolder holder) {
    // TODO Auto-generated method stub

    try {
        mCamera.setPreviewDisplay(mSurfHolder);

        Camera.CameraInfo camInfo = new Camera.CameraInfo();
        Camera.getCameraInfo(0, camInfo);

        int rotation =
            mActivity.getWindowManager().
                getDefaultDisplay().getRotation();
        int degrees = 0;
        switch (rotation) {
        case Surface.ROTATION_0:
            degrees = 0; break;
        case Surface.ROTATION_90:
            degrees = 90; break;
        case Surface.ROTATION_180:
            degrees = 180; break;
        case Surface.ROTATION_270:
            degrees = 270; break;
```

```
            }

                int result;
                result = (camInfo.orientation - degrees + 360) % 360;
                mCamera.setDisplayOrientation(result);

                mCamera.startPreview();

                Camera.Parameters camParas = mCamera.getParameters();
                if (camParas.getFocusMode().equals(
                        Camera.Parameters.FOCUS_MODE_AUTO) ||
                    camParas.getFocusMode().equals(
                        Camera.Parameters.FOCUS_MODE_MACRO))
                    mCamera.autoFocus(onCamAutoFocus);
                else
                    Toast.makeText(getContext(), "照相机不支持自动对焦！",
                            Toast.LENGTH_SHORT)
                        .show();
        } catch (Exception e) {
            // TODO Auto-generated catch block
            Toast.makeText(getContext(), "照相机启始错误！",
                    Toast.LENGTH_LONG)
                .show();
        }
    }

    @Override
    public void surfaceDestroyed(SurfaceHolder holder) {
        // TODO Auto-generated method stub

    }

    Camera.AutoFocusCallback onCamAutoFocus = new Camera.
    AutoFocusCallback() {

        @Override
        public void onAutoFocus(boolean success, Camera camera) {
            // TODO Auto-generated method stub
            Toast.makeText(getContext(), "自动对焦！",
                    Toast.LENGTH_SHORT)
                .show();
        }

    };
}
```

针对以上程序代码说明如下：

1. 首先我们定义一个 Camera 对象用来储存使用的摄影机，SurfaceHolder 是 SurfaceView

类需要用到的对象，Activity 对象是用来取得屏幕的旋转角度，以便让摄影机的拍摄界面能够直立显示。

2. 在 CameraPreview 类的构造方法中我们设置好 SurfaceHolder 对象，并如前一小节的说明设置好 Surface 类型。

3. 我们自定义一个 set()方法让外部程序设置所使用的 Activity 对象和 Camera 对象。

4. 在 surfaceCreated()这个 callback 方法中先指定 Camera 对象所使用的 SurfaceView，接下来一连串的程序代码就是取得 Activity 界面的旋转角度，然后计算出 Camera 对象正确的图像显示角度，接着调用 startPreview()开始显示实时的拍摄界面。

5. 另外我们示范如何利用 Camera 对象的 getParameters()方法取得摄影机的对焦功能。当调用 autoFocus()设置自动对焦时，可以传入一个 callback 函式让系统调用。

Step 5 开启 src/(组件路径名称)文件夹中的主程序文件，在程序编辑窗口中右击，选择 Source > Override/Implement Methods…，从弹出的方法清单对话框中勾选 onResume()、onPause()、onCreateOptionsMenu()和 onOptionsItemSelected()，完成后按下 OK 按钮，以下分别说明每一个方法中的工作：

1. 在 onCreate()方法中建立用来当成程序界面的 CameraPreview 对象，然后调用 setContentView()将它设置成为程序界面。

2. 系统运行完 onCreate()方法后会运行 onResume()，在 onResume()中我们取得 Camera 对象，并将它和程序本身的 Activity 传给 CameraPreview 对象。传入这些对象的目的如上一个步骤中的说明。

3. 当程序即将暂停运行时会调用 onPause()，此时我们停止 Camera 的运行并释放它的使用权。

4. 在 onCreateOptionsMenu()和 onOptionsItemSelected()中则是利用菜单的方式建立"照相"和"显示照片"两项功能让用户单击。如果使用者单击"照相"则运行 takePicture()进行拍照，同时传入相关的 callback 函数，有关这些 callback 函式的功能请参考程序代码中的批注。如果用户单击"显示照片"，则利用 Intent 对象请求外部程序帮忙显示照片。

详细的程序代码如下：

```
package …

import …

public class Main extends Activity {

    private static final int MENU_TAKE_PICTURE = Menu.FIRST,
            MENU_SHOW_PICTURE = Menu.FIRST + 1;
```

```java
    private Camera mCamera;
    private CameraPreview mCamPreview;

    /** Called when the activity is first created. */
    @Override
    public void onCreate(Bundle savedInstanceState) {
        super.onCreate(savedInstanceState);

        getWindow().setFormat(PixelFormat.TRANSLUCENT);

        // 如果是手机程序,启用下列程序代码让摄影机界面使用整个屏幕
//        requestWindowFeature(Window.FEATURE_NO_TITLE);
//        getWindow().setFlags(WindowManager.LayoutParams.FLAG_FULLSCREEN,
//                             WindowManager.LayoutParams.FLAG_FULLSCREEN);

        mCamPreview = new CameraPreview(this);
        setContentView(mCamPreview);
    }

    @Override
    protected void onResume() {
        // TODO Auto-generated method stub

        mCamera = Camera.open();
        mCamPreview.set(this, mCamera);

        super.onResume();
    }

    @Override
    protected void onPause() {
        // TODO Auto-generated method stub

        mCamera.stopPreview();

        mCamera.release();
        mCamera = null;

        super.onPause();
    }

    @Override
    public boolean onCreateOptionsMenu(Menu menu) {
        // TODO Auto-generated method stub

        menu.add(0, MENU_TAKE_PICTURE, 0, "照相");
        menu.add(0, MENU_SHOW_PICTURE, 0, "显示照片");
```

```java
        return super.onCreateOptionsMenu(menu);
    }

    @Override
    public boolean onOptionsItemSelected(MenuItem item) {
        // TODO Auto-generated method stub

        switch (item.getItemId()) {
        case MENU_TAKE_PICTURE:
            mCamera.takePicture(camShutterCallback, camRawDataCallback,
                    camJpegCallback);
            break;
        case MENU_SHOW_PICTURE:
            Intent it = new Intent(Intent.ACTION_VIEW);
            File file = new File("/sdcard/photo.jpg");
            it.setDataAndType(Uri.fromFile(file), "image/");
            startActivity(it);
            break;
        }

        return super.onOptionsItemSelected(item);
    }

    ShutterCallback camShutterCallback = new ShutterCallback() {
        public void onShutter() {
            // 通知使用者已完成拍照，例如发出一个声音
        }
    };

    PictureCallback camRawDataCallback = new PictureCallback() {
        public void onPictureTaken(byte[] data, Camera camera) {
            // 用来接收原始的图像数据
        }
    };

    PictureCallback camJpegCallback = new PictureCallback() {
        public void onPictureTaken(byte[] data, Camera camera) {
            // 用来接收压缩成 jpeg 格式的图像数据

            FileOutputStream outStream = null;
            try {
                outStream = new FileOutputStream("/sdcard/photo.jpg");
                outStream.write(data);
                outStream.close();
            } catch (IOException e) {
                Toast.makeText(Main.this, "图像文件储存错误！",
                        Toast.LENGTH_SHORT)
                        .show();
```

```
            }
            mCamera.startPreview();
        }
    };
}
```

完成这个程序项目之后，我们必须将它安装到实体手机或平板电脑上才能运行，相关的操作方式请读者参考 UNIT 8 中的说明。

UNIT 76 录像程序

Android 版本	1.X	2.X	3.X	4.X
适用性	★	★	★	★

在前一个单元中我们已经学会如何使用 Camera 对象和 SurfaceView 对象让程序可以实时显示摄像机所拍摄的图像，并且利用 Camera 对象的 takePicture() 方法完成拍照，但是如果要让程序能够录像，Camera 对象就力有未逮，我们必须再度请出 UNIT 73 中介绍的 MediaRecorder 类，加上它才能够完成录像功能。

76-1　Camera 和 MediaRecorder 通力合作

　　Camera 对象和 MediaRecorder 对象都需要使用实体摄像机，但是它们在录像的过程中所扮演的角色并不相同。Camera 对象是负责提供预览的功能，也就是在开始录像之前显示摄像机所拍摄的图像。当用户启动录像功能之后，便要换成 MediaRecorder 上场，Camera 对象必须将实体摄像机的使用权释放出来让 MediaRecorder 使用，以取得摄像机的拍摄界面。

　　另外 Camera 和 MediaRecorder 也都要用到 SurfaceView 对象，以显示摄像机拍摄的实时图像。当程序在预览的状态时，Camera 对象会将摄像机的拍摄界面显示在 SurfaceView。当启动录像功能之后，Camera 对象就要释放 SurfaceView 的使用权，让 MediaRecorder 可以显示正在录制的界面，图 76-1 是录像程序的功能架构图。

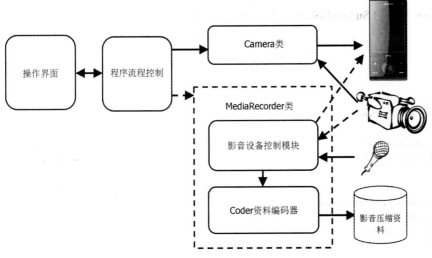

图 76-1 录像程序的功能架构图

76-2 在接口布局文件中建立 SurfaceView

如果读者回想前一个单元的摄像机界面显示方式,我们是根据建立一个继承 SurfaceView 的类,再将该类的对象设置给 Camera 并当成程序的界面。这种方式让这个 SurfaceView 独占整个屏幕,如果我们希望在程序界面中同时显示 SurfaceView 和其他控制组件,像是按钮,就必须换成在接口布局文件中建立 SurfaceView 组件,这样就可以在程序的操作界面中加入其他接口组件,如以下范例:

```
<?xml version="1.0" encoding="utf-8"?>
<LinearLayout xmlns:android="http://schemas.android.com/apk/res/android"
    android:orientation="vertical"
    android:layout_width="match_parent"
    android:layout_height="match_parent"
    >
<SurfaceView android:id="@+id/camPreview"
    android:layout_width="(设置适合的宽度)"
    android:layout_height="(设置适合的高度)"
    android:layout_gravity="center_horizontal"
    />
```
-
- (其他接口组件)
-

```
</LinearLayout>
```

采用以上方式建立 SurfaceView 时,必须让主程序类实现 SurfaceHolder.Callback 接口,再将主程序类传给 SurfaceView 对象的 SurfaceHolder 成为 callback 函数如下：

```java
package ...

import ...

public class Main extends Activity
    implements SurfaceHolder.Callback {

    private SurfaceView mCamPreview;
    private SurfaceHolder mSurfHolder;

    @Override
    public void onCreate(Bundle savedInstanceState) {
        super.onCreate(savedInstanceState);
        setContentView(R.layout.main);

        setupViewComponent();
    }

    private void setupViewComponent() {
        mCamPreview = (SurfaceView) findViewById(R.id.camPreview);
        mSurfHolder = mCamPreview.getHolder();
        mSurfHolder.addCallback(this);
            ...
    }

    public void surfaceCreated(SurfaceHolder holder) {

    }

    public void surfaceChanged(SurfaceHolder holder, int format, int width,
            int height) {

    }

    public void surfaceDestroyed(SurfaceHolder holder) {

    }
}
```

在了解录像程序的运行架构和相关对象的用法之后，接下来让我们利用一个程序项目来学习建立录像程序的完整步骤。

76-3　范例程序

录像程序需要使用实体摄像机，因此当完成程序的开发之后，必须将它安装到实体手机或平板电脑才能运行（参考 UNIT 8 中的说明），以下是完成录像程序的步骤，图 76-2 是程序在实体手机的运行界面：

图 76-2　录像程序在实体手机的运行界面

Step 1　建立一个新的 Android 程序项目，项目的属性设置依照之前的惯例即可。

Step 2　从 Eclipse 左边的项目检查窗口中开启程序功能描述文件 AndroidManifest.xml，加入使用摄像机的相关设置，以及录音和写入 sd card 的功能。

```
<?xml version="1.0" encoding="utf-8"?>
<manifest ...>
  ...
  <uses-permission android:name="android.permission.WRITE_EXTERNAL_STORAGE" />
  <uses-permission android:name="android.permission.RECORD_AUDIO" />
  <uses-permission android:name="android.permission.CAMERA" />
  <uses-feature android:name="android.hardware.camera" />
    <uses-feature android:name="android.hardware.camera.autofocus" />

  <application ...>
      ...
  </application>
</manifest>
```

Step 3 在 Eclipse 左边的项目检查窗口中展开此项目的 res/layout 文件夹，开启接口布局文件 main.xml，参考前面的说明建立<SurfaceView>组件标签。在这个范例中我们将利用菜单的方式来启动和停止录像，因此不需要在接口布局文件中加入其他组件。如果读者想要换成使用按钮的方式操作，可以自行加入<Button>组件标签并视需要修改程序代码。

```xml
<?xml version="1.0" encoding="utf-8"?>
<LinearLayout xmlns:android="http://schemas.android.com/apk/res/android"
    android:orientation="vertical"
    android:layout_width="match_parent"
    android:layout_height="match_parent"
    >
<SurfaceView android:id="@+id/camPreview"
    android:layout_width="match_parent"
    android:layout_height="match_parent"
    android:layout_gravity="center_horizontal"
    />
</LinearLayout>
```

Step 4 在 Eclipse 左边的项目检查窗口中展开此项目的 src/(组件路径名称)文件夹，开启主程序文件，在这个主程序中我们必须完成下列工作：

1．实现 SurfaceHolder.Callback 接口以提供 SurfaceView 对象的 callback 函数。
2．取得接口布局文件中的 SurfaceView 组件并设置它的 callback 函式。
3．建立 Camera 对象以提供录像前的预览功能。
4．建立 MediaRecorder 对象以提供录制视频的功能。

有关 SurfaceView、Camera 和 MediaRecorder 对象的用法都已经在前面的单元介绍过，这个单元的程序项目只是将它们整合在一起，完成后的程序代码如下（操作提示：开始编辑程序代码之前先叫出程序编辑窗口的快捷菜单，然后单击 Source > Override/Implement Methods…加入 onResume()、onPause()、onCreateOptionsMenu()和 onOptionsItemSelected()等四个方法）：

```java
package ...

import ...

public class Main extends Activity
    implements SurfaceHolder.Callback {

    private static final int MENU_START_RECORDING = Menu.FIRST,
                    MENU_STOP_RECORDING = Menu.FIRST + 1,
                    MENU_VIEW_RECORDING = Menu.FIRST + 2;

    private final String mFileName = "recorded_video.3gp";
```

拍照、录音、录像与多媒体播放　PART 14

```java
    private Camera mCamera;
    private MediaRecorder mRecorder;
    private SurfaceView mCamPreview;
    private SurfaceHolder mSurfHolder;
    private boolean mRecording = false;
    private int mRotateDegree;

    @Override
    public void onCreate(Bundle savedInstanceState) {
        super.onCreate(savedInstanceState);

        // 如果要在手机上运行，可以启用下列程序代码
        // 让拍摄的图像占满整个屏幕
//      requestWindowFeature(Window.FEATURE_NO_TITLE);
//      getWindow().setFlags(WindowManager.LayoutParams.FLAG_FULLSCREEN,
//                  WindowManager.LayoutParams.FLAG_FULLSCREEN);
//      setRequestedOrientation(ActivityInfo.SCREEN_ORIENTATION_PORTRAIT);

        setContentView(R.layout.main);

        setupViewComponent();
    }

    @Override
    protected void onResume() {
        // TODO Auto-generated method stub

        mRecorder = new MediaRecorder();
        mCamera = Camera.open();

        super.onResume();
    }

    @Override
    protected void onPause() {
        // TODO Auto-generated method stub

        mCamera.stopPreview();

        mCamera.release();
        mCamera = null;

        super.onPause();
    }

    private void setupViewComponent() {
        mCamPreview = (SurfaceView) findViewById(R.id.camPreview);
        mSurfHolder = mCamPreview.getHolder();
```

```java
            mSurfHolder.addCallback(this);
            mSurfHolder.setType(SurfaceHolder.SURFACE_TYPE_PUSH_BUFFERS);
        }

        private void startRecording() {
            try {
                mRecorder.setCamera(mCamera);
                mRecorder.setAudioSource(MediaRecorder.AudioSource.MIC);
                mRecorder.setVideoSource(MediaRecorder.VideoSource.DEFAULT);
                mRecorder.setOutputFormat(MediaRecorder.OutputFormat.MPEG_4);
                mRecorder.setAudioEncoder(MediaRecorder.AudioEncoder.AMR_WB);
                mRecorder.setVideoEncoder(MediaRecorder.VideoEncoder.MPEG_4_SP);
                mRecorder.setOutputFile("/sdcard/" + mFileName);
                mRecorder.setPreviewDisplay(mSurfHolder.getSurface());
                mRecorder.setOrientationHint(mRotateDegree);
                mRecorder.prepare();
                mRecorder.start();
            }
            catch (Exception e) {
                Toast.makeText(Main.this, "录像错误！", Toast.LENGTH_LONG)
                    .show();
            }
        }

        public void surfaceCreated(SurfaceHolder holder) {
            try {
                mCamera.setPreviewDisplay(mSurfHolder);

                Camera.CameraInfo camInfo = new
                        Camera.CameraInfo();
                Camera.getCameraInfo(0, camInfo);

                int rotation = getWindowManager().getDefaultDisplay().
                getRotation();
                int degrees = 0;
                switch (rotation) {
                case Surface.ROTATION_0:
                    degrees = 0; break;
                case Surface.ROTATION_90:
                    degrees = 90; break;
                case Surface.ROTATION_180:
                    degrees = 180; break;
                case Surface.ROTATION_270:
                    degrees = 270; break;
                }

                mRotateDegree = (camInfo.orientation - degrees +
                360) % 360;
```

```java
            mCamera.setDisplayOrientation(mRotateDegree);

            mCamera.startPreview();

                Camera.Parameters camParas = mCamera.getParameters();
                if (camParas.getFocusMode().equals(Camera.
                    Parameters.FOCUS_MODE_AUTO) ||
                    camParas.getFocusMode().equals(Camera.Paramete
                    rs.FOCUS_MODE_MACRO))
                        mCamera.autoFocus(null);
                else
                    Toast.makeText(this, "照相机不支持自动对焦！",
                    Toast.LENGTH_SHORT)
                        .show();
        } catch (Exception e) {
            // TODO Auto-generated catch block
            Toast.makeText(this, "照相机启始错误！",
            Toast.LENGTH_LONG)
                .show();
        }
    }

    public void surfaceChanged(SurfaceHolder holder, int format,
    int width,
            int height) {
    }

    public void surfaceDestroyed(SurfaceHolder holder) {
        if (mRecording) {
            mRecorder.stop();
            mRecording = false;
        }
        mRecorder.release();
        finish();
    }

    @Override
    public boolean onCreateOptionsMenu(Menu menu) {
        // TODO Auto-generated method stub

        menu.add(0, MENU_START_RECORDING, 0, "开始录像");
        menu.add(0, MENU_STOP_RECORDING, 0, "停止录像");
        menu.add(0, MENU_VIEW_RECORDING, 0, "播放影片");

        return super.onCreateOptionsMenu(menu);
    }

    @Override
```

```java
        public boolean onOptionsItemSelected(MenuItem item) {
            // TODO Auto-generated method stub
            switch (item.getItemId()) {
            case MENU_START_RECORDING:
                mRecording = true;
                mCamera.stopPreview();
                mCamera.unlock();
                startRecording();
                break;
            case MENU_STOP_RECORDING:
                mRecording = false;
                mRecorder.stop();

                try {
                    mCamera.reconnect();
                    mCamera.startPreview();
                } catch (Exception e) {
                    // TODO Auto-generated catch block
                    Toast.makeText(this, "Camera 启始错误！
", Toast.LENGTH_LONG)
                        .show();
                }

                break;
            case MENU_VIEW_RECORDING:
                Intent it = new Intent();
                it.setClass(Main.this, PlayVideo.class);
                it.putExtra("FILE_NAME", mFileName);
                startActivity(it);
                break;
            }

            return super.onOptionsItemSelected(item);
        }
    }
```

另外在整合 Camera 对象和 MediaRecorder 对象时必须注意以下几点：

1. 要开始启动 MediaRecorder 的录像功能以前，必须先运行 Camera 对象的 stopPreview()释放程序界面的 SurfaceView，以及运行 unlock()释放摄像机的控制权。
2. 必须将 Camera 对象传给 MediaRecorder（调用 MediaRecorder 的 setCamera()）。
3. 在操作 MediaRecorder 对象的过程中，必须确实遵守图 73-2 的流程，否则运行时会出现例外错误。
4. 由于手机和平板电脑的摄像机拍摄方向和用户观看图像的角度可能会有不一致的情况，我们可以利用 MediaRecorder 的 setOrientationHint()方法指定影片播放时的旋转角度。

针对设置 MediaRecorder 的影音压缩格式，Android SDK 技术文件提供表 76-1 的建议。

表 76-1　Android SDK 技术文件建议的影音压缩格式设置

	标准设置，SD（低品质）	标准设置，SD（高质量）	高分辨率，HD （并非所有设备都支持）
影片压缩格式	H.264 Baseline Profile	H.264 Baseline Profile	H.264 Baseline Profile
影片分辨率	176×144 px	480×360 px	1280×720 px
每秒帧数 （frame rate）	12 fps	30 fps	30 fps
影片比特率 （bitrate）	56 Kb/s	500 Kb/s	2 Mb/s
声音压缩格式	AAC-LC	AAC-LC	AAC-LC
声道数目	1 (mono)	2 (stereo)	2 (stereo)
声音比特率 （bitrate）	24 Kb/s	128 Kb/s	192 Kb/s

Step 5 新增一个继承 Activity 的新类，我们将利用它来建立播放影片的程序，这个新类可以取名为 PlayVideo，它的程序代码和 UNIT 74 的程序范例很类似，基本上就是从接收到的 Intent 对象中取出要播放的影片档名称，再利用 VideoView 和 MediaController 完成播放的动作，这个新类的接口布局文件和程序代码如下：

- 界面布局文件 play_video.xml：

```
<?xml version="1.0" encoding="utf-8"?>
<LinearLayout xmlns:android="http://schemas.android.com/apk/res/android"
    android:orientation="vertical"
    android:layout_width="match_parent"
    android:layout_height="match_parent"
    >
<VideoView android:id="@+id/videoView"
    android:layout_width="match_parent"
    android:layout_height="match_parent"
    android:layout_centerInParent="true" />
</LinearLayout>
```

- 程序文件 PlayVideo.java：

```
package ...

import ...

public class PlayVideo extends Activity
    implements MediaPlayer.OnErrorListener,
MediaPlayer.OnCompletionListener {
```

```java
    private VideoView mVideoView;

    /** Called when the activity is first created. */
    @Override
    public void onCreate(Bundle savedInstanceState) {
    super.onCreate(savedInstanceState);
        setContentView(R.layout.play_video);

        setupViewComponent();
    }

    @Override
    protected void onResume() {
        // TODO Auto-generated method stub
        super.onResume();

        mVideoView.start();
    }

    @Override
    public void onCompletion(MediaPlayer mp) {
        // TODO Auto-generated method stub

        Toast.makeText(this, "播放完毕！ ", Toast.LENGTH_LONG)
            .show();
    }

    @Override
    public boolean onError(MediaPlayer mp, int what, int extra) {
        // TODO Auto-generated method stub
        Toast.makeText(this, "运行错误！ ", Toast.LENGTH_LONG)
            .show();

        return true;
    }

    private void setupViewComponent() {
        mVideoView = (VideoView)findViewById(R.id.videoView);
        MediaController mediaController = new MediaController(this);
        mVideoView.setMediaController(mediaController);
        mVideoView.setOnCompletionListener(this);
        mVideoView.setOnErrorListener(this);

        String sVideoFileName = getIntent().getStringExtra
          ("FILE_NAME");
        Uri uri = Uri.parse("/sdcard/" + sVideoFileName);
        mVideoView.setVideoURI(uri);
    }
}
```

PART 15　WebView 与网页处理

UNIT 77　WebView 的网页浏览功能
UNIT 78　自己打造网页浏览器
UNIT 79　JavaScript 和 Android 程序代码之间的调用

Android 版本	1.X	2.X	3.X	4.X
适用性	★	★	★	★

UNIT 77
WebView 的网页浏览功能

网络联机是智能手机和平板电脑的必备功能，连上网络就等于建立了一条通往全世界的虚拟道路，可以随手取得各种实时信息、查询想要了解的相关问题、以及使用五花八门的服务。为了能够满足用户的需求，各种网络相关的应用程序不断地推陈出新，其中使用频率最高的莫过于网页浏览程序（Web Browser），因此学习网络程序设计最重要的主题之一就是了解如何开发具有网页浏览功能的程序。

如果读者曾经设计过网页，就会知道目前网页技术的复杂性。网页大约在 1994 年开始出现，最初只能显示静态的数据，经过将近 20 年的发展，现在已经能够做到非常完整的互动功能，包括现在快速兴起的"云端运算"（Cloud Computing），都必须使用 Web 相关的技术。Web 的强大功能意味着它背后技术的高度复杂性，如果我们要自己从头打造一个网页解译程序，那会是一个非常艰巨的任务。还好 Android 系统内建一个 WebView 类，它就是一个现成的网页解译和显示组件，我们可以利用它来开启指定的网页。

> **补充说明**
>
> WebView 类内部是使用 WebKit 网页处理引擎，WebKit 是由 Apple 开发并公开让大家使用，包括 Google 的 Chrome 和 Apple 的 Safari 都是使用 WebKit 引擎，Microsoft 的 IE 则是使用 Trident 引擎，Firefox 是使用 Gecko 引擎。

77-1　WebView 的用法

WebView 对于网页数据的解译和浏览提供非常完整的支持，而且它的用法就像我们前面学过的接口组件一样简单，使用 WebView 的基本步骤如下：

Step 1 在程序的接口布局文件中建立一个<WebView>组件标签并设置好相关的属性如下：

```xml
<?xml version="1.0" encoding="utf-8"?>
<LinearLayout xmlns:android="http://schemas.android.com/apk/res/android"
    android:orientation="vertical"
    android:layout_width="match_parent"
    android:layout_height="match_parent"
    >
<WebView android:id="@+id/webView"
    android:layout_width="match_parent"
    android:layout_height="match_parent"
    />
</LinearLayout>
```

Step 2 在程序的功能描述文件 AndroidManifest.xml 中设置使用网络功能。

```xml
<?xml version="1.0" encoding="utf-8"?>
<manifest …>
    <uses-permission android:name="android.permission.INTERNET" />

    <application …>
        …
    </application>
</manifest>
```

Step 3 在程序代码中定义一个 WebView 类型的对象，然后调用 findViewById()从程序的接口取得 WebView 组件存入此 WebView 对象。

Step 4 如果需要修改 WebView 对象的设置，可以调用 WebView 对象的 getSettings()取得 WebSettings 对象，并视需要进行修改。

Step 5 调用 WebView 对象的其他方法设置相关功能。

Step 6 调用 WebView 对象的 loadUrl()开启指定的网页。

以下是步骤三到步骤六的程序代码范例：

```java
WebView mWebView;
mWebView = (WebView)findViewById(R.id.webView);

WebSettings webSettings = mWebView.getSettings();
webSettings.setJavaScriptEnabled(true); // 开启 Java Script 解译功能

mWebView.loadUrl(mEdtUrl.getText().toString());
```

77-2 范例程序

我们利用以上介绍的 WebView 类实现一个可以浏览网页的程序，它的运行界面如图 77-1 所示，程序界面的上方有一个 EditText 组件可以让用户输入网址，输入完毕后按下右边的"开启网址"按钮就会在下方的 WebView 组件中显示网页。这个程序项目的接口描述文件和程序代码如下，在接口描述文件中我们使用二层的 LinearLayout 让按钮和 Text 组件排列在 WebView 组件的上方，另外提醒读者记得在程序功能描述文件 AndroidManifest.xml 中设置使用网络功能。如果用户单击网页中的超链接，程序会另外启动 Android 内建的网页浏览器开启新的网址，如图 77-2 所示，这是 WebView 对象内定的作法，不过我们可以改变它，另外也可以做到像一般网页浏览器的"回上一页"和"到下一页"的切换功能，这些进阶的控制我们留到下一个单元再做介绍。

图 77-1　具备网页浏览功能的程序

■ 界面布局文件：

```
<?xml version="1.0" encoding="utf-8"?>
<LinearLayout xmlns:android="http://schemas.android.com/apk/res/android"
    android:orientation="vertical"
    android:layout_width="match_parent"
    android:layout_height="match_parent"
    >
<LinearLayout
    android:orientation="horizontal"
    android:layout_width="match_parent"
    android:layout_height="wrap_content"
    >
```

图 77-2　按下网页中的超链接后程序自动启动内建的网页浏览器

```xml
<TextView
    android:layout_width="wrap_content"
    android:layout_height="wrap_content"
    android:text="网址："
    android:textSize="20sp"
    />
<EditText android:id="@+id/edtUrl"
    android:layout_width="500dp"
    android:layout_height="wrap_content"
    />
<Button android:id="@+id/btnOpenUrl"
    android:layout_width="wrap_content"
    android:layout_height="wrap_content"
    android:text="开启网址"
    />
</LinearLayout>
<WebView android:id="@+id/webView"
    android:layout_width="match_parent"
    android:layout_height="match_parent"
    />
</LinearLayout>
```

■　程序文件：

```
package …

import …

public class Main extends Activity {
```

```java
private Button mBtnOpenUrl;
private EditText mEdtUrl;
private WebView mWebView;

/** Called when the activity is first created. */
@Override
public void onCreate(Bundle savedInstanceState) {
    super.onCreate(savedInstanceState);
    setContentView(R.layout.main);
    setupViewComponent();
}

private void setupViewComponent() {
    mBtnOpenUrl = (Button)findViewById(R.id.btnOpenUrl);
    mEdtUrl = (EditText)findViewById(R.id.edtUrl);
    mWebView = (WebView)findViewById(R.id.webView);

    WebSettings webSettings = mWebView.getSettings();
    webSettings.setJavaScriptEnabled(true);

    mBtnOpenUrl.setOnClickListener(btnClickLisOpenUrl);
}

private Button.OnClickListener btnClickLisOpenUrl = new
Button.OnClickListener() {
    public void onClick(View v) {
        mWebView.loadUrl(mEdtUrl.getText().toString());
    }
};
}
```

Android 版本	1.X	2.X	3.X	4.X
适用性	★	★	★	★

UNIT 78
自己打造网页浏览器

在学会了 WebView 的基本用法之后，这个单元我们将进一步利用 WebView 来开发一个网页浏览器。这个网页浏览器将具备完整的网页切换和状态显示功能，它的运行界面如图 78-1 所示，其中的网址列可以让使用者输入想要开启的网址，在开启网页的过程中会在上方的标题栏显示"正在下载网页…"的信息并在屏幕的右上角显示等待循环，"停止"按钮也会启动让用户单击。随着使用者切换网页的过程，"回上一页"和"到下一页"按钮会视操作情况自动启动让使用者操作。程序也会提供放大和缩小网页的控制按钮（图 78-1 右下方，当按下网页区域并进行拖曳动作后才会显示），用户可以放大网页的内容以方便阅读。其实这些网页浏览的相关功能都是 WebView 内建的操作方法，程序只要在适当的时机调用这些方法就可以完成网页的控制和切换。因此在开始建立网页浏览器之前，我们先学习更多操作 WebView 的方法。

图 78-1　自行建立的网页浏览器

78-1　WebView 的网页操作方法

在上一个单元的实现范例中我们特别说明一个情况，就是当使用者单击 WebView 网页中的超链接时，程序会自动启动其他的网页浏览器来开启超链接的网页。这种内定的运行方式是为了安全性的考虑，因为在 WebView 中我们可以启动 JavaScript 的运行功能，如果有恶意的网页在 WebView 中开启，并且程序中又有和 JavaScript 连结的程序代码（下一个单元将介绍实现方法），那么这个恶意的网页便能够运行我们的程序，进而机窃取个人资料或是危害系统安全。当然我们可以根据程序代码的控制，让特定的网页在 WebView 中开启而不必启动其他的网页浏览器，这需要借助 WebClient 对象，我们将在下一小节中介绍。以下我们先学习 WebView 的网页控制方法：

1. goBack()
   回到前一个浏览过的网页，这是所有网页浏览器的必备功能，在开启不同网页的过程中，WebView 会自动记录浏览的顺序，只要调用 goBack()就可以回到前一个网页。goBack()方法通常会和 canGoBack()以及 getUrl()方法搭配使用。程序可以先运行 canGoBack()检查是否有前一个网页存在，例如如果已经回到第一个网页，就不再有前一个网页，这时候就不要启用"回上一页"按钮。如果还有前一个网页，而且使用者又按下"回上一页"按钮，这时候程序便运行 goBack()，然后调用 getUrl()取得对应的网址并显示在网址列。

2. goForward()
   到下一个浏览过的网页，这也是网页浏览器的基本功能，它和 goBack()是互相对应的方法，在实际的程序中也会和 canGoForward()以及 getUrl()方法搭配使用。

3. stopLoading()
   停止正在下载的网页。

4. reload()
   重新下载网页。例如开启的网页出现错误时，可以按下"更新"按钮运行 WebView 的 reload()方法重新下载网页数据。

5. setSupportZoom()和 setBuiltInZoomControls()
   设置 WebView 的网页缩放功能，调用这两个方法并传入 true 就可以启用网页的缩放功能以方便阅读。

关于程序和 WebView 之间的关系，以及上述的 WebView 内部的方法请读者参考图 78-2。在开发网页浏览器的时候我们需要考虑什么时候应该运行 canGoBack()和 canGoForward()，以便更新"回上一页"和"到下一页"按钮的状态，答案是当 WebView 中的网页变更的时候，又例如我们必须在开始下载网页和完成下载的这一段时间才启用"停止"按钮。有关网页下载状态的控制牵涉 WebViewClient 和 WebChromeClient 这两个对象，因此我们必须学会这两个对象的用法。

PART 15　WebView 与网页处理

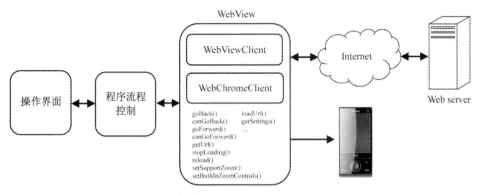

图 78-2　程序和 WebView 的运行架构

78-2　设置 WebViewClient 和 WebChromeClient

WebViewClient 和 WebChromeClient 是在 WebView 中运行的对象，它们的功能是负责下载网页数据，再利用 callback 函数的方式通知主程序运行结果。例如以下程序代码就是在 WebView 中建立一个 WebViewClient 对象，如此一来当用户单击 WebView 网页中的超链接时，就会直接在 WebView 中开启，而不会像上一个单元的程序必须启动其他的网页浏览器。

```
WebView mWebView = (WebView)findViewById(R.id.webView);
mWebView.setWebViewClient(new WebViewClient());
```

上述的程序代码是最简单的作法，它无法让程序和 WebViewClient 对象进行互动，也就是说缺乏 callback 函数，以致于无法通知主程序它的运行结果。为了能够让主程序和 WebViewClient 对象之间能够互动，我们可以在程序项目中建立一个继承 WebViewClient 的新类，假设我们将它取名为 MyWebViewClien，然后利用程序编辑窗口的快捷菜单中的 Source > Override/Implement Methods…，叫出方法清单对话框，从中勾选以下程序代码范例中的 shouldOverrideUrlLoading()、onPageStarted()、onPageFinished()和 onReceivedError()四个 callback 方法，另外我们加入一个自定义的方法 setupViewComponent()如下：

```
package …

import …

public class MyWebViewClient extends WebViewClient {

    @Override
    public boolean shouldOverrideUrlLoading(WebView view, String url) {
        // TODO Auto-generated method stub
        // 当使用者单击 WebView 网页中的超链接时运行
```

613

```java
            return true;
        }

        @Override
        public void onPageStarted(WebView view, String url, Bitmap favicon) {
            // TODO Auto-generated method stub
            //当开始下载网页时运行
            super.onPageStarted(view, url, favicon);
        }

        @Override
        public void onPageFinished(WebView view, String url) {
            // TODO Auto-generated method stub
            //当完成网页下载时运行
            super.onPageFinished(view, url);
        }

        @Override
        public void onReceivedError(WebView view, int errorCode,
                String description, String failingUrl) {
            // TODO Auto-generated method stub
            //当下载网页出现错误时运行
            super.onReceivedError(view, errorCode, description, failingUrl);
        }

        public MyWebViewClient setupViewComponent(…) {
            // 设置运行时需要用到的对象
            …
            return this;
        }
    }
```

这四个 callback 函数的运行时机如程序代码中的批注，setupViewComponent()方法是用来设置网页下载的过程中需要更新的接口组件，像是"停止"按钮。以上是我们在开发网页浏览器的过程中会用到的函式，除此之外还有许多其他的 callback 函数会在不同的时间点运行，读者可以查阅 Android SDK 技术文件中有关 WebViewClient 类的说明。建立好这个新类之后，就可以在主程序中产生一个此类的对象，然后调用 WebView 的 setWebViewClient()完成设置。

除了 WebViewClient 之外，还有一个 WebChromeClient 对象也和网页下载有关，我们可以利用它获得网页下载的进度。由于只需要用到它的 onProgressChanged()方法，我们可以直接将建立 WebChromeClient 对象的程序代码写在主程序文件，而不需要另外建立 WebChromeClient 类的程序文件，以下是在主程序中设置 WebViewClient 和 WebChromeClient 对象的范例。在了解 WebView 的功能和 WebViewClient 以及 WebChromeClient 之间的运行方式之后，我们就可以开始实现网页浏览器。

```
mWebView.setWebViewClient(new MyWebViewClient()
                .setupViewComponent(…));
mWebView.setWebChromeClient(new WebChromeClient() {
        public void onProgressChanged(WebView view, int progress) {
            // Activity 和 WebViews 的进度值使用不同的数值范围，
            // 因此必须乘上 100，当到达 100%后进度列会自动消失
            setProgress(progress * 100);
        }
});
```

78-3 范例程序

在本单元一开始的时候我们就已经介绍过网页浏览器的运行界面，它的操作方式和功能如同前面的说明，每一个按钮的状态都会随着操作过程适当地改变。在网页下载的时候，标题栏会显示"正在下载网页…"，当按下"回上一页"和"到下一页"按钮时，网址列的内容也会自动更新，以下是建立网页浏览器程序项目的详细步骤：

Step 1 建立一个新的 Android 程序项目，项目的属性设置依照之前的惯例即可。

Step 2 在程序的功能描述文件 AndroidManifest.xml 中设置使用网络功能。

```xml
<?xml version="1.0" encoding="utf-8"?>
<manifest …>
    <uses-permission android:name="android.permission.INTERNET" />

    <application …>
        …
    </application>
</manifest>
```

Step 3 在 Eclipse 左边的项目检查窗口中展开此项目的 res/layout 文件夹，开启接口布局文件 main.xml 然后编辑如下，我们使用二层的 LinearLayout 让所有按钮和 Text 组件排列在 WebView 组件的上方：

```xml
<?xml version="1.0" encoding="utf-8"?>
<LinearLayout xmlns:android="http://schemas.android.com/apk/res/android"
    android:orientation="vertical"
    android:layout_width="match_parent"
    android:layout_height="match_parent">
<LinearLayout
    android:orientation="horizontal"
    android:layout_width="match_parent"
    android:layout_height="wrap_content">
<Button android:id="@+id/btnGoBack"
    android:layout_width="wrap_content"
```

```xml
        android:layout_height="wrap_content"
        android:text="回上一页"
        android:enabled="false"/>
    <Button android:id="@+id/btnGoForward"
        android:layout_width="wrap_content"
        android:layout_height="wrap_content"
        android:text="到下一页"
        android:enabled="false"/>
    <Button android:id="@+id/btnReload"
        android:layout_width="wrap_content"
        android:layout_height="wrap_content"
        android:text="更新"
        android:enabled="false"/>
    <Button android:id="@+id/btnStop"
        android:layout_width="wrap_content"
        android:layout_height="wrap_content"
        android:text="停止"
        android:enabled="false"/>
    <TextView
        android:layout_width="wrap_content"
        android:layout_height="wrap_content"
        android:textSize="20sp"
        android:text="网址："/>
    <EditText android:id="@+id/edtUrl"
        android:layout_width="300dp"
        android:layout_height="wrap_content"
        android:singleLine="true"/>
    <Button android:id="@+id/btnOpenUrl"
        android:layout_width="wrap_content"
        android:layout_height="wrap_content"
        android:text="开启网址"/>
</LinearLayout>
<WebView android:id="@+id/webView"
    android:layout_width="match_parent"
    android:layout_height="match_parent"/>
</LinearLayout>
```

Step 4 新增一个继承 WebViewClient 的新类，我们可以将它取名为 MyWebViewClient，它的程序代码架构和功能如同前一个小节的说明，在这个实现范例中，必须将"回上一页"、"到下一页"、"更新"和"停止"四个按钮，以及 Activity 和 WebView 对象传给 MyWebViewClient 以便在网页下载的过程中更新按钮的状态并且在标题栏显示信息，程序代码如下：

```
package …

import …

public class MyWebViewClient extends WebViewClient {
```

```java
    private Activity mActivity;
    private Button mBtnGoBack,
                   mBtnGoForward,
                   mBtnStop,
                   mBtnReload;
    private WebView mWebView;

    @Override
    public boolean shouldOverrideUrlLoading(WebView view, String url) {
        // TODO Auto-generated method stub
        if (Uri.parse(url).getHost().indexOf("google") >= 0) {
            return false;
        }

        Intent it = new Intent(Intent.ACTION_VIEW, Uri.parse(url));
        mActivity.startActivity(it);
        return true;
    }

    @Override
    public void onPageStarted(WebView view, String url, Bitmap favicon) {
        // TODO Auto-generated method stub

        // 手机程序可以在标题栏显示横式进度列
        // 平板电脑只能显示环状等待循环
//      mActivity.setProgressBarVisibility(true);
        mActivity.setProgressBarIndeterminateVisibility(true);
        mActivity.setTitle("正在下载网页...");
        mBtnReload.setEnabled(false);
        mBtnStop.setEnabled(true);
        super.onPageStarted(view, url, favicon);
    }

    @Override
    public void onPageFinished(WebView view, String url) {
        // TODO Auto-generated method stub

        // 平板电脑程序 - 隐藏环状等待循环
        mActivity.setProgressBarIndeterminateVisibility(false);
        mActivity.setTitle(R.string.app_name);
        mBtnReload.setEnabled(true);
        mBtnStop.setEnabled(false);

        if (mWebView.canGoBack())
            mBtnGoBack.setEnabled(true);
        else
            mBtnGoBack.setEnabled(false);
```

```java
            if (mWebView.canGoForward())
                mBtnGoForward.setEnabled(true);
            else
                mBtnGoForward.setEnabled(false);

            super.onPageFinished(view, url);
        }

        @Override
        public void onReceivedError(WebView view, int errorCode,
                String description, String failingUrl) {
            // TODO Auto-generated method stub

            mBtnReload.setEnabled(true);
            mBtnStop.setEnabled(false);

            Toast.makeText(mActivity, "开启网页错误：" + failingUrl
                    + description, Toast.LENGTH_LONG)
                .show();
            super.onReceivedError(view, errorCode, description, failingUrl);
        }

        public MyWebViewClient setupViewComponent(Activity act,
                                                  WebView webView,
                                                  Button btnGoBack,
                                                  Button btnGoForward,
                                                  Button btnReload,
                                                  Button btnStop) {
            mActivity = act;
            mWebView = webView;
            mBtnGoBack = btnGoBack;
            mBtnGoForward = btnGoForward;
            mBtnReload = btnReload;
            mBtnStop = btnStop;

            return this;
        }
    }
```

在 shouldOverrideUrlLoading()方法中，我们先检查要开启的网址是否是属于 google 的网页，如果是的话就直接 return false 让网页直接在 WebView 中开启。如果不是属于 google 的网页，则利用 Intent 对象要求外部的网页浏览器开启该网址。利用这种方式就可以筛选出可信任的网页，让它们直接在 WebView 中开启，以避免前面解释过的安全问题。

另外手机程序和平板电脑程序的标题所显示的进度列类型也不相同，手机程序的标题可以显示具有百分比的进度列，但是平板电脑的标题只能显示环状等待循环，因此程序中我们特别加上批注说明，其他有关按钮状态的切换和错误信息的提示请读者直接参考程序代码就能了解。

Step 5 在 Eclipse 左边的项目检查窗口中展开此项目的 src/(组件路径名称)文件夹并开启主程序文件，我们让主程序类实现 OnClickListener 接口，然后依照程序编辑窗口的错误提示和修正建议功能，加入 onClick()方法，这样我们就可以将主程序类的对象传给按钮当作 OnClickListener。另外我们在 onCreate()方法中调用 requestWindowFeature()方法启用标题栏的进度显示功能，在 setupViewComponent()方法中则是依照前面的说明设置好 WebView 对象，这样网页便能正常运行，以下是主程序类的程序代码：

```
package …

import …

public class Main extends Activity
    implements OnClickListener {

    private Button mBtnOpenUrl,
                mBtnGoBack,
                mBtnGoForward,
                mBtnStop,
                mBtnReload;
    private EditText mEdtUrl;
    private WebView mWebView;

    /** Called when the activity is first created. */
    @Override
    public void onCreate(Bundle savedInstanceState) {
        super.onCreate(savedInstanceState);

        // 手机程序可以在标题栏显示横式进度列
        // 平板电脑只能显示环状等待循环
//      requestWindowFeature(Window.FEATURE_PROGRESS);
        requestWindowFeature(Window.FEATURE_INDETERMINATE_PROGRESS);
        setContentView(R.layout.main);

        // 平板电脑程序 - 隐藏环状等待循环
        setProgressBarIndeterminateVisibility(false);

        setupViewComponent();
    }

    private void setupViewComponent() {
        mBtnOpenUrl = (Button)findViewById(R.id.btnOpenUrl);
        mBtnGoBack = (Button)findViewById(R.id.btnGoBack);
        mBtnGoForward = (Button)findViewById(R.id.btnGoForward);
```

```java
mBtnStop = (Button)findViewById(R.id.btnStop);
mBtnReload = (Button)findViewById(R.id.btnReload);
mEdtUrl = (EditText)findViewById(R.id.edtUrl);
mWebView = (WebView)findViewById(R.id.webView);

// 使用自定义的 MyWebViewClient 可以筛选在程序中
// 浏览的网页，或是启动外部的浏览器
mWebView.setWebViewClient(new MyWebViewClient()
        .setupViewComponent(this,
                            mWebView,
                            mBtnGoBack,
                            mBtnGoForward,
                            mBtnReload,
                            mBtnStop));
mWebView.setWebChromeClient(new WebChromeClient() {
    public void onProgressChanged(WebView view, int progress) {
        // Activity 和 WebViews 的进度值使用不同的代表值，
        // 所以必须乘上 100，当到达 100%进度列会自动消失
        setProgress(progress * 100);
    }
});
WebSettings webSettings = mWebView.getSettings();
webSettings.setJavaScriptEnabled(true);
webSettings.setSupportZoom(true);
webSettings.setBuiltInZoomControls(true);

mBtnOpenUrl.setOnClickListener(this);
mBtnGoBack.setOnClickListener(this);
mBtnGoForward.setOnClickListener(this);
mBtnStop.setOnClickListener(this);
mBtnReload.setOnClickListener(this);
}

@Override
public void onClick(View v) {
    // TODO Auto-generated method stub
    switch (v.getId()) {
    case R.id.btnOpenUrl:
        mWebView.loadUrl(mEdtUrl.getText().toString());
        break;
    case R.id.btnGoBack:
        mWebView.goBack();
        mEdtUrl.setText(mWebView.getUrl());
        break;
    case R.id.btnGoForward:
        mWebView.goForward();
        mEdtUrl.setText(mWebView.getUrl());
        break;
```

```
            case R.id.btnReload:
                mWebView.reload();
                break;
            case R.id.btnStop:
                mWebView.stopLoading();

                // 平板电脑程序 - 隐藏环状等待循环
                setProgressBarIndeterminateVisibility(false);
                setTitle(R.string.app_name);
                mBtnReload.setEnabled(true);
                mBtnStop.setEnabled(false);
                break;
        }
    }
}
```

完成这个程序项目之后便可以运行程序测试网页浏览功能,读者可以仔细观察每一个按钮的状态变化,如果要启用网页缩放控制列,可以按住屏幕并进行拖曳便会自动显示。

Android 版本	1.X	2.X	3.X	4.X
适用性	★	★	★	★

UNIT 79

JavaScript 和 Android 程序代码之间的调用

网页中的 JavaScript 程序代码是由 WebView 组件负责解释和运行，Android 程序代码则是在自己的 UI thread 中运行，二者看来互不相干，可是 Android 系统却为二者提供了互通的机制，让在 WebView 中运行的 JavaScript 可以调用 Android 程序中的方法，而 Android 程序也可以调用 WebView 中的 JavaScript 的 function。这个功能让程序开发者可以设计出功能更强大的程序，但是要提醒读者在运用这些功能的背后必须留意程序安全上的问题，避免恶意的网页通过程序代码侵犯个人隐私或危害系统安全。

79-1 从 JavaScript 调用 Android 程序代码

以下是让 WebView 中的 JavaScript 能够调用 Android 程序代码的步骤：

Step 1 在 Android 程序项目中新增一个类，这个类中的方法是专门提供给 WebView 中的 JavaScript 调用，例如我们可以将这个类取名为 JavaScriptCallFunc，它的内容如下：

```
public class JavaScriptCallFunc {

    public void func1(...) {
        // Android 程序要运行的程序代码
        ...
    }
```

```
    public void func2(…) {
        // Android 程序要运行的程序代码
        …
    }
}
```

Step 2 在程序代码中启用 WebView 的 JavaScript 运行功能,并建立一个上述类的对象作为程序代码和 JavaScript 之间的接口,这个建立接口的动作是利用 WebView 的 addJavascriptInterface()方法来完成,它需要两个参数,第一个就是上述类的对象,第二个则是指定这个接口在 JavaScript 中使用的名称,以下范例是将这个名称指定为 Android。

```
WebView mWebView = (WebView)findViewById(R.id.webView);

WebSettings webSettings = mWebView.getSettings();
webSettings.setJavaScriptEnabled(true);

mWebView.addJavascriptInterface(new JavaScriptCallFunc(), "Android");
```

Step 3 建立一个 html 网页文件,在网页的 JavaScript 程序代码中利用前一个步骤设置的接口名称调用 Android 程序代码中的方法,如以下范例:

```html
<html>

<head>
<META http-equiv="Content-Type" content="text/html; charset=UTF-8">
<title>My web page</title>
</head>

<body>
…（网页的内容）

<script type="text/javascript">
function callAndroidFunc1(…) {
    Android.Func1(…);
}

function callAndroidFunc2(…) {
    Android.Func2(…);
}
</script>
</body>

</html>
```

这个网页文件可以储存在程序项目的 assets 文件夹中,再利用 WebView 的 loadUrl("file:///android_asset/（网页文件名）.html")将此网页文件加载 WebView 中运行。如果网页文件中有中文字,必须仿照以上范例加入<META>标签指定使用 UTF-8 编码,并将此网页文件改成使用 UTF-8 编码的方式储存。操作的方式是在 Eclipse 左边的项目检查窗口中，用鼠标右击该网页文件，再从快捷菜单中选择 Properties 就会出现图 79-1 的对话框。在对话框右下方找到 Text file encoding 字段，单击 Other 再从下拉式对话框中选择 UTF-8，然后按下 Apply 按钮，最后按下 OK 按钮。

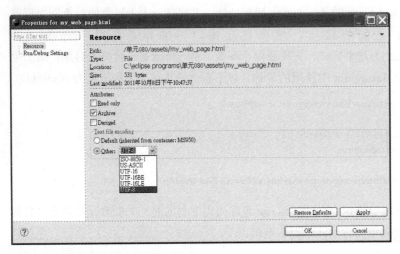

图 79-1　配置文件案储存的编码方式

79-2　从 Android 程序代码调用 JavaScript 的 function

要完成这项工作非常简单，假设我们已经将以下的 html 文件加载到 WebView 中运行：

```
<html>

<head>
<META http-equiv="Content-Type" content="text/html; charset=UTF-8">
<title>My web page</title>
</head>

<body>
…（网页的内容）

<script type="text/javascript">
function funcDoSomething(…) {
    …
}
</script>
</body>

</html>
```

其中有一个名为 funcDoSomething() 的 JavaScript 函数。如果要在 Android 程序中调用这个函数，只要利用 WebView 的 loadUrl() 方法，并指定 javascript:funcDoSomething() 即可，如以下程序代码：

    mWebView.loadUrl("javascript:funcDoSomething()");

79-3　使用 WebView 的 loadData()

截至目前为止我们都是使用 WebView 的 loadUrl() 来加载网页文件，其实 WebView 还提供了其他不同的网页加载方式，其中 loadData() 也是很常用的一种，它让我们可以直接在程序代码中建立一个符合 html 格式的字符串，然后将它直接输入给 WebView 完成解译和显示。在调用 loadData() 的时候必须指定字符串的类型和编码方式，一般是使用 UTF-8 编码以支持各种语言，以下是程序代码范例：

```
String sHtml = null;
try {
sHtml = URLDecoder.decode(
        "<html>" +
        "<META http-equiv=\"Content-Type\" content=\"text/html; charset=UTF-8\">" +
        "<body>这是由程序代码建立的网页。</body>" +
        "</html>", "utf-8");
} catch (Exception e) {
// TODO Auto-generated catch block
e.printStackTrace();
}

mWebView.loadData(sHtml, "text/html", "utf-8");
```

为了得到 UTF-8 编码的字符串，我们利用 URLDecoder 类的 decode() 方法来建立字符串，并指定第二个参数为"utf-8"。建立好 html 格式的字符串后就可以调用 WebView 的 loadData() 传入该字符串，然后指定字符串类型为"text/html"，编码方式为"utf-8"，这样就可以在 WebView 中显示这个 html 网页。接下来我们根据实现一个程序项目帮助读者了解完整的程序架构和建立过程。

79-4　范例程序

以下建立的程序项目将包含前面介绍的三种 WebView 的功能，程序在平板电脑的运行界面如图 79-2 所示。首先按下"加载网页"按钮开启储存在程序项目 assets 文件夹中的网页文件，该网页中包含一个"调用 Android 程序显示 Toast"的按钮和二个 JavaScript function。按下网页中的"调用 Android 程序显示 Toast"按钮时，网页中的 JavaScript 会调用 Android 程序代码显示一个 Toast 信息，如图 79-2 所示。如果按下程序的"在网页中显示图片"按钮，程序代码会调用网页中的

JavaScript 显示一个图片文件。如果按下"用程序代码建立网页"按钮，则会显示一个由程序代码建立的网页，如图 79-3 所示。以下是完成这个程序项目的详细步骤：

图 79-2　JavaScript 和 Android 程序代码之间的调用

图 79-3　用程序代码产生的网页

Step 1　建立一个新的 Android 程序项目，项目的属性设置依照之前的惯例即可。

Step 2　如果想要让程序能够开启 Internet 的网页，必须在程序功能描述文件 AndroidManifest.xml 中设置开启网络功能（如同上一个单元的范例程序）。不过以这个程序范例来说，并不需要连到 Internet，因此可以不用设置。

Step 3　新增一个类作为 JavaScript 调用 Android 程序代码的接口，这个类只需要继承内定的

java.lang.Object 类即可，然后在该类的程序代码中建立要提供给 JavaScript 调用的方法，以下是完成后的程序代码：

```java
package …

import …

public class JavaScriptCallFunc {
    Context mContext;

    JavaScriptCallFunc(Context c) {
        mContext = c;
    }

    public void showToastMsg(String s) {
        Toast.makeText(mContext, s, Toast.LENGTH_LONG)
            .show();
    }
}
```

我们把这个类取名为 JavaScriptCallFunc，它有一个构造方法用来传入主程序的运行环境以提供 Toast 信息使用，另外有一个 showToastMsg() 方法用来提供给网页中的 JavaScript 调用。

Step 4 开启程序项目的接口布局文件，将它编辑如下，其中包含三个按钮和一个 WebView 组件。我们利用二层 LinearLayout 的排列方式，让按钮排列在程序界面的上方，按钮的功能如同前面的说明。

```xml
<?xml version="1.0" encoding="utf-8"?>
<LinearLayout xmlns:android="http://schemas.android.com/apk/res/android"
    android:orientation="vertical"
    android:layout_width="match_parent"
    android:layout_height="match_parent"
    >
<LinearLayout
    android:orientation="horizontal"
    android:layout_width="match_parent"
    android:layout_height="wrap_content"
    >
<Button android:id="@+id/btnLoadHtml"
    android:layout_width="wrap_content"
    android:layout_height="wrap_content"
    android:text="载入网页"
    />
<Button android:id="@+id/btnShowImage"
    android:layout_width="wrap_content"
    android:layout_height="wrap_content"
```

```xml
        android:text="在网页中显示图片"
        />
    <Button android:id="@+id/btnBuildHtml"
        android:layout_width="wrap_content"
        android:layout_height="wrap_content"
        android:text="用程序代码建立网页"
        />
</LinearLayout>
<WebView android:id="@+id/webView"
    android:layout_width="match_parent"
    android:layout_height="match_parent"
    />
</LinearLayout>
```

Step 5 新增一个 html 网页文件,请读者在 Eclipse 左边的项目检查窗口中用鼠标右击 assets 文件夹,然后选择 New > File,并在对话框中的 File name 字段输入网页文件名称,文件名只能用小写英文字和底线字符,例如我们可以取名为 my_web_page.html,完成后按下 Finish 按钮。

Step 6 这个新增的网页文件会自动开启在程序编辑窗口中,不过它是以浏览器模式运行,因此请将它关闭,然后在 Eclipse 左边的项目检查窗口中用鼠标右击这个网页文件,选择 Open With > Text Editor,就可以进入文字编辑模式,然后输入以下网页的内容:

```html
<html>

<head>
<META http-equiv="Content-Type" content="text/html; charset=UTF-8">
<title>My web page</title>
</head>

<body>
<input type="button" value="调用 Android 程序显示 Toast"
        onClick="callAndroidShowToast('由 JavaScript 调用')" />
<br>
<img id="img_demo"/>

<script type="text/javascript">
function callAndroidShowToast(msg) {
    Android.showToastMsg(msg);
}

function showImage() {
    document.getElementById("img_demo").src="file:///
android_asset/android.jpg";
}
</script>
```

```
</body>
</html>
```

网页中包含一个"调用 Android 程序显示 Toast"的按钮和两个名为 callAndroidShowToast()和 showImage()的 JavaScript function。当按下该按钮时会运行 callAndroidShowToast()并传入要显示的信息字符串，callAndroidShowToast()内部会调用 Android 程序代码显示 Toast 信息。showImage()则是提供给后续建立的 Android 程序代码调用使用。

Step 7 开启主类程序文件，首先我们让它实现 OnClickListener 接口，以便将它设置给 Button 对象当成是 OnClickListener。在 setupViewComponent()方法中我们设置好所有的接口组件，包括启动 WebView 对象的 JavaScript 运行功能和建立一个让 JavaScript 调用的接口，我们将这个接口取名为 Android。最后在 onClick()方法中判断用户按下的按钮并运行对应的程序代码，这些程序代码的功能如同本单元前面的说明，以下是完整的程序文件内容：

```
package …

import …

public class Main extends Activity
        implements OnClickListener {

    private Button mBtnLoadHtml,
                mBtnShowImage,
                mBtnBuildHtml;
    private WebView mWebView;

    /** Called when the activity is first created. */
    @Override
    public void onCreate(Bundle savedInstanceState) {
        super.onCreate(savedInstanceState);
        setContentView(R.layout.main);

        setupViewComponent();
    }

    private void setupViewComponent() {
        mBtnLoadHtml = (Button)findViewById(R.id.btnLoadHtml);
        mBtnShowImage = (Button)findViewById(R.id.btnShowImage);
        mBtnBuildHtml = (Button)findViewById(R.id.btnBuildHtml);
        mWebView = (WebView)findViewById(R.id.webView);

        WebSettings webSettings = mWebView.getSettings();
        webSettings.setJavaScriptEnabled(true);
```

```java
        mWebView.addJavascriptInterface(new JavaScriptCallFunc(this),
        "Android");

        mBtnLoadHtml.setOnClickListener(this);
        mBtnShowImage.setOnClickListener(this);
        mBtnBuildHtml.setOnClickListener(this);
    }

    @Override
    public void onClick(View v) {
        // TODO Auto-generated method stub

        switch (v.getId()) {
        case R.id.btnLoadHtml:
            mWebView.loadUrl("file:///android_asset/my_web_page.html");
            break;
        case R.id.btnShowImage:
            mWebView.loadUrl("javascript:showImage()");
            break;
        case R.id.btnBuildHtml:
            String sHtml = null;
            try {
              sHtml = URLDecoder.decode(
                    "<html>" +
                    "<META http-equiv=\"Content-Type\" " +
                    "content=\"text/html;" +
                    "charset=UTF-8\">" +
                    "<body>这是由程序代码建立的网页。</body>" +
                    "</html>", "utf-8");
            } catch (Exception e) {
                // TODO Auto-generated catch block
                e.printStackTrace();
            }

            mWebView.loadData(sHtml, "text/html", "utf-8");
            break;
        }
    }
}
```

我们利用三个单元介绍 WebView 的相关功能，从这些解说和实现范例读者应该可以体会到 WebView 的强大功能，但这还不是 WebView 的全部，往后读者如果遇到需要处理网页资料的时候，第一个念头就是先寻找 WebView 的相关用法，多半都会有您想要的答案。

PART 16 开发 NFC 程序

UNIT 80　NFC 程序设计
UNIT 81　把数据写入 NFC tag
UNIT 82　NFC 的进阶用法

Android 版本	1.X	2.X	3.X	4.X
适用性		★	★	★

UNIT 80

NFC 程序设计

NFC（Near Field Communication）是所谓的"近场通讯"，它是一种新兴的手机应用，运行方式是让手机靠近一个含有 NFC tag 的设备或是小贴片，以取得或传送数据，如图 80-1 所示。它可以用在手机付款、个人资料交换、商品数据撷取等，其中最基本的用法是所谓的 NFC Data Exchange Format，简称 NDEF，它具有下列两种功能：

1. 从 NFC tag 读取 NDEF 数据。
2. 从一个 NFC 设备传送 NDEF 数据到另一个 NFC 设备。

当 Android 系统侦测到 NFC tag 时，它会利用 Tag Dispatch 机制来进行处理，Tag Dispatch 的过程牵涉 NFC tag 资料的解析和建立 Intent 对象，以下我们将从头开始好好地检查一下这整个过程。

图 80-1　NFC 的运行模式

① 从 Android 2.3.3（API level 10）开始才完整支持 NFC 功能。

80-1　Android 处理 NFC tag 数据的方式

要了解 Android 如何处理 NFC tag 中的资料，必须先对 NFC tag 的数据格式有一些了解。在 NFC tag 中 NDEF 数据是封装在一个 NFC message 里面，NFC message 内部可以有一个或多个 record（参考图 80-1），第一个 record 含有以下信息：

1. TNF（Type Name Format）
   决定 Type 字段的格式，TNF 可以是表 80-1 的几种设置值。

2. Type
   TNF 字段的值决定此字段的数据格式，例如如果 TNF 的值是 TNF_WELL_KNOWN，这个字段就是 RTD（Record Type Definition），RTD 的值如表 80-2 所示。

3. ID
   record 的 ID 编号，通常不需要设置。

4. Payload
   用来储存实际的数据，NFC message 中的数据可以分开储存在多个 record 中。

表 80-1　NFC record 的 TNF 字段的设置值

TNF 的设置值	说明
TNF_ABSOLUTE_URI	Type 字段是 URI 数据
TNF_EMPTY	Android 系统会以 ACTION_TECH_DISCOVERED 的方式处理
TNF_EXTERNAL_TYPE	Type 字段是 URN 类型的 URI，Android 会将它的格式转成 vnd.android.nfc://ext/<domain_name>:<service_name>
TNF_MIME_MEDIA	Type 字段是描述 MIME 的类型
TNF_UNCHANGED	Android 系统会以 ACTION_TECH_DISCOVERED 的方式处理
TNF_UNKNOWN	Android 系统会以 ACTION_TECH_DISCOVERED 的方式处理
TNF_WELL_KNOWN	Type 字段是 RTD

表 80-2　当 TNF 的值是 TNF_WELL_KNOWN 时，Type 字段的设置值

Type 字段的值（RTD）	说明
RTD_ALTERNATIVE_CARRIER	Android 系统会以 ACTION_TECH_DISCOVERED 的方式处理
RTD_HANDOVER_CARRIER	Android 系统会以 ACTION_TECH_DISCOVERED 的方式处理
RTD_HANDOVER_REQUEST	Android 系统会以 ACTION_TECH_DISCOVERED 的方式处理
RTD_HANDOVER_SELECT	Android 系统会以 ACTION_TECH_DISCOVERED 的方式处理
RTD_SMART_POSTER	Payload 字段是 URI
RTD_TEXT	MIME 类型是 text/plain
RTD_URI	Payload 字段是 URI

当 Android 手机或平板电脑的屏幕被解锁之后，Android 系统便开始连续不断地侦测 NFC tag（使用者可以利用手机或平板电脑的系统设置取消这项功能）。当侦测到 NFC tag 时，Android 系统开始分析从 NFC tag 中取得的数据。首先利用 TNF 和 Type 两个字段决定数据是否属于 MIME 或是 URI 类型。如果是的话代表这个 NFC tag 是属于 NDEF，接着取出实际的数据，再将它包装成一个 ACTION_NDEF_DISCOVERED 类型的 Intent 对象，然后搜寻目前安装的程序，找出谁能够处理这个 Intent 对象，然后启动该程序把 Intent 对象交给它处理。如果 Android 系统分析 NFC tag 的结果发现不是 NDEF，便会建立一个 ACTION_TECH_DISCOVERED 类型的 Intent 对象，其中包含一个 Tag 对象和从 NFC tag 中取出的数据。

例如 NFC message 中的第一个 record 的 TNF 是 TNF_ABSOLUTE_URI，则代表 Type 字段就是 URI，此时 Android 会取出该 URI 连同 Payload 中的数据一起包装成一个 ACTION_NDEF_DISCOVERED 类型的 Intent 对象。如果 NFC message 的第一个 record 的 TNF 是 TNF_UNKNOWN，则 Android 会建立一个 ACTION_TECH_DISCOVERED 类型的 Intent 对象。当 Android 根据 NFC tag 中的数据建立好 Intent 对象后，便开始搜寻可以处理该 Intent 的程序，并根据下列步骤进行后续操作：

Step 1 如果是 ACTION_NDEF_DISCOVERED 类型的 Intent 而且有程序可以进行处理，就启动该程序，然后把 Intent 传给它，如果找不到可以处理的程序就进行下一个步骤。

Step 2 重新建立一个 ACTION_TECH_DISCOVERED 类型的 Intent，然后寻找可以处理它的程序。如果有找到就启动该程序然后把 Intent 传给它，如果找不到可以处理的程序就进行下一个步骤。

Step 3 重新建立一个 ACTION_TAG_DISCOVERED 类型的 Intent，然后寻找可以处理它的程序。如果有找到就启动该程序然后把 Intent 传给它，如果找不到可以处理的程序就放弃处理这个 NFC tag 的数据。

以上的处理流程请参考图 80-2。

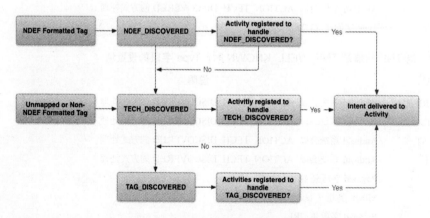

图 80-2　Android 系统处理 NFC tag 的流程（Tag Dispatch System）

80-2　开发 NFC 程序

要让程序能够处理 NFC tag 的数据必须先在程序功能描述文件 AndroidManifest.xml 中完成以下设置：

```xml
<?xml version="1.0" encoding="utf-8"?>
<manifest ...>
   ...
<uses-sdk android:minSdkVersion="10" />
<uses-permission android:name="android.permission.NFC" />
<uses-feature android:name="android.hardware.nfc" android:required="true" />

    <application ...>
        …
    </application>
</manifest>
```

Android 从 2.3.1（API level 9）开始支持 NFC，但是最初只能使用 ACTION_TAG_DISCOVERED 的形式，直到 2.3.3 的版本（API level 10）才支持完整的 NFC 读写功能，而且 Android 4.0（API level 14）以后新增以 Android Beam 的方式使用 NFC，因此 android:minSdkVersion 属性必须依照程序使用的 NFC 功能进行设置。另外在 <uses-feature …> 标签中是要求程序所安装的手机或平板电脑必须具有 NFC 功能，这个设置也可以省略，换成在程序中调用 getDefaultAdapter() 侦测 NFC 功能，下一个单元有实现的程序代码范例。

接下来是设置程序能够处理的 NFC Intent 的形式，这其实是利用 UNIT 40 介绍的 Intent Filter 技术。例如要让程序能够处理 ACTION_NDEF_DISCOVERED 类型的 Intent，而且数据是属于 MIME 的 text 形式，就必须在程序功能描述文件中加入以下设置：

```xml
<?xml version="1.0" encoding="utf-8"?>
<manifest ...>
    …
    <application ...>
        <activity ...>
            <intent-filter>
                <action android:name="android.nfc.action.NDEF_DISCOVERED"/>
                <category android:name="android.intent.category.DEFAULT"/>
                <data android:mimeType="text/plain" />
            </intent-filter>
        </activity>

    </application>
</manifest>
```

如果程序要处理的数据是属于 URI 类型，而且内容是 http://developer.android.com/index.html，就必须设置为：

```
…（同前一个范例）
<intent-filter>
    <action android:name="android.nfc.action.NDEF_DISCOVERED"/>
    <category android:name="android.intent.category.DEFAULT"/>
    <data android:scheme="http"
              android:host="developer.android.com"
              android:pathPrefix="/index.html" />
</intent-filter>
…（同前一个范例）
```

如果要让程序能够处理 ACTION_TECH_DISCOVERED 类型的 Intent，必须先在程序项目的 res 文件夹中新增一个 xml 子文件夹，并且在里头建立一个 xml 文件，例如可以将它取名为 nfc_tech_list.xml，然后编辑它的内容如下：

```
<resources xmlns:xliff="urn:oasis:names:tc:xliff:document:1.2">
    <tech-list>
        <tech>android.nfc.tech.NfcA</tech>
        <tech>android.nfc.tech.Ndef</tech>
    </tech-list>
    <tech-list>
        <tech>android.nfc.tech.NfcB</tech>
        <tech>android.nfc.tech.Ndef</tech>
    </tech-list>
</resources>
```

这个文件定义程序能够处理的 NFC tag 形式，以这个范例而言代表程序能够处理 NfcA 搭配 Ndef 的形式，或者是 NfcB 搭配 Ndef 的形式，然后在程序功能描述文件中加入下列设置：

```
<?xml version="1.0" encoding="utf-8"?>
<manifest …>
    …
    <application …>
        <activity>
            ...
            <intent-filter>
                    <action android:name="android.nfc.action.TECH_DISCOVERED"/>
            </intent-filter>

            <meta-data android:name="android.nfc.action.TECH_DISCOVERED"
                       android:resource="@xml/nfc_tech_list" />
            ...
        </activity>
    </application>
</manifest>
```

如果要让程序能够处理 ACTION_TAG_DISCOVERED 类型的 Intent，则必须在程序功能描述文件中加入以下设置：

```
<intent-filter>
    <action android:name="android.nfc.action.TAG_DISCOVERED"/>
</intent-filter>
```

程序收到 NFC Intent 之后必须从中取出数据，一般来说数据会以下列两种方式储存在 Intent 中：

1. EXTRA_TAG
   代表是一个 Tag 类型的对象。
2. EXTRA_NDEF_MESSAGES
   代表这是 NDEF 类型的数据。

以下程序代码范例先检查接收到的 NFC Intent 是否为 ACTION_NDEF_DISCOVERED 类型，如果是的话再从中取出 EXTRA_NDEF_MESSAGES 形式的数据：

```
public void onResume() {
    super.onResume();
    ...
    if (NfcAdapter.ACTION_NDEF_DISCOVERED.equals(getIntent().getAction())) {
        Parcelable[] rawMsgs = intent.getParcelableArrayExtra
        (NfcAdapter.EXTRA_NDEF_MESSAGES);
        if (rawMsgs != null) {
            msgs = new NdefMessage[rawMsgs.length];
            for (int i = 0; i < rawMsgs.length; i++) {
                msgs[i] = (NdefMessage) rawMsgs[i];
            }
        }
    }
    //处理 msgs[]数组中的数据
    ...
}
```

如果要取得 NFC Intent 中的 Tag 对象可以利用下列程序代码：

```
Tag tag = intent.getParcelableExtra(NfcAdapter.EXTRA_TAG);
```

Android 版本	1.X	2.X	3.X	4.X
适用性		★[②]	★[②]	★

UNIT 81
把资料写入 NFC tag

根据上一个单元的说明，NFC tag 中的数据是储存在 record 中，而且表 80-1 列出各种类型的 record，如果程序中需要建立 record，可以参考下列程序代码：

1. 建立 TNF_ABSOLUTE_URI 类型的 record。

    ```
    NdefRecord nfcRecord = new NdefRecord(NdefRecord.TNF_ABSOLUTE_URI,
        "http://developer.android.com/index.html".getBytes(Charset.forName
          ("US-ASCII")),
        new byte[0], new byte[0]);
    ```

2. 建立 TNF_MIME_MEDIA 类型的 record。

    ```
    NdefRecord nfcRecord = new NdefRecord(NdefRecord.TNF_MIME_MEDIA,
        "application/com.example.android.beam".getBytes(Charset.forName
          ("US-ASCII")),
        new byte[0], "Beam me up, Android!".getBytes(Charset.forName
          ("US-ASCII")));
    ```

3. 建立 TNF_WELL_KNOWN 类型的 record 并储存 RTD_TEXT 类型的数据。

    ```
    byte[] langBytes = locale.getLanguage().getBytes(Charset.forName("US-ASCII"));
    Charset utfEncoding = encodeInUtf8 ? Charset.forName("UTF-8") : Charset.forName("UTF-16");
    byte[] textBytes = payload.getBytes(utfEncoding);
    int utfBit = encodeInUtf8 ? 0 : (1 << 7);
    char status = (char) (utfBit + langBytes.length);
    byte[] data = new byte[1 + langBytes.length + textBytes.length];
    ```

[①] 从 Android 2.3.3（API level 10）开始才完整支持 NFC 功能，Android Beam 和 AAR 功能必须在 Android 4.0（API level 14）版本以上才能使用。

[②] Android Beam 和 AAR 功能必须在 Android 4.0（API level 14）版本以上才能使用。

```
data[0] = (byte) status;
System.arraycopy(langBytes, 0, data, 1, langBytes.length);
System.arraycopy(textBytes, 0, data, 1 + langBytes.length, textBytes.length);
NdefRecord nfcRecord = new NdefRecord(NdefRecord.TNF_WELL_KNOWN,
NdefRecord.RTD_TEXT, new byte[0], data);
```

4. 建立 TNF_WELL_KNOWN 类型的 record 并储存 RTD_URI 类型的数据。

```
byte[] uriField = "example.com".getBytes(Charset.forName("US-ASCII"));
byte[] payload = new byte[uriField.length + 1];
//加 1 是为了 URI 的前导符
byte payload[0] = 0x01;      //在 URI 前面加上 http://www.
System.arraycopy(uriField, 0, payload, 1, uriField.length);
//在 payload 后面加上 URI
NdefRecord nfcRecord = new NdefRecord(
NdefRecord.TNF_WELL_KNOWN, NdefRecord.RTD_URI, new byte[0], payload);
```

5. 建立 TNF_EXTERNAL_TYPE 类型的 record。

```
byte[] payload;
...（把资料存到 payload 中）
NdefRecord nfcRecord = new NdefRecord(
    NdefRecord.TNF_EXTERNAL_TYPE, "example.com:externalType", new byte[0], payload);
```

准备好 NFC record 之后就可以建立一个 NFC message 将 record 包装起来：

```
NdefMessage msg = new NdefMessage( new NdefRecord[] { nfcRecord });
```

如果要在 NFC message 中放入多个 record，可以参考以下程序代码：

```
NdefRecord nfcRecord1 = new NdefRecord(…);
NdefRecord nfcRecord2 = new NdefRecord(…);
NdefRecord nfcRecord3 = new NdefRecord(…);
NdefMessage msg = new NdefMessage( new NdefRecord[] { nfcRecord1, nfcRecord2, nfcRecord3 });
```

81-1　Android 4.X 的 Android Application Record（AAR）

AAR 是 Android 4.0 以后的版本新加入的功能，它可以把程序名称（包含组件路径名称）写入 NFC message 中，当 Android 收到 NFC message 时，会先在里头寻找 AAR，如果有找到，便直接启动 AAR 指定的程序让它处理 NFC message 中的资料。如果系统目前没有安装该程序，Android 会自动连到 Android Market 网站让用户下载该程序。

当程序要将 NFC message 写入 NFC tag 或传给另一个 Android 设备时，可以在里头加入 AAR，相关程序代码请参考以下范例：

```
NdefMessage msg = new NdefMessage(
    new NdefRecord[] { nfcRecord,    // nfcRecord 是前面建立好的 record
```

```
        NdefRecord.createApplicationRecord("(程序的组件路径和名称)")
        // 建立 AAR
});
```

81-2　Android Beam

　　Android 4.0 以后的版本新增 Beam 的功能让程序可以通过 NFC 的方式把数据传给另一个 Android 设备，有下列两种方式可以让程序运行 Beam 功能：
1. 调用 setNdefPushMessage() 传入建立好的 NdefMessage 对象，当要接收的 Android 设备靠近时，数据就会自动传送过去。
2. 调用 setNdefPushMessageCallback() 设置一个 call back 函数，当要接收的 Android 设备靠近时，Android 系统会调用这个 call back 函数，我们在该函数中建立要传送的 message 并完成传送的动作。

　　使用 Beam 功能时要注意传送端和接收端的设备都必须在 unlocked 状态。由于接收到含有 AAR 的 NFC message 时，系统会直接运行 AAR 指定的程序，因此不需要在程序功能描述文件中设置 Intent Filter 的信息。不过 Intent Filter 可以和 Beam 功能同时使用，让程序可以处理多种类型的 NFC message。以下是利用 Beam 传送数据的程序代码范例：

```
package tw.android;

import ...

public class Beam extends Activity implements CreateNdefMessageCallback {
    NfcAdapter mNfcAdapter;
    TextView textView;

    @Override
    public void onCreate(Bundle savedInstanceState) {
        super.onCreate(savedInstanceState);
        setContentView(R.layout.main);
        TextView textView = (TextView) findViewById(R.id.textView);
        // 检查是否具有 NFC 功能
        mNfcAdapter = NfcAdapter.getDefaultAdapter(this);
        if (mNfcAdapter == null) {
            Toast.makeText(this, "没有 NFC 功能", Toast.LENGTH_LONG).show();
            finish();
            return;
        }

        mNfcAdapter.setNdefPushMessageCallback(this, this);
    }
```

```java
@Override
public NdefMessage createNdefMessage(NfcEvent event) {
    String text = ("要传送的数据\n\n" +
            "传送时间: " + System.currentTimeMillis());
    NdefMessage msg = new NdefMessage(
            new NdefRecord[] { createMimeRecord(
                    "application/com.example.android.beam", text.getBytes()),
                    NdefRecord.createApplicationRecord("tw.android.beam")
    });
    return msg;
}

@Override
public void onResume() {
    super.onResume();
    // 检查是否因为 NFC Beam 才启动
    if (NfcAdapter.ACTION_NDEF_DISCOVERED.equals(getIntent().getAction())) {
        processIntent(getIntent());
    }
}

@Override
public void onNewIntent(Intent intent) {
    // 在 onResume()之前运行,设置 intent
    setIntent(intent);
}

void processIntent(Intent intent) {
    // 处理 NDEF Message
    textView = (TextView) findViewById(R.id.textView);
    Parcelable[] rawMsgs = intent.getParcelableArrayExtra(
            NfcAdapter.EXTRA_NDEF_MESSAGES);
    // only one message sent during the beam
    NdefMessage msg = (NdefMessage) rawMsgs[0];
    // record 0 contains the MIME type, record 1 is the AAR, if present
    textView.setText(new String(msg.getRecords()[0].getPayload()));
}

public NdefRecord createMimeRecord(String mimeType, byte[] payload) {
    // 建立一个 MIME type 在 NDEF record 中使用
    byte[] mimeBytes = mimeType.getBytes(Charset.forName("US-ASCII"));
    NdefRecord mimeRecord = new NdefRecord(
            NdefRecord.TNF_MIME_MEDIA, mimeBytes, new byte[0], payload);
    return mimeRecord;
}
}
```

Android 版本	1.X	2.X	3.X	4.X
适用性		★[①]	★	★

UNIT 82
NFC 的进阶用法

通常 NFC tag 会利用 NDEF 格式储存数据，当 Android 接收到 NDEF 格式的 message 时，它可以自动解析其中的资料，但是如果遇到非 NDEF 格式的 NFC tag 时，Android 会建立 ACTION_TECH_DISCOVERED 类型的 Intent。如果程序收到这种 Intent 就代表必须自己解析 NFC tag 中的数据。相关的操作类是定义在 android.nfc.tech 组件中如表 82-1，程序可以调用 getTechList() 取得 tag 支持的格式，再利用 TagTechnology 对象和相关类来进行处理。如果要让程序能够处理 ACTION_TECH_DISCOVERED 类型的 Intent，必须完成以下步骤：

表 82-1 android.nfc.tech 组件中的类

类	说明
TagTechnology	这是一个 Interface，所有 Tag Technology 的类都必须实现这个 Interface
NfcA	NFC-A (ISO 14443-3A)的型式
NfcB	NFC-B (ISO 14443-3B)的型式
NfcF	NFC-F (JIS 6319-4)的型式
NfcV	NFC-V (ISO 15693)的型式
IsoDep	ISO-DEP (ISO 14443-4)的型式
Ndef	用来处理 NDEF 格式的数据
NdefFormatable	可以用 NDEF 格式来处理
MifareClassic	MIFARE 的型式
MifareUltralight	MIFARE Ultralight 的型式

Step 1 依照 UNIT 80 的说明编辑程序功能描述文件 AndroidManifest.xml。

[①] 从 Android 2.3.3（API level 10）开始才完整支持 NFC 功能。

Step 2 当程序收到 ACTION_TECH_DISCOVERED 类型的 Intent 时，取出其中的 Tag 对象。

```
Tag tagFromIntent = intent.getParcelableExtra(NfcAdapter.EXTRA_TAG);
```

Step 3 调用 getTechList()决定 Tag 的型式，然后运行对应的 get()方法取得 TagTechnology 对象。

以下程序代码示范如何利用 android.nfc.tech 组件中的 MifareUltralight 类对 NFC tag 进行数据读写：

```java
public class MifareUltralightTagTester {

    private static final String TAG = MifareUltralightTagTester.class.getSimpleName();

    public void writeTag(Tag tag, String tagText) {
        MifareUltralight ultralight = MifareUltralight.get(tag);
        try {
            ultralight.connect();
            ultralight.writePage(4, "abcd".getBytes(Charset.forName("US-ASCII")));
            ultralight.writePage(5, "efgh".getBytes(Charset.forName("US-ASCII")));
            ultralight.writePage(6, "ijkl".getBytes(Charset.forName("US-ASCII")));
            ultralight.writePage(7, "mnop".getBytes(Charset.forName("US-ASCII")));
        } catch (IOException e) {
            Log.e(TAG, "IOException while closing MifareUltralight...", e);
        } finally {
            try {
                ultralight.close();
            } catch (IOException e) {
                Log.e(TAG, "IOException while closing MifareUltralight...", e);
            }
        }
    }

    public String readTag(Tag tag) {
        MifareUltralight mifare = MifareUltralight.get(tag);
        try {
            mifare.connect();
            byte[] payload = mifare.readPages(4);
            return new String(payload, Charset.forName("US-ASCII"));
        } catch (IOException e) {
            Log.e(TAG, "IOException while writing MifareUltralight message...", e);
        } finally {
            if (mifare != null) {
                try {
                    mifare.close();
                }
                catch (IOException e) {
                    Log.e(TAG, "Error closing tag...", e);
```

```
                    }
                }
            }
            return null;
        }
    }
```

82-1 让运行中的程序优先处理 NFC Intent

如果要让目前运行中的程序优先处理 NFC 的 Intent（称为 foreground dispatch），必须依照下列步骤修改程序：

Step 1 在 onCreate() 中建立一个 PendingIntent 对象，然后注册 Intent Filter 和可以处理的 tag 型式。

Step 2 加入 onPause()、onResume() 和 onNewIntent() 方法对 NFC Intent 进行适当的处理。

详细的作法请参考以下程序代码范例，其中的 **addDataType()** 方法是用来设置程序要处理的数据格式，**techListsArray** 是 String 类型的二维数组用来指定程序可以处理的 tag 型式。

■ 主程序文件：

```
package …

import …

public class DashboardActivity extends Activity {

    NFCForegroundUtil nfcForegroundUtil = null;
    private TextView info;

    @Override
    public void onCreate(Bundle savedInstanceState) {
        super.onCreate(savedInstanceState);
        setContentView(R.layout.main);
        info = (TextView)findViewById(R.id.info);

        nfcForegroundUtil = new NFCForegroundUtil(this);
    }

    public void onPause() {
        super.onPause();
        nfcForegroundUtil.disableForeground();
    }

    public void onResume() {
        super.onResume();
        nfcForegroundUtil.enableForeground();
```

```java
        if (!nfcForegroundUtil.getNfc().isEnabled())
        {
            Toast.makeText(getApplicationContext(), "Error!",
            Toast.LENGTH_LONG).show();
            startActivity(new Intent(android.provider.Settings.ACTION_
            WIRELESS_SETTINGS));
        }
    }

    public void onNewIntent(Intent intent) {
        Tag tag = intent.getParcelableExtra(NfcAdapter.EXTRA_TAG);
        info.setText(NFCUtil.printTagDetails(tag));
    }
}
```

- NFCForegroundUtil 类程序文件：

```java
public class NFCForegroundUtil {

    private NfcAdapter nfc;
    private Activity activity;
    private IntentFilter intentFiltersArray[];
    private PendingIntent intent;
    private String techListsArray[][];

    public NFCForegroundUtil(Activity activity) {
        super();
        this.activity = activity;
        nfc = NfcAdapter.getDefaultAdapter(activity.getApplicationContext());

        intent = PendingIntent.getActivity(activity, 0, new Intent(activity,
            activity.getClass()).addFlags(Intent.FLAG_ACTIVITY_SINGLE_TOP), 0);

        IntentFilter ndef = new IntentFilter(NfcAdapter.ACTION_NDEF_DISCOVERED);

        try {
            ndef.addDataType("text/plain");
        } catch (MalformedMimeTypeException e) {
            throw new RuntimeException("Unable to speciy */* Mime Type", e);
        }
        intentFiltersArray = new IntentFilter[] { ndef };

        techListsArray = new String[][] {
            new String[] { NfcA.class.getName(), NfcB.class.getName() },
            new String[] {NfcV.class.getName()} };
    }
```

```java
    public void enableForeground()
    {
        Log.d("demo", "Foreground NFC dispatch enabled");
        nfc.enableForegroundDispatch(activity, intent, intentFiltersArray,
            techListsArray);
    }

    public void disableForeground()
    {
        Log.d("demo", "Foreground NFC dispatch disabled");
        nfc.disableForegroundDispatch(activity);
    }

    public NfcAdapter getNfc() {
        return nfc;
    }
}
```

附录　本书光盘内容与使用说明

本书光盘内容包含书上所有的范例程序源代码和补充说明文件,有关光盘内容的储存路径和使用说明请参阅以下表格。

文件夹名称	子文件夹名称	说明
范例程序源代码	每个子文件夹对应到不同单元的范例程序项目	请先将程序项目的文件夹复制到 Eclipse 的 workspace 文件夹中,再利用 Eclipse 主选单的 File > Import 功能加载程序项目
其他文件	无	补充说明文件